Lecture Notes in Computer Science 6605

Commenced Publication in 1973
Founding and Former Series Editors:
Gerhard Goos, Juris Hartmanis, and Jan van Leeuwen

W0193064

Advanced Research in Computing and Software Science

Subline of Lectures Notes in Computer Science

Parosh Aziz Abdulla
K. Rustan M. Leino (Eds.)

Tools and Algorithms for the Construction and Analysis of Systems

17th International Conference, TACAS 2011
Held as Part of the Joint European Conferences
on Theory and Practice of Software, ETAPS 2011
Saarbrücken, Germany, March 26–April 3, 2011
Proceedings

 Springer

Volume Editors

Parosh Aziz Abdulla
University of Uppsala
Dept. of Information Technology
751 05 Uppsala, Sweden
E-mail: parosh@it.uu.se

K. Rustan M. Leino
Microsoft Research
Redmond, WA 98052, USA
E-mail: leino@microsoft.com

ISSN 0302-9743   e-ISSN 1611-3349
ISBN 978-3-642-19834-2   e-ISBN 978-3-642-19835-9
DOI 10.1007/978-3-642-19835-9
Springer Heidelberg Dordrecht London New York

Library of Congress Control Number: 2011922620

CR Subject Classification (1998): F.3, D.2, C.2, D.3, D.2.4, C.3

LNCS Sublibrary: SL 1 – Theoretical Computer Science and General Issues

Typesetting: Camera-ready by author, data conversion by Scientific Publishing Services, Chennai, India

Printed on acid-free paper

Springer is part of Springer Science+Business Media (www.springer.com)

Foreword

ETAPS 2011 was the 14th instance of the European Joint Conferences on Theory and Practice of Software. ETAPS is an annual federated conference that was established in 1998 by combining a number of existing and new conferences. This year it comprised the usual five sister conferences (CC, ESOP, FASE, FOSSACS, TACAS), 16 satellite workshops (ACCAT, BYTECODE, COCV, DICE, FESCA, GaLoP, GT-VMT, HAS, IWIGP, LDTA, PLACES, QAPL, ROCKS, SVARM, TERMGRAPH, and WGT), one associated event (TOSCA), and seven invited lectures (excluding those specific to the satellite events).

The five main conferences received 463 submissions this year (including 26 tool demonstration papers), 130 of which were accepted (2 tool demos), giving an overall acceptance rate of 28%. Congratulations therefore to all the authors who made it to the final programme! I hope that most of the other authors will still have found a way of participating in this exciting event, and that you will all continue submitting to ETAPS and contributing to make of it the best conference on software science and engineering.

The events that comprise ETAPS address various aspects of the system development process, including specification, design, implementation, analysis and improvement. The languages, methodologies and tools which support these activities are all well within its scope. Different blends of theory and practice are represented, with an inclination towards theory with a practical motivation on the one hand and soundly based practice on the other. Many of the issues involved in software design apply to systems in general, including hardware systems, and the emphasis on software is not intended to be exclusive.

ETAPS is a confederation in which each event retains its own identity, with a separate Programme Committee and proceedings. Its format is open-ended, allowing it to grow and evolve as time goes by. Contributed talks and system demonstrations are in synchronised parallel sessions, with invited lectures in plenary sessions. Two of the invited lectures are reserved for 'unifying' talks on topics of interest to the whole range of ETAPS attendees. The aim of cramming all this activity into a single one-week meeting is to create a strong magnet for academic and industrial researchers working on topics within its scope, giving them the opportunity to learn about research in related areas, and thereby to foster new and existing links between work in areas that were formerly addressed in separate meetings.

ETAPS 2011 was organised by the *Universität des Saarlandes* in cooperation with:

- ▷ European Association for Theoretical Computer Science (EATCS)
- ▷ European Association for Programming Languages and Systems (EAPLS)
- ▷ European Association of Software Science and Technology (EASST)

It also had support from the following sponsors, which we gratefully thank: DFG Deutsche Forschungsgemeinschaft; AbsInt Angewandte Informatik GmbH; Microsoft Research; Robert Bosch GmbH; IDS Scheer AG / Software AG; T-Systems Enterprise Services GmbH; IBM Research; gwSaar Gesellschaft für Wirtschaftsförderung Saar mbH; Springer-Verlag GmbH; and Elsevier B.V.

The organising team comprised:

General Chair:	*Reinhard Wilhelm*
Organising Committee:	*Bernd Finkbeiner, Holger Hermanns* (chair),
	Reinhard Wilhelm, Stefanie Haupert-Betz,
	Christa Schäfer
Satellite Events:	*Bernd Finkbeiner*
Website:	*Hernán Baró Graf*

Overall planning for ETAPS conferences is the responsibility of its Steering Committee, whose current membership is:

Vladimiro Sassone (Southampton, Chair), Parosh Abdulla (Uppsala), Gilles Barthe (IMDEA-Software), Lars Birkedal (Copenhagen), Michael O'Boyle (Edinburgh), Giuseppe Castagna (CNRS Paris), Marsha Chechik (Toronto), Sophia Drossopoulou (Imperial College London), Bernd Finkbeiner (Saarbrücken) Cormac Flanagan (Santa Cruz), Dimitra Giannakopoulou (CMU/NASA Ames), Andrew D. Gordon (MSR Cambridge), Rajiv Gupta (UC Riverside), Chris Hankin (Imperial College London), Holger Hermanns (Saarbrücken), Mike Hinchey (Lero, the Irish Software Engineering Research Centre), Martin Hofmann (LMU Munich), Joost-Pieter Katoen (Aachen), Paul Klint (Amsterdam), Jens Knoop (Vienna), Barbara König (Duisburg), Shriram Krishnamurthi (Brown), Juan de Lara (Madrid), Kim Larsen (Aalborg), Rustan Leino (MSR Redmond), Gerald Luettgen (Bamberg), Rupak Majumdar (Los Angeles), Tiziana Margaria (Potsdam), Ugo Montanari (Pisa), Luke Ong (Oxford), Fernando Orejas (Barcelona), Catuscia Palamidessi (INRIA Paris), George Papadopoulos (Cyprus), David Rosenblum (UCL), Don Sannella (Edinburgh), João Saraiva (Minho), Helmut Seidl (TU Munich), Tarmo Uustalu (Tallinn), and Andrea Zisman (London).

I would like to express my sincere gratitude to all of these people and organisations, the Programme Committee Chairs and members of the ETAPS conferences, the organisers of the satellite events, the speakers themselves, the many reviewers, all the participants, and Springer for agreeing to publish the ETAPS proceedings in the ARCoSS subline.

Finally, I would like to thank the Organising Chair of ETAPS 2011, Holger Hermanns and his Organising Committee, for arranging for us to have ETAPS in the most beautiful surroundings of Saarbrücken.

January 2011	Vladimiro Sassone
	ETAPS SC Chair

Preface

This volume contains the proceedings of the 17th International Conference on Tools and Algorithms for the Construction and Analysis of Systems (TACAS 2011). TACAS 2011 took place in Saarbrücken, Germany, March 28–31, 2011, as part of the 14th European Joint Conferences on Theory and Practice of Software (ETAPS 2011), whose aims, organization, and history are presented in the foreword of this volume by the ETAPS Steering Committee Chair, Vladimiro Sassone.

TACAS is a forum for researchers, developers, and users interested in rigorously based tools and algorithms for the construction and analysis of systems. The conference serves to bridge the gaps between different communities that share common interests in tool development and its algorithmic foundations. The research areas covered by such communities include, but are not limited to, formal methods, software and hardware verification, static analysis, programming languages, software engineering, real-time systems, communications protocols, and biological systems. The TACAS forum provides a venue for such communities at which common problems, heuristics, algorithms, data structures, and methodologies can be discussed and explored. TACAS aims to support researchers in their quest to improve the usability, utility, flexibility, and efficiency of tools and algorithms for building systems. Tool descriptions and case studies with a conceptual message, as well as theoretical papers with clear relevance for tool construction, are all encouraged. The specific topics covered by the conference include, but are not limited to, the following: specification and verification techniques for finite and infinite-state systems, software and hardware verification, theorem proving and model checking, system construction and transformation techniques, static and run-time analysis, abstraction techniques for modeling and validation, compositional and refinement-based methodologies, testing and test-case generation, analytical techniques for safety, security, or dependability, analytical techniques for real-time, hybrid, or stochastic systems, integration of formal methods and static analysis in high-level hardware design or software environments, tool environments and tool architectures, SAT and SMT solvers, and applications and case studies.

TACAS traditionally considers two types of papers: research papers and tool demonstration papers. Research papers are full-length papers that contain novel research on topics within the scope of the TACAS conference and have a clear relevance for tool construction. Tool demonstration papers are shorter papers that give an overview of a particular tool and its applications or evaluation. TACAS 2011 received a total of 112 submissions including 24 tool demonstration papers and accepted 32 papers of which 10 papers were tool demonstration papers. Each submission was evaluated by at least three reviewers. After a six-week reviewing process, the program selection was carried out in a two-week electronic Program

Committee meeting. We believe that the committee deliberations resulted in a strong technical program. One highlight is the quantity and quality of the tool papers submitted to the conference and accepted for presentation.

Gerard J. Holzmann, Jet Propulsion Laboratory, California Institute of Technology, USA, gave the unifying ETAPS 2011 invited talk on "Reliable Software Development: Analysis-Aware Design." Andreas Podelski, University of Freiburg, Germany, gave the TACAS 2011 invited talk on "Transition Invariants and Transition Predicate Abstraction for Program Termination". The abstracts of the talks are included in this volume.

As TACAS 2011 Program Committee Co-chairs, we would like to thank the authors of all submitted papers, the Program Committee members, and all the referees for their invaluable contribution in guaranteeing such a strong technical program. We also thank the EasyChair system for hosting the conference submission and program selection process and automating much of the proceedings generation process. We would like to express our appreciation to the ETAPS Steering Committee and especially its Chair, Vladimiro Sassone, as well as the Organizing Committee for their efforts in making ETAPS 2011 such a successful event.

January 2011

Parosh Aziz Abdulla
K. Rustan M. Leino

Conference Organization

Steering Committee

Ed Brinksma	ESI and University of Twente (The Netherlands)
Rance Cleaveland	University of Maryland and Fraunhofer USA Inc. (USA)
Kim G. Larsen	Aalborg University (Denmark)
Bernhard Steffen	Technical University Dortmund (Germany)
Lenore Zuck	University of Illinois at Chicago (USA)

Program Chairs

Parosh A. Abdulla	Uppsala University (Sweden)
K. Rustan M. Leino	Microsoft Research (USA)

Program Committee

Nikolaj Bjørner	Microsoft Research (USA)
Ahmed Bouajjani	LIAFA, University of Paris 7 (France)
Patricia Bouyer-Decitre	LSV, CNRS and ENS Cachan (France)
Alessandro Cimatti	Istituto per la Ricerca Scientifica e Tecnologica (Italy)
Rance Cleaveland	University of Maryland and Fraunhofer USA Inc. (USA)
Thierry Coquand	Chalmers University (Sweden)
Giorgio Delzanno	Università di Genova (Italy)
Javier Esparza	Technische Universität München (Germany)
Orna Grumberg	Technion - Israel Institute of Technology (Israel)
Peter Habermehl	LIAFA University Paris 7 (France)
Reiner Hähnle	Chalmers University of Technology (Sweden)
Naoki Kobayashi	Tohoku University (Japan)
Kim G. Larsen	Aalborg University (Denmark)
Rupak Majumdar	Max Planck Institute for Software Systems (Germany)
Panagiotis Manolios	Northeastern University (USA)
Richard Mayr	University of Edinburgh (UK)
Doron Peled	Bar Ilan University (Israel)
Anna Philippou	University of Cyprus (Cyprus)

C.R. Ramakrishnan	University at Stony Brook (USA)
Xavier Rival	INRIA, ENS Paris (France)
Natasha Sharygina	University of Lugano (Switzerland)
Armando Solar-Lezama	MIT (USA)
Bernhard Steffen	Technical University Dortmund (Germany)
Tomáš Vojnar	Brno University of Technology (Czech Republic)
Verena Wolf	Saarland University (Germany)
Lenore Zuck	University of Illinois at Chicago (USA)

External Reviewers

Markus Aderhold
Francesco Alberti
Aleksandr Andreychenko
Mohamed Faouzi Atig
Simon Baeumler
Christel Baier
Jiri Barnat
Sharon Barner
Ananda Basu
Prithwish Basu
Nathalie Bertrand
Julien Bertrane
Michel Bidoit
Jasmin Christian Blanchette
Bernard Boigelot
Matthew Bolton
Marius Bozga
Marco Bozzano
Tomas Brazdil
Thomas Brihaye
Vaclav Brozek
Roberto Bruttomesso
Richard Bubel
Peter Buchholz
Sebastian Burckhardt
Franck Cassez
Ivana Cerna
Harsh Raju Chamarthi
Yu-Fang Chen
Alvin Cheung
Lorenzo Clemente
Christopher Conway
Katherine Coons

Pepijn Crouzen
Pieter Cuijpers
Mads Dam
Alexandre David
Cezara Dragoi
Michael Emmi
Constantin Enea
Uli Fahrenberg
Jean-Christophe Filliatre
Shaked Flur
Vojtech Forejt
Guy Gallasch
Pierre Ganty
Samir Genaim
Patrice Godefroid
Eugene Goldberg
Andreas Griesmayer
Alberto Griggio
Ashutosh Gupta
Serge Haddad
Ernst Moritz Hahn
Keijo Heljanko
Jane Hillston
Lukas Holik
Florian Horn
Falk Howar
Radu Iosif
Malte Isberner
Yoshinao Isobe
Susmit Jha
Barbara Jobstmann
Manu Jose
Line Juhl

Kenneth Yrke Jørgensen
Vineet Kahlon
Gal Katz
Christian Kern
Filip Konecny
Hillel Kugler
Shashi Kumar
Marta Kwiatkowska
Peter Lammich
Kai Lampka
Axel Legay
Jerome Leroux
Stefan Leue
Shuhao Li
Ann Lillieström
Florian Lonsing
Michael Luttenberger
Claude Marché
Yael Meller
Maik Merten
Marco Mesiti
Roland Meyer
Linar Mikeev
Marius Mikučionis
Wojciech Mostowski
Sergio Mover
Iman Narasamdya
Stefan Naujokat
Daniel Neider
Johannes Neubauer
Mads Chr. Olesen
Jörg Olschewski
Gabriele Paganelli
Vasilis Papavasileiou
Gennaro Parlato
Hans-Jörg Peter
Ricardo Peña
Andre Platzer
Isabelle Puaut
Hongyang Qu
Yusi Ramadian
Gianna Reggio
Ahmed Rezine
Marina Ribaudo

Enric Rodriguez Carbonell
Simone Fulvio Rollini
Marco Roveri
Pritam Roy
Philipp Ruemmer
Andrey Rybalchenko
Oliver Rüthing
Yaniv Sa'ar
Indranil Saha
Arnaud Sangnier
Viktor Schuppan
Ondrej Sery
K.C. Shashidhar
Sarai Sheinvald
Mihaela Sighireanu
Rishabh Singh
Oleg Sokolsky
David Spieler
Scott Stoller
Ofer Strichman
Kohei Suenaga
Damian Sulewski
Andrzej Tarlecki
Tino Teige
Claus Thrane
Anthony To
Stefano Tonetta
Tayssir Touili
Yih-Kuen Tsay
Aliaksei Tsitovich
Antti Valmari
Boudewijn van Dongen
Laurent Vigneron
Yakir Vizel
Christian von Essen
Björn Wachter
Xiaoyang Wang
Zheng Wang
Sam Weber
Zhilei Xu
Eran Yahav
Kuat Yessenov
Gianluigi Zavattaro

Table of Contents

Reliable Software Development: Analysis-Aware Design
(Invited Talk) 1
   Gerard J. Holzmann

Transition Invariants and Transition Predicate Abstraction for Program
Termination (Invited Talk) 3
   Andreas Podelski and Andrey Rybalchenko

Memory Models and Consistency

Sound and Complete Monitoring of Sequential Consistency for Relaxed
Memory Models 11
   Jabob Burnim, Koushik Sen, and Christos Stergiou

Compositionality Entails Sequentializability 26
   Pranav Garg and P. Madhusudan

Litmus: Running Tests against Hardware 41
   Jade Alglave, Luc Maranget, Susmit Sarkar, and Peter Sewell

Invariants and Termination

Canonized Rewriting and Ground AC Completion Modulo Shostak
Theories 45
   Sylvain Conchon, Evelyne Contejean, and Mohamed Iguernelala

Invariant Generation in Vampire 60
   Kryštof Hoder, Laura Kovács, and Andrei Voronkov

Enforcing Structural Invariants Using Dynamic Frames 65
   Diego Garbervetsky, Daniel Gorín, and Ariel Neisen

Loop Summarization and Termination Analysis 81
   Aliaksei Tsitovich, Natasha Sharygina,
   Christoph M. Wintersteiger, and Daniel Kroening

Timed and Probabilistic Systems

Off-Line Test Selection with Test Purposes for Non-deterministic
Timed Automata 96
   Nathalie Bertrand, Thierry Jéron, Amélie Stainer, and Moez Krichen

Quantitative Multi-objective Verification for Probabilistic Systems 112
  Vojtěch Forejt, Marta Kwiatkowska, Gethin Norman,
  David Parker, and Hongyang Qu

Efficient CTMC Model Checking of Linear Real-Time Objectives........ 128
  Benoît Barbot, Taolue Chen, Tingting Han,
  Joost-Pieter Katoen, and Alexandru Mereacre

Interpolations and SAT-Solvers

Efficient Interpolant Generation in Satisfiability Modulo Linear Integer
Arithmetic 143
  Alberto Griggio, Thi Thieu Hoa Le, and Roberto Sebastiani

Generalized Craig Interpolation for Stochastic Boolean Satisfiability
Problems 158
  Tino Teige and Martin Fränzle

Specification-Based Program Repair Using SAT...................... 173
  Divya Gopinath, Muhammad Zubair Malik, and Sarfraz Khurshid

Optimal Base Encodings for Pseudo-Boolean Constraints 189
  Michael Codish, Yoav Fekete, Carsten Fuhs, and
  Peter Schneider-Kamp

Learning

Predicate Generation for Learning-Based Quantifier-Free Loop
Invariant Inference 205
  Yungbum Jung, Wonchan Lee, Bow-Yaw Wang, and Kwangkuen Yi

Next Generation LearnLib 220
  Maik Merten, Bernhard Steffen, Falk Howar, and Tiziana Margaria

Model Checking

Applying CEGAR to the Petri Net State Equation 224
  Harro Wimmel and Karsten Wolf

Biased Model Checking Using Flows................................ 239
  Muralidhar Talupur and Hyojung Han

S-TALiRo: A Tool for Temporal Logic Falsification for Hybrid
Systems 254
  Yashwanth Annapureddy, Che Liu, Georgios Fainekos, and
  Sriram Sankaranarayanan

Games and Automata

GAVS+: An Open Platform for the Research of Algorithmic Game
Solving 258
  *Chih-Hong Cheng, Alois Knoll, Michael Luttenberger, and
  Christian Buckl*

Büchi Store: An Open Repository of Büchi Automata 262
  *Yih-Kuen Tsay, Ming-Hsien Tsai, Jinn-Shu Chang, and
  Yi-Wen Chang*

QUASY: Quantitative Synthesis Tool 267
  *Krishnendu Chatterjee, Thomas A. Henzinger,
  Barbara Jobstmann, and Rohit Singh*

Unbeast: Symbolic Bounded Synthesis 272
  Rüdiger Ehlers

Verification (I)

Abstractions and Pattern Databases: The Quest for Succinctness and
Accuracy 276
  Sebastian Kupferschmid and Martin Wehrle

The ACL2 Sedan Theorem Proving System 291
  *Harsh Raju Chamarthi, Peter Dillinger, Panagiotis Manolios, and
  Daron Vroon*

Probabilistic Systems

On Probabilistic Parallel Programs with Process Creation and
Synchronisation 296
  Stefan Kiefer and Dominik Wojtczak

Confluence Reduction for Probabilistic Systems 311
  Mark Timmer, Mariëlle Stoelinga, and Jaco van de Pol

Model Repair for Probabilistic Systems 326
  *Ezio Bartocci, Radu Grosu, Panagiotis Katsaros,
  C.R. Ramakrishnan, and Scott A. Smolka*

Verification (II)

Boosting Lazy Abstraction for SystemC with Partial Order
Reduction 341
  Alessandro Cimatti, Iman Narasamdya, and Marco Roveri

Modelling and Verification of Web Services Business Activity
Protocol 357
    Anders P. Ravn, Jiří Srba, and Saleem Vighio

CADP 2010: A Toolbox for the Construction and Analysis of
Distributed Processes 372
    Hubert Garavel, Frédéric Lang, Radu Mateescu, and Wendelin Serwe

GameTime: A Toolkit for Timing Analysis of Software 388
    Sanjit A. Seshia and Jonathan Kotker

Author Index 393

Reliable Software Development: Analysis-Aware Design

Gerard J. Holzmann

Laboratory for Reliable Software,
Jet Propulsion Laboratory, California Institute of Technology,
Pasadena, CA 91109, USA
http://lars-lab.jpl.nasa.gov/

Abstract. The application of formal methods in software development does not have to be an *all-or-nothing* proposition. Progress can be made with the introduction of relatively unobtrusive techniques that simplify analysis. This approach is meant replace traditional *analysis-agnostic* coding with an *analysis-aware* style of software development.

Software verification efforts are often stumped by the complexity of not just the analysis itself, but also the preparations that have to be made to enable it. This holds especially if the code base is large, and written in a more traditional programming language with limited builtin protection. When asked to analyze, for instance, the embedded software of an automobile in a high-profile study of the potential software causes for unintended acceleration incidents, our first challenge is generally in the preparations phase. Very similar challenges can exist in the analysis of mission-critical flight software for space missions. Some of the more time-consuming obstacles in these efforts can be avoided, though, if code is designed explicitly with the possibility of independent analysis in mind.

To give a, perhaps overly simple, example of the wisdom or restricting access to shared data in a multi-tasking system: if valuables are stored in the open in a yard, and some are damaged or missing, the analyst will generally have a difficult problem finding out what happened. If they are stored in a locked room and the same thing happens, the job of finding out what happened is reduced to finding out who had access to the key. The analogy to software will be clear: taking even simple precautions can have a large effect.

The adoption of somewhat stronger analysis-aware coding principles can make a notable difference in the types of guarantees that one can give about a large software system, especially when formal methods related tools are used such as static analyzers [10], logic model checkers [6], or provers such as VCC [3]. As one example, reconstructing where state information is stored in a complex system can be one of the hardest obstacles in the application of model-driven verification techniques [7]. The task becomes almost trivial if the global state information is co-located in memory, or placed in a single data structure.

There are many other relatively benign tactics that can be adopted to make code safer and more thoroughly verifiable. As one other example, the integration of complex code with simple aspect-oriented annotations [1] can support

P.A. Abdulla and K.R.M. Leino (Eds.): TACAS 2011, LNCS 6605, pp. 1–2, 2011.

the mechanical generation of code instrumentations that can help a verifier extract design information, or build visualizations of dynamic data structures and message flows that can guide a verification effort.

The benefit of assertions is also well-known [2,9]. A richer set of assertions can be used though [5,4]. As an example, two types of inline *temporal assertions* can be used in combination with the FeaVer model extractor [5]: response and precedence assertions. A response assertion, $assert\_r(e)$, expresses the LTL property $\Diamond e$), stating that a condition e must hold within a finite number of steps from the point in the code where the assertion is placed. A precedence assertion, $assert\_p(e1, e2)$, expresses the LTL property $(e1 \; U \; e2)$, which states that condition $e1$ must hold at least until, within a finite number of steps, condition $e2$ also holds. Temporal assertions can be used to derive property automata for model checkers, as in the FeaVer system [5], or to generate runtime monitors for use in runtime verification, as in the TimeRover system [4].

Finally, in the analysis of complex systems, exhaustive proof will often be beyond reach. This is not necessarily fatal. As Sanjoy Mahajan, a theoretical physicist at Caltech, noted, the attempt to solve complex problems with rigorous methods can lead to *rigor mortis*, which can only be avoided by breaking some of the rules. In our case this can mean using randomized proof techniques and massively-parallel search techniques [8], which can be remarkably effective.

Acknowledgments. The research described in this paper was carried out at the Jet Propulsion Laboratory, California Institute of Technology, under a contract with the National Aeronautics and Space Administration.

References

1. http://en.wikipedia.org/wiki/Aspect-oriented_programming
2. Clarke, L.A., Rosenblum, D.: A historical perspective on runtime assertion checking in software development. ACM SIGSOFT Software Eng. Notes 31(3) (May 2006)
3. Cohen, E., Dahlweid, M., et al.: VCC: A Practical System for Verifying Concurrent C. In: Berghofer, S., Nipkow, T., Urban, C., Wenzel, M. (eds.) TPHOLs 2009. LNCS, vol. 5674, pp. 23–42. Springer, Heidelberg (2009)
4. Drusinsky, D.: Temporal Rover. In: Havelund, K., Penix, J., Visser, W. (eds.) SPIN 2000. LNCS, vol. 1885, pp. 323–330. Springer, Heidelberg (2000)
5. Holzmann, G.J., Smith, M.H.: FeaVer 1.0 User Guide, Bell Laboratories Technical Report, 64 pages (2000), http://cm.bell-labs.com/cm/cs/what/modex/
6. Holzmann, G.J.: The Spin Model Checker: Primer and Reference Manual. Addison-Wesley, Reading (2004)
7. Holzmann, G.J., Joshi, R., Groce, A.: Model driven code checking. Automated Software Eng. Journal 15(3-4), 283–297 (2008)
8. Holzmann, G.J., Joshi, R., Groce, A.: Swarm Verification Techniques. IEEE Trans. on Software Eng. (to appear, 2011)
9. Kudrjavets, G., Nagappan, N., Ball, T.: Assessing the relationship between software assertions and code quality: an empirical investigation. Microsoft Technical Report, MSR_TR-2006-54, 17 pages (2006)
10. http://spinroot.com/static/

Transition Invariants and Transition Predicate Abstraction for Program Termination

Andreas Podelski[1] and Andrey Rybalchenko[2]

[1] University of Freiburg
[2] Technische Universität München

Abstract. Originally, the concepts of transition invariants and transition predicate abstraction were used to formulate a proof rule and an abstraction-based algorithm for the verification of liveness properties of concurrent programs under fairness assumptions. This note applies the two concepts for proving termination of sequential programs. We believe that the specialized setting exhibits the underlying principles in a more direct way.

1 Introduction

Transition invariants allow one to combine several ranking functions into a single termination argument. Transition predicate abstraction automates the computation of transition invariants using automated theorem proving techniques. Together, transition invariants and transition predicate abstraction overcome critical deficiencies of the classical proof method for program termination. The classical method for proving program termination is based on the construction of a single ranking function for the entire program. This construction cannot be supported by the abstraction of a program into a finite-state program (each finite-state program with n states will contain a loop to accomodate executions with length greater than n).

Transition invariants and transition predicate abstraction were introduced in [3] and [4], respectively (we refer to [3,4] for the discussion of related work). Here, we use a uniform setting in order to present the two concepts together (as it was done in the earlier technical report [1]). Originally, the concepts of transition invariants and transition predicate abstraction were used to formulate a proof rule and an abstraction-based algorithm for the verification of liveness properties of concurrent programs under fairness assumptions. This note applies the two concepts for proving termination of sequential programs. The purpose of this note is to provide a short, direct, and comprehensive access to the underlying principles.

2 Preliminaries

We abstract away from a particular programming language and use transition relations to describe programs. To further simplify the presentation, our

P.A. Abdulla and K.R.M. Leino (Eds.): TACAS 2011, LNCS 6605, pp. 3–10, 2011.

```
11: y := read_int();
12: while (y > 0) {
      y := y-1;
    }
```

$$\rho_1 : pc = \ell_1 \wedge pc' = \ell_2$$
$$\rho_1 : pc = \ell_2 \wedge pc' = \ell_2 \wedge y > 0 \wedge y' = y - 1$$

$$T_1 : pc = \ell_1 \wedge pc' = \ell_2$$
$$T_2 : y > 0 \wedge y' < y$$

Fig. 1. Program ANY-Y contains unbounded non-determinism at line 11. The union of binary relations ρ_1 and ρ_2 is the transition relation of the program. Termination of ANY-Y cannot be proved with ranking functions ranging over the set of natural numbers (the initial rank must be at least the ordinal ω). The union of binary relations T_1 and T_2 is a transition invariant for ANY-Y. The binary relations T_1 and T_2 are both well-founded (T_1 does not have a chain longer than 1; T_2 does not have a chain longer than the value of y in its starting element). Thus we have a disjunctively well-founded transition invariant for ANY-Y, which proves the termination of ANY-Y.

definition of programs does not specify a particular set of initial states. We assume that the program can be started from any state.

Definition 1 (Transition-based program). *We define a* program *as a triple*

$$P = (\Sigma, \mathcal{T}, \rho),$$

consisting of:

- *a set of* states Σ,
- *a finite set of* transitions $\mathcal{T}$, *which can be thought of as labels of program statements, and*
- *a function ρ which assigns to each transition a binary* transition relation *over states,*

$$\rho_\tau \subseteq \Sigma \times \Sigma, \qquad for\ \tau \in \mathcal{T}.$$

The transition relation *of P, denoted R_P, comprises the transition relations ρ_τ of all transitions $\tau \in \mathcal{T}$, i.e.,*

$$R_P = \bigcup_{\tau \in \mathcal{T}} \rho_\tau.$$

A program state is a valuation of program variables, including the program counter. An assertion over program variables denotes a set of program states, while an assertion over program variables and their primed versions denotes a binary relation over program states. We identify sets and binary relations by assertions denoting them.

See Figure 1 for an example of a program and its formal representation in terms of its transition relations.

A program P is *terminating* if its transition relation R_P is well-founded. This means that the relation R_P does not have an infinite chain, i.e., an infinite sequence

$$s_1, s_2, s_3, \ldots$$

where each pair of successive states (s_i, s_{i+1}) is contained in the relation R_P.

3   Disjunctively Well-Founded Transition Invariants

In this section we give a brief description of terminology and results of [3] restricted to termination ([3] also deals with general liveness properties and fairness). We write r^+ to denote the transitive closure of a relation r.

Definition 2 (Transition invariant). *Given a program $P = (\Sigma, T, \rho)$, a transition invariant T is a binary relation over states T that contains the program's transition relation R_P^+, i.e.,*

$$R_P^+ \subseteq T .$$

Definition 3 (Disjunctively well-founded relation). *A relation T is disjunctively well-founded if it is a finite union of well-founded relations:*

$$T = T_1 \cup \cdots \cup T_n .$$

Theorem 1 (Proof rule for termination). *A program P is terminating if and only if there exists a disjunctively well-founded transition invariant for P.*

Proof. "Only if" ($\Rightarrow$) is trivial: if P is terminating, then both R_P and R_P^+ are well-founded. Choose $n = 1$ and $T_1 = R_P^+$.

"If" ($\Leftarrow$): we show that if P is *not* terminating and $T_1 \cup \cdots \cup T_n$ is a transition invariant, then some T_i is not well-founded. Nontermination of P means there exists an infinite computation:

$$s_0 \xrightarrow{\tau_1} s_1 \xrightarrow{\tau_2} s_3 \xrightarrow{\tau_3} \ldots$$

Let a choice function f satisfy

$$f(k, \ell) \in \{ \, T_i \mid (s_k, s_\ell) \in T_i \, \}$$

for $k, \ell \in \mathbb{N}$ with $k < \ell$. (The condition $R_P^+ \subseteq T_1 \cup \cdots \cup T_n$ implies that f exists, but does not define it uniquely.) Define equivalence relation $\simeq$ on f's domain by

$$(k, \ell) \simeq (k', \ell') \text{ if and only if } f(k, \ell) = f(k', \ell')$$

Relation $\simeq$ is of finite index since the set of T_i's is finite. By Ramsey's Theorem there exists an infinite sequence of natural numbers $k_1 < k_2 < \ldots$ and fixed $m, n \in \mathbb{N}$ such that

$$(k_i, k_{i+1}) \simeq (m, n) \qquad \text{for all } i \in \mathbb{N}.$$

Hence $(s_{k_i}, s_{k_{i+1}}) \in T_{f(m,n)}$ for all i. This is a contradiction: $T_{f(m,n)}$ is not well-founded. □

The proof of Theorem 1 uses a weak version of Ramsey's theorem. This version states that every infinite complete graph that is colored with finitely many colors contains a monochrome infinite *path* (as opposed to a monochrome infinite *complete subgraph*, in the strong version of Ramsey's theorem).

```
l1: while (x => 0) {
       y := 1;
l2:    while (y < x) {
          y := y+1;
       }
       x := x-1;
    }
```

$$\rho_1 : pc = \ell_1 \wedge pc' = \ell_2 \wedge x \geq 0 \wedge x' = x \wedge y' = 1$$
$$\rho_2 : pc = \ell_2 \wedge pc' = \ell_2 \wedge y < x \wedge x' = x \wedge y' = y + 1$$
$$\rho_3 : pc = \ell_2 \wedge pc' = \ell_1 \wedge y \geq x \wedge x' = x - 1 \wedge y' = y$$

$$T_1 : pc = \ell_1 \wedge pc' = \ell_2$$
$$T_2 : pc = \ell_2 \wedge pc' = \ell_1$$
$$T_3 : x \geq 0 \wedge x' < x$$
$$T_4 : x - y > 0 \wedge x' - y' < x - y$$

Fig. 2. Program BUBBLE contains a nested loop. Termination of BUBBLE is classically shown with the *lexicographic* ranking function $\langle x, x - y \rangle$ defined by the pair of the ranking functions x and $x - y$. The disjunctively well-founded transition invariant shown (the union of binary relations $T_1, \ldots, T_4$) does not prescribe an order between the two ranking functions x and $x - y$ (which have T_3 and T_4 as the corresponding *ranking relations*).

```
1: while (x > 0 && y > 0) {
      if (read_int()) {
         (x, y) := (x-1, x);
      } else {
         (x, y) := (y-2, x+1);
      }
   }
```

$$\rho_1 : pc = pc' = \ell \wedge x > 0 \wedge y > 0 \wedge$$
$$x' = x - 1 \wedge y' = x$$
$$\rho_2 : pc = pc' = \ell \wedge x > 0 \wedge y > 0 \wedge$$
$$x' = y - 2 \wedge y' = x + 1$$

$$T_1 : x > 0 \wedge x' < x$$
$$T_2 : y > 0 \wedge y' < y$$
$$T_3 : x + y > 0 \wedge x' + y' < x + y$$

Fig. 3. Program CHOICE contains a non-determinstic choice between two simultaneous assignment statements in the loop body. The disjunctively well-founded transition invariant shown (the union of binary relations $T_1, \ldots, T_3$) presents the ranking relations for the three ranking functions x, y, and $x + y$, none of which by itself suffices to prove terminations.

See Figures 1, 2, 3, and 4 for examples of disjunctively well-founded transition invariants.

As a consequence of the above theorem, we can prove termination of a program P as follows. We compute a disjunctively well-founded superset of the transitive closure of the transition relation of the program P, i.e., we construct a finite number of well-founded relations $T_1, \ldots, T_n$ whose union covers R_P^+. We need to show that the inclusion $R_P^+ \subseteq T_1 \cup \cdots \cup T_n$ indeed holds, and we need to show that each of the relations $T_1, \ldots, T_n$ is indeed well-founded. Transition predicate abstraction can be used to obtain an automation of the three steps, as shown in the next section.

```
1: while (x > 0 && y > 0) {
      if (read_int()) {
        x := x-1;
        y := read_int();
      } else {
        y := y-1;
      }
   }
```

$$\rho_1 : pc = pc' = \ell \wedge x > 0 \wedge y > 0 \wedge$$
$$x' = x - 1$$
$$\rho_2 : pc = pc' = \ell \wedge x > 0 \wedge y > 0 \wedge$$
$$x' = x \wedge y' = y - 1$$

$$T_1 : x \geq 0 \wedge x' < x$$
$$T_2 : y > 0 \wedge y' < y$$

Fig. 4. The program XORY shown contains a non-deterministic choice between two simultaneous assignment statements in the loop body. The first one decrements x and erases y (assigns a non-deterministic value to y). The second decrements y. The validity of the disjunctively well-founded transition invariant shown (the union of the ranking relations for the two ranking functions x and y) is shown by transition predicate abstraction.

4   Transition Predicate Abstraction (TPA)

A transition predicate is a binary relation over program states. Transition predicate abstraction [4] is a method to compute transition invariants, just as predicate abstraction is a method to compute invariants.

Definition 4 (Set of abstract transitions $\mathcal{T}_{\mathcal{P}}^{\#}$). *Given the set of transition predicates $\mathcal{P}$, the set of abstract transitions $\mathcal{T}_{\mathcal{P}}^{\#}$ is the set that contains for every subset of transition predicates $\{p_1, \ldots, p_m\} \subseteq \mathcal{P}$ the conjunction of these transition predicates, i.e.,*

$$\mathcal{T}_{\mathcal{P}}^{\#} = \{p_1 \wedge \ldots \wedge p_m \mid 0 \leq m \text{ and } p_i \in \mathcal{P} \text{ for } 1 \leq i \leq m\} \ .$$

The set of abstract transitions $\mathcal{T}_{\mathcal{P}}^{\#}$ is closed under intersection, and it contains the assertion *true* (the empty intersection, corresponding to the case $m = 0$), which denotes the set of all pairs of program states.

Example 1. Consider the following set of transition predicates.

$$\mathcal{P} = \{x' = x, x' < x, y' < y\}$$

The set of abstract transitions $\mathcal{T}_{\mathcal{P}}^{\#}$ is

$$\{\mathsf{true}, x' = x, x' < x, y' < y, x' = x \wedge y' < y, x' < x \wedge y' < y, \mathsf{false}\} \ .$$

The abstract transition written as true is the set of all state pairs $\Sigma \times \Sigma$ and is the empty conjunction of transition predicates. The abstract transition written as false is the empty relation; e.g., the conjunction of $x = x'$ and $x > x'$ is false.

We next define a function that assigns to a binary relation T over states the least (wrt. inclusion) abstract transition that is a superset of T.

Definition 5 (Abstraction function α). *A set of transition predicates $\mathcal{P}$ defines the* abstraction function

$$\alpha : 2^{\Sigma \times \Sigma} \to \mathcal{T}_{\mathcal{P}}^{\#}$$

which assigns to a relation $r \subseteq \Sigma \times \Sigma$ the smallest abstract transition that is a superset of r, i.e.,

$$\alpha(r) = \bigwedge \{p \in \mathcal{P} \mid r \subseteq p\}.$$

We note that α is extensive. i.e., the inclusion

$$r \subseteq \alpha(r)$$

holds for any binary relation over states $r \subseteq \Sigma \times \Sigma$.

Example 2. After taking the transition predicates $x > 0$ and $y > 0$, which leave the primed variables unconstrained, into consideration the application of the abstraction function α to the transition relations ρ_1 and ρ_2 of the program in Figure 4 results in the following abstract transitions.

$$\alpha(\rho_1) = x > 0 \wedge y > 0 \wedge x' < x$$
$$\alpha(\rho_2) = x > 0 \wedge y > 0 \wedge x' = x \wedge y' < y$$

We next present an algorithm that uses the abstraction α to compute (a set of abstract transitions that represents) a transition invariant. The algorithm terminates because the set of abstract transitions $\mathcal{T}_{\mathcal{P}}^{\#}$ is finite.

Algorithm 1 (TPA).
Transition invariants via transition predicate abstraction.

    Input: *program $P = (\Sigma, \mathcal{T}, \rho)$*
               set of transition predicates $\mathcal{P}$
               abstraction α defined by $\mathcal{P}$ (according to Def. 5)
    Output: *set of abstract transitions $P^{\#} = \{T_1, \ldots, T_n\}$*
               such that $T_1 \cup \cdots \cup T_n$ is a transition invariant

    $P^{\#} := \{\alpha(\rho_\tau) \mid \tau \in \mathcal{T}\}$
    repeat
        $P^{\#} := P^{\#} \cup \{\alpha(T \circ \rho_\tau) \mid T \in P^{\#},\, \tau \in \mathcal{T},\, T \circ \rho_\tau \neq \emptyset\}$
    until *no change*

Our notation $P^{\#}$ for the set of abstract transitions computed by the TPA algorithm stems from [4]. There, $P^{\#}$ is called an abstract transition program. In contrast to [4] we do not consider edges between the abstract transitions, since they only needed for keeping track of fairness assumption when proving fair termination and liveness properties.

Theorem 2 (TPA). *Let $P^{\#} = \{T_1, \ldots, T_n\}$ be the set of abstract transitions computed by Algorithm* TPA. *If every abstract relation $T_1, \ldots, T_n$ is well-founded, then program P is terminating.*

Proof. The union of the abstract relations $T_1 \cup \cdots \cup T_n$ is a transition invariant. If every abstract relation $T_1, \ldots, T_n$ is well-founded, the union $T_1 \cup \cdots \cup T_n$ is a disjunctively well-founded transition invariant and by Theorem 1 the program P is terminating. $\square$

Example 3. Consider the program P in Figure 4 and the set of transition predicates $\mathcal{P}$ in Example 1. The output of Algorithm TPA is

$$P^{\#} = \{x > x', \quad x = x' \wedge y > y'\}$$

Both abstract transitions in $P^{\#}$ are well-founded. Hence P is terminating. In fact, it is sufficient to show the well-foundedness of the (simpler) binary relations T_1 and T_2 given in Figure 4. The transition invariant given there is valid because it contains the one defined by $P^{\#}$.

Each abstract transition in $P^{\#}$ (the representation of a transition invariant computed by the transition-predicate abstraction-based algorithm) is a conjunction of transition predicates. Thus it corresponds to a conjunction $g \wedge u$ of a *guard* formula g which contains only unprimed variables, and an *update* formula u which contains primed variables, for example $x > 0 \wedge x > x'$. Thus it denotes the transition relation of a *simple* while program of the form `while g { u }`, for example, `while (x > 0) { assume(x > x'); x := x' }`. The well-foundedness of the abstract transition is thus equivalent to the termination of the simple while program. Since we have fast and complete procedures that find ranking functions for such programs [2], we can automate also the third step of transition invariant-based termination proofs, as outlined in the previous section.

5 Conclusion

We have presented disjunctively well-founded transition invariants as the basis of a new proof rule for program termination, and transition predicate abstraction as the basis of its automation. As a result, we obtain the foundation for a new class of automatic methods for proving program termination.

Acknowledgements. Discussions with Neil Jones and Chin Soon Lee started this work. We thank Amir Pnueli for inspiration and encouragement, and for suggesting the terminology of disjunctively well-founded transition invariants.

References

1. Podelski, A., Rybalchenko, A.: Software model checking of liveness properties via transition invariants. Technical Report MPI-I-2003-2-004, Max-Planck-Institut für Informatik (December 2003),
   http://domino.mpi-inf.mpg.de/internet/reports.nsf/NumberView/2003-2-004

2. Podelski, A., Rybalchenko, A.: A complete method for the synthesis of linear rank-ing functions. In: Steffen, B., Levi, G. (eds.) VMCAI 2004. LNCS, vol. 2937, pp. 239–251. Springer, Heidelberg (2004)
3. Podelski, A., Rybalchenko, A.: Transition invariants. In: LICS 2004: Proceedings of the 19th Annual IEEE Symposium on Logic in Computer Science, Washington, DC, USA, pp. 32–41. IEEE Computer Society Press, Los Alamitos (2004)
4. Podelski, A., Rybalchenko, A.: Transition predicate abstraction and fair termina-tion. In: POPL 2005: Proceedings of the ACM SIGPLAN-SIGACT Symposium on Principles of Programming Languages, vol. 32, pp. 132–144. ACM, New York (2005)

Sound and Complete Monitoring of Sequential Consistency for Relaxed Memory Models

Jabob Burnim, Koushik Sen, and Christos Stergiou

EECS Department, University of California, Berkeley
{jburnim,ksen,chster}@cs.berkeley.edu

Abstract. We present a technique for verifying that a program has no executions violating sequential consistency (SC) when run under the relaxed memory models Total Store Order (TSO) and Partial Store Order (PSO). The technique works by monitoring sequentially consistent executions of a program to detect if similar program executions could fail to be sequentially consistent under TSO or PSO. We propose novel monitoring algorithms that are sound and complete for TSO and PSO—if a program can exhibit an SC violation under TSO or PSO, then the corresponding monitor can detect this on some SC execution. The monitoring algorithms arise naturally from the operational definitions of these relaxed memory models, highlighting an advantage of viewing relaxed memory models operationally rather than axiomatically. We apply our technique to several concurrent data structures and synchronization primitives, detecting a number of violations of sequential consistency.

1 Introduction

Programmers writing concurrent software often assume that the underlying memory model is sequentially consistent. However, sequential consistency strongly constrains the ordering of memory operations, which can make it difficult to achieve high performance in commodity microprocessors [9,20]. Thus, to enable increased concurrency and performance, processors often provide a *relaxed* memory model. Unfortunately, working with relaxed memory models often requires subtle and difficult reasoning [9,20].

Nevertheless, developers of high-performance concurrent programs, such as lock-free data-structures and synchronization libraries, often use regular load and store operations, atomic compare-and-swap-like primitives, and explicit data races instead of locks to increase performance. Concurrency bugs are notoriously hard to detect and debug; relaxed memory models make the situation even worse.

Recently, there has been great interest in developing techniques for the verification and analysis of concurrent programs under relaxed memory models [9,20,13,3,15,4,2,10]. In a promising and practical approach for such verification, Burckhardt and Musuvathi [4] argued that programmers, despite using ad-hoc synchronization, expect their program to be sequentially consistent. They proposed SOBER, which monitors sequentially consistent executions to detect violations of sequential consistency (SC). A key observation made in their work

P.A. Abdulla and K.R.M. Leino (Eds.): TACAS 2011, LNCS 6605, pp. 11–25, 2011.

is that, for the Total Store Order (TSO) [23] memory model (which is quite similar to that of the x86 architecture [21]), if a program execution under TSO violates sequential consistency (SC), then this fact can be detected by examining some sequentially consistent execution of the program. Therefore, if run-time monitoring is combined with a traditional model checker, which explores all sequentially consistent executions of the program, then all violations of SC under TSO can be detected. Burckhardt and Musuvathi [4] use an axiomatic definition of SC and TSO to derive the SOBER monitoring algorithm.

In this paper, we develop two novel monitoring algorithms for detecting violations of sequential consistency (SC) under relaxed memory models Total Store Order (TSO) [23] and Partial Store Order (PSO) [23]. Each algorithm, when monitoring a sequentially consistent execution of a program, simulates a similar TSO or PSO execution, reporting if this similar execution can ever violate sequential consistency. We prove both monitors *sound*—if they report a warning, then the monitored program can violate SC under TSO or PSO—and *complete*—if a program can violate SC under TSO or PSO, then the corresponding monitor can detect this fact by examining some sequentially consistent execution.

Rather than working with axiomatic definitions of these relaxed memory models, as [4] does, we derive our algorithms from *operational* definitions of TSO and PSO. We show that this alternate approach naturally leads to fundamentally different monitoring algorithms, with several advantages over SOBER.

One advantage of our operational approach is that our monitoring algorithms follow simply from the operational definitions of TSO and PSO. While monitoring algorithms based on axiomatic definitions require the design of complex vector clocks, in addition to the standard vector clocks to track the traditional happens-before relation, our approach can directly "run" the operational memory model. Thus, we were easily able to develop a monitoring algorithm for PSO, in addition to TSO—such an extension is unknown for the SOBER algorithm.

Another advantage of our algorithms is that they have a run-time complexity of $O(N \cdot P)$ when monitoring a sequentially consistent execution with N shared memory operations and P processes. This complexity is an improvement over the run-time complexity $O(N \cdot P^2)$ of the SOBER algorithm. We see this improvement in run-time complexity because we do not need to maintain additional vector clocks for TSO and PSO.

Further, in developing our monitoring algorithms, we discovered a bug in the SOBER algorithm—in its axiomatic definition of TSO—which makes the algorithm incomplete. We believe that this bug is quite subtle. The same bug is also present [21] in the 2007 Intel® 64 Architecture Memory Order White Paper. In this paper, we identify and correct the error [8] in the SOBER algorithm arising from a too-strict axiomatic definition of the TSO memory model.

We have implemented our monitoring algorithms for C programs in THRILLE [16] and, combined with THRILLE's preemption-bounded model checking [19], have applied our monitors to two mutual exclusion algorithms and several concurrent data structures. Our experiments show that we can detect sequential consistency violations under TSO or PSO in all of these benchmarks.

Other Related Work. There have been many efforts to verify or check concurrent programs on relaxed memory models [9,20,13,3,15,4,2,10]. Some of these techniques [13,3] encode a program and the underlying memory model as a constraint system and use a constraint solver to find bugs. Other techniques [9,20,15] explicitly explore the state space of a program to find bugs.

Recently, [10] proposed an adversarial memory for testing at run-time if a data race in a Java program could be harmful under Java's memory model.

2   Preliminaries

We consider a parallel program P to consist of a number of *parallel threads*, each with a thread-local state, that communicate though a *shared memory*. The *execution* of a parallel program consists of a sequence of program *steps*. In each step, one of the program threads *issues* a shared memory *operation*—e.g., a read or write of a shared variable—and then updates its local state based on its current local state and anything returned by the memory operation—e.g., the value read for a shared variable.

Below, we will give operational definitions of relaxed memory models TSO and PSO as abstract machines. These abstract machine definitions are designed to be simple for a programmer to understand and to reason about, not to exactly describe the internal structure or operation of real hardware processors. For example, our operational definitions contain no caches or mechanisms for ensuring cache coherency. At the same time, these operational definitions are faithful in that they allow exactly the program behaviors allowed by more traditional axiomatic definitions of TSO and PSO.

2.1   Programming Model

Let *Proc* be set of all program processes or thread identifiers and *Value* be the set of possible program values. Then, we define the set *Event* of shared memory operations, or *events*, to consist of all:

- Stores $\mathbf{st}(p, a, v)$ by process $p \in Proc$ to address $a \in Adr$ of $v \in Value$.
- Loads $\mathbf{ld}(p, a)$ by process $p \in Proc$ of address $a \in Adr$.
- Atomic operations $\mathbf{atm}(p, a, f)$ by $p \in Proc$ on $a \in Adr$, which store $f(v)$ to address a after reading $v \in Value$ from a, where $f : Value \rightarrow Value$.
  This operations models atomic shared memory primitives such as compare-and-swap (CAS), fetch-and-add, test-and-set, etc.[1]

Note that we need not explicitly include a memory fence operation. In our operational models, memory barriers can be simulated by atomic operations $\mathbf{atm}(p, a, f)$, which restrict reordering or delaying of earlier memory ops.

---

[1] For example, a CAS(a, old, new) by process p is modeled by $\mathbf{atm}(p, a, f)$, where $f = (\lambda x. \text{ if } x = old \text{ then } new \text{ else } x)$. The CAS "succeeds" when it reads value old, and "fails" otherwise.

We denote the process p and address a of an event $e \in Event$ by $p(e)$ and $a(e)$, respectively.

We now formalize programs as independent processes with local state and communicating only via the above shared memory operations. We abstract unnecessary details such as the source language, control flow, and structure of the process-local state. We define a program P to be a tuple $(s_0, next, update)$ where:

- Function $s_0 : Proc \rightarrow \Sigma$ is the initial program state, mapping process p to its thread-local state $s_0(p) \in \Sigma$, where Σ is the set of all possible local states.
- Partial function $next : Proc \times \Sigma \rightarrow Event$ indicates the next *memory operation* or *event*, $next(p, \sigma)$, that process p issues when in local state σ. If $next$ is undefined for (p, σ), denoted $next(p, \sigma) = \bot$, then process p is complete. Program P terminates when $next(p, s(p)) = \bot$ for all p.
- Function $update : Proc \times \Sigma \times Value \rightarrow \Sigma$ indicates the new local state $update(p, \sigma, v)$ of process p after it receives response v to op $next(p, \sigma)$.

We similarly formalize a memory model MM as a labeled transition system with initial state m_0 and with transitions labeled by events from $Event$, paired with memory model responses. We also allow memory model MM to have a set of labeled internal transitions τ_{MM}, to model any internal nondeterminism of the memory model. Then, an *execution* of program P under memory model MM is a sequence of labeled transitions:

$$(s_0, m_0) \xrightarrow{l_1} (s_1, m_1) \xrightarrow{l_2} \cdots \xrightarrow{l_n} (s_n, m_n)$$

where each transition $(s_{i-1}, m_{i-1}) \xrightarrow{l_i} (s_i, m_i)$ is either labeled by an ordered pair $(e_i, r_i) \in Event \times Value$, in which case:

- $e_i = next(p(e_i), s_{i-1}(p(e_i)))$
- $s_i(p(e_i)) = update(p(e_i), s_{i-1}(p(e_i)), r_i)$, where r_i is the value returned for e_i, which is $\bot$ for stores.
- $s_i(p') = s_{i-1}(p')$ for $p' \neq p(e_i)$
- $m_{i-1} \xrightarrow{(e_i, r_i)} m_i$ is a transition in MM

or is labeled by an internal transition from τ_{MM}, in which case:

- $s_i = s_{i-1}$, and $m_{i-1} \xrightarrow{l_i} m_i$ is a transition in MM

In this model, there are two sources of nondeterminism in a program execution: (1) The thread schedule—i.e., at each step, which process p executes a transition, and (2) the internal nondeterminism of the memory model.

3   Operational Memory Models

We now give our operational definitions for three memory models: sequential consistency (SC) and relaxed memory models TSO and PSO. Fundamentally, these definitions are equivalent to operational definitions given by other researchers for SC [14,5,2,18], TSO [14,5,2,21,18], and PSO [14,2,18].

In our presentation, we aim for definitions that provide a simple and easy to understand model for a programmer. We present each memory model as a library or module with an internal state, representing the abstract state of a shared memory, and with methods **store**(p, a, v), **load**(p, a), and **atomic**(p, a, f) by which a program interacts with the shared memory. The memory model executes all such methods atomically—i.e. one at a time and uninterrupted. Additionally, memory models TSO and PSO each have an *internal* method **store**^{c}. Each memory model is permitted to nondeterministically call this internal method, with any arguments, whenever no other memory model method is being executed.

A note about connecting these definitions to our formalism in Section 2:

- $m \xrightarrow{\mathbf{ld}(p,a),v} m$ when **load**(p, a), run in memory model state m, returns v. (Note that **load**(p, a) does not modify the memory model state m.)
- $m \xrightarrow{\mathbf{st}(p,a,v),\perp} m'$ when **store**(p, a, v), run in state m, yields state m'.
- $m \xrightarrow{\mathbf{atm}(p,a,f),v} m'$ when **atomic**(p, a, f), run in m, returns v and yields state m'.

Sequential Consistency (SC). Our operational definition of SC [17] is given in Figure 1. The SC abstract machine simply models shared memory as an array m mapping addresses to values, reading from and writing to this array on loads, stores, and atomic operations.

Total Store Order (TSO). Our definition of TSO [23] is given in Figure 2. In addition to modeling shared memory as array m, mapping addresses to values, the TSO abstract machine has a FIFO *write buffer* $B[p]$ for each process p.

We omit a proof that our operational definition is equivalent to more traditional axiomatic ones. Our model is similar to the operational definitions in [4] and [21], both of which are proved equivalent to axiomatic definitions of TSO. Conceptually, the per-process write buffers allow stores to be reordered or delayed past later loads, while ensuring that each process's stores become globally visible in the order in which they are performed. And there is a total order on all stores—the order in which stores *commit*—that respects the program order.

Partial Store Order (PSO). Our operational definition of PSO [23] is given in Figure 3. Our PSO abstract machine is very similar to that of TSO, except that pending writes are stored in per-process and *per-address* write buffers. Internal method **store**$^c_{PSO}(p, a)$ commits the oldest pending store *to address a* and **atomic**$_{PSO}(p, a, f)$ commits/flushes only pending stores *to address a*.

---

m[*Adr*] : *Val*

store$_{SC}(p, a, v)$:
    $m[a] := v$

load$_{SC}(p, a)$:
    return $m[a]$

atomic$_{SC}(p, a, f)$:
    $ret := m[a]$
    $m[a] := f(m[a])$
    return ret

Fig. 1. Operational Model of Sequential Consistency (SC)

m[Adr] : Val
B[$Proc$] : **FIFOQueue** of (Adr, Val)

store$_{\mathrm{TSO}}(p, a, v)$:
  $B[p].addLast(a, v)$

store$_{\mathrm{TSO}}^{c}(p)$:
  if not $B[p].empty()$:
    $(a, v) := B[p].removeFirst()$
    $m[a] := v$

load$_{\mathrm{TSO}}(p, a)$:
  if $B[p].contains((a, *))$:
    $(a, v) :=$ last element $(a, *)$ of $B[p]$
    return v
  else:
    return $m[a]$

atomic$_{\mathrm{TSO}}(p, a, f)$:
  while not $B[p].empty()$:
    store$_{\mathrm{TSO}}^{c}(p)$
  $ret := m[a]$
  $m[a] := f(m[a])$
  return ret

Fig. 2. Operational Model of TSO

m[Adr] : Val
B[$Proc$][Adr] : **FIFOQueue** of Val

store$_{\mathrm{PSO}}(p, a, v)$:
  $B[p][a].addLast(v)$

store$_{\mathrm{PSO}}^{c}(p, a)$:
  if not $B[p][a].empty()$:
    $m[a] := B[p][a].removeFirst()$

load$_{\mathrm{PSO}}(p, a)$:
  if not $B[p][a].empty()$:
    return $B[p][a].getLast()$

  else:
    return $m[a]$

atomic$_{\mathrm{PSO}}(p, a, f)$:
  while not $B[p][a].empty()$:
    store$_{\mathrm{PSO}}^{c}(p, a)$
  $ret := m[a]$
  $m[a] := f(m[a])$
  return ret

Fig. 3. Operational Model of PSO

4   Violations of Sequential Consistency

In this section, we formally define what it means for a program to have a *violation* of sequential consistency under a relaxed memory model. In Section 5, we will give monitoring algorithms for detecting such violations under TSO and PSO by examining only the SC executions of a program.

4.1   Execution Traces

If a program exhibits some behavior in a TSO or PSO execution, we would like to say that the behavior is not sequentially consistent if there is no execution under SC exhibiting the same behavior. We will define a *trace* of an SC, TSO, or PSO execution to capture this notion of the behavior of a program execution.

Following [12], [4], and [6], we formally define a trace of an execution of a program P to be a tuple $(E, \rightarrow_p, src, \rightarrow_{st})$, where

- $E \subseteq Event$ is the set of shared memory operations or events in the execution.
- For each process $p \in Proc$, relation $\rightarrow_p \subseteq E \times E$ is a total order on the events $e \in E$ from process p—i.e. with $p(e) = p$. In particular, $e \rightarrow_p e'$ iff $p(e) = p(e')$ and e is before e' in the execution. Thus, $\rightarrow_p$ does not relate events from different processes.

Relation $\rightarrow_p$ is called the *program order relation*, and $e \rightarrow_p e'$ indicates that process p issued operation e before issuing e'.

- For each load or atomic operation $e \in E$, partial function $src : Event \rightarrow Event$ indicates the store or atomic operation $src(e) \in Event$ from which e reads its value. $src(e) = \bot$ indicates that e got its value from no store, instead reading the initial default value in the shared memory. Note that $a(src(e)) = a(e)$ whenever $src(e)$ is defined.
- For each address a, relation $\rightarrow_{st} \subseteq E \times E$ is a total order on the stores and atomic operations on a in the execution. In particular, $e \rightarrow_{st} e'$ iff $a(e) = a(e')$ and e becomes globally visible before e' in the execution. Thus, $\rightarrow_{st}$ does not relate events on different addresses.

  Memory models SC, TSO, and PSO all guarantee the existence, for each address, of such a total order in which the writes to that address become globally visible[2]. Note that not all relaxed memory models guarantee the existence of such an ordering.

Note that we can only define a trace for a *complete* TSO or PSO execution—that is, one in which every store has become globally visible or, in the language of our abstract models, committed. If multiple processes have pending writes to the same address, then the execution does not specify an order on those writes and thus does not define a total $\rightarrow_{st}$ relation.

4.2   Sequential Consistency and the Happens-Before Relation

We define a trace of a program P to be sequentially consistent only if it arises from some sequentially consistent execution of P:

Definition 1. *A trace* $T = (E, \rightarrow_p, src, \rightarrow_{st})$ *of a program* P *is* sequentially consistent *iff there exists some execution of* P *under SC with trace* T.

This definition is not very convenient for showing that a trace is *not* sequentially consistent. Thus, following [22,4], we give an axiomatic characterization of SC traces by defining relations $\rightarrow_c$ and $\rightarrow_{hb}$ on the events of a trace:

Definition 2. *Let* $(E, \rightarrow_p, src, \rightarrow_{st})$ *be a trace. Events* $e, e' \in E$ *are related by the* **conflict-order relation**, *denoted* $e \rightarrow_c e'$, *iff* $a(e) = a(e')$ *and one of the following holds:*

- e *is a write*, e' *is a read, and* $e = src(e')$,
- e *is a write*, e' *is a write, and* $e \rightarrow_{st} e'$
- e *is a read*, e' *is a write, and either* $src(e) = \bot$ *or* $src(e) \rightarrow_{st} e'$

Definition 3. *For a trace* $(E, \rightarrow_p, src, \rightarrow_{st})$, *the* happens-before relation *is defined as the union of the program-order and conflict-order relations on* E—*i.e.* $\rightarrow_{hb} = (\rightarrow_p \cup \rightarrow_c)$. *We refer to the reflexive transitive closure of the happens-before relation as* $\rightarrow_{hb}^*$.

---

[2] This property is closely related to *store atomicity* [1]. TSO and PSO do not technically have store atomicity, however, because a process's loads can see the values of the process's earlier stores before those stores become globally visible.

As observed in [22] and [4], a trace is sequentially consistent iff its $\rightarrow_{hb}$ relation is acyclic:

Proposition 1. *Let $T = (E, \rightarrow_p, src, \rightarrow_{st})$ be a trace of an execution of program P. Trace T is sequentially consistent iff relation $\rightarrow_{hb}$ is acyclic on E.*

We define a sequential consistency *violation* as a program execution with a non-sequentially-consistent trace:

Definition 4. *A program P has a* violation of sequential consistency *under relaxed memory model TSO (resp. PSO) iff there exists some TSO (resp. PSO) execution of P with trace T such that T is not sequentially consistent.*

5 Monitoring Algorithms

We describe our monitoring algorithms for TSO and PSO in this section. We suppose here that we are already using a model checker to explore and to verify the correctness of all sequentially consistent executions of program P. A number of existing model checkers [19,16] explore the sequentially consistent, interleaved schedules of a parallel program.

Figures 4 and 5 list our monitor algorithms for TSO and PSO. We present our algorithms as *online* monitors—that is, for an SC execution $(s_0, m_0) \xrightarrow{e_1, r_1} \cdots \xrightarrow{e_n, r_n} (s_n, m_n)$ of some program P, we run **monitor**$_{TSO}$ (respectively **monitor**$_{PSO}$) on the execution by:

- Initializing the internal state B and **last** as described in Figure 4 (resp. 5).
- Then, for each step $e_i \in Event$ in the execution, from e_1 to e_n, we call **monitor**$_{TSO}(e_i)$.
- If any of these calls signals "SC Violation", then P has a sequential consistency violation under TSO (resp., PSO). Note that we do not stop our monitoring algorithm after detecting the first violation. Thus, we find all such violations of sequential consistency along each SC execution.

Conceptually, each monitor algorithm works by simulating a TSO (respectively PSO) version of the given SC execution. Array B simulates the FIFO write buffers in a TSO (resp. PSO) execution, but buffers the pending store *events* rather than just the pending values to be written. Lines 16–25 update these write buffers, queueing a store when the SC execution performs a store and flushing the buffers when the SC execution performs an atomic operation.

But when should this simulated TSO (resp. PSO) execution "commit" these pending stores, making them globally visible? Lines 10–12 commit these pending stores as late as possible while still ensuring that the simulated TSO (resp. PSO) execution has the same trace and $\rightarrow_{hb}$-relation as the SC execution. This is achieved by, just before simulating operation e by process p on address a, committing all pending stores to address a from any other process—these pending stores *happen before* e in the SC execution, so they must be committed to ensure they *happen before* e in the simulated TSO (resp. PSO) execution. Note that, in the TSO monitor, this may commit pending stores to other addresses, as well.

```
1  B[Proc] : FIFOQueue of (Adr, Event)      1  B[Proc][Adr] : FIFOQueue of Event
2  prev[Proc] : Event initialized to ⊥      2  prev[Proc] : Event initialized to ⊥
3                                            3
4  monitor_TSO(e):                           4  monitor_PSO(e):
5  // Could simulation have →_hb−cycle?      5  // Could simulation have →_hb−cycle?
6  if ∃(a(e), e′) ∈ B[p(e′)] with            6  if ∃e′ ∈ B[p(e′)][a(e)] with
       p(e′) ≠ p(e) ∧ e′ →*_hb prev[p(e)]:        p(e′) ≠ p(e) ∧ e′ →*_hb prev[p(e)]:
7     signal "SC Violation"                  7     signal "SC Violation"
8                                            8
9  // Ensure equivalence to SC.              9  // Ensure equivalence to SC.
10 while ∃(a(e), ∗) ∈ B[p′] with p′ ≠ p(e): 10 while ∃p′ ≠ p(e) with not
                                                          B[p′][a(e)].empty():
11    B[p′].removeFirst()                    11    B[p′][a(e)].removeFirst()
12    emit st^c(p′)                          12    emit st^c(p′, a(e))
13                                            13
14 // Execute e in TSO simulation.           14 // Execute e in PSO simulation.
15 prev[p(e)] := e                           15 prev[p(e)] := e
16 if e =st(p, a, v):                        16 if e =st(p, a, v):
17    B[p].addLast(a, e)                      17    B[p][a].addLast(e)
18    emit st(p, a, v)                       18    emit st(p, a, v)
19 else if e =ld(p, a):                      19 else if e =ld(p, a):
20    emit ld(p, a)                           20    emit ld(p, a)
21 else if e =atm(p, a, f):                   21 else if e =atm(p, a, f):
22    while not B[p].empty():                 22    while not B[p][a].empty():
23       B[p].removeFirst()                    23       B[p][a].removeFirst()
24       emit st^c(p)                          24       emit st^c(p, a)
25    emit atm(p, a, f)                        25    emit atm(p, a, f)
```

Fig. 4. Monitoring algorithm for TSO   **Fig. 5.** Monitoring algorithm for PSO

But first, Line 6 of $\textbf{monitor}_{TSO}(e)$, resp. $\textbf{monitor}_{PSO}(e)$, checks if we can create a violation of sequential consistency by executing memory operation e in the TSO (resp. PSO) simulation *before* committing any pending and conflicting stores. That is, suppose e is an operation on address a by process p, and in our simulated TSO (resp. PSO) execution there is a pending store $e′$ to a by process $p′ \neq p$. In the TSO (resp. PSO) execution, we can force $e \to_c e′$ by executing e (and committing e, if it is a store) before committing $e′$. Further, suppose that $e′$ satisfies the rest of the condition at Line 6. That is, $e′ \to^*_{hb} \textbf{prev}[p]$—in the trace of the SC execution, event $e′$ *happens before* the event $\textbf{prev}[p]$ issued by process p just before e. Then, as proved in Theorem 1, in the trace of the simulated TSO (resp. PSO) execution we will have $e′ \to^*_{hb} \textbf{prev}[p] \to_p e \to_c e′$. This is a cycle in the $\to_{hb}$-relation, indicating that the simulated TSO (resp. PSO) execution is not sequentially consistent.

In order to track the classic $\to^*_{hb}$ relation on the trace of the SC execution that we are monitoring, we use a well-known vector clock algorithm. The algorithm has a time complexity of $O(N \cdot P)$ on a trace/execution of length N with P processes. A short description of the algorithm can be found in, e.g., [11].

Theorem 1. *Algorithms **monitor**$_{TSO}$ and **monitor**$_{PSO}$ are sound monitoring algorithms for TSO and PSO, respectively. That is, whenever either reports a violation of sequential consistency given an SC execution of a program P, then P really has a violation of sequential consistency under TSO (resp. PSO).*

Theorem 2. *Algorithms **monitor**$_{TSO}$ and **monitor**$_{PSO}$ are complete for TSO and PSO, respectively. That is, if program P has a violation of sequential consistency under TSO (resp. PSO), then there exists some SC execution of P on which **monitor**$_{TSO}$ (resp. **monitor**$_{PSO}$) reports an SC violation.*

Theorem 3. *On a sequentially consistent execution of length N on P processes, monitoring algorithms **monitor**$_{TSO}$ and **monitor**$_{PSO}$ run in time $O(N \cdot P)$.*

We sketch the proofs of these results in the Appendix. Complete proofs can be found in our accompanying technical report.

6   Comparison to SOBER

Our work is inspired by SOBER [4], a previous monitoring algorithm that detects program executions under TSO that violate sequential consistency by examining only SC executions. SOBER is derived from the axiomatic characterization of relaxed memory model TSO, while we work from operational definitions of TSO and PSO. There are four key differences between our work and SOBER.

First, we give monitor algorithms for detecting sequential consistency violations under both TSO and the more relaxed PSO memory model, while SOBER detects only violations under TSO.

Second, the run-time complexity of our algorithms is $O(N \cdot P)$, where P is the number of processors and N is the length of the monitored SC execution. This is an improvement over the complexity $O(N \cdot P^2)$ of the SOBER algorithm. The additional factor of $O(P)$ in SOBER is from a vector clock algorithm to maintain the *relaxed happens-before relation*, which axiomatically defines the behaviors legal under TSO. In contrast, when working from our operational definitions for TSO and PSO, there is no need for such additional vector clocks.

Third, the SOBER monitoring algorithm is more sensitive than our **monitor**$_{TSO}$. That is, for some programs there exist individual sequentially consistent executions for which SOBER will report the existence of an SC violation while **monitor**$_{TSO}$ will not. However, this does not affect the completeness of our monitoring algorithms—for such programs, there will always exist some other SC execution on which **monitor**$_{TSO}$ will detect and report a violation of sequential consistency. In our experimental evaluation, this reduced sensitivity does not seem to hinder our ability to find violations of sequential consistency when combining our monitors with preemption-bounded model checking [19].

Fourth, we believe that working with operational definitions for relaxed memory models TSO and PSO is both simpler than working with axiomatic definitions and leads to more natural and intuitive monitoring algorithms. As evidence for this belief, we note that we have discovered [8] a subtle error in the axiomatic definition of TSO given in [4], which leads SOBER to fail to detect some real violations of sequential consistency under TSO. This error has been confirmed [7] both by the authors of [4] and by [18]. We discuss this error, and how to correct it, in our accompanying technical report.

7   Experimental Evaluation

In order to experimentally evaluate our monitor algorithms, we have implemented **monitor**$_{TSO}$ and **monitor**$_{PSO}$ on top of the THRILLE [16] tool for model checking, testing, and debugging parallel C programs.

In our experiments, we use seven benchmarks. The names and sizes of these benchmarks are given in Columns 1 and 2 of Tables 1 and 2. Five are implementations of concurrent data structures taken from [3]: msn, a non-blocking queue, ms2, a two-lock queue, lazylist, a list-based concurrent set, harris, a non-blocking set, and snark, a double-ended queue (dequeue). The other two benchmarks are implementations of Dekker's algorithm and Lamport's bakery algorithm for mutual exclusion. Previous research [3,4] has demonstrated that the benchmarks have sequential consistency violations under relaxed memory models without added memory fences. For each of the benchmarks we have manually constructed a test harness.

In our experimental evaluation, we combine our monitoring algorithms with THRILLE's *preemption-bounded* [19] model checking. That is, we run **monitor**$_{TSO}$ and **monitor**$_{PSO}$ on all sequentially consistent executions of each benchmark with a bounded number of preemptive context switches. This verification is not complete—because we do not apply our monitor algorithms to *every* SC execution, we may miss possible violations of SC under TSO or PSO. We evaluate only whether our monitoring algorithms are effective in finding real violations of SC when combined with a systematic but incomplete model checker.

We run two sets of experiments, one with a preemption bound of 1 and the other with a preemption bound of 2. Columns 3 and 4 list the number of parallel interleavings explored and the total time taken with a preemption bound of 1 (Table 1) and a bound of 2 (Table 2). The cost of running our unoptimized monitor implementations on every sequentially consistent execution adds an overhead of roughly 20% to THRILLE's model checking for the data structure benchmarks and about 100% to the mutual exclusion benchmarks.

Table 1. Experimental evaluation of **monitor**$_{TSO}$ and **monitor**$_{PSO}$ on all interleavings with up to 1 preemption

bench	LOC	# inter-leavings	total time	distinct violations under TSO	distinct violations under PSO
dekker	20	79	6.4	3	5
bakery	30	197	42.4	3	4
msn	80	92	7.5	0	3
ms2	80	123	11.0	0	2
harris	160	161	18.8	0	2
lazylist	120	139	14.0	0	4
snark	150	172	15.4	0	4

Table 2. Experimental evaluation on all interleavings with up to 2 preemptions

bench	LOC	# inter-leavings	total time	distinct violations under TSO	distinct violations under PSO
dekker	20	1714	180.0	9	11
bakery	30	13632	3992.4	3	4
msn	80	2300	196.1	0	3
ms2	80	3322	300.0	0	2
harris	160	5646	661.7	0	2
lazylist	120	4045	428.4	0	4
snark	150	6510	609.9	0	10

Rather than report every single parallel interleaving on which one of our monitor algorithms signaled a violation, we group together violations caused by the same pair of operations e and e'. We say that a violation is *caused by* e and e' when $\mathbf{monitor}_{TSO}(e)$ or $\mathbf{monitor}_{PSO}(e)$ is the call on which a violation is signaled, and e' is the conflicting memory access identified in the condition at Line 6. For such a violation, e' *happens before* the event $\mathbf{prev}[p(e)]$ just before e in process $p(e)$, but event e also *happens before* e' because we delay store e' until after e completes in the violating TSO or PSO execution.

Columns 4 and 5 of Tables 1 and 2 list the number of such distinct violations of sequential consistency found under TSO and PSO in our experiments. Note that we find no violations of sequential consistency under TSO for any of the data structure benchmarks. Their use of locks and compare-and-swap operations appear to be sufficient to ensure sequential consistency under TSO. On the other hand, we find violations of sequential consistency for all benchmarks under PSO.

Acknowledgments. We would like to thank Krste Asanović, Pallavi Joshi, Chang-Seo Park, and our anonymous reviewers for their valuable comments. This research supported in part by Microsoft (Award #024263) and Intel (Award #024894) funding and by matching funding by U.C. Discovery (Award #DIG07-10227), by NSF Grants CNS-0720906, CCF-1018729, and CCF-1018730, and by a DoD ND-SEG Graduate Fellowship. Additional support comes from Par Lab affiliates National Instruments, NEC, Nokia, NVIDIA, Samsung, and Sun Microsystems.

References

1. Arvind, A., Maessen, J.W.: Memory model = instruction reordering + store atomicity. In: ISCA 2006: Proceedings of the 33rd Annual International Symposium on Computer Architecture, pp. 29–40. IEEE Computer Society, Los Alamitos (2006)
2. Atig, M.F., Bouajjani, A., Burckhardt, S., Musuvathi, M.: On the verification problem for weak memory models. In: The 36th Annual ACM SIGPLAN-SIGACT Symposium on Principles of Programming Languages, POPL (2010)
3. Burckhardt, S., Alur, R., Martin, M.M.K.: CheckFence: checking consistency of concurrent data types on relaxed memory models. In: ACM SIGPLAN Conference on Programming Language Design and Implementation (2007)

4. Burckhardt, S., Musuvathi, M.: Effective program verification for relaxed memory models. In: Gupta, A., Malik, S. (eds.) CAV 2008. LNCS, vol. 5123, pp. 107–120. Springer, Heidelberg (2008)
5. Burckhardt, S., Musuvathi, M.: Effective program verification for relaxed memory models. Tech. Rep. MSR-TR-2008-12, Microsoft Research (2008)
6. Burckhardt, S., Musuvathi, M.: Memory model safety of programs. In $(EC)^2$: Workshop on Exploting Concurrency Efficiently and Correctly (2008)
7. Burckhardt, S., Musuvathi, M.: Personal communcation (2010)
8. Burnim, J., Sen, K., Stergiou, C.: Sound and complete monitoring of sequential consistency in relaxed memory models. Tech. Rep. UCB/EECS-2010-31, EECS Department, University of California, Berkeley (March 2010), http://www.eecs.berkeley.edu/Pubs/TechRpts/2010/EECS-2010-31.html
9. Dill, D.L., Park, S., Nowatzyk, A.G.: Formal specification of abstract memory models. In: Symposium on Research on Integrated Systems (1993)
10. Flanagan, C., Freund, S.N.: Adversarial memory for detecting destructive races. In: ACM SIGPLAN Conference on Programming Language Design and Implementation, PLDI (2010)
11. Flanagan, C., Godefroid, P.: Dynamic partial-order reduction for model checking software. In: Proc. of the 32nd Symposium on Principles of Programming Languages (POPL 2005), pp. 110–121 (2005)
12. Gibbons, P., Korach, E.: The complexity of sequential consistency. In: Fourth IEEE Symposium on Parallel and Distributed Processing. pp. 317–235 (1992)
13. Gopalakrishnan, G., Yang, Y., Sivaraj, H.: QB or not QB: An efficient execution verification tool for memory orderings. In: Alur, R., Peled, D.A. (eds.) CAV 2004. LNCS, vol. 3114, pp. 401–413. Springer, Heidelberg (2004)
14. Higham, L., Kawash, J., Verwaal, N.: Weak memory consistency models. part i: Definitions and comparisons. Tech. Rep. 97/603/05, Department of Computer Science, The University of Calgary (1998)
15. Huynh, T.Q., Roychoudhury, A.: Memory model sensitive bytecode verification. FMSD 31(3), 281–305 (2007)
16. Jalbert, N., Sen, K.: A trace simplification technique for effective debugging of concurrent programs. In: The 18th ACM SIGSOFT International Symposium on the Foundations of Software Engineering (SIGSOFT 2010/FSE-18) (2010)
17. Lamport, L.: How to make a multiprocessor computer that correctly executes multiprocess programs. IEEE Trans. Comput. 28(9), 690–691 (1979)
18. Mador-Haim, S., Alur, R., Martin, M.M.: Generating litmus tests for contrasting memory consistency models. In: Touili, T., Cook, B., Jackson, P. (eds.) CAV 2010. LNCS, vol. 6174, pp. 273–287. Springer, Heidelberg (2010)
19. Musuvathi, M., Qadeer, S.: Iterative context bounding for systematic testing of multithreaded programs. In: PLDI 2007: Proceedings of the 2007 ACM SIGPLAN Conference on Programming Language Design and Implementation. ACM, New York (2007)
20. Park, S., Dill, D.L.: An executable specification, analyzer and verifier for RMO (relaxed memory order). In: ACM Symposium on Parallel Algorithms and Achitectures, pp. 34–41. ACM Press, New York (1995)
21. Sewell, P., Sarkar, S., Owens, S., Nardelli, F.Z., Myreen, M.O.: x86-TSO: a rigorous and usable programmer's model for x86 multiprocessors. Commun. ACM 53(7), 89–97 (2010)
22. Shasha, D., Snir, M.: Efficient and correct execution of parallel programs that share memory. ACM Trans. Program. Lang. Syst. 10(2), 282–312 (1988)
23. SPARC International. The SPARC architecture manual (v. 9). Prentice-Hall, Englewood Cliffs (1994)

A   Soundness and Completeness Proof Sketches

We sketch here proofs of the soundness and completeness of our monitoring algorithms. Complete proofs can be found in our accompanying technical report.

Theorem 1. *Algorithms* **monitor**$_{TSO}$ *and* **monitor**$_{PSO}$ *are sound monitoring algorithms for TSO and PSO, respectively.*

Proof (sketch). Let $(\sigma_0, m_0) \xrightarrow{e_1,r_1} \cdots \xrightarrow{e_n,r_n} (\sigma_n, m_n)$ be an SC execution of a program P such that **monitor**$_{TSO}$ (resp., **monitor**$_{PSO}$), on this execution, first signals "Sequential Consistency Violation" on event e_n. We prove that P has an execution under TSO (resp., PSO) with a non-sequentially-consistent trace.

Observe that, during its execution, **monitor**$_{TSO}$ (resp., **monitor**$_{PSO}$) emits labels from *Event* and τ_{TSO} (resp., τ_{PSO}). We can show:

(1) That the events emitted during the first $n-1$ calls to **monitor**$_{TSO}$ (resp., **monitor**$_{PSO}$) form a TSO (resp., PSO) execution of program P. Further, this execution has the same trace as the SC execution.
    We can show that the emitted $\mathbf{st}^c$ transitions ensure that the loads and atomic operations in the TSO (resp., PSO) execution see the same values as in the SC execution. Thus, program P behaves identically.
(2) That operation e_n can be performed in this TSO (resp., PSO) execution in a way that creates a trace with a cycle in its $\to_{hb}$-relation.
    In the emitted TSO (resp., PSO) execution, the event e' in the condition at Line 6 is still pending. Thus, we can perform e (and commit it, if it is a store) before e' commits, so that $e \to_c e'$. And $e' \to_{hb}^* \mathbf{prev}[p(e)]$ because the emitted execution has the same trace as the SC execution $e_1, \ldots, e_{n-1}$. Thus, we create a TSO (resp., PSO) execution with happens-before cycle:

$$e' \to_{hb}^* \mathbf{prev}[p(e)] \to_p e \to_c e'$$

Theorem 2. *Algorithms* **monitor**$_{TSO}$ *and* **monitor**$_{PSO}$ *are complete monitoring algorithms for TSO and PSO, respectively.*

Proof (sketch). Suppose $(s_0, m_0) \xrightarrow{l_1} \cdots \xrightarrow{l_n} (s_n, m_n)$ is a TSO (resp., PSO) execution of program P with a trace $(E, \to_p, src, \to_{st})$ that is not sequentially consistent. Recall that each l_i is either a memory event from *Event* or an internal transition $\mathbf{st}^c(p)$ for TSO or $\mathbf{st}^c(p,a)$ for PSO.

We can obtain shorter TSO (resp., PSO) executions of P by removing some *Event*-labeled transition from the execution $l_1, \ldots, l_n$, as well as possibly removing corresponding $\mathbf{st}^c$ transitions. For example, we can safely remove the last *Event* issued by any process p, even if it is not last in the execution, as long as it does not write a value that is read by a later operation.

We use this freedom to construct a shorter TSO (resp., PSO) execution that is a *minimal* violation of sequential consistency. That is, if any further *Event*'s are removed, the trace of the execution becomes sequentially consistent.

On this minimally-violating execution, consider the *Event*'s that can be safely removed—i.e. their removal leaves a valid TSO (resp., PSO) execution, but this execution has an SC trace. For such a safely-removable e, let $last(e)$ denote the last write to $a(e)$ to become globally visible in the TSO (resp., PSO) execution, not including e itself. In the TSO (resp., PSO) execution $e \to_c last$, but we would have $last(e) \to_{hb}^* e$ in the SC execution in which e is removed and then run at the end. We can show that, for at least one of the these e, no other event comes after $last(e)$ in any SC trace and also forces $last(e)$ to be committed before event e executes. Thus, $\mathbf{monitor}_{TSO/PSO}(e)$ reports a violation on this execution.

B   Complexity of Monitoring Algorithms

Lemma 1. *During the execution of $\mathbf{monitor}_{TSO}$ (respectively, $\mathbf{monitor}_{PSO}$), for each address $a \in Adr$, at any given time at most one process $p \in Proc$ will have pending stores to address a in its write buffer $B[p]$ (resp., $B[p][a]$).*

Proof (by induction). Initially, before any calls to $\mathbf{monitor}_{TSO}$ (resp., $\mathbf{monitor}_{PSO}$), the lemma clearly holds.

Suppose the lemma holds after k calls to $\mathbf{monitor}_{TSO}$ (resp., $\mathbf{monitor}_{PSO}$), and let $\mathbf{monitor}_{TSO}(e)$ or $\mathbf{monitor}_{PSO}(e)$ be the $(k+1)$-st call. If e is not a store, or if $e = \mathbf{st}(p, a, v)$ where no other process $p' \neq p$ has any pending stores to a, then the lemma clearly holds during and after the call.

Suppose instead that $e = \mathbf{st}(p, a, v)$ and process $p' \neq p$ is the only process with pending stores to address a. Then, in $\mathbf{monitor}_{TSO}$ (resp., $\mathbf{monitor}_{PSO}$), the while-loop at Lines 10–12 commits all such pending stores by p' before Line 17 adds a pending store to a to a write buffer of process p.

By Lemma 1, in the condition at Line 6 in $\mathbf{monitor}_{TSO}(e)$ and $\mathbf{monitor}_{PSO}(e)$ at most one processor p' can have pending writes to $a(e)$.

We further observe that $\mathbf{monitor}_{TSO}$ and $\mathbf{monitor}_{PSO}$ remain complete if, at Line 6, we check this condition with only the last (i.e. most recent) pending store to a in $B[p']$ or $B[p'][a]$. See the proof of Theorem 2 for details.

Theorem 3. *On a sequentially consistent execution of length N on P processes, monitoring algorithms $\mathbf{monitor}_{TSO}$ and $\mathbf{monitor}_{PSO}$ run in time $O(N \cdot P)$.*

Proof. We show that each call $\mathbf{monitor}_{TSO/PSO}(e)$ runs in amortized $O(P)$ time, yielding $O(N \cdot P)$ total time. As mentioned in the previous section, updating the vector clocks to maintain the happens-before relation on the monitored SC execution requires $O(P)$ time per call to $\mathbf{monitor}_{TSO/PSO}(e)$.

For each address $a \in Adr$, we track the single process p which has pending stores to a. Further, for a we maintain a pointer into the **FIFOQueue** of this process to the last (i.e. most recent) pending store to a. We can maintain these two pieces of per-address information in $O(1)$ time per call to $\mathbf{monitor}_{TSO/PSO}(e)$. Using this information, checking the condition at Line 6 requires $O(1)$ time to find the last pending store to $a(e)$ and $O(1)$ time to check $e' \to_{hb}^* e$. The condition at Line 10 can similarly be checked in $O(1)$.

Finally, the total number of iterations of the while-loops at Lines 10 and 22, across all N calls to $\mathbf{monitor}_{TSO/PSO}(e)$, cannot exceed $O(N)$ as we buffer no more than N writes.

Compositionality Entails Sequentializability

Pranav Garg and P. Madhusudan

University of Illinois at Urbana-Champaign

Abstract. We show that any concurrent program that is amenable to compositional reasoning can be effectively translated to a sequential program. More precisely, we give a reduction from the verification problem for concurrent programs against safety specifications to the verification of sequential programs against safety specifications, where the reduction is parameterized by a set of auxiliary variables A, such that the concurrent program *compositionally* satisfies its specification using auxiliary variables A iff the sequentialization satisfies its specification. Existing sequentializations for concurrent programs work only for underapproximations like bounded context-switching, while our sequentialization has the salient feature that it can *prove* concurrent programs entirely correct, as long as it has a compositional proof. The sequentialization allows us to use sequential verification tools (including deductive verification tools and predicate abstraction tools) to analyze and prove concurrent programs correct. We also report on our experience in the deductive verification of concurrent programs by proving their sequential counterparts using the program verifier BOOGIE.

Keywords: concurrent programs, compositional verification, sequentialization.

1   Introduction

Sequentializing concurrent programs has been a topic of recent research. Given a concurrent program with a safety specification, we would like to reduce the problem of verifying the concurrent program to the verification of a sequential program. Moreover, and most importantly, we seek a sequential program that does not simply simulate the global evolution of the concurrent program as that would be quite complex and involve taking the product of the local state-spaces of the processes. Instead, we seek a sequential program that tracks a *bounded* number of copies of the local and shared variables, where the bound is independent of the number of parallel components.

The appeal of sequentialization is that it allows using the existing class of sequential verification tools to verify concurrent programs. A large number of sequential verification techniques and tools, like deductive verification, abstraction-based model-checking, and static dataflow analysis immediately come into play when a sequentialization is possible.

Of course, such sequentializations are not possible for all concurrent programs and specifications. In fact, in the presence of recursion and when variables have

P.A. Abdulla and K.R.M. Leino (Eds.): TACAS 2011, LNCS 6605, pp. 26–40, 2011.

bounded domains, concurrent verification is undecidable while sequential verification is decidable, which proves that an effective sequentialization is in general impossible.

The currently known sequentializations have hence focussed on capturing *under-approximations* of concurrent programs. Lal and Reps [13] showed that given a concurrent program with finitely many threads and a bound k, the problem of checking whether the concurrent program is safe on all executions that involve only k context-rounds can be reduced to the verification of a sequential program. A *lazy* sequentialization for bounded context rounds that ensures that the sequential program explores only states reachable by the concurrent program was defined by La Torre et al [10]. A sequentialization for unboundedly many threads and bounded round-robin rounds of context-switching is also known [11]. Lahiri, Qadeer and Rakamarić have used the sequentialization of Lal and Reps to check concurrent C-programs by unrolling loops in the sequential program a bounded number of times, and subjecting them to deductive SMT-solver based verification [12,8].

In this paper, we show a general sequentializability result that is not restricted to under-approximations. We show that any concurrent program with finitely many threads can *always* be sequentialized provided there exists a *compositional proof* of correctness of the concurrent program. More precisely, we show that given a concurrent program C with assertions and a set of *auxiliary* variables A, there is a sequentialization of it, $S_{C,A}$ with assertions, and that C can be shown to compositionally satisfy its assertions by exposing the auxiliary variables A if and only if $S_{C,A}$ satisfies its assertions. The notion of C compositionally satisfying its assertions using auxiliary variables A is defined semantically, and intuitively captures the *rely-guarantee proofs* pioneered by Jones [9]. Rely-guarantee proofs of concurrent programs are very standard, and perhaps the best known compositional verification technique for concurrent programs. In these proofs, *auxiliary variables* can be seen as local states that get exposed in order to build a compositional rely-guarantee proof.

Compositional proofs of programs may not always exist, and since our sequentialization only produces sequential programs that are precise when a compositional proof exists over the fixed auxiliary variables A, proving its sequentialization correct can be seen as a sound but incomplete mechanism for verifying the concurrent program. Note that our sequentialization *does not* require the compositional proof to be given; it is only parameterized by the auxiliary variables A. In fact, if the sequential program is correct, then we show that the concurrent program is always correct. Conversely, if the concurrent program is correct and has a compositional proof using variables A, then we show that the sequential program is guaranteed to be correct as well.

The salient aspect of our sequentialization is that it can be used to prove concurrent programs *entirely* correct, as opposed to checking underapproximations of it. Moreover, though our sequentializations are sound but incomplete, we believe they are useful on most practical applications since concurrent programs often have compositional proofs. Our result also captures the *cost* of sequentialization (i.e. the number of variables in the sequentialization) as directly proportional to the number of auxiliary variables that are required to build a compositional proof. Concurrent programs that are "loosely coupled" often require only a small number of auxiliary variables to be exposed, and hence admit efficient sequentializations.

We also describe our experience in applying our sequentialization to prove a suite of concurrent programs entirely correct by using deductive verification of their sequentializations. More precisely, we wrote rely-guarantee proof annotations for some concurrent programs by formulating the rely and guarantee conditions, the loop invariants, and pre- and post-conditions for every function. We then sequentialized the concurrent program and also *transformed* the rely-guarantee proof annotations to corresponding proof annotations on the sequential program. As we show, in this translation, rely and guarantee conditions naturally get transformed to pre- and post-conditions of methods, while loop-invariants and pre- and post-conditions get translated to loop invariants and pre- and post-conditions in the sequential program. Then, using an automatic sequential program verifier BOOGIE, we verified the sequentializations correct. BOOGIE takes our programs with the proof annotations, generates verification conditions, and discharges them using an automatic theorem prover (SMT solver).

The above use of sequentialization for deductive verification is not the best use of our sequentializations, as given rely-guarantee proofs, simpler techniques for statically verifying them are known [6]. However, our sequentializations can be applied even when the rely-guarantee proofs are not known, provided the sequential verification tool is powerful to prove it correct. Indeed, we have also used the sequentialization followed by an automatic predicate-abstraction tool (SLAM [2]) to prove some concurrent programs correct.

In summary, the result presented in this paper shows a surprising connection between compositional proofs and sequentializability. We believe that this constitutes a fundamental theoretical understanding of when concurrent programs are efficiently sequentializable, and offers the first efficient sequentializations that work without underapproximation restrictions, enabling us to verify concurrent program entirely using sequentializations.

Related work: Thread-modular verification [6,7] is in fact precisely the same as compositional verification á la Jones, but has been adapted to both model-checking [7] and extended static checking [6]. Our result can be hence seen as showing how thread-modular verification of concurrent programs can be reduced to pure sequential verification. There has also been work on using counter-example guided predicate-abstraction and refinement for rely-guarantee reasoning [4], and building rely-guarantee interfaces using *learning* [3,1].

2   A Compositional Abstract Semantics for Programs

We define a *non-standard compositional semantics* for concurrent programs, different from the traditional semantics, in order to capture when a parallel composition of programs can be *argued compositionally to satisfy a specification*. This semantics is parameterized by a set of auxiliary variables, and is the semantic analog of *compositional rely-guarantee proofs* pioneered by Jones [9].

Let us fix two processes P_1 and P_2, working concurrently, with local variables L_1 and L_2 respectively, and a set of shared variables S (assume L_1, L_2 and S are pairwise disjoint, without loss of generality). For any set of (typed) variables V, let Val_V denote the set of valuations of V to their respective data-domains (data-domains are finite or countably infinite). For any $u \in Val_V$, let $u \downarrow V'$ denote the valuation u restricted to the variables in $V \cap V'$. We extend this notation to sets of valuations, $U \downarrow V'$. Also, for any $u \in Val_V$ and $u' \in Val_{V'}$, where $V \cap V' = \emptyset$, let $u \cup u'$ denote the unique valuation in $Val_{V \cup V'}$ that extends u and u' to $V \cup V'$.

Let $Init \subseteq (Val_{L_1} \times Val_{L_2} \times Val_S)$ be the set of initial global configurations of $P_1 || P_2$. Let $\delta_1 \subseteq (Val_{L_1} \times Val_S \times Val_{L_1} \times Val_S)$ and $\delta_2 \subseteq (Val_{L_2} \times Val_S \times Val_{L_2} \times Val_S)$ be the local transition relations of P_1 and P_2, respectively.

The natural (interleaving) semantics of $P_1 || P_2$ is, of course, defined by the function $\delta \subseteq (Val_{L_1} \times Val_{L_2} \times Val_S \times Val_{L_1} \times Val_{L_2} \times Val_S)$, where $\delta(l_1, l_2, s, l_1', l_2', s')$ holds iff $\delta_1(l_1, s, l_1', s')$ holds and $l_2' = l_2$, or $\delta_2(l_2, s, l_2', s')$ holds and $l_1' = l_1$. The set of *reachable states* according to this relation, *Reach*, is defined as the set of global states that can be reached from the initial state.

Let us now define the non-standard compositional semantics of $P_1 || P_2$. This definition is parameterized by a set of *auxiliary variables* $A \subseteq L_1 \cup L_2$.

Definition 1. *The semantics of the compositional semantics of parallel composition with respect to the set of auxiliary variables A, denoted $P_1 ||_A P_2$, is defined using the* four *sets:*

$$R_1 \subseteq (Val_{L_1} \times Val_S \times Val_{A \cap L_2}),$$
$$R_2 \subseteq (Val_{L_2} \times Val_S \times Val_{A \cap L_1}),$$
$$Guar_1, Guar_2 \subseteq (Val_S \times Val_A \times Val_S \times Val_A),$$

which are defined as the least *sets that satisfy the following conditions:*

a) **Initialization:**
 - R_1 *contains the set* $\{(l_1, s, t) \mid l_1 \cup s \cup t \in Init \downarrow (L_1 \cup A \cup S)\}$.
 - R_2 *contains the set* $\{(l_2, s, t) \mid l_2 \cup s \cup t \in Init \downarrow (L_2 \cup A \cup S)\}$.
b) **Transitions of P_1:** *If* $(l_1, s, t) \in R_1$ *and* $\delta_1(l_1, s, l_1', s')$ *holds, then*
 - **Local update:** $(l_1', s', t) \in R_1$.
 - **Update to guarantee:** $(s, l_1 \downarrow A \cup t, s', l_1' \downarrow A \cup t) \in Guar_1$.
c) **Transitions of P_2:** *If* $(l_2, s, t) \in R_2$ *and* $\delta_2(l_2, s, l_2', s')$ *holds, then*
 - **Local update:** $(l_2', s', t) \in R_2$
 - **Update to guarantee:** $(s, l_2 \downarrow A \cup t, s', l_2' \downarrow A \cup t) \in Guar_2$.
d) **Interference:**
 - *If* $(l_1, s, t) \in R_1$ *and* $(s, l_1 \downarrow A \cup t, s', t') \in Guar_2$, *then* $(l_1, s', t' \downarrow L_2) \in R_1$.
 - *If* $(l_2, s, t) \in R_2$ *and* $(s, l_2 \downarrow A \cup t, s', t') \in Guar_1$, *then* $(l_2, s', t' \downarrow L_1) \in R_2$.

The set of reachable states *according to the* non-standard compositional semantics *with respect to the set of auxiliary variables A is defined as*

$$Reach_A = \{(l_1, s, l_2) \mid (l_1, s, l_2 \downarrow A) \in R_1 \text{ and } (l_2, s, l_1 \downarrow A) \in R_2\}. \qquad \square$$

Intuitively, under the compositional semantics, we track independently the view of P_1 (and P_2) using valuations of its local variables, shared variables, and the subset of the other process's local variables declared to be auxiliary (using the sets R_1 and R_2). Furthermore, we keep the set of *guarantee* transition-relations $Guar_1$ and $Guar_2$ that summarize what transitions P_1 and P_2 can take, but restricted to the auxiliary and shared variables only. The guarantee-relation of P_1 is used to update the view of P_2 (i.e. R_2), and vice versa. The crucial aspect of the definition above is that it *ignores* the correlation between local variables of P_1 and P_2 that are not defined to be auxiliary variables. The computation of $P_1||_A P_2$ hence proceeds mostly locally, with updates using the guarantee relation of the other process (which affects shared and auxiliary variables only), and is combined in the end to get the set of globally reachable configurations.

It is not hard to see that $Reach \subseteq Reach_A$, for any A. Hence, the compositional semantics is an *over-approximation* of the set of reachable states of the program, and proving that a program is safe under the compositional semantics is sufficient to prove that the program is safe. Moreover, when the auxiliary variables include all local variables (including the program counter and local call stack), the compositional semantics coincides with the natural semantics.

The above definitions and rules can be generalized to k processes running in parallel, and we can define the compositional semantics $P_1||_A P_2||_A \cdots ||_A P_n$ where A is subset of local variables of each process.

The rely-guarantee proof method of Jones

The rely-guarantee method of Jones [9] essentially builds compositional rely-guarantee *proofs* using a similar abstraction. Given *sequential programs* P_1 and P_2, and a pre-condition *pre* and a post-condition *post* for $P_1||P_2$, the rely-guarantee proof technique over a set of auxiliary variables A involves providing a pair of tuples, $(pre_1, post_1, rely_1, guar_1)$ and $(pre_2, post_2, rely_2, guar_2)$, where pre_1, $post_1$, pre_2, and $post_2$ are unary predicates defining subsets of states, and $rely_1$, $guar_1$, $rely_2$, $guar_2$ are binary relations defining transformations of the shared variables and the auxiliary variables A. The meaning of the tuple for P_1 is that, when P_1 is started with a state satisfying pre_1 and in an environment that could change the auxiliary variables and shared variables allowed by $rely_1$, P_1 would make transitions that accord to $guar_1$, and if it terminates, will satisfy $post_1$ at the exit. An analogous meaning holds for P_2. Note that $rely_1$, $rely_2$, $guar_1$ and $guar_2$ are defined over shared variables and the auxiliary variables. The programs P_1 and P_2 are proved to satisfy these conditions using a *local* proof by considering each P_i interacting with a general environment satisfying $rely_i$; in particular invariants of P_i needed to establish the Hoare-style proof of P_i should be *invariant* or stable with respect to $rely_i$.

The following proof rule can then be used to prove partial correctness of $P_1||P_2$:

$$\frac{guar_1 \Rightarrow rely_2, \qquad guar_2 \Rightarrow rely_1,}{P \models (pre, post_1, rely_1, guar_1), \qquad Q \models (pre, post_2, rely_2, guar_2)}$$

$$P||Q \models (pre, post_1 \wedge post_2)$$

The rely-guarantee method works also for *nested* parallellism compositionally; see [9,16] for details.

It is easy to see that a compositional rely-guarantee proof of $P_1||P_2$, over a set of auxiliary variables A, is really a proof that the compositional semantics of $P_1||_A P_2$ is correct. Note that if $P_1||_A P_2$ is correct, it does not imply a rely-guarantee proof exists, however, as proofs have limitations of the logical syntax used to write the rely and guarantee conditions, and hence do not always exist.

The main result
We can now state the main result of this paper. We show that, given a parallel composition of sequential programs $P_1||P_2||\ldots||P_n$ with assertions, and a set of auxiliary variables A, we can build a sequential program S with assertions such that S has the following properties:

- At any point, the scope of S contains at most one copy of the local variables of a *single* process P_i, three copies of the auxiliary variables, and at most three copies of the shared variables.
- The compositional semantics of $P_1||_A P_2||_A \ldots ||_A P_n$ with respect to the auxiliary variables A satisfies its assertions iff S satisfies its assertions.
- If S satisfies its assertions, then $P_1||P_2||\ldots||P_n$ also satisfies its assertions.

The first remark above says that the sequentialized program has less variables in scope than the naive *product* of the individual processes; the sequentialization intuitively simulates the processes *separately*, keeping track of only an extra copy of auxiliary variables and shared variables. Second, the sequentialization is a *precise* reduction of the verification problem, provided the concurrent program can be proved compositionally (i.e. if the auxiliary variables are sufficient to make the compositional semantics of the program be assertion-failure free). Finally, the sequential program is an over-approximation of the behaviors of the parallel program for *any* set of auxiliary variables, and hence proving it correct proves the parallel program correct.

The above result will be formalized in the sequel (see Theorem 1) for a class of parallel programs that has sequential recursive functions, but with no thread creation or dynamic memory allocation (the result can be extended to dynamic data-structures but will require mechanisms to cache heap-structures and compare them for equality). Our main theorem hence states that any parallel program that is amenable to compositional reasoning can be sequentialized, where the number of new variables added in the sequentialization grows with the number of auxiliary variables required to prove the program correct. We utilize the sequentialization result in one verification context, namely deductive verification, to build a compositional deductive verification tool for concurrent programs using the sequential verifier BOOGIE.

3 A High Level Overview of the Sequentialization

In this section, we give a brief overview of our sequentialization. For ease of explanation, let us consider a concurrent program consisting of two processes P_1

and P_2. Let A be the set of auxiliary variables and assume that the compositional semantics of the concurrent program is correct with respect to A.

Assume we had functions $G_1(s^*, a^*)$ (and $G_2(s^*, a^*)$) that *somehow* takes a shared and auxiliary state (s^*, a^*) and non-deterministically returns all states (s, a) such that $(s^*, a^*, s, a) \in Guar_1$ (respectively $Guar_2$), where $Guar_1$ and $Guar_2$ are as in Definition 1. Then we could write a function that computes the states reachable by P_1 according to the compositional semantics (i.e. R_1 in Definition 1) using the following code:

```
while(*) {
  if (*) then
    <<simulate a transition of P1>>
  else
    (s,a) := G2(s,a);
  fi
}
return (s,a);
```

In other words, we could write a sequential program that returns precisely the states in R_1, by interleaving simulations of P_1 with calls to G_2 to compute interference according to $Guar_2$ (see Definition 1). We can similarly implement the sequential code that explores R_2 using calls to G_1.

Note that on two *successive* calls to $G_2()$, there is no preservation of the local states of P_2, *except its variables declared to be auxiliary*. However, we do *not* have to preserve the exact local state of P_2 as we are not simulating the natural semantics of the program, but only its compositional semantics with respect to auxiliary variables A. This is the crux of the argument as to why we can sequentially compute R_1 without simultaneously tracking all the local variables of P_2.

Now, turning to the function G_2 (and G_1), consider Definition 1 again, and notice that, given (s^*, a^*), in order to compute (s, a) such that $(s^*, a^*, s, a) \in Guar_2$, we must essentially be able to find a local state l_2 such that $(l_2, s^*, a^* \downarrow L_1) \in R_2$ where $l_2 \downarrow A = a^* \downarrow L_2$, and then we can take its transitive closure with respect to δ_2. We can hence write G_2 using the following sequential code:

```
G2(s*,a*) {
    <<initialize variables of P2>>
    while(*) {
        if (*) then
            <<simulate a transition of P2>>
        else
            (s,a) := G1(s,a);
        fi
    }
    assume (local and shared state is consistent with s*,a*);
    while(*) {
        <<simulate a transition of P2>>
    }
    return (s,a);
}
```

Intuitively, G_2 starts with the *initial* state of P_2 and sets about recomputing a state (l_2, s_2) that is compatible with its given input (s^*, a^*) (i.e. with $s_2 = s^*$ and $l_2 \downarrow A = a^* \downarrow L_2$). It does this by essentially running the code for R_2 (i.e. by simulating P_2 and calling G_1). Once it has found such a state, it simulates P_2 for a while longer, and returns the resulting state.

We hence get four procedures that compute R_1, R_2, $Guar_1$ and $Guar_2$, respectively, with mutually recursive calls between the functions computing $Guar_1$ and $Guar_2$. The correctness of the sequential programs follow readily from Definition 1, as it is a direct encoding of that computation. Our sequentialization transformation essentially creates these functions G_1 and G_2. However, since our program cannot have statements like "simulate a transition of P2", we perform a syntactic transformation of the concurrent code into a sequential code, where control code is inserted between statements of the concurrent program in order to define the functions G_1 and G_2. This complication combined with the handling of recursive functions makes the translation quite involved; however, the above explanation captures the crux of the construction.

4   Sequential and Concurrent Programs

Our language for concurrent programs consists of a parallel composition of recursive sequential programs. Variables in our programs are defined over integer and Boolean domains. The syntax of programs is defined by the following grammar:

$$\begin{aligned}
\langle conc\text{-}pgm \rangle &::= \langle decl \rangle^* \langle pgm\text{-}list \rangle \\
\langle pgm\text{-}list \rangle &::= \langle pgm\text{-}list \rangle \parallel \langle pgm\text{-}list \rangle \mid \langle pgm \rangle \\
\langle pgm \rangle &::= \langle decl \rangle^* \langle proc \rangle^* \\
\langle proc \rangle &::= f(\overline{x}) \text{ begin } \langle decl \rangle^* \langle stmt \rangle \text{ end} \\
\langle stmt \rangle &::= \langle stmt \rangle; \langle stmt \rangle \mid \text{ skip } \mid \overline{x} := expr(\overline{x}) \mid \overline{x} := f(\overline{y}) \mid \\
&\qquad f(\overline{x}) \mid \text{ return } \overline{x} \mid \text{assume } b\text{-}expr \mid \text{ assert } b\text{-}expr \mid \\
&\qquad \text{if } b\text{-}expr \text{ then } \langle stmt \rangle \text{ else } \langle stmt \rangle \text{ fi} \mid \\
&\qquad \text{while } b\text{-}expr \text{ do } \langle stmt \rangle \text{ od} \mid \text{ atomic } \{\langle stmt \rangle\} \\
\langle decl \rangle &::= \text{int } \langle var\text{-}list \rangle; \mid \text{ bool } \langle var\text{-}list \rangle; \\
\langle var\text{-}list \rangle &::= \langle var\text{-}list \rangle, \langle var\text{-}list \rangle \mid \langle literal \rangle
\end{aligned}$$

A concurrent program consists of k sequential program components $P_1...P_k$ (for some k) communicating with each other through shared variables $\mathcal{S}$. These shared variables are declared in the beginning of the concurrent program (we assume integer variables are initialized to 0 and Boolean variables to *false*). Each sequential component consists of a procedure called *main* and a list of other procedures. The control flow for all sequential components P_i starts in the corresponding *main* procedure, which we call $main_i$. The *main* for all sequential components has *zero* arguments and no return value.

Each procedure is a declaration of local variables followed by a sequence of statements, where statements can be simultaneous assignments, function calls (call-by-value) that take in multiple parameters and return multiple values, conditionals, while loops, assumes, asserts, atomic, and return statements. In the above syntax, $\bar{x}$ represents a vector of variables. We allow non-determinism in our programs; boolean constants are *true*, *false* and $*$, where $*$ evaluates non-deterministically to *true* or *false*.

The safety specifications for both concurrent and sequential programs are expressed in our language as assert statements. The semantics of an assume statement is slightly different. If the value of the boolean expression (*b-expr*) evaluates to *true*, then the assume behaves like a skip. Otherwise, if the boolean expression evaluates to *false*, the program silently terminates. Synchronization and atomicity are achieved by the *atomic* construct. All the statements enclosed in the atomic block are executed without any interference by the other processes. Locks can be simulated in our syntax by modeling a lock l as an integer variable 1 and by modeling P_i acquiring l using the code:

```
atomic { assume(l=0); l := i;}
```

and modeling the release with the code:

```
atomic { if (l=i) then l := 0;}
```

We assume programs do not have nested atomic blocks.

The syntax of sequential programs is the same as the syntax of concurrent programs except that we disallow the parallel composition operator ($\|$) and the *atomic* construct.

5   The Sequentialization

In this section, we describe our sequentialization for concurrent programs and argue its correctness.

Let us fix a concurrent program with shared variables S and auxiliary variables A; we assume auxiliary variables are *global* in each thread P_i. Let the concurrent program be composed of k sequential components.

The sequential program corresponding to the concurrent program will have a new function *main*, and additionally, as explained in Section 3, will have a procedure G_i for each sequential component P_i of the concurrent program that semantically captures the guarantee $Guar_i$ of P_i. The procedure G_i takes a shared state $(\bar{s}^*)$ and auxiliary state $(\bar{a}^*)$ as input and returns $(\bar{s}, \bar{a})$ such that $(\bar{s}^*, \bar{a}^*, \bar{s}, \bar{a}) \in Guar_i$. Finally, each $G_i()$ is formed using procedures that are obtained by transforming the process P_i (using the function $\tau_i[]$ shown below that essentially *inserts* the interference code $\mathcal{I}_i$ shown in Figure 1 between the statements of P_i).

The shared variables and auxiliary variables are modeled as global variables in the sequential program. Furthermore, we have an extra copy of the shared and auxiliary variables ($\bar{s}^*$ and $\bar{a}^*$) that are used to pass shared and auxiliary states between the processes $G_i()$. We also have a copy of shared and auxiliary variables ($\bar{s}'$ and $\bar{a}'$) that are declared to be *local* in each procedure to store a shared and auxiliary state and restore it after a call to a function $G_j()$ to compute interference. Besides these, the sequential program also uses global Boolean variables z and *term*; intuitively, z is used to keep track of when the shared and auxiliary state $\overline{s}^*$ and $\overline{a}^*$ has been reached and *term* (for *terminate*) is used to signal that $G_i()$ has finished computing and wants to return the value.

- Global variable declarations are:
  ```
  // insert declaration for s̄,ā,s̄*,ā* as global variables
  decl bool term, z;
  ```

- The function main() is defined as: `main() begin G_1() end`

- Each function $G_i()$ is defined as below:
  ```
  G_i() begin
      z := false;   term := false;
      s̄* := s̄;   ā* := ā;
      // insert code to initialize s̄,ā
      main_i();
      assume (term = true);
      return
  end
  ```

- The function τ_i that transforms the program for P_i is defined as:
  - $\tau_i[\texttt{f}(\overline{x})\ \texttt{begin}\ decl\ stmt\ \texttt{end}] =$
    ```
    f(x̄) begin  decl
        // insert declaration of s̄',ā' as local variables.
    ```
    $\tau_i[stmt]$
    ```
    end
    ```
  - $\tau_i[\mathcal{S}_1;\ \mathcal{S}_2] = \tau_i[\mathcal{S}_1];\ \tau_i[\mathcal{S}_2]$
  - $\tau_i[\mathcal{S}] = \mathcal{I}_i;\ \mathcal{S}$ where $\mathcal{S}$ is an assignment, skip, assume, assert, a function call or a return statement.
  - $\tau_i[\texttt{while}\ b\text{-}expr\ \texttt{do}\ \mathcal{S}\ \texttt{od}] = \mathcal{I}_i;\ \texttt{while}\ b\text{-}expr\ \texttt{do}\ \tau_i[\mathcal{S}];\ \mathcal{I}_i\ \texttt{od}$
  - $\tau_i[\texttt{if}\ b\text{-}expr\ \texttt{then}\ \mathcal{S}_1\ \texttt{else}\ \mathcal{S}_2\ \texttt{fi}] =$
    $\mathcal{I}_i;\ \texttt{if}\ b\text{-}expr\ \texttt{then}\ \tau_i[\mathcal{S}_1]\ \texttt{else}\ \tau_i[\mathcal{S}_2]\ \texttt{fi}$
  - $\tau_i[\texttt{atomic}\ \{\mathcal{S}\}] = \mathcal{I}_i;\ \mathcal{S}$

The procedure *main* in the sequential program simply calls the method G_1.

The procedure G_i is obtained from the corresponding program component P_i by a simple transformation. At a high level, this procedure first copies the incoming shared and auxiliary state into the variables $\overline{s}^*$ and $\overline{a}^*$. It then computes a local state of P_i which is consistent with the state $(\overline{s}^*, \overline{a}^*)$ (at which point z turns to *true*), and then non-deterministically simulates the transitions of P_i from this local state, to return a reachable shared state $\overline{s}$ and auxiliary state $\overline{a}$. Every time G_i is called, it starts from its initial state, and simulates P_i, interleaving it with the control code $\mathcal{I}_i$ given in Figure 1.

The interference code $\mathcal{I}_i$ (Figure 1) keeps track of whether the incoming state $(\overline{s}^*, \overline{a}^*)$ has been reached through a boolean variable z which is initialized to *false*. If z is *false* (i.e. the state $(\overline{s^*}, \overline{a^*})$ has not been reached), then before any transition of P_i, the control code can non-determinsitically choose to invoke its environment (in doing so, in order to preserve its input $\overline{s^*}, \overline{a^*}$, it stores them in a local state and restores them after the call returns and restores its variables z and *term* to *false*).

When the state $(\overline{s^*}, \overline{a^*})$ is reached, z can be non-deterministically set to *true*, from which point no interference code G_j can be called, and only local computation proceeds, till non-deterministically the program decides to terminate by setting *term* to *true*. Once *term* is *true*, the code pops the control-stack all the way back to reach the function G_i which then returns to its caller, returning the new state in $(\overline{s}, \overline{a})$. Note that the state $z = false$, $term = false$ corresponds to the first while loop in the code for the guarantee G_2 in Section 3. Similarly, setting *term* to *true* corresponds to the termination of the second while loop. We conclude this section by stating our main theorem:

```
if(term = true) then return fi
if(!z & *) then
  while(*) do
    // call G_1
    if(*) then
      s' := s*;   a' := a*;
      G_1();
      z := false;
      term := false;
      s* := s';   a* := a'
    fi
    Similarly call G_2 ... G_k except G_i
  od
fi
if(!z & s = s* & a = a* & *) then
  z := true fi
if(z & *) then
  term := true; return
fi
```

Fig. 1. The interference control code $\mathcal{I}_i$

Theorem 1. *Let C be a concurrent program with auxiliary variables A (assumed global), and let $S_{C,A}$ be its sequentialization with respect to A. Then the compositional semantics of C with respect to the auxiliary variables A has no reachable state violating any of its assertions iff $S_{C,A}$ violates none of its assertions.* □

Illustration of a sequentialization

Figure 2 shows a concurrent program consisting of two threads, say P_1 and P_2. The program consists of a shared variable x whose initial value is *zero*. Both the threads *atomically* increment the value of x. Let $A = \{pc1, pc2\}$ be the auxiliary variables capturing the control position in the respective processes and let the initial value of these variables also be *zero*. In general, new auxiliary variables may be needed for performing compositional proofs; these new variables are *written to* but never read from in the program, and hence do not affect the semantics of the original program; see [9,15].

It can be easily seen that the compositional semantics of this program with respect to the auxiliary variables A is correct. Figure 3 shows the sequential program obtained from the sequentialization of this concurrent program with respect to these auxiliary variables. Our result allows us to verify the concurrent program in Figure 2 by verifying the correctness of its sequentialization with respect to the auxiliary variables A, shown in Figure 3.

```
                        decl int x, pc1, pc2;

          main_1()                     ||    main_2()
          begin                        ||    begin
            atomic {                   ||      atomic {
              x := x + 1;              ||        x := x + 2;
              pc1 := 1                 ||        pc2 := 1
            }                          ||      }
            assert(spec₁);             ||      assert(spec₂);
            return                     ||      return
          end                          ||    end
```

$spec_1$: pc1 = 1 && ((pc2 = 0 && x = 1) || (pc2 = 1 && x = 3))
$spec_2$: pc2 = 1 && ((pc1 = 0 && x = 2) || (pc1 = 1 && x = 3))

Fig. 2. An example program

```
decl int x, pc1, pc2;
decl int x*, pc1*, pc2*;
decl bool z, term;
main begin
    G_1();
    return
end
_____

Iᵢ: if(term = true) then
      return fi
    if(!z & *) then
      x', pc1', pc2' :=
         x*, pc1*, pc2*;
      G_{3-i}();
      z, term:=false, false;
      x*, pc1*, pc2* :=
         x', pc1', pc2'
    fi
    if(x = x* & pc1 = pc1* &
    pc2 = pc2* & *) then
      z := true
    fi
    if(z & *) then
      term := true; return
    fi
```

```
G_1() begin
  z, term:=false, false;
  x*, pc1*, pc2* :=
          x, pc1, pc2
  x, pc1, pc2 := 0, 0, 0;
  main_1();
  assume(term = true);
  return
end

main_1() begin
  decl int x', pc1', pc2';
  I₁
  x := x + 1;
  pc1 := 1;
  I₁
  assert(spec₁);
  I₁
  return
end
```

```
G_2() begin
  z, term:=false, false;
  x*, pc1*, pc2*:=
          x, pc1, pc2
  x, pc1, pc2:=0, 0, 0;
  main_2();
  assume(term = true);
  return
end

main_2() begin
  decl int x', pc1', pc2';
  I₂
  x := x + 2;
  pc2 := 1;
  I₂
  assert(spec₂);
  I₁
  return
end
```

Fig. 3. The sequential program obtained from the concurrent program in Figure 2

6   Experience

The sequentialization used in this paper can be used to verify a concurrent program using *any* sequential verification tool. This includes tools based on abstract interpretation and predicate abestraction, those based on bounded model-checking, as well as those based on deductive-verification based extended static checking.

Deductive verification: We used the sequentialization for proving concurrent programs using deductive verification. Given a concurrent program and its Jones-style rely-guarantee proof annotations (pre, post, rely, guar, and loop-invariants), we sequentialized it with respect to the auxiliary variables and syntactically transformed the user-provided proof annotations to obtain the proof annotations (pre-conditions and post-conditions of methods and loop invariants) of the sequential program. In general, the pre-condition of method G_i asserts that "*term = false*" and its post-condition asserts that the guarantee $guar_i$ is true across the function (if *term* is *true*). Furthermore, the pre-conditions and post-conditions of every function in P_i gets translated to pre-conditions and post-conditions in its sequentialization with the extra condition that $guar_i$ is *true* when *term* is equal to *true*.

These annotated sequential programs were fed to the sequential verifier BOO-GIE that generates verification conditions that are in turn solved by an SMT solver (Z3 in this case). If the sequential program is proved correct, it proves the correctness of the original concurrent program. Note that though similar static extended checking techniques are known for the rely-guarantee method [6], our technique allows one to use just a sequential verifier like BOOGIE to prove the program correct, and requires no other decision problems to be solved (the checking in [6], for example, requires a separate call to a theorem prover to check guarantees are reflexive and transitive, etc.)

We used this technique to prove correct the following set of concurrent programs: X++ (Figure 2), Lock [7], Peterson's mutual exclusion algorithm, the Bakery protocol, ArrayIndexSearch [9], GCD [5], and a simplified version of a Windows NT Bluetooth driver. Lock is a simple example program consisting of two threads that modify a shared variable after acquiring a lock; the safety condition in the example asserts that these modifications cannot occur concurrently. The program ArrayIndexSearch finds the *least index* of an array such that the value at that index satisfies a given predicate, and consists of two threads, one that searches odd indices and the other that searches even indices, and communicate on a shared variable *index* that is always kept updated to the current least index value. GCD is a concurrent version of Euclid's algorithm for computing the greatest common divisor of any two numbers; here the two concurrent threads update the pair of integers.

The Windows NT bluetooth driver is a *parameterized program* (i.e. has an unbounded number of threads). It consists of two types of threads: there is one stopper-thread and an unbounded number of adder-threads. A stopper calls a procedure to halt the driver, while an adder calls a procedure to perform I/O in the driver. The I/O is successfully handled if the driver is not stopped while it executes. The program, though small, has an intricate global invariant that requires a shared variable to reflect the number of active adder threads.

Table 1. Experimental Results. Evaluated on Intel dual-core 1.6GHz, 1Gb RAM.

Programs	Concurrent pgm			Sequential pgm		
	#Threads	#LOC	#Lines of annotations	#LOC	#Lines of annotations	Time
X++	2	38	5	113	6	8s
Lock	2	50	9	184	10	122s
Peterson	2	52	35	232	36	145s
Bakery	2	55	8	147	13	18s
ArrayIndexSearch	2	74	17	222	21	126s
GCD	2	78	23	279	29	869s
Bluetooth	unbdd	69	20	276	55	107s

Table 1 gives the experimental results[1]. For each program, we report the number of threads in the concurrent program, the number of lines of code in the concurrent program and its sequentialization, the number of lines of annotations in both the concurrent program (which includes rely/guarantee annotations and loop invariants) and its sequentialization, and the time taken by BOOGIE to verify the sequentialized program.

BOOGIE was able to verify the correctness of all our programs. All these programs except the Windows NT bluetooth driver consist of two threads and are sequentialized as detailed in Section 5. The Bluetooth driver is an example of a parameterized program running any number of instances of the adder threads. In our sequentialization, we model the environment consisting of all the adder threads together with a single procedure. If we keep track of the number of adders at a particular program location (counter abstraction [14]) and expose these auxiliary variables, it turns out that the device driver can be proved correct under compositional semantics. We used this rely-guarantee proof, sequentialized the program, and used BOOGIE to prove the Bluetooth driver correct in its full generality.

Predicate abstraction: We have also used our sequentialization followed by the predicate-abstraction tool SLAM [2] to prove programs automatically correct. In this case, we need no annotations and just the set of auxiliary variables. We were able to automatically prove the correctness of the programs X++, Lock, Peterson and the Bakery protocol, in negligible time.

7   Future Directions

One drawback of our sequentialization is that it creates recursive programs, even when the concurrent program has no recursion. It is hence natural to ask whether there is a sequentialization that does not introduce recursion. We know indeed of such a sequentialization, where the process G_i does not start from the *initial* state to match the state given to it, but rather "jumps" directly to a state consistent with the state given to it. This, of course, does not capture compositional semantic

---

[1] Experiments available at http://www.cs.uiuc.edu/$\sim$garg11/tacas11

reasoning precisely, and seems to be an over-approximation of it. Evaluating the effectiveness of this translation is an interesting future direction.

Finally, since our sequentialization captures the concurrent program without under-approximations, utilizing the sequentialization followed by techniques such as static analysis, predicate-abstraction, and even bounded model-checking, would be interesting directions to pursue.

Acknowledgements. This work is partially supported by NSF grants #0747041 and #1018182.

References

1. Alur, R., Madhusudan, P., Nam, W.: Symbolic compositional verification by learning assumptions. In: Etessami, K., Rajamani, S.K. (eds.) CAV 2005. LNCS, vol. 3576, pp. 548–562. Springer, Heidelberg (2005)
2. Ball, T., Rajamani, S.K.: The SLAM project: debugging system software via static analysis. In: POPL, pp. 1–3. ACM, New York (2002)
3. Cobleigh, J.M., Giannakopoulou, D., Păsăreanu, C.S.: Learning assumptions for compositional verification. In: Garavel, H., Hatcliff, J. (eds.) TACAS 2003. LNCS, vol. 2619, pp. 331–346. Springer, Heidelberg (2003)
4. Cohen, A., Namjoshi, K.S.: Local proofs for global safety properties. In: Damm, W., Hermanns, H. (eds.) CAV 2007. LNCS, vol. 4590, pp. 55–67. Springer, Heidelberg (2007)
5. Feng, X.: Local rely-guarantee reasoning. In: POPL, pp. 315–327. ACM, New York (2009)
6. Flanagan, C., Freund, S.N., Qadeer, S.: Thread-modular verification for shared-memory programs. In: Le Métayer, D. (ed.) ESOP 2002. LNCS, vol. 2305, pp. 262–277. Springer, Heidelberg (2002)
7. Flanagan, C., Qadeer, S.: Thread-modular model checking. In: Ball, T., Rajamani, S.K. (eds.) SPIN 2003. LNCS, vol. 2648, pp. 213–224. Springer, Heidelberg (2003)
8. Ghafari, N., Hu, A.J., Rakamarić, Z.: Context-bounded translations for concurrent software: An empirical evaluation. In: van de Pol, J., Weber, M. (eds.) SPIN 2010. LNCS, vol. 6349, pp. 227–244. Springer, Heidelberg (2010)
9. Jones, C.B.: Tentative steps toward a development method for interfering programs. ACM Trans. Program. Lang. Syst. 5(4), 596–619 (1983)
10. La Torre, S., Madhusudan, P., Parlato, G.: Reducing context-bounded concurrent reachability to sequential reachability. In: Bouajjani, A., Maler, O. (eds.) CAV 2009. LNCS, vol. 5643, pp. 477–492. Springer, Heidelberg (2009)
11. La Torre, S., Madhusudan, P., Parlato, G.: Sequentializing parameterized programs (2010), http://www.cs.uiuc.edu/~madhu/seqparam.pdf
12. Lahiri, S.K., Qadeer, S., Rakamarić, Z.: Static and precise detection of concurrency errors in systems code using SMT solvers. In: Bouajjani, A., Maler, O. (eds.) CAV 2009. LNCS, vol. 5643, pp. 509–524. Springer, Heidelberg (2009)
13. Lal, A., Reps, T.W.: Reducing concurrent analysis under a context bound to sequential analysis. In: Gupta, A., Malik, S. (eds.) CAV 2008. LNCS, vol. 5123, pp. 37–51. Springer, Heidelberg (2008)
14. Lubachevsky, B.D.: An approach to automating the verification of compact parallel coordination programs. Acta Inf. 21, 125–169 (1984)
15. Owicki, S.S., Gries, D.: An axiomatic proof technique for parallel programs. Acta Inf. 6, 319–340 (1976)
16. Xu, Q., de Roever, W.P., He, J.: The rely-guarantee method for verifying shared variable concurrent programs. Formal Asp. Comput. 9(2), 149–174 (1997)

Litmus: Running Tests against Hardware[*]

Jade Alglave[1,3], Luc Maranget[1], Susmit Sarkar[2], and Peter Sewell[2]

[1] INRIA
[2] University of Cambridge
[3] Oxford University

Abstract. Shared memory multiprocessors typically expose subtle, poorly understood and poorly specified relaxed-memory semantics to programmers. To understand them, and to develop formal models to use in program verification, we find it essential to take an empirical approach, testing what results parallel programs can actually produce when executed on the hardware. We describe a key ingredient of our approach, our litmus tool, which takes small 'litmus test' programs and runs them for many iterations to find interesting behaviour. It embodies various techniques for making such interesting behaviour appear more frequently.

1 Introduction

Modern shared memory multiprocessors do not actually provide the sequentially consistent (SC) memory semantics [Lam79] typically assumed in concurrent program verification. Instead, they provide a *relaxed memory model*, arising from optimisations in multiprocessor hardware, such as store buffering and instruction reordering (relaxed-memory behaviour can also arise from compiler optimisations). For example, in hardware with store buffers, the program below (in pseudo-code on the left and x86 assembly on the right) can end with 0 in both r_0 and r_1 on x86, a result not possible under SC:

Shared: x, y, initially zero		X86 SB (* Store Buffer test *)
Thread-local: r_0, r_1		{ x=0; y=0; }
Proc 0	Proc 1	P0 | P1 ;
$y \leftarrow 1$	$x \leftarrow 1$	MOV [y],\$1 | MOV [x],\$1 ;
$r_0 \leftarrow x$	$r_1 \leftarrow y$	MOV EAX,[x] | MOV EAX,[y] ;
Finally: is $r_0 = 0$ and $r_1 = 0$ possible?		exists (0:EAX=0 /\ 1:EAX=0)

The actual relaxed memory model exposed to the programmer by a particular multiprocessor is often unclear. Many models are described only in informal prose documentation [int09, pow09], which is often ambiguous, usually incomplete [SSS+10, AMSS10], and sometimes unsound (forbidding behaviour that is observable in reality) [SSS+10]. Meanwhile, researchers have specified various formal models for relaxed memory, but whether they accurately capture the subtleties of actual processor implementations is usually left unexamined. In

---

[*] We acknowledge funding from EPSRC grants EP/F036345, EP/H005633, and EP/H027351, from ANR project parsec (ANR-06-SETIN-010), and from INRIA associated team MM.

P.A. Abdulla and K.R.M. Leino (Eds.): TACAS 2011, LNCS 6605, pp. 41–44, 2011.
© Springer-Verlag Berlin Heidelberg 2011

contrast, we take a firmly empirical approach: testing what current implementations actually provide, and use the test results to inform the building of models. This is in the spirit of Collier's early work on ARCHTEST [Col92], which explores various violations of SC, but which does not deal with many complexities of modern processors, and also does not easily support testing new tests.

Much interesting memory model behaviour already shows up in small, but carefully crafted, concurrent programs operating on shared memory locations, "litmus tests". Given a specified initial state, the question for each test is what final values of registers and memory locations are permitted by actual hardware. Our litmus tool takes as input a litmus file, as on the right above, and runs the program within a test harness many times. On one such run of a million executions, it produced the result below, indicating that the result of interest occurred 34 times.

```
Positive: 34, Negative: 999966
Condition exists (0:EAX=0 /\ 1:EAX=0) is validated
```

The observable behaviour of a typical multiprocessor arises from an extremely complex (and commercially confidential) internal structure, and is highly non-deterministic, dependent on details of timing and the processors' internal state. Black-box testing cannot be guaranteed to produce all permitted results in such a setting, but with careful design the tool does generate interesting results with reasonable frequency.

2 High Level Overview

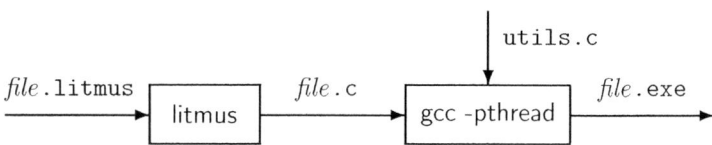

Our litmus tool takes as input small concurrent programs in x86 or Power assembly code (*file*.litmus). It accepts symbolic locations (such as x and y in our example), and symbolic registers. The tool then translates the program *file*.litmus into a C source file, encapsulating the program as inline assembly in a test harness. The C file is then compiled by gcc into executables which can be run on the machine to perform checks. The translation process performs some simple liveness analysis (to properly identify registers read and trashed by inline assembly), and some macro expansions (macros for lock acquire and release are translated to packaged assembly code).

The test harness initialises the shared locations, and then spawns threads (using the POSIX pthread library) to run the various threads within a loop. Each thread does some mild synchronization to ensure the programs run roughly at the same time, but with some variability so that interesting behaviour can show up. In the next section we describe various ways in which the harness can be adjusted, so that results of interest show up more often.

The entire program consists of about 10,000 lines of Objective Caml, plus about 1,000 lines of C. The two phases can be separated, allowing translated C files to be transferred to many machines. It is publicly distributed as a part of the diy tool suite, available at `http://diy.inria.fr`, with companion user documentation. litmus has been run successfully on Linux, Mac OS and AIX [AMSS10].

3   Test Infrastructure and Parameters

Users can control various parameters of the tool, which impact efficiency and outcome variability, sometimes dramatically.

Test repetition. To benefit from parallelism and stress the memory subsystem, given a test consisting of t threads $P_0, \ldots, P_{t-1}$, we run $n = \max(1, a/t)$ identical test *instances* concurrently on a machine with a cores. Each of these tests consists in repeating r times the sequence of creating t threads, collectively running the litmus test s times, then summing the produced outcomes in an histogram.

Thread assignment. We first fork t POSIX threads $T_0, \ldots T_{t-1}$ for executing $P_0, \ldots, P_{t-1}$. We can control which thread executes which code with the *launch mode*: if *fixed* then T_k executes P_k; if *changing* (the default) the association between POSIX and test threads is random. In our experience, the launch mode has a marginal impact, except when affinity is enabled—see *Affinity* below.

Accessing memory cells. Each thread executes a loop of size s. Loop iteration number i executes the code of one test thread and saves the final contents of its observed registers in arrays indexed by i; a memory location x in the `.litmus` source corresponds to an array cell. The access to this array cell depends on the *memory mode*. In *direct mode* the array cell is accessed directly as $x[i]$; hence cells are accessed sequentially and false sharing effects are likely. In *indirect mode* (the default) the array cell is accessed by a shuffled array of pointers, giving a much greater variability of outcomes. If the (default) *preload mode* is enabled, a preliminary loop of size s reads a random subset of the memory locations accessed by P_k, also leading to a greater outcome variability.

Thread synchronisation. The iterations performed by the different threads T_k may be unsynchronised, synchronised by a pthread-based barrier, or synchronised by busy-wait loops. Absence of synchronisation is of marginal interest when t exceeds a or when $t = 2$. Pthread-based barriers are slow and in fact offer poor synchronisation for short code sequences. Busy-waiting synchronisation is thus the preferred technique and the default.

Affinity. Affinity is a scheduler property binding software (POSIX) threads to given hardware *logical processor*. The latter may be single cores or, on machines with hyper-threading (x86) or simultaneous multi threading (SMT, Power) each core may host several logical processors.

   We allocate logical processors test instance by test instance (parameter n) and then POSIX thread by POSIX thread, scanning the logical processors sequence left-to-right by steps of the specified *affinity increment*. Suppose a logical processors sequence $P = 0, 1, \ldots, A - 1$ (the default on a machine with A logical processors available) and an increment i: we allocate (modulo A) first the processor 0, then i, then $2i$, *etc.* If we reach 0 again, we allocate the processor 1 and then increment again. Thereby, all the processors in the sequence will get allocated to different threads naturally, provided of course that less than A threads are scheduled to run.

4   The Impact of Test Parameters

Test parameters can have a large impact on the frequency of interesting results. Our tests are non-deterministic and parallel, and the behaviours of interest arise from specific microarchitectural actions at specific times. Thus the observed frequency is quite sensitive to the machine in question and to its operating system, in addition to the specific test itself.

Let us run the SB test from the introduction with various combinations of parameters on a lightly loaded Intel Core 2 Duo. There is one interesting outcome here, and we graph the frequency of that outcome arising per second below against the logarithm of the iteration size s. Note that only the orders of magnitude are significant, not the precise numbers, for a test of this nature.

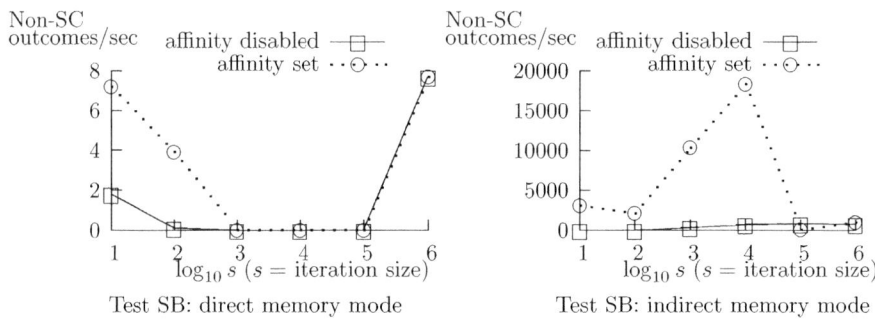

Test SB: direct memory mode   Test SB: indirect memory mode

We obtain the best results with indirect memory mode and affinity control, and 10^4 iterations per thread creation. These settings depend on the characteristics of the machine and scheduler, and we generally find such combinations of parameters remain good on the same testbed, even for different tests.

References

[AMSS10]   Alglave, J., Maranget, L., Sarkar, S., Sewell, P.: Fences in weak memory models. In: Touili, T., Cook, B., Jackson, P. (eds.) CAV 2010. LNCS, vol. 6174, pp. 258–272. Springer, Heidelberg (2010)
[Col92]   Collier, W.W.: Reasoning About Parallel Architectures. Prentice-Hall, Englewood Cliffs (1992)
[int09]   Intel 64 and IA-32 Architectures Software Developer's Manual, vol. 3A, rev. 30 (March 2009)
[Lam79]   Lamport, L.: How to Make a Correct Multiprocess Program Execute Correctly on a Multiprocessor. IEEE Trans. Comput. 46(7), 779–782 (1979)
[pow09]   Power ISA Version 2.06 (2009)
[SSS+10]   Sewell, P., Sarkar, S., Owens, S., Zappa Nardelli, F., Myreen, M.O.: x86-TSO: A rigorous and usable programmer's model for x86 multiprocessors. Communications of the ACM 53(7), 89–97 (2010) (Research Highlights)

Canonized Rewriting and Ground AC Completion Modulo Shostak Theories⋆

Sylvain Conchon, Evelyne Contejean, and Mohamed Iguernelala

LRI, Univ Paris-Sud, CNRS, Orsay F-91405
INRIA Saclay – Ile-de-France, ProVal, Orsay, F-91893

Abstract. AC-completion efficiently handles equality modulo associative and commutative function symbols. When the input is ground, the procedure terminates and provides a decision algorithm for the word problem. In this paper, we present a modular extension of ground AC-completion for deciding formulas in the combination of the theory of equality with user-defined AC symbols, uninterpreted symbols and an arbitrary signature disjoint Shostak theory X. Our algorithm, called AC(X), is obtained by augmenting in a modular way ground AC-completion with the canonizer and solver present for the theory X. This integration rests on canonized rewriting, a new relation reminiscent to normalized rewriting, which integrates canonizers in rewriting steps. AC(X) is proved sound, complete and terminating, and is implemented to extend the core of the ALT-ERGO theorem prover.

Keywords: decision procedure, associativity and commutativity, rewriting, AC-completion, SMT solvers, Shostak's algorithm.

1 Introduction

Many mathematical operators occurring in automated reasoning such as union and intersection of sets, or boolean and arithmetic operators, satisfy the following associativity and commutativity (AC) axioms

$$\forall x.\forall y.\forall z.\ u(x, u(y, z)) = u(u(x, y), z) \quad \text{(A)}$$
$$\forall x.\forall y.\ u(x, y) = u(y, x) \quad \text{(C)}$$

Automated AC reasoning is known to be difficult. Indeed, the mere addition of these two axioms to a prover will usually glut it with plenty of useless equalities which will strongly impact its performances[1]. In order to avoid this drawback, built-in procedures have been designed to efficiently handle AC symbols. For instance, SMT-solvers incorporate dedicated decision procedures for some *specific* AC symbols such as arithmetic or boolean operators. On the contrary, algorithms found in resolution-based provers such as AC-completion allow a powerful *generic* treatment of user-defined AC symbols.

⋆ Work partially supported by the French ANR project ANR-08-005 Decert.
[1] Given a term t of the form $u(c_1, u(c_2, \ldots, u(c_n, c_{n+1})\ldots))$, the axiomatic approach may have to explicitly handle the $(2n)!/n!$ terms equivalent to t.

P.A. Abdulla and K.R.M. Leino (Eds.): TACAS 2011, LNCS 6605, pp. 45–59, 2011.
© Springer-Verlag Berlin Heidelberg 2011

Given a finite word problem $\bigwedge_{i \in I} s_i = t_i \vdash s = t$ where the function symbols are either uninterpreted or AC, AC-completion attempts to transform the conjunction $\bigwedge_{i \in I} s_i = t_i$ into a finitely terminating, confluent term rewriting system R whose reductions preserve identity. The rewriting system R serves as a decision procedure for validating $s = t$ modulo AC: the equation holds if and only if the normal forms of s and t w.r.t R are equal modulo AC. Furthermore, when its input contains only ground equations, AC-completion terminates and outputs a convergent rewriting system [16].

Unfortunately, AC reasoning is only a part of the automated deduction problem, and what we really need is to decide formulas combining AC symbols and other theories. For instance, in practice, we are interested in deciding finite ground word problems which contain a mixture of uninterpreted, interpreted and AC function symbols, as in the following assertion

$$
\begin{aligned}
&u(a, c_2 - c_1) \approx a \wedge u(e_1, e_2) - f(b) \approx u(d, d) \wedge \\
&d \approx c_1 + 1 \wedge e_2 \approx b \wedge u(b, e_1) \approx f(e_2) \wedge c_2 \approx 2 * c_1 + 1
\end{aligned} \quad \vdash a \approx u(a, 0)
$$

where u is an AC symbol, $+, -, *$ and the numerals are from the theory of linear arithmetic, f is an uninterpreted function symbol and the other symbols are uninterpreted constants. A combination of AC reasoning with linear arithmetic and the free theory $\mathcal{E}$ of equality is necessary to prove this formula. Linear arithmetic is used to show that $c_2 - c_1 = c_1 + 1$ so that (i) $u(a, c_1 + 1) = a$ follows by congruence. Independently, $e_2 = b$ and $d = c_1 + 1$ imply (ii) $u(c_1 + 1, c_1 + 1) = 0$ by congruence, linear arithmetic and commutativity of u. AC reasoning can finally be used to conclude that (i) and (ii) imply that $u(a, c_1 + 1, c_1 + 1)$ is equal to both a and $u(a, 0)$.

There are two main methods for combining decision procedures for disjoint theories. First, the Nelson-Oppen approach [18] is based on a variable abstraction mechanism and the exchange of equalities between shared variables. Second, the Shostak's algorithm [21] extends a congruence closure procedure with theories equipped with canonizers and solvers, $i.e.$ procedures that compute canonical forms of terms and solve equations, respectively. While ground AC-completion can be easily combined with other decision procedures by the Nelson-Oppen method, it cannot be directly integrated in the Shostak's framework since it actually does not provide a solver for the AC theory.

In this paper, we investigate the integration of Shostak theories in ground AC-completion. We first introduce a new notion of rewriting called *canonized* rewriting which adapts normalized rewriting to cope with canonization. Then, we present a modular extension of ground AC-completion for deciding formulas in the combination of the theory of equality with user-defined AC symbols, uninterpreted symbols and an arbitrary signature disjoint Shostak theory X. The main ideas of our integration are to substitute standard rewriting by canonized rewriting, using a global canonizer for AC and X, and to replace the equation orientation mechanism found in ground AC-completion with the solver for X.

AC-completion has been studied for a long time in the rewriting community [15,20]. A generic framework for combining completion with a generic built-in equational theory E has been proposed in [10]. Normalized completion [17] is designed to use a modified rewriting relation when the theory E is equivalent to the union of the AC theory and a convergent rewriting system $\mathcal{S}$. In this setting, rewriting steps are only performed on $\mathcal{S}$-normalized terms. AC(X) can be seen as an adaptation of ground normalized completion to efficiently handle the theory E when it is equivalent to the union of the AC theory and a Shostak theory X. In particular, $\mathcal{S}$-normalization is replaced by the application of the canonizer of X. This modular integration of X allows us to reuse proof techniques of ground AC-completion [16] to show the correctness of AC(X).

Kapur [11] used ground completion to demystify Shostak's congruence closure algorithm and Bachmair *et al.* [3] compared its strategy with other ones into an abstract congruence closure framework. While the latter approach can also handle AC symbols, none of these works formalized the integration of Shostak theories into (AC) ground completion.

Outline. Section 2 recalls standard ground AC completion. Section 3 is devoted to Shostak theories and global canonization. Section 4 presents the AC(X) algorithm and illustrates its use through an example. The correctness of AC(X) is sketched in Section 5 and experimental results are presented in Section 6. Conclusion and future works are presented in Section 7.

2   Ground AC-Completion

In this section, we first briefly recall the usual notations and definitions of [1,7] for term rewriting modulo AC. Then, we give the usual set of inference rules for ground AC-completion procedure and we illustrate its use through an example.

Terms are built from a signature $\Sigma = \Sigma_{AC} \uplus \Sigma_{\mathcal{E}}$ of AC and uninterpreted symbols, and a set of variables $\mathcal{X}$ yielding the term algebra $\mathcal{T}_\Sigma(\mathcal{X})$. The range of letters $a \dots f$ denotes uninterpreted symbols, u denotes an AC function symbol, s, t, l, r denote terms, and x, y, z denote variables. Viewing terms as trees, subterms within a term s are identified by their positions. Given a position p, $s|_p$ denotes the subterm of s at position p, and $s[r]_p$ the term obtained by replacement of $s|_p$ by the term r. We will also use the notation $s(p)$ to denote the symbol at position p in the tree, and the root position is denoted by Λ. Given a subset Σ' of Σ, a subterm $t|_p$ of t is a Σ'-alien of t if $t(p) \notin \Sigma'$ and p is minimal *w.r.t* the prefix word ordering[2]. We write $\mathcal{A}_{\Sigma'}(t)$ the multiset of Σ'-aliens of t.

A substitution is a partial mapping from variables to terms. Substitutions are extended to a total mapping from terms to terms in the usual way. We write $t\sigma$ for the application of a substitution σ to a term t. A well-founded quasi-ordering [6] on terms is a reduction quasi-ordering if $s \preceq t$ implies $s\sigma \preceq t\sigma$ and $l[s]_p \preceq l[t]_p$, for any substitution σ, term l and position p. A quasi-ordering $\preceq$ defines an equivalence relation $\simeq$ as $\preceq \cap \succeq$ and a partial ordering $\prec$ as $\preceq \cap \not\succeq$.

An equation is an unordered pair of terms, written $s \approx t$. The variables contained in an equation, if any, are understood as being universally quantified. Given a set of equations E, the equational theory of E, written $=_E$, is the set of

---

[2] Notice that according to this definition, a variable may be a Σ'-alien.

equations that can be obtained by reflexivity, symmetry, transitivity, congruence and instances of equations in E^3. The word problem for E consists in determining if, given two ground terms s and t, the equation $s \approx t$ is in $=_E$, denoted by $s =_E t$. The word problem for E is ground when E contains only ground equations. An equational theory $=_E$ is said to be *inconsistent* when $s =_E t$, for *any* s and t.

A rewriting rule is an oriented equation, usually denoted by $l \rightarrow r$. A term s rewrites to a term t at position p by the rule $l \rightarrow r$, denoted by $s \rightarrow^p_{l \rightarrow r} t$, iff there exists a substitution σ such that $s|_p = l\sigma$ and $t = s[r\sigma]_p$. A rewriting system R is a set of rules. We write $s \rightarrow_R t$ whenever there exists a rule $l \rightarrow r$ of R such that s rewrites to t by $l \rightarrow r$ at some position. A normal form of a term s w.r.t to R is a term t such that $s \rightarrow^*_R t$ and t cannot be rewritten by R. The system R is said to be *convergent* whenever any term s has a unique normal form, denoted $s\downarrow_R$, and does not admit any infinite reduction. Completion [12] aims at converting a set E of equations into a convergent rewriting system R such that the sets $=_E$ and $\{s \approx t \mid s\downarrow_R = t\downarrow_R\}$ coincide. Given a suitable reduction ordering on terms, it has been proved that completion terminates when E is ground [14].

Rewriting modulo AC. Let $=_{AC}$ be the equational theory obtained from the set:

$$AC = \bigcup_{u \in \Sigma_{AC}} \{\, u(x,y) \approx u(y,x),\ u(x, u(y,z)) \approx u(u(x,y),z) \,\}$$

In general, given a set E of equations, it has been shown that no suitable reduction ordering allows completion to produce a convergent rewriting system for $E \cup AC$. When E is ground, an alternative consists in in-lining AC reasoning both in the notion of rewriting step and in the completion procedure.

Rewriting modulo AC is directly related to the notion of matching modulo AC as shown by the following example. Given a rule $u(a, u(b, c)) \rightarrow t$, we would like the following reductions to be possible:

$$(1)\ f(u(c, u(b, a)), d) \rightarrow f(t, d) \qquad (2)\ u(a, u(c, u(d, b))) \rightarrow u(t, d)$$

Associativity and commutativity of u are needed in (1) for the subterm $u(c, u(b, a))$ to match the term $u(a, u(b, c))$, and in (2) for the term $u(a, u(c, u(d, b)))$ to be seen as $u(u(a, u(b, c)), d)$, so that the rule can be applied. More formally, this leads to the following definition.

Definition 1 (Ground rewriting modulo AC). *A term s rewrites to a term t modulo AC at position p by the rule $l \rightarrow r$, denoted by $s \rightarrow^p_{AC \backslash l \rightarrow r} t$, iff (1) $s|_p =_{AC} l$ and $t = s[r]_p$ or (2) $l(\Lambda) = u$ and there exists a term s' such that $s|_p =_{AC} u(l, s')$ and $t = s[u(r, s')]_p$*

In order to produce a convergent rewriting system, ground AC-completion requires a well-founded reduction quasi-ordering $\preceq$ total on ground terms with an underlying equivalence relation which coincides with $=_{AC}$. Such an ordering will be called a total ground AC-reduction ordering.

---

[3] The equational theory of the free theory of equality $\mathcal{E}$, defined by the empty set of equations, is simply denoted $=$.

$$\textbf{Trivial} \quad \frac{\langle\ E \cup \{s \approx t\}\ \mid\ R\ \rangle}{\langle\ E\ \mid\ R\ \rangle}\ s =_{AC} t$$

$$\textbf{Orient} \quad \frac{\langle\ E \cup \{s \approx t\}\ \mid\ R\ \rangle}{\langle\ E\ \mid\ R \cup \{s \rightarrow t\}\ \rangle}\ t \prec s$$

$$\textbf{Simplify} \quad \frac{\langle\ E \cup \{s \approx t\}\ \mid\ R\ \rangle}{\langle\ E \cup \{s' \approx t\}\ \mid\ R\ \rangle}\ s \rightarrow_{AC \backslash R} s'$$

$$\textbf{Compose} \quad \frac{\langle\ E\ \mid\ R \cup \{l \rightarrow r\}\ \rangle}{\langle\ E\ \mid\ R \cup \{l \rightarrow r'\}\ \rangle}\ r \rightarrow_{AC \backslash R} r'$$

$$\textbf{Collapse} \quad \frac{\langle\ E\ \mid\ R \cup \{g \rightarrow d, l \rightarrow r\}\ \rangle}{\langle\ E \cup \{l' \approx r\}\ \mid\ R \cup \{g \rightarrow d\}\ \rangle}\ \begin{cases} l \rightarrow_{AC \backslash g \rightarrow d} l' \\ g \prec l\ \lor\ (g \simeq l \land d \prec r) \end{cases}$$

$$\textbf{Deduce} \quad \frac{\langle\ E\ \mid\ R\ \rangle}{\langle\ E \cup \{s \approx t\}\ \mid\ R\ \rangle}\ s \approx t \in \texttt{headCP}(R)$$

Fig. 1. Inference rules for ground AC-completion

The inference rules for ground AC-completion are given in Figure 1. The rules describe the evolution of the state of a procedure, represented as a configuration $\langle\ E \mid R\ \rangle$, where E is a set of ground equations and R a ground set of rewriting rules. The initial state is $\langle\ E_0 \mid \emptyset\ \rangle$ where E_0 is a given set of ground equations. **Triv**ial removes an equation $u \approx v$ from E when u and v are equal modulo AC. **Orient** turns an equation into a rewriting rule according to a given total ground AC-reduction ordering $\preceq$. R is used to rewrite either side of an equation (**Simplify**), and to reduce right hand side of rewriting rules (**Com**pose). Given a rule $l \rightarrow r$, **Coll**apse either reduces l at an inner position, or replaces l by a term smaller than r. In both cases, the reduction of l to l' may influence the orientation of the rule $l' \rightarrow r$ which is added to E as an equation in order to be re-oriented. Finally, **Deduce** adds equational consequences of rewriting rules to E. For instance, if R contains two rules of the form $u(a, b) \rightarrow s$ and $u(a, c) \rightarrow t$, then the term $u(a, u(b, c))$ can either be reduced to $u(s, c)$ or to the term $u(t, b)$. The equation $u(s, c) \approx u(t, b)$, called *critical pair*, is thus necessary for ensuring convergence of R. Critical pairs of a set of rules are computed by the following function (a^μ stands for the maximal term w.r.t. size enjoying the assertion):

$$\texttt{headCP}(R) = \left\{\ u(b, r') \approx u(b', r)\ \middle|\ \begin{array}{l} l \rightarrow r \in R,\ \ l' \rightarrow r' \in R \\ \exists a^\mu : l =_{AC} u(a^\mu, b)\ \land\ l' =_{AC} u(a^\mu, b') \end{array}\ \right\}$$

Example. To get a flavor of ground AC-completion, consider a modified version of the assertion given in the introduction, where the arithmetic part has been removed (and uninterpreted constant symbols renamed for the sake of simplicity)

$$u(a_1, a_4) \approx a_1, u(a_3, a_6) \approx u(a_5, a_5), a_5 \approx a_4, a_6 \approx a_2 \vdash a_1 \approx u(a_1, u(a_6, a_3))$$

The precedence $a_1 \prec_p \cdots \prec_p a_6 \prec_p u$ defines an AC-RPO ordering on terms [19] which is suitable for ground AC-completion. The table in Figure 2 shows the application steps of the rules given in Figure 1 from an initial configuration $\langle \{u(a_1, a_4) \approx a_1, u(a_3, a_6) \approx u(a_5, a_5), a_5 \approx a_4, a_6 \approx a_2\} \mid \emptyset \rangle$ to a final configuration $\langle \emptyset \mid R_f \rangle$, where R_f is the set of rewriting rules $\{1, 3, 5, 7, 10\}$. It can be checked that $a_1 \downarrow_{R_f}$ and $u(a_1, u(a_6, a_3)) \downarrow_{R_f}$ are identical.

1	$\mathbf{u(a_1, a_4) \rightarrow a_1}$	**Ori** $u(a_1, a_4) \approx a_1$
2	$u(a_3, a_6) \rightarrow u(a_5, a_5)$	**Ori** $u(a_3, a_6) \approx u(a_5, a_5)$
3	$\mathbf{a_5 \rightarrow a_4}$	**Ori** $a_5 \approx a_4$
4	$u(a_3, a_6) \rightarrow u(a_4, a_4)$	**Com** 2 and 3
5	$\mathbf{a_6 \rightarrow a_2}$	**Ori** $a_6 \approx a_2$
6	$u(a_3, a_2) \approx u(a_4, a_4)$	**Col** 4 and 5
7	$\mathbf{u(a_4, a_4) \rightarrow u(a_3, a_2)}$	**Ori** 6
8	$u(a_1, a_4) \approx u(a_1, u(a_3, a_2))$	**Ded** from 1 and 7
9	$a_1 \approx u(a_1, u(a_3, a_2))$	**Sim** 8 by 1
10	$\mathbf{u(a_1, u(a_3, a_2)) \rightarrow a_1}$	**Ori** 9

Fig. 2. Ground AC-completion example

3 Shostak Theories and Global Canonization

In this section, we recall the notions of canonizers and solvers underlying Shostak theories and show how to obtain a global canonizer for the combination of the theories $\mathcal{E}$ and AC with an arbitrary signature disjoint Shostak theory X.

From now on, we assume given a theory X with a signature Σ_X. A canonizer for X is a function $\mathtt{can_X}$ that computes a unique normal form for every term such that $s =_X t$ iff $\mathtt{can_X}(s) = \mathtt{can_X}(t)$. A solver for X is a function $\mathtt{solve_X}$ that solves equations between Σ_X-terms. Given an equation $s \approx t$, $\mathtt{solve_X}(s \approx t)$ either returns a special value $\bot$ when $s \approx t \cup X$ is inconsistent, or an equivalent substitution. A Shostak theory X is a theory with a canonizer and a solver which fulfill some standard properties given for instance in [13].

Our combination technique is based on the integration of a Shostak theory X in ground AC-completion. From now on, we assume that terms are built from a signature Σ defined as the union of the disjoint signatures Σ_{AC}, $\Sigma_{\mathcal{E}}$ and Σ_X. We also assume a total ground AC-reduction ordering $\preceq$ defined on $\mathcal{T}_\Sigma(\mathcal{X})$ used later on for completion. The combination mechanism requires defining both a global canonizer for the union of $\mathcal{E}$, AC and X, and a wrapper of $\mathtt{solve_X}$ to handle heterogeneous equations. These definitions make use of a global one-to-one mapping $\alpha : \mathcal{T}_\Sigma \rightarrow \mathcal{X}$ (and its inverse mapping ρ) and are based on a variable abstraction mechanism which computes the *pure* Σ_X-part $[\![t]\!]$ of a heterogeneous term t as follows:

$$[\![t]\!] = f([\![s]\!]) \text{ when } t = f(s) \text{ and } f \in \Sigma_X \qquad \text{and} \qquad [\![t]\!] = \alpha(t) \text{ otherwise}$$

The canonizer for AC defined in [9] is based on flattening and sorting techniques which simulate associativity and commutativity, respectively. For instance, the term $u(u(u'(c, b), b), c)$ is first flattened to $u(u'(c, b), b, c)$ and then sorted[4] to get

---

[4] For instance, using the AC-RPO ordering based on the precedence $b \prec_p c \prec_p u'$.

the term $u(b, c, u'(c, b))$. It has been formally proved that this canonizer solves the word problem for AC [5]. However, this definition implies a modification of the signature Σ_{AC} where arity of AC symbols becomes variadic. Using such canonizer would impact the definition of AC-rewriting given in Section 2. In order to avoid such modification we shall define an equivalent canonizer that builds degenerate trees instead of flattened terms. For instance, we would expect the normal form of $u(u(u'(c, b), b), c)$ to be $u(b, u(c, u'(c, b)))$. Given a signature Σ which contains Σ_{AC} and any total ordering $\trianglelefteq$ on terms, we define can_{AC} by:

$$
\begin{aligned}
\mathrm{can}_{AC}(x) &= x && \text{when } x \in \mathcal{X} \\
\mathrm{can}_{AC}(f(\boldsymbol{v})) &= f(\mathrm{can}_{AC}(\boldsymbol{v})) && \text{when } f \notin \Sigma_{AC} \\
\mathrm{can}_{AC}(u(t_1, t_2)) &= u(s_1, u(s_2, \ldots, u(s_{n-1}, s_n) \ldots)) \\
&\quad \text{where } t_i' = \mathrm{can}_{AC}(t_i) \text{ for } i \in [1, 2] \\
&\quad \text{and } \{\!\{s_1, \ldots, s_n\}\!\} = \mathcal{A}_{\{u\}}(t_1') \cup \mathcal{A}_{\{u\}}(t_2') \\
&\quad \text{and } s_i \trianglelefteq s_{i+1} \text{ for } i \in [1, n-1] \text{when } u \in \Sigma_{AC}
\end{aligned}
$$

We can easily show that can_{AC} enjoys the standard properties required for a canonizer. The proof that can_{AC} solves the word problem for AC follows directly from the one given in [5].

Using the technique described in [13], we define our global canonizer can which combines can_X with can_{AC} as follows:

$$
\begin{aligned}
\mathrm{can}(x) &= x && \text{when } x \in \mathcal{X} \\
\mathrm{can}(f(\boldsymbol{v})) &= f(\mathrm{can}(\boldsymbol{v})) && \text{when } f \in \Sigma_{\mathcal{E}} \\
\mathrm{can}(u(s, t)) &= \mathrm{can}_{AC}(u(\mathrm{can}(s), \mathrm{can}(t))) && \text{when } u \in \Sigma_{AC} \\
\mathrm{can}(f_X(\boldsymbol{v})) &= \mathrm{can}_X(f_X([\![\mathrm{can}(\boldsymbol{v})]\!]))\rho && \text{when } f_X \in \Sigma_X
\end{aligned}
$$

Again, the proofs that can solves the word problem for the union $\mathcal{E}$, AC and X and enjoys the standard properties required for a canonizer are similar to those given in [13]. The only difference is that can_{AC} directly works on the signature Σ, which avoids the use of a variable abstraction step when canonizing a mixed term of the form $u(t_1, t_2)$ such that $u \in \Sigma_{AC}$.

Using the same mappings α, ρ and the abstraction function, the wrapper $\mathtt{solve}$ can be easily defined by:

$$
\mathtt{solve}(s \approx t) = \begin{cases} \bot & \text{if } \mathtt{solve}_X([\![s]\!] \approx [\![t]\!]) = \bot \\ \{\, x_i\rho \to t_i\rho \,\} & \text{if } \mathtt{solve}_X([\![s]\!] \approx [\![t]\!]) = \{x_i \approx t_i\} \end{cases}
$$

In order to ensure termination of AC(X), the global canonizer and the wrapper must be compatible with the ordering $\preceq$ used by AC-completion, that is:

Lemma 1. $\forall t \in \mathcal{T}_\Sigma,\ \mathrm{can}(t) \preceq t$
$\forall s, t \in \mathcal{T}_\Sigma,\ \text{if } \mathtt{solve}(s \approx t) = \bigcup\{p_i \to v_i\} \text{ then } v_i \prec p_i$

We can prove that the above properties hold when the theory X enjoys the following local compatibility properties:

Axiom 1. $\forall t \in \mathcal{T}_\Sigma,\ \mathrm{can}_X([\![t]\!]) \preceq [\![t]\!]$
$\forall s, t \in \mathcal{T}_\Sigma,\ \text{if } \mathtt{solve}_X([\![s]\!] \approx [\![t]\!]) = \bigcup\{x_i \approx t_i\} \text{ then } t_i\rho \prec x_i\rho$

To fulfil this axiom, AC-reduction ordering can be chosen as an AC-RPO order-
ing [19] based on a precedence relation $\prec_p$ such that $\Sigma_X \prec_p \Sigma_\mathcal{E} \cup \Sigma_{AC}$. From
now on, we assume that X is locally compatible with $\preceq$.

Example. To solve the equation $u(a,b) + a \approx 0$, we use the abstraction $\alpha =
\{u(a,b) \mapsto x, \ a \mapsto y\}$ and call $\texttt{solve}_X$ on $x + y \approx 0$. Since $a \prec u(a,b)$, the only
solution which fulfills the axiom above is $\{x \approx -y\}$. We apply ρ and get the set
$\{u(a,b) \to -a\}$ of rewriting rules.

4   Ground AC-Completion Modulo X

In this section, we present the $\mathsf{AC(X)}$ algorithm which extends the ground AC-
completion procedure given in Section 2. For that purpose, we first adapt the
notion of ground AC-rewriting to cope with canonizers. Then, we show how
to refine the inference rules given in Figure 1 to reason modulo the equational
theory induced by a set E of ground equations and the theories $\mathcal{E}$, AC and X.

4.1   Canonized Rewriting

From rewriting point of view, a canonizer behaves like a convergent rewriting system: it gives an effective way of computing normal forms. Thus, a natural way for
integrating $\texttt{can}$ in ground AC-completion is to extend normalized rewriting [17].

Definition 2. *Let* $\texttt{can}$ *be a canonizer. A term s* $\texttt{can}$*-rewrites to a term t at
position p by the rule $l \to r$, denoted by $s \leadsto^p_{l \to r} t$, iff*

$$s \to^p_{AC \backslash l \to r} t' \qquad \text{and} \qquad \texttt{can}(t') = t$$

Example. Using the usual canonizer $\texttt{can}_A$ for linear arithmetic and the rule
$\gamma : u(a,b) \to a$, the term $f(a + 2 * u(b,a))$ $\texttt{can}_A$-rewrites to $f(3 * a)$ by $\leadsto_\gamma$ as
follows: $f(a + 2 * u(b,a)) \to_{AC \backslash \gamma} f(a + 2 * a)$ and $\texttt{can}_A(f(a + 2 * a)) = f(3 * a)$.

Lemma 2. $\forall s, t. \quad s \leadsto_{l \to r} t \implies s =_{AC,X,l \approx r} t$

4.2   The $\mathsf{AC(X)}$ Algorithm

The first step of our combination technique consists in replacing the rewriting
relation found in completion by canonized rewriting. This leads to the rules of
$\mathsf{AC(X)}$ given in Figure 3. The state of the procedure is a pair $\langle\, E \mid R \,\rangle$ of equations
and rewriting rules. The initial configuration is $\langle\, E_0 \mid \emptyset \,\rangle$ where E_0 is supposed to
be a set of equations between canonized terms. Since $\mathsf{AC(X)}$'s rules only involve
canonized rewriting, the algorithm maintains the invariant that terms occurring
in E and R are in canonical forms. **Tri**vial thus removes an equation $u \approx v$
from E when u and v are syntactically equal. A new rule **Bottom** is used to
detect inconsistent equations. Similarly to normalized completion, integrating
the global canonizer $\texttt{can}$ in rewriting is not enough to fully extend ground AC-
completion with the theory X: in both cases the orientation mechanism has
to be adapted . Therefore, the second step consists in integrating the wrapper
$\texttt{solve}$ in the **Ori**ent rule. The other rules are much similar to those of ground
AC-completion except that they use the relation $\leadsto_R$ instead of $\to_{AC \backslash R}$.

$$\text{Trivial } \frac{\langle\, E \cup \{s \approx t\} \mid R \,\rangle}{\langle\, E \mid R \,\rangle} \; s = t \qquad \text{Bottom } \frac{\langle\, E \cup \{s \approx t\} \mid R \,\rangle}{\perp} \; \texttt{solve}(s,t) = \perp$$

$$\text{Orient } \frac{\langle\, E \cup \{s \approx t\} \mid R \,\rangle}{\langle\, E \mid R \cup \texttt{solve}(s,t) \,\rangle} \; \texttt{solve}(s,t) \neq \perp$$

$$\text{Simplify } \frac{\langle\, E \cup \{s \approx t\} \mid R \,\rangle}{\langle\, E \cup \{s' \approx t\} \mid R \,\rangle} \; s \leadsto_R s' \qquad \text{Compose } \frac{\langle\, E \mid R \cup \{l \to r\} \,\rangle}{\langle\, E \mid R \cup \{l \to r'\} \,\rangle} \; r \leadsto_R r'$$

$$\text{Collapse } \frac{\langle\, E \mid R \cup \{g \to d,\, l \to r\} \,\rangle}{\langle\, E \cup \{l' \approx r\} \mid R \cup \{g \to d\} \,\rangle} \quad \begin{cases} l \leadsto_{g \to d} l' \\ g \prec l \;\lor\; (g \simeq l \land d \prec r) \end{cases}$$

$$\text{Deduce } \frac{\langle\, E \mid R \,\rangle}{\langle\, E \cup \{s \approx t\} \mid R \,\rangle} \; s \approx t \in \texttt{headCP}(R)$$

Fig. 3. Inference rules for ground AC-completion modulo X

1	$u(a, c_2 - c_1) \to a$	Ori $u(a, c_2 - c_1) \approx a$
2	$u(e_1, e_2) \to u(d, d) + f(b)$	Ori $u(e_1, e_2) - f(b) \approx u(d, d)$
3	$\mathbf{d} \to \mathbf{c_1 + 1}$	Ori $d \approx c_1 + 1$
4	$u(e_1, e_2) \to u(c_1 + 1, c_1 + 1) + f(b)$	Com 2 and 3
5	$\mathbf{e_2} \to \mathbf{b}$	Ori $e_2 \approx b$
6	$u(b, e_1) \approx u(c_1 + 1, c_1 + 1) + f(b)$	Col 4 and 5
7	$u(b, e_1) \to u(c_1 + 1, c_1 + 1) + f(b)$	Ori $u(b, e_1) \approx u(c_1 + 1, c_1 + 1) + f(b)$
8	$u(c_1 + 1, c_1 + 1) + f(b) \approx f(b)$	Sim $u(b, e_1) \approx f(e_2)$ by 5 and 7
9	$\mathbf{u(c_1 + 1, c_1 + 1)} \to \mathbf{0}$	Ori $u(c_1 + 1, c_1 + 1) + f(b) \approx f(b)$
10	$\mathbf{u(b, e_1)} \to \mathbf{f(b)}$	Com 7 and 9
11	$\mathbf{c_2} \to \mathbf{2 * c_1 + 1}$	Ori $c_2 \approx 2 * c_1 + 1$
12	$u(a, c_1 + 1) \approx a$	Col 1 and 11
13	$\mathbf{u(a, c_1 + 1)} \to \mathbf{a}$	Ori $u(a, c_1 + 1) \approx a$
14	$u(0, a) \approx u(a, c_1 + 1)$	Ded from 9 and 13
15	$u(0, a) \approx a$	Sim 14 by 13
16	$\mathbf{u(0, a)} \to \mathbf{a}$	Ori 15

Fig. 4. AC(X) on the running example

Example. We illustrate AC(X) on the example given in the introduction:

$$u(a, c_2 - c_1) \approx a \land u(e_1, e_2) - f(b) \approx u(d, d) \land$$
$$d \approx c_1 + 1 \land e_2 \approx b \land u(b, e_1) \approx f(e_2) \land c_2 \approx 2 * c_1 + 1 \quad \vdash a \approx u(a, 0)$$

The table given in Figure 4 shows the application of the rules of AC(X) on the example when X is instantiated by linear arithmetic. We use an AC-RPO ordering based on the precedence $1 \prec_p 2 \prec_p a \prec_p b \prec_p c_1 \prec_p c_2 \prec_p d \prec_p e_1 \prec_p e_2 \prec_p f \prec_p u$. The procedure terminates and produces a convergent rewriting system $R_f = \{3, 5, 9, 10, 11, 13, 16\}$. Using R_f, we can check that a and $u(a, 0)$ can-rewrite to the same normal form.

5   Correctness

As usual, in order to enforce correctness, we cannot use any (unfair) strategy. We say that a strategy is *strongly fair* when no possible application of an inference rule is infinitely delayed and **Ori**ent is only applied over fully reduced terms.

Theorem 1. *Given a set E of ground equations, the application of the rules of AC(X) under a strongly fair strategy terminates and either produces $\perp$ when $E \cup AC \cup X$ is inconsistent, or yields a final configuration $\langle\ \emptyset \mid R\ \rangle$ such that:*
$$\forall s, t \in \mathcal{T}_\Sigma.\ s =_{E, AC, X} t \iff \mathrm{can}(s)\wr_R = \mathrm{can}(t)\wr_R$$

The proof[5] is based on three intermediate theorems, stating respectively soundness, completeness and termination. In the following, we shall consider a fixed run of the completion procedure,

$$\langle\ E_0 \mid \emptyset\ \rangle \to \langle\ E_1 \mid R_1\ \rangle \to \ldots \to \langle\ E_n \mid R_n\ \rangle \to \langle\ E_{n+1} \mid R_{n+1}\ \rangle \to \ldots$$

starting from the initial configuration $\langle\ E_0 \mid \emptyset\ \rangle$. We denote R_∞ (resp. E_∞) the set of all encountered rules $\bigcup_n R_n$ (resp. equations $\bigcup_n E_n$) and $\mathcal{R}_\omega$ (resp. E_ω) the set of persistent rules $\bigcup_n \bigcap_{i \geq n} R_i$ (resp. equations $\bigcup_n \bigcap_{i \geq n} E_i$).

5.1   Soundness

Soundness is ensured by the following invariant:

Theorem 2. *For any configuration $\langle\ E_n \mid R_n\ \rangle$ reachable from $\langle\ E_0 \mid \emptyset\ \rangle$,*
$$\forall\ s,\ t,\quad (s, t) \in E_n \cup R_n \implies s =_{AC, X, E_0} t$$

Proof. The invariant obviously holds for the initial configuration and is preserved by all the inference rules. The rules **Simp**lify, **Comp**ose, **Coll**apse and **Ded**uce preserve the invariant since for any rule $l \to r$, if $l =_{AC, X, E_0} r$, for any term s rewritten by $\leadsto_{l \to r}$ into t, then $s =_{AC, X, E_0} t$. If **Ori**ent is used to turn an equation $s \approx t$ into a set of rules $\{p_i \to v_i\}$, by definition of solve, $p_i = x_i \rho$ and $v_i = t_i \rho$, where $\mathrm{solve}_X(\llbracket s \rrbracket \approx \llbracket t \rrbracket) = \{x_i \approx t_i\}$. By soundness of solve_X $x_i =_{X, \llbracket s \rrbracket \approx \llbracket t \rrbracket} t_i$. An equational proof of $x_i =_{X, \llbracket s \rrbracket \approx \llbracket t \rrbracket} t_i$ can be instantiated by ρ, yielding an equational proof $p_i =_{X, s \approx t} v_i$. Since by induction $s =_{AC, X, E_0} t$ holds, we get $p_i =_{AC, X, E_0} v_i$.

In the rest of this section, we assume that the strategy is strongly fair. This implies in particular that $\mathrm{headCP}(R_\omega) \subseteq E_\infty$, $E_\omega = \emptyset$ and R_ω is inter-reduced, that is none of its rules can be collapsed or composed by another one. We also assume that $\perp$ is not encountered, otherwise, termination is obvious.

5.2   Completeness

Completeness is established by using a variant of the technique introduced by Bachmair *et al.* in [2] for proving completeness of completion. It transforms a

---

[5] All the details of the proof can be found in a research report [4].

proof between two terms which is not under a suitable form into a smaller one, and the smallest proofs are the desired ones. The proofs we are considering are made of elementary steps, either equational steps, with AC, X and E_∞, or rewriting steps, with R_∞ and the additional (possibly infinite) rules $R_{\text{can}} = \{t \to \text{can}(t) \mid \text{can}(t) \neq t\}$. Rewriting steps with R_∞ can be either $\leadsto_{R_\infty}$ or $\to_{R_\infty}$[6].

The measure of a proof is the multiset of the elementary measures of its elementary steps. The measure of an elementary step takes into account the number of terms which are in a canonical form in an elementary proof: the canonical weight of a term t, $w_{\text{can}}(t)$ is equal to 0 if $\text{can}(t) =_{AC} t$ and to 1 otherwise. Notice that if $w_{\text{can}}(t) = 1$, then $\text{can}(t) \prec t$, and if $w_{\text{can}}(t) = 0$, then $\text{can}(t) \simeq t$. The measure of an elementary step between t_1 and t_2 performed thanks to:

- an equation is equal to $(\{t_1, t_2\}, \text{-}, \text{-}, \text{-}, \text{-})$
- a rule $l \to r \in R_\infty$ is equal to $(\{t_1\}, 1, w_{\text{can}}(t_1) + w_{\text{can}}(t_2), l, r)$ if $t_1 \leadsto_{l \to r} t_2$ or $t_1 \to_{l \to r} t_2$.
- a rule of R_{can} is equal to $(\{t_1\}, 0, w_{\text{can}}(t_1) + w_{\text{can}}(t_2), t_1, t_2)$ if $t_1 \to_{R_{\text{can}}} t_2$.

As usual the measure of a step $s \leftarrow t$ is the measure of $t \to s$. Elementary steps are compared lexicographically using the multiset extension of $\preceq$ for the first component, the usual ordering over natural numbers for the components 2 and 3, and $\preceq$ for last ones. Since $\preceq$ is an AC-reduction ordering, the ordering defined over proofs is well-founded.

The general methodology is to show that a proof which contains some unwanted elementary steps can be replaced by a proof with a strictly smaller measure. Since the ordering over measures is well-founded, there exists a minimal proof, and such a minimal proof is of the desired form.

Lemma 3. *A proof containing*

- *an elementary step $\longleftrightarrow_{s \approx t}$, where $s \approx t \in AC \cup X \cup E_\infty$,*
- *or an elementary rewriting step truly of the form $\longrightarrow_{R_\infty}$ or $\longleftarrow_{R_\infty}$, or $\leadsto_{l \to r}$ or $\leadsto_{r \leftarrow l}$, where $l \to r \in R_\infty \setminus R_\omega$*
- *or a peak $s \leftarrow_{R_{\text{can}}} t \to_{R_{\text{can}}} s'$, $s \leadsto_{R_\omega} t \to_{R_\omega} s'$, or $s \leadsto_{R_\omega} t \longrightarrow_{R_{\text{can}}} s'$*

is not minimal.

Theorem 3. *If s and t are two terms such that $s \longleftrightarrow^*_{AC, X, E_\infty, R_\infty} s'$ then $\text{can}(s) \downarrow_{R_\omega} = \text{can}(t) \downarrow_{R_\omega}$.*

Proof. If s and s' are equal modulo $\longleftrightarrow^*_{AC, X, E_\infty, R_\infty}$, so are $\text{can}(s)$ and $\text{can}(s')$. By the above lemma, a minimal proof between $\text{can}(s)$ and $\text{can}(s')$ is necessary of the form $\text{can}(s)(\leadsto_{R_\omega} \cup \to_{R_{\text{can}}})^* (\leftharpoonup_{R_\omega} \cup \leftarrow_{R_{\text{can}}})^* \text{can}(s')$. This sequence of steps can also be seen as $\text{can}(s) \to^*_{R_{\text{can}}} (\leadsto_{R_\omega} \to^*_{R_{\text{can}}})^* (\leftarrow^*_{R_{\text{can}}} \leftharpoonup_{R_\omega})^* \leftarrow^*_{R_{\text{can}}} \text{can}(s')$. By definition, $\to_{R_{\text{can}}}$ cannot follow a $\leadsto_{R_\omega}$-step, and $\text{can}(s)$ and $\text{can}(s')$ cannot be reduced by $\to_{R_{\text{can}}}$, hence the wanted result.

---

[6] Here, $s \longrightarrow_{R_\infty} t$ actually means $s \longrightarrow_{AC \setminus R_\infty} t'$ and $t = \text{can}_{AC}(t')$.

5.3   Termination

We shall prove that, under a strongly fair strategy, R_ω is finite and obtained in a finite time (by cases on the head function symbol of the rule's left-hand side), and then we show that R_ω will clean up the next configurations and the completion process eventually halts on $\langle\, \emptyset \mid R_\omega \,\rangle$. In order to make our case analysis on rules, and to prove the needed invariants, we define several sets of terms (assuming without loss of generality that $E_0 = \mathsf{can}(E_0)$):

$$T_0 = \{t \mid \exists t_0, e_1, e_2 \in \mathcal{T}_\Sigma(\mathcal{X}), e_1 \approx e_2 \in E_0 \ \text{ and } \ t_0 = e_i|_p \ \text{ and } \ t_0 \leadsto^*_{R_\infty} t\}$$
$$T_{0\mathsf{X}} = T_0 \cup \{f_\mathsf{X}(t_1, \ldots, t_n) \mid f_\mathsf{X} \in \Sigma_\mathsf{X} \ \text{ and } \ \forall i, t_i \in T_{0\mathsf{X}}\}$$
$$T_1 = \{t \mid t \in T_0 \ \text{ and } \ \forall p, t|_p \in T_{0\mathsf{X}}\}$$
$$T_2 = \{u(t_1, \ldots, t_n) \mid u \in \Sigma_{AC} \ \text{ and } \ \forall i, t_i \in T_1\}$$

T_0 is the set of all terms and subterms in the original problem as well as their reducts by R_∞. The set $T_{0\mathsf{X}}$ moreover contains terms with X-aliens in T_0. T_1 is the set of terms that can be introduced by X from terms of T_0 (by solving or canonizing). T_2 is a superset of the terms built by critical pairs.

We first establish by structural induction over terms that:

$$\forall \gamma, t, s, \ \gamma \in R_\infty \cap T_j^2 \wedge t \in T_i \wedge t \leadsto_\gamma s \implies s \in T_i, \ \text{for } i, j = 1, 2$$

Then, by induction over n, we show that any configuration $\langle\, E_n \mid R_n \,\rangle$ accessible from $\langle\, E_0 \mid \emptyset \,\rangle$ after n steps is such that $E_n \cup R_n \subseteq T_1^2 \cup T_2^2$.

The fact that R_∞ is finitely branching is a corollary of

Lemma 4. *If $l \to r_n$ is created at step n in R_n and $l \to r_m$ at step m in R_m, with $n < m$, then r_m is a reduct of r_n by $\leadsto_{R_\infty}$.*

The proof of this lemma is by induction over the length of the derivation, and by a case analysis over the applied inference rule.

Theorem 4. *Under a strongly fair strategy, AC(X) terminates.*

By the above properties, $\mathcal{R}_\omega$ can be divided in $\mathcal{R}_\omega \cap T_1^2$ and $\mathcal{R}_\omega \cap T_2^2$. $\mathcal{R}_\omega \cap T_1^2$ is finite, since all its left-hand sides are reducts of a finite number of terms by R_∞ which is well-founded and finitely branching. $\mathcal{R}_\omega \cap T_2^2$ is finite by using the same argument as in the ground AC-completion proof, based on the Higman's lemma. Hence $\mathcal{R}_\omega$ is finite and obtained in a finite number of steps, that is, there exists n such that $\mathcal{R}_\omega \subseteq R_n$. Then $\mathcal{R}_\omega$ will clean the rest of E_n, and the newly generated critical pairs will be discarded as trivial ones.

6   Experimental Results

AC(X) has been implemented in the ALT-ERGO [8] theorem prover. In this section, we show that this extension has strong impact both on performances and memory allocation *w.r.t.* an axiomatic approach. For that purpose, we benchmarked our implementation and compared its performances with state-of-the-art

SMT solvers (Z3 v2.8, CVC3 v2.2, Simplify v1.5.4). All measures were obtained on a laptop running Linux equipped with a $2.58\,GHz$ dual-core Intel processor and with $4\,Gb$ main memory. Provers were given a time limit of five minutes for each test and memory limitation was managed by the system. The results are given in seconds; we write TO for *timeout* and OM for *out of memory*.

Our test suite is made of formulas which are valid in the combination of the theory of linear arithmetic $\mathcal{A}$, the free theory of equality[7] $\mathcal{E}$ and a small part of the theory of sets defined by the symbols $\cup$, $\subseteq$, the singleton constructor $\{\cdot\}$, and the following set of axioms:

$$
AC \begin{cases} Assoc: & \forall x,y,z.\ \ x \cup (y \cup z) \approx (x \cup y) \cup z \\ Commut: & \forall x,y.\ \ \ x \cup y \approx y \cup x \end{cases}
$$

$$
S \begin{cases} SubTrans: & \forall x,y,z.\ \ x \subseteq y \to y \subseteq z \to x \subseteq z \\ SubSuper: & \forall x,y,z.\ \ x \subseteq y \to x \subseteq y \cup z \\ SubUnion: & \forall x,y,z.\ \ x \subseteq y \to x \cup z \subseteq y \cup z \\ SubRefl: & \forall x.\ \ x \subseteq x \end{cases}
$$

In order to get the most accurate information from our benchmarks, we classify formulas in three categories according to the subset of axioms needed to prove their validity. We use the standard mathematical notation $\bigcup_{i=1}^{d} a_i$ for the terms of the form $a_1 \cup (a_2 \cup (\cdots \cup a_d)) \cdots)$ and we write $\bigcup_{i=1}^{d} a_i; b$ for terms of the form $a_1 \cup (a_2 \cup (\cdots \cup (a_d \cup b))) \cdots)$. Formulas in the first category are of the form:

$$
\bigwedge_{p=1}^{n} (\{e\} \cup \bigcup_{i=1}^{d} a_i^p) \approx b^p\ \to\ \underbrace{\bigwedge_{p=1}^{n} \bigwedge_{q=p+1}^{n} \bigcup_{i=d}^{1} a_i^p; b^q \approx \bigcup_{i=d}^{1} a_i^q; b^p}_{G}
$$

and proving their validity only requires the theory $\mathcal{E}$ and the AC properties of $\cup$. The second category contains formulas additionally involving linear arithmetic:

$$
\bigwedge_{p=1}^{n} (\{t_p - p\} \cup \bigcup_{i=1}^{d} a_i^p) \approx b^p \wedge \bigwedge_{p=1}^{n-1} t_p + 1 \approx t_{p+1} \to G
$$

Formulas in the third category involve the $\subseteq$ symbol and are of the form:

$$
\bigwedge_{p=1}^{n} \bigcup_{i=1}^{d} \{e_i^p\} \approx b^p \wedge \bigcup_{i=d}^{d} \{e + e_i^p\} \approx c^p \wedge e \approx 0 \to \bigwedge_{p=1}^{n} c^p \subseteq (b^p \cup \{e_d^p\}) \cup \{e\}
$$

In order to prove their validity, we additionally need some axioms of S. The results of the benchmarks are shown in Fig. 5, Fig. 6 and Fig 7. The first column contains the results for Alt-Ergo when we *explicitly* declare $\cup$ as an AC symbol and remove the AC axioms from the problem. In the second column, we do not take advantage of AC(X) and keep the AC axioms in the context. Figures 5 and 6 show that, contrary to the axiomatic approach, built-in AC reasoning is little sensitive to the depth of terms: given a number n of equations, the running time is proportional to the depth d of terms. However, we notice a slowdown when n increases. This is due to the fact that AC(X) has to process a quadratic number of critical pairs generated from the equations in the hypothesis. From Fig. 7, we notice that ALT-ERGO with AC(X) performs better than the other provers.

---

[7] These two theories are built-in for all SMT solvers we used for our benchmarks.

n-d	AC(X)	A-E	z3	CVC3	SIMP.
3-3	**0.01**	0.19	0.22	0.40	0.18
3-6	**0.01**	32.2	OM	132	OM
3-12	**0.01**	TO	OM	OM	OM
6-3	**0.01**	11.2	1.10	13.2	2.20
6-6	**0.02**	TO	OM	OM	OM
6-12	**0.02**	TO	OM	OM	OM
12-3	**0.16**	TO	5.64	242	11.5
12-6	**0.24**	TO	OM	OM	OM
12-12	**0.44**	TO	OM	OM	OM

Fig. 5. AC + $\mathcal{E}$

n-d	AC(X)	A-E	z3	CVC3	SIMP.
3-3	**0.00**	1.10	0.03	0.11	0.19
3-6	**0.00**	TO	3.67	4.21	OM
3-12	**0.00**	TO	OM	OM	OM
6-3	**0.02**	149	0.10	2.26	2.22
6-6	**0.02**	TO	17.7	99.3	OM
6-12	**0.04**	TO	OM	OM	OM
12-3	**0.27**	TO	0.35	44.5	11.2
12-6	**0.40**	TO	76.7	TO	OM
12-12	**0.72**	TO	OM	OM	OM

Fig. 6. AC + $\mathcal{E}$ + $\mathcal{A}$

n-d	AC(X)	A-E	z3	CVC3	SIMP.
3-3	**0.02**	3.16	0.09	10.2	OM
3-6	**0.04**	TO	60.6	OM	OM
3-12	**0.12**	TO	OM	OM	OM
6-3	**0.07**	188	0.18	179	OM
6-6	**0.12**	TO	TO	OM	OM
6-12	**0.66**	TO	OM	OM	OM
12-3	**0.20**	TO	0.58	OM	OM
12-6	**0.43**	TO	TO	OM	OM
12-12	**1.90**	TO	OM	OM	OM

Fig. 7. AC + $\mathcal{E}$ + $\mathcal{A}$ + $\mathcal{S}$

The main reason is that its instantiation mechanism is not spoiled by the huge number of intermediate terms the other provers generate when they instantiate the AC axioms.

7   Conclusion and Future Works

We have presented a new algorithm AC(X) which efficiently combines, in the ground case, the AC theory with a Shostak theory X and the free theory of equality. Our combination consists in a tight embedding of the canonizer and the solver for X in ground AC-completion. The integration of the canonizer relies on a new rewriting relation, reminiscent to normalized rewriting, which interleaves canonization and rewriting rules. We proved correctness of AC(X) by reusing standard proof techniques. Completeness is established thanks to a proofs' reduction argument, and termination follows the lines of the proof of ground AC-completion where the finitely branching result is adapted to account for the theory X.

AC(X) has been implemented in the ALT-ERGO theorem prover. The first experiments are very promising and show that a built-in treatment of AC, in the combination of the free theory of equality and a Shostak theory, is more efficient than an axiomatic approach. Although effective, the integration of AC(X) in ALT-ERGO fails to prove the formula

$$(\forall x, y, z.P((x \cup y) \cup z)) \wedge b \approx c \cup d \rightarrow P(a \cup b)$$

since the trigger for the internal quantified formula (the term $(x \cup y) \cup z)$) does not match the term $a \cup b$, even when exploiting the ground equation $b = c \cup d$ which allows to match the term $a \cup (c \cup d)$). Introducing explicitly the AC axioms for $\cup$ would allow the matcher to generate the ground term $(a \cup c) \cup d$ that could be matched. However, as shown by our benchmarks, too many terms are generated with these axioms in general. In order to fix this problem, we intend to extend the pattern-matching algorithm of ALT-ERGO to exploit both ground equalities and properties of AC symbols. In the near future, we also plan to extend AC(X) to handle the AC theory with unit or idempotence. This will be a first step towards a decision procedure for a substantial part of the finite sets theory.

References

1. Baader, F., Nipkow, T.: Term Rewriting and All That. Cambridge University Press, Cambridge (1998)
2. Bachmair, L., Dershowitz, N., Hsiang, J.: Orderings for equational proofs. In: Proc. 1st LICS, Cambridge, Mass., pp. 346–357 (June 1986)
3. Bachmair, L., Tiwari, A., Vigneron, L.: Abstract congruence closure. Journal of Automated Reasoning 31(2), 129–168 (2003)
4. Conchon, S., Contejean, E., Iguernelala, M.: Canonized Rewriting and Ground AC Completion Modulo Shostak Theories. Research Report 1538, LRI (December 2010)
5. Contejean, E.: A certified AC matching algorithm. In: van Oostrom, V. (ed.) RTA 2004. LNCS, vol. 3091, pp. 70–84. Springer, Heidelberg (2004)
6. Dershowitz, N.: Orderings for term rewriting systems. Theoretical Computer Science 17(3), 279–301 (1982)
7. Dershowitz, N., Jouannaud, J.-P.: Rewrite systems. In: van Leeuwen, J. (ed.) Handbook of Theoretical Computer Science, pp. 243–320. North-Holland, Amsterdam (1990)
8. Conchon, S., Contejean, E., Bobot, F., Lescuyer, S., Iguernelala, M.: The Alt-Ergo theorem prover, http://alt-ergo.lri.fr
9. Hullot, J.-M.: Associative commutative pattern matching. In: Proc. 6th IJCAI, Tokyo, vol. I, pp. 406–412 (August 1979)
10. Jouannaud, J.-P., Kirchner, H.: Completion of a set of rules modulo a set of equations. SIAM Journal on Computing 15(4) (November 1986)
11. Kapur, D.: Shostak's congruence closure as completion. In: Comon, H. (ed.) RTA 1997. LNCS, vol. 1232. Springer, Heidelberg (1997)
12. Knuth, D.E., Bendix, P.B.: Simple word problems in universal algebras. In: Leech, J. (ed.) Computational Problems in Abstract Algebra, pp. 263–297. Pergamon Press, Oxford (1970)
13. Krstić, S., Conchon, S.: Canonization for disjoint unions of theories. Information and Computation 199(1-2), 87–106 (2005)
14. Lankford, D.S.: Canonical inference. Memo ATP-32, University of Texas at Austin (March 1975)
15. Lankford, D.S., Ballantyne, A.M.: Decision procedures for simple equational theories with permutative axioms: Complete sets of permutative reductions. Memo ATP-37, University of Texas, Austin, Texas, USA (August 1977)
16. Marché, C.: On ground AC-completion. In: Book, R.V. (ed.) RTA 1991. LNCS, vol. 488. Springer, Heidelberg (1991)
17. Marché, C.: Normalized rewriting: an alternative to rewriting modulo a set of equations. Journal of Symbolic Computation 21(3), 253–288 (1996)
18. Nelson, G., Oppen, D.C.: Simplification by cooperating decision procedures. ACM Trans. on Programming, Languages and Systems 1(2), 245–257 (1979)
19. Nieuwenhuis, R., Rubio, A.: A precedence-based total AC-compatible ordering. In: Kirchner, C. (ed.) RTA 1993. LNCS, vol. 690. Springer, Heidelberg (1993)
20. Peterson, G.E., Stickel, M.E.: Complete sets of reductions for some equational theories. J. ACM 28(2), 233–264 (1981)
21. Shostak, R.E.: Deciding combinations of theories. J. ACM 31, 1–12 (1984)

Invariant Generation in Vampire*

Kryštof Hoder[1], Laura Kovács[2], and Andrei Voronkov[1]

[1] University of Manchester
[2] TU Vienna

Abstract. This paper describes a loop invariant generator implemented in the theorem prover Vampire. It is based on the symbol elimination method proposed by two authors of this paper. The generator accepts a program written in a subset of C, finds loops in it, analyses the loops, generates and outputs invariants. It also uses a special consequence removal mode added to Vampire to remove invariants implied by other invariants. The generator is implemented as a standalone tool, thus no knowledge of theorem proving is required from its users.

1   Introduction

In [9] a new *symbol elimination* method of loop invariant generation was introduced. The method is based on the following ideas. Suppose we have a loop L with a set of (scalar and array) variables V. The set V defines the *language* of L. We extend the language L to a richer language L' by a number of functions and predicates. For every scalar variable v of the loop we add to L' a unary function $v(i)$ which denotes the value of v after i iterations of L, and similar for array variables. Thus, all loop variables are considered as functions of the loop counter. Further, we add to L' so-called *update predicates* expressing updates to arrays as formulas depending on the loop counter. After that we automatically generate a set P of first-order properties of the loop in the language L'. These properties are valid properties of the loop, yet they are not loop invariants since they use the extended language L'.

Note that any logical consequence of P that only contains variables in L is also a loop invariant. Thus, we are interested in finding logical consequences of formulas in P expressed in L. To this end, we run a first-order theorem prover using a saturation algorithm on P in such a way that it tries to derive formulas in L. To obtain a saturation algorithm specialised to efficiently derive consequences in L, we enhanced the theorem prover Vampire [7] by so-called *colored proofs* and a *symbol elimination mode*. In colored proofs, some (predicate and/or function) symbols are declared to have colors, and every proof inference can use symbols of at most one color.

As reported in [9], we tested Vampire on several benchmarks for invariant generation. It was shown that symbol elimination can infer complex properties with quantifier alternations. Symbol elimination thus provides new perspectives in automating program verification, since such invariants could not be automatically derived by other methods.

---

* Kryštof Hoder is supported by the School of Computer Science at the University of Manchester. Laura Kovács is supported by an FWF Hertha Firnberg Research grant (T425-N23). Andrei Voronkov is partially supported by an EPSRC grant. This research was partly supported by Dassault Aviation.

P.A. Abdulla and K.R.M. Leino (Eds.): TACAS 2011, LNCS 6605, pp. 60–64, 2011.
© Springer-Verlag Berlin Heidelberg 2011

As the method is new, its practical power and limitations are not well-understood. The main obstacle to its experimental evaluation lies in the fact that program analysis and generation of input for symbol elimination by a separate tool is error-prone and requires full knowledge of our invariant generation method. The tool described in this paper was designed with the purpose of creating a *standalone tool implementing invariant generation* by symbol elimination. Vampire can still be used for symbol elimination only, so that the program analysis is done by another tool (for example, for experiments with variations of the method).

The purpose of this paper is to describe the program analyser and loop invariant generator of Vampire, their implementation and use. We do not overview Vampire itself.

Related work. Reasoning about loop invariants is a challenging and widely studied research topic. We overview only some papers most closely related to our tool.

Automatic loop invariant generation is described in a number of papers, including [3,12,5,10,4,6]. In [12] loop invariants are inferred by predicate abstraction over a set of a priori defined predicates, while [4] employs constraint solving over invariant templates. Input predicates in conjunction with interpolation are used to infer invariants in [10]. Unlike these works, we require no used guidance in providing input templates and/or predicates. User guidance is also not required in [3,5], but invariants are derived using abstract interpretation [3,5] or symbolic computation [6]. However, these approaches can only infer universally quantified invariants, whereas we can also derive invariants with quantifier alternation.

Our work is also related to first-order theorem proving [11,13,8]. These works implement superposition calculi, with a limited support for theories. However, only Vampire implements colored proofs and consequence removal essential for the symbol elimination method.

A more complex and general framework for program analysis is given in, e.g., [1,2]. Whereas in [1,2] theorem proving is integrated in a program analysis environment, we integrate program analysis in a theorem proving framework. Although our approach at the moment is limited to the analysis of a restricted class of loops, we are able to infer richer and more complex quantified invariants than [1,2]. Combining our method with other techniques for verification and invariant generation is left for further work.

2   Invariant Generation in Vampire: Overview

To create an integrated environment for invariant generation, we implemented a simple program analyser and several new features in Vampire. The workflow of the invariant generation process is given in Figure 1.

The analyser itself comprises about 4,000 lines of C++ code (all Vampire is written in C++). In addition to the analyser, we had to extend formulas and terms with if-then-else and let-in constructs, implement colored proofs, automatic theory loading, and the consequence removal mode. All together making Vampire into an invariant generator required about 12,000 lines of code. Currently, the analyser only generates loop properties for symbol elimination, but we plan to use it in the future for a more powerful integration of program analysis and theorem proving.

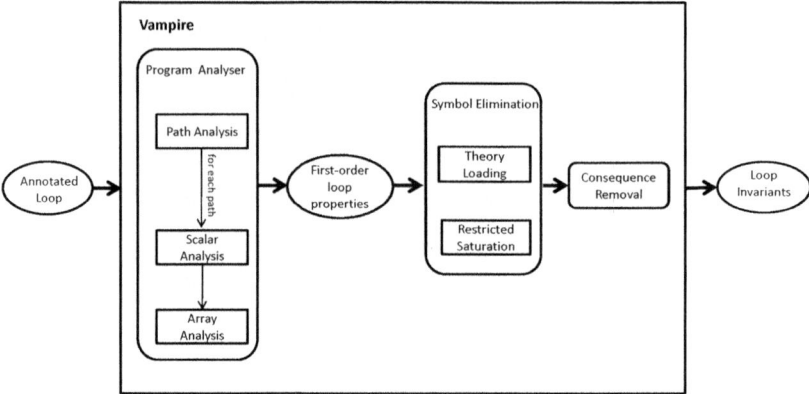

Fig. 1. Program Analysis and Invariant Generation in Vampire

Program analysis.   The program analysis part works as follows. First, it extracts all loops from the input program. It ignores nested loops and performs the following steps on every non-nested loop.

1. Find all loop variables.
2. Classify them into variables updated by the loop and constant variables.
3. Find counters, that is, updated scalar variables that are only incremented or decremented by constant values. Note that expressions used as array indexes in loops are typically counters.
4. Save properties of counters.
5. Generate update predicates of updated array variables and save their properties.
6. Save the formulas corresponding to the transition relation of the loop.
7. Generate a symbol elimination task for Vampire.

The input to the analyser is a program written in a subset of C. The subset consists of scalar integer variables, array variables, arithmetical expressions, assignments, conditionals and while-do loops. Nested loops are not yet handled.

Symbol Elimination and Theory Loading. The program analyser generates a set of first-order loop properties and information about which symbols should be eliminated. A (predicate and/or function) symbol is to be eliminated in Vampire whenever it is specified to have some color. The next phase of invariant generation runs symbol elimination on the set of formulas generated by the analyser. Before doing symbol elimination, Vampire checks which theory symbols (such as integer addition) are used and loads axioms relevant to these theory symbols. Theory symbols have no color in Vampire. After theory loading, Vampire runs a saturation algorithm on the theory axioms and the formulas generated by its analyser. A special term ordering is used to ensure that symbol elimination is effective and efficient.

Consequence Removal. The result of the symbol elimination phase is a set of loop invariants. This set is sometimes too large. For example, it is not unusual that Vampire generates over a hundred invariants in less than a second.

An analysis of these invariants shows that some invariants are concise and natural for humans, while some other invariants look artificial (this does not mean they are not interesting and/or not useful). It is typically the case that the generated set of invariants contains many invariants implied by other invariants in the set.

The next phase of invariant generation prunes the generated set by removing the implied invariants. Checking whether each generated invariant is implied by all other invariants is too inefficient. To remove them efficiently, we implemented a special consequence removal mode. The output of the tool is the set of all non-removed invariants.

Implementation and Availability. We implemented our approach to invariant generation in Vampire. The new version of Vampire is available from `http://www.vprover.org`. The current version of Vampire runs under Windows, Linux and MacOS.

Experiments. We evaluated invariant generation in Vampire using two benchmark suites: (1) challenging loops taken from [3,12], and (2) a collection of 38 loops taken from programs provided by Dassault Aviation. We used a computer with a 2GHz processor and 2GB RAM and ran experiments using Vampire version 0.6. The symbol elimination phase was run with a 1 second time limit and the consequence removal phase with a 20 seconds time limit.

For all the examples the program analyser took essentially no time. It turned out that symbol elimination in one second can produce a large amount of invariants, ranging from one to hundreds. Consequence removal normally deletes about 80% of all invariants.

3   Conclusion

It is not unusual that program analysers call theorem provers or contain theorem provers as essential parts. Having a program analyser as part of a theorem prover is less common. We implemented an extension of Vampire by program analysis tools, which resulted in a standalone automatic loop invariant generator. Our tool derives logically complex invariants, strengthening the state-of-the-art in reasoning about loops.

References

1. Barnett, M., Leino, K.R.M., Schulte, W.: The Spec# Programming System: An Overview. In: Barthe, G., Burdy, L., Huisman, M., Lanet, J.-L., Muntean, T. (eds.) CASSIS 2004. LNCS, vol. 3362, pp. 49–69. Springer, Heidelberg (2005)
2. Correnson, L., Cuoq, P., Puccetti, A., Signoles, J.: Frama-C User Manual. In: CEA LIST (2010)
3. Gopan, D., Reps, T.W., Sagiv, S.: A Framework for Numeric Analysis of Array Operations. In: Proc. of POPL, pp. 338–350 (2005)
4. Gupta, A., Rybalchenko, A.: InvGen: An efficient invariant generator. In: Bouajjani, A., Maler, O. (eds.) CAV 2009. LNCS, vol. 5643, pp. 634–640. Springer, Heidelberg (2009)
5. Halbwachs, N., Peron, M.: Discovering Properties about Arrays in Simple Programs. In: Proc. of PLDI, pp. 339–348 (2008)

6. Henzinger, T.A., Hottelier, T., Kovács, L., Rybalchenko, A.: Aligators for arrays (Tool paper). In: Fermüller, C.G., Voronkov, A. (eds.) LPAR-17. LNCS (LNAI), vol. 6397, pp. 348–356. Springer, Heidelberg (2010)
7. Hoder, K., Kovacs, L., Voronkov, A.: Interpolation and Symbol Elimination in Vampire. In: Proc. of IJCAR, pp. 188–195 (2010)
8. Korovin, K.: iProver – An Instantiation-Based Theorem Prover for First-Order Logic (System Description). In: Armando, A., Baumgartner, P., Dowek, G. (eds.) IJCAR 2008. LNCS (LNAI), vol. 5195, pp. 292–298. Springer, Heidelberg (2008)
9. Kovács, L., Voronkov, A.: Finding Loop Invariants for Programs over Arrays Using a Theorem Prover. In: Chechik, M., Wirsing, M. (eds.) FASE 2009. LNCS, vol. 5503, pp. 470–485. Springer, Heidelberg (2009)
10. McMillan, K.L.: Quantified Invariant Generation Using an Interpolating Saturation Prover. In: Ramakrishnan, C.R., Rehof, J. (eds.) TACAS 2008. LNCS, vol. 4963, pp. 413–427. Springer, Heidelberg (2008)
11. Schulz, S.: System description: E 0.81. In: Basin, D., Rusinowitch, M. (eds.) IJCAR 2004. LNCS (LNAI), vol. 3097, pp. 223–228. Springer, Heidelberg (2004)
12. Srivastava, S., Gulwani, S.: Program Verification using Templates over Predicate Abstraction. In: Proc. of PLDI, pp. 223–234 (2009)
13. Weidenbach, C., Schmidt, R.A., Hillenbrand, T., Rusev, R., Topic, D.: System description: SPASS version 3.0. In: Pfenning, F. (ed.) CADE 2007. LNCS (LNAI), vol. 4603, pp. 514–520. Springer, Heidelberg (2007)

Enforcing Structural Invariants Using Dynamic Frames

Diego Garbervetsky, Daniel Gorín, and Ariel Neisen

Departamento de Computación, FCEyN, Universidad de Buenos Aires
{diegog,dgorin,aneisen}@dc.uba.ar

Abstract. The theory of dynamic frames is a promising approach to handle the so-called *framing problem*, that is, giving a precise characterizations of the locations in the heap that a procedure may modify.

In this paper, we show that the machinery used for dynamic frames may be exploited even further. In particular, we use it to check that implementations of abstract data types maintain certain structural invariants that are very hard to express with usual means, including being acyclic (like non-circular linked lists and trees) and having a unique path between nodes (like in a tree).

The idea is that regions in this formalism over-approximate the set of reachable objects. We can then maintain this structural invariants by including special preconditions in assignments, of the kind that can be verified by state-of-the-art SMT-based tools. To test this approach we modified the verifier for the Dafny programming language in a suitable way and were able to enforce these invariants in non-trivial examples.

1 Introduction

A typical procedure specification describes the effects of the procedure over its arguments. However, in a context involving pointers, this information is not enough to enable the verification (neither manual nor automatic) of conformance of an implementation to its specification –at least not in a modular way [9].

What is missing in typical specifications is a precise characterization of the locations in the program heap that can be safely assumed to be left untouched by an operation. The problem of formally describing such locations is known as the *frame problem*. The theory of *dynamic frames* [5] is, perhaps, one of the most promising proposals on how to address this problem in a practical way. Recent implementations have shown that verification of programs containing dynamic frame specifications is feasible using state-of-the-art tools [7,4,8,14].

To use dynamic frames basically means to equip each value o in the heap with a *specification variable* or *ghost field* (i.e., a field that can be used in the program specification but not in the program text) that represents the collection of heap locations that must be affected in order to make the observable value of o change.[1] This attribute is usually called *representation region*. Intuitively, if o_1

---

[1] More precisely, a value can be equipped with more than one such attributes, which allows for more precise specifications, but for the general case, one is enough.

P.A. Abdulla and K.R.M. Leino (Eds.): TACAS 2011, LNCS 6605, pp. 65–80, 2011.

and o_2 have disjoint representation regions, one can guarantee that modifying o_1 will not inadvertently alter o_2 too.

Now, this article is *not* about framing the scope of the effects of a procedure call. Instead, our starting point is the simple observation that, in practice, the representation region of o roughly corresponds to the set of locations that are *reachable* from o by chasing pointers. This means that in a setting with dynamic frames the developer is required to provide, for the sake of framing, information about the heap that might be extremely useful for the verification of heap-related properties such as reachability, shape-analysis, etc. We are therefore interested in the question of how this information can be effectively exploited for such tasks.

In this paper we give a first step in that direction: we use the machinery of dynamic frames to verify that (the graph induced by the points-to relation of) the internal representation of an *abstract data type* satisfies certain structural properties, such as *acyclicity* or *tree-shapedness*. This kind of structural properties constitute the invariant of countless data-structures and at the same time they tend to be tricky to maintain properly, leading to very subtle bugs. In addition, these are properties that cannot even be expressed using a formula of plain first-order logic and, therefore, are not very amenable to the automated techniques employed in contract-based verification.

The paper is structured as follows. In §2 we briefly introduce the dynamic frames methodology and define a simple language with region inference that serves as a basis for the developments in later sections. In §3, for instance, we show how to extend it with a class qualifier for acyclity and in §4 with another one for tree-shapedness. In §5 we focus on improving the precision of the techniques and in §6 we present some preliminary results using a prototype implementation. We conclude in §7 and §8 discussing related work and future directions.

2   Dynamic Frames

We begin by fixing notations and terminology. We assume an infinite set *Ref* of *references* (or *locations*) and a set *Val* of *values*, which we assume *indexed tuples* of references (we shall call them *records* or *objects*). We also assume an infinite set *Var* of *program variables*. As usual, all these sets are mutually disjoint.

A *store* (or *heap*) is a *finite* and *injective* mapping σ_h from *Ref* to *Val*,[2] while an *environment* is a *finite* mapping σ_e from *Var* to *Ref*. A *state* is then a pair $\sigma = \langle \sigma_h, \sigma_e \rangle$ and we will typically write both $\sigma(\iota)$ and $\sigma(v)$ for $\iota \in Ref$ and $v \in Var$ meaning $\sigma_h(\iota)$ and $\sigma_h(\sigma_e(v))$ respectively. That is, $\sigma(x)$ denotes the value of x, where x can be either a program variable or a reference. In the same vein, for f a field of an indexed tuple, we will write $\sigma(\iota.f)$ and $\sigma(v.f)$ for $\sigma(\sigma(\iota).f)$ and $\sigma(\sigma(v).f)$. We refer to the domain of σ_h as the references *used in* σ. Finally, we say that an object o_i *reaches* an object o_j in a state σ if either $i = j$ or for some field f, $\sigma(o_i.f)$ reaches o_j in σ (if the state σ is clear from context, we say simply "o_i reaches o_j").

---

[2] In very concrete terms, σ_h maps *memory locations* to *objects* and the injectiveness condition witnessess the fact that different locations refer to different objects.

The idea behind the dynamic frames theory is to use finite sets of references, called *regions*, to *frame* a value. Intuitively, a region ρ frames a reference ι in state σ if in any other state τ that coincides with σ in the value of the objects denoted in ρ, $\sigma(\iota)$ and $\tau(\iota)$ must coincide too. It is easy to see that if ρ frames ι then it must be the case that $\iota \in \rho$.[3]

Now, if we know that ρ frames ι, and we also have that region π is the set of references touched by a procedure invocation and we can show that ρ and π are disjoint (denoted $\rho \parallel \pi$), then we can assert that the invocation did not change the value of ι (this is the Value Preservation Theorem in [5]).

The dynamic frames methodology can therefore be summed up as the requirement to assign to each object in the store a region that frames it, called its *representation region*. Representation regions can then appear in procedure specifications, e.g., a precondition may require that two objects have disjoint representation regions –which means they are not aliased; a postcondition may indicate that an object's representation region grew only by the addition of fresh references (cf. *swinging pivots* [9]), which preserves disjointness of regions, etc.

2.1 Inferring Regions Automatically

In order to illustrate dynamic frames with an example and provide a basis for the developments in the following sections, we introduce next a very small programming language. Its syntax, shown in Fig. 1, is loosely based on Dafny's [7]. We assume the language to be statically typed, but we will not give the formal typing rules since this is standard.

```
Member   ::=  Field | Method
 Field   ::=  var f : C
Method   ::=  method m(x_1 : C_1, ..., x_n : C_n) returns (y : C)
                 requires α(this, x_1, ..., x_n, σ);
                 ensures β(this, x_1, ..., x_n, y, σ, τ);
                 modifies ρ(this, x_1, ..., x_n, σ);
                 { Stmt }
  Stmt   ::=  v := Expr | v := new C | v_t.f := v_s | v_t := v_s.m(v_1, ..., v_n)
         |    if (Expr == null){Stmt} else {Stmt} | Stmt; Stmt
   Ref   ::=  v | this | null
  Expr   ::=  Ref | Ref.f
```

Fig. 1. Syntax of a language with support for dynamic frame annotations

Similarly, we leave the language for method contracts unspecified and simply take it to be a form of many-sorted first-order language. In an *ensures* clause, the free variables σ and τ in $\beta(this, x_1, ..., x_n, y, \sigma, \tau)$ both have sort *state* and correspond to the states before and after the execution of the method, respectively. In a *requires* clause, only the state variable σ may occur free. That said, we will sometimes write specifications in a sugared form, as in Fig. 2, and leave the unsugaring to the reader. Notice that in sugared *ensure* clauses, variable x corresponds to the term $\tau_e(x)$, while **old**(x) denotes $\sigma_e(x)$.

---

[3] We are excluding the degenerate case where there is only one possible value for ι.

```
method returnCopy(x,y) returns (z)
    modifies x;                       // ⤳ {σₑ(x)}
    requires x ≠ null ∧ y ≠ null;     // ⤳ σₑ(x) ≠ null ∧ σₑ(y) ≠ null
    ensures x.f = old(y).f ∧ z = old(x); // ⤳ τ(x.f) = σ(y.f) ∧ τ(z) = σ(x)
```

Fig. 2. A typical procedure specification and the unsugared version (commented)

The *modifies clause* is an expression of sort *set of references*, where only a state variable σ occurs free. An example of such an expression would be:

$$\{\sigma_e(\mathbf{this})\} \cup reg(\sigma(x.\mathsf{f})) \tag{1}$$

which we will normally write in its sugared form $\{\mathbf{this}\} \cup reg(x.\mathsf{f})$. Notice that *reg* is used to denote the representation region of an object, which every object has. The representation region is an attribute of an object, therefore $o_1 = o_2$ implies $reg(o_1) = reg(o_2)$. It is not a field accessible from the program text, though; in fact two objects may have $reg(o_1) \neq reg(o_2)$ while $o_1.\mathsf{f} = o_2.\mathsf{f}$ for every field f. We use notation $o_1 \approx o_2$ for this form of *structural* equality that takes into account only the value of these fields (of course, $o_1 = o_2$ implies $o_1 \approx o_2$). As we will see next, it is possible to have an execution from σ to τ such that for some reference ι, $\sigma_h(\iota) \approx \tau_h(\iota)$ (i.e., ι is not "touched") while $\sigma_h(\iota) \neq \tau_h(\iota)$ (e.g., $reg(\sigma_h(\iota)) \neq reg(\tau_h(\iota))$ because some reference reachable from ι in σ was modified in τ). For conciseness, we may write $reg_\sigma(x)$ for $reg(\sigma(x))$.

We will impose some sanity conditions on valid states. For instance, to simplify definitions, we want to assume that the null reference ø denotes a special *null object* with an empty representation region. More importantly, the representation region of an object o must include the set of objects reachable from o. We can express these conditions as a *state invariant* $I_s(\sigma)$ using the following:

$$\operatorname{im} \sigma_e \subseteq \operatorname{dm} \sigma_h \wedge \text{ø} \in \operatorname{dm} \sigma_h \wedge reg_\sigma(\text{ø}) = \emptyset \wedge \forall f \cdot \sigma(\text{ø}.f) = \text{ø} \tag{2}$$

$$\forall \iota \in \operatorname{dm} \sigma_h \cdot (\iota \neq \text{ø} \Rightarrow \iota \in reg_\sigma(\iota) \wedge \forall f \cdot reg_\sigma(\iota.f) \subseteq reg_\sigma(\iota) \subseteq \operatorname{dm} \sigma_h) \tag{3}$$

Fig. 3 presents the interesting cases of the semantics of this language, in an axiomatic form. To minimize boilerplate we will consistently use the constructions $Pre^{\mathbf{P}}(\alpha_1(\sigma), \ldots, \alpha_n(\sigma))$ and $Pst_{\mathbf{Q}}^{\mathbf{P}}(\beta_1(\sigma, \tau), \ldots, \beta_m(\sigma, \tau))$, where the α_i and β_i are the relevant parts of the pre and post-conditions. They correspond, respectively to $\forall \sigma \cdot (\mathbf{P}(\sigma) \Rightarrow (I_s(\sigma) \wedge \alpha_1(\sigma) \wedge \cdots \wedge \alpha_n(\sigma)))$ and $\forall \sigma, \tau \cdot ((\mathbf{P}(\sigma) \wedge \beta_1(\sigma, \tau) \wedge \cdots \wedge \beta_m(\sigma, \tau)) \Rightarrow \mathbf{Q}(\sigma, \tau))$.

Unlike Dafny, the semantics dictate the way in which the representation regions of the objects in the heap are updated. Consider, for example, rule NEW. Firstly, it requires $\mathbf{P}$ to be strong enough to ensure the state invariant holds at the pre-state σ. Next, it requires that $\mathbf{Q}$ must hold whenever $\mathbf{P}$ was satisfied by σ and the post-state τ differs from σ only on the value of variable v (see definition of $\triangleright$ below), which corresponds to a *fresh* reference (i.e., not occurring in σ) and refers to an object that *is the only member of its representation region*.

The STORE rule indicates that an assignment may *modify* the representation region of the target v_t: whatever might be reachable from v_s (i.e., its representation region) is added to what may be reachable from v_t. But this is not the whole story: in the statement $x.f := y;\ y.f := z$, after the second assignment, x's representation region needs to be adjusted too. As we will see next, the state modification operator $\triangleright$ takes care of this also.

Let ν be a set of variables and let ρ be a set of locations; the predicate $\sigma \triangleright^{\nu}_{\rho} \tau$ is then the conjunction of the following formulas:

$$I_s(\tau) \wedge \mathrm{dm}\,\sigma_e \subseteq \mathrm{dm}\,\tau_e \wedge \mathrm{dm}\,\sigma_h \subseteq \mathrm{dm}\,\tau_h \tag{4}$$

$$\forall v \in (\mathrm{dm}\,\sigma_e \setminus \nu) \cdot \sigma_e(v) = \tau_e(v) \wedge \forall \iota \in (\mathrm{dm}\,\sigma_h \setminus \rho) \cdot \sigma(\iota) \approx \tau(\iota) \tag{5}$$

$$\forall \iota \in \mathrm{dm}\,\sigma_h \cdot reg_\tau(\iota) = reg_\sigma(\iota) \cup regs_\tau(\rho \cap reg_\sigma(\iota)) \tag{6}$$

where $regs_\sigma(\rho) = \bigcup_{\kappa \in \rho} reg_\sigma(\kappa)$. Clearly (4) is just a basic state sanity condition; (5) indicates that ν and ρ characterize the variables and locations that may have changed; while (6) propagates updates in representation regions. Notice that for $\rho \parallel reg_\sigma(\iota)$, (5) and (6) guarantee that $\sigma(\iota) = \tau(\iota)$. We write $\triangleright^\nu$ and $\triangleright_\rho$ for $\triangleright^\nu_\emptyset$ and $\triangleright^\emptyset_\rho$ respectively.

$$\text{NEW}\quad \frac{\models Pre^{\mathbf{P}}()\\ \models Pst^{\mathbf{P}}_{\mathbf{Q}}(\sigma \triangleright^{\{v\}} \tau, \tau_e(\mathsf{v}) \in (\mathrm{dm}\,\tau_h \setminus \mathrm{dm}\,\sigma_h), reg_\tau(\mathsf{v}) = \{\tau_e(\mathsf{v})\})}{\{\mathbf{P}\}\ \mathsf{v} := \mathbf{new}\ \mathsf{C}\ \{\mathbf{Q}\}}$$

$$\text{READ}\quad \frac{\models Pre^{\mathbf{P}}(\sigma(\mathsf{v}_s) \neq \sigma(\emptyset))\\ \models Pst^{\mathbf{P}}_{\mathbf{Q}}(\sigma \triangleright^{\{v_t\}} \tau, \tau(\mathsf{v}_t) = \sigma(\mathsf{v}_s.\mathsf{f}))}{\{\mathbf{P}\}\ \mathsf{v}_t := \mathsf{v}_s.\mathsf{f}\ \{\mathbf{Q}\}}$$

$$\text{STORE}\quad \frac{\models Pre^{\mathbf{P}}(\sigma(\mathsf{v}_t) \neq \sigma(\emptyset))\\ \models Pst^{\mathbf{P}}_{\mathbf{Q}}(\sigma \triangleright_{\{\sigma_e(v_t)\}} \tau, \tau(\mathsf{v}_t) \approx \sigma(\mathsf{v}_t)[\mathsf{f} \mapsto \sigma(\mathsf{v}_s)], reg_\tau(\mathsf{v}_t) = reg_\sigma(\mathsf{v}_t) \cup reg_\sigma(\mathsf{v}_s))}{\{\mathbf{P}\}\ \mathsf{v}_t.\mathsf{f} := \mathsf{v}_s\ \{\mathbf{Q}\}}$$

$$\text{CALL}\quad \frac{\models Pre^{\mathbf{P}}(\sigma(\mathsf{v}_s) \neq \sigma(\emptyset), \alpha(\mathsf{v}_s, \mathsf{v}_1, \ldots, \mathsf{v}_n, \sigma))\\ \models Pst^{\mathbf{P}}_{\mathbf{Q}}(\sigma \triangleright^{\{v_t\}}_{\rho(\mathsf{v}_s, \mathsf{v}_1, \ldots, \mathsf{v}_n, \sigma)} \tau, \beta(\mathsf{v}_s, \mathsf{v}_0, \ldots, \mathsf{v}_n, \mathsf{v}_t, \sigma, \tau))}{\{\mathbf{P}\}\ \mathsf{v}_t := \mathsf{v}_s.\mathsf{m}(\mathsf{v}_1, ..., \mathsf{v}_n)\ \{\mathbf{Q}\}}$$

Fig. 3. Semantic rules for the language of Fig. 1 (fragment)

In rule CALL, α, β and ρ correspond to the *requires*, *ensures* and *modifies* clauses of method m, respectively. Therefore, the scope of the effects of the method call is *framed* by $\rho(\mathsf{v}_s, \mathsf{v}_1, \ldots, \mathsf{v}_n, \sigma)$, since the post-state τ must satisfy $\sigma \triangleright^{\{v_t\}}_{\rho(\mathsf{v}_s, \mathsf{v}_1, \ldots, \mathsf{v}_n, \sigma)} \tau$.

Given these semantic clauses, one can derive program acceptance rules in a straightforward way: a class C is accepted if all its methods are accepted; and a method declaration of the form

> **method** $m(x_1 : \mathsf{C}_1, \ldots, x_n : \mathsf{C}_n)$ **returns** $(z: \mathsf{D})$
> **requires** $\alpha(this, x_1, \ldots, x_n, \sigma)$;
> **ensures** $\beta(this, x_1, \ldots, x_n, z, \sigma, \tau)$;
> **modifies** $\rho(this, x_1, \ldots, x_n, \sigma)$;
> $\{\ S\ \}$

is accepted if, for $\{v_1, \ldots, v_k\}$ the local variables in S, we have:

$$\{\alpha(this, x_1, \ldots, x_n, \sigma)\} S \{\sigma \rhd_{\rho(this, x_1, \ldots, x_n, \sigma)}^{\{v_1, \ldots, v_k\}} \wedge \beta(this, x_1, \ldots, x_n, z, \sigma, \tau)\} \quad (7)$$

Of course, in order to decide acceptance one needs to resort to some sort of automated reasoner, like is done with the Dafny verifying-compiler. Verification this way can be seen as a form of typing.

It is not hard to prove that programs that are thus accepted behave well with respect to the frame conditions of the **modifies** clauses. For this, one needs to prove a stronger result, namely, that representation regions over-approximate *reachability* (i.e., if o_1 reaches o_2 in σ then $o_2 \in reg(o_1)$), which follows from the fact that this condition is preserved according to the semantics. Formal definitions and proofs can be found in [12]

As a final remark, notice that according to our rules (including condition (6)), representation regions are *monotonic* in the sense that no reference is ever removed –even those that may be no longer reachable are kept. This is suboptimal: one may easily end up in a scenario with two objects o_1 and o_2 such that $reg(o_1) \nparallel reg(o_2)$ although no location reachable from one is reachable by the other. In Kassios's original formulation [5] this was not the case; but it required higher-order logic and inductive reasoning, which is just too hard for state-of-the-art automated reasoners. Our presentation can be seen as a compromise between precision and automatic verifiability.

Example 1. Consider the declaration of a class *List* in Fig. 4. We are interested only in the framing specification and will ignore the functional part.

```
class List {                                    method concat(l : List)
  method add(d : Data)                            requires reg(this)‖reg(l);
    modifies reg(this);                           modifies reg(this);
    ensures                                       ensures
      (reg(old(this)) ∪ reg(d)) ⊆ reg(this);        reg(this) = reg(old(this)) ∪ reg(l);
    ensures                                       ensures ...
      fresh(reg(this)\(reg(old(this)) ∪ reg(d)));  }
    ensures ...
```

Fig. 4. A simple *List* type interface

Method *add* modifies the list to append a new element and its **modifies** clause states that the set of references reachable from **this** can be affected by this method. Therefore, we need to specify the effect only for those locations. The specification says that after executing *add* every reachable object remains reachable and, in addition, d will be reachable too. It also says that it will not introduce aliasing with other existing objects by declaring that any other object reachable from **this** will be fresh[4].

---

[4] A fresh object will most probably correspond to a newly allocated node that will hold the data; but this is an implementation detail and nothing else need to be said about it in the interface.

Similarly, method *concat* declares that after its execution **this** will also reach the objects reachable from list l. The combination of the **requires** and **modifies** clauses also guarantees that l will not be mutated.

3   Verifiably Acyclic Data Structures

In this section we will show how the language of §2.1 can be easily adapted to support verifiably correct *acyclic data structures*. These are ubiquitous in computer science, typically implemented using a node type with a recursive reference. Fig. 5 shows two structures that can be built using this type of nodes. The one on the right, though, does not correspond to what one expects from a linked list since it contains a cycle. An incorrect implementation of a linked list that allows such instances to be built may lead to bugs that are very hard to track-down.

Fig. 5. Structures l_1 and l_2 are built with linked list nodes; l_2 is not *acyclic*

The fact that no node in a linked list should participate in a cycle can be seen as part of the class invariant of the list type. One would be tempted to include this requirement as part of the class invariant of the type and use standard techniques to verify that it holds at the end of every procedure call [1,3]. Regrettably, no first-order logic formula can express this condition (this is a straightforward consequence of the compactness theorem, see, e.g., [12] for more details), which makes this approach currently unfeasible.

The idea we will explore here is to treat this requirement as a *strong* form of class invariant, which must hold at every point of the method execution. We will see that exploiting representation regions makes it feasible to guarantee that this condition is preserved.

3.1   A Characterization of *acyclicity*

Suppose we implement the class *List* of Fig. 4 using a linked list and want to enforce its acyclicity. We propose to extend the syntax of the language of §2.1 with a special *class qualifier* "**acyclic**" that allows us to write declarations as the one in Fig. 6. The intended meaning is that any object of a class qualified as **acyclic** satisfies a strong class invariant. The exact invariant deserves some considerations, though. We shall say that o_i *occurs in a cycle* in σ whenever for some f, $\sigma(o_i.\text{f})$ reaches o_i in σ. The most intuitive definition would be, perhaps, to stipulate that an object o of a class tagged as **acyclic** satisfies the invariant "o does not reach an object that occurs in a cycle". This would be too strong in practice. Consider for example an object n of class *LLNode* (Fig. 6); to fulfill this invariant we might as well demand T to be qualified with **acyclic** too. But

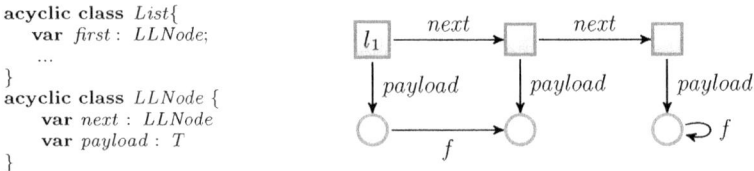

```
acyclic class List{
    var first : LLNode;
    ...
}
acyclic class LLNode {
    var next : LLNode
    var payload : T
}
```

Fig. 6. Declaration of an *acyclic* linked list node

this would impose a big restriction on the type of the payload. In other words, a linked list node should not put demands on the internal representation of the payload.

The invariant we will use, instead, is that if o is of a class qualified as **acyclic** then "o cannot occur in a cycle made of objects of *acyclic classes*". That is, o may occur in a cycle as long as some object in the cycle is not tagged as acyclic. To avoid confusion, we will term the cycles that this invariant forbids "invalid cycles". We believe this weaker notion of acyclicity constitutes a good compromise in practice. For instance, observe that if one is given a while-loop where every inductive variable is a reference to an object qualified as **acyclic** that is not mutated during the cycle (as it would typically be the case in algorithms traversing the internal representation of an abstract datatype implemented using an acyclic structure), then if every iteration of the loop can be shown to terminate, the loop cannot hang.

3.2 Preserving the Acyclicity Invariant

We want to guarantee that an acyclic object remains acyclic after the execution of any statement (that is, assuming the pre-condition of the statement holds). What we will exploit is the fact that $reg(o)$ over-approximates the set of objects that o may reach.

The first thing to observe is that only assignments and method invocations can introduce cycles; and among assignments, only *store* instructions $v_t.f = v_s$ do. Notice, furthermore that if v_t or v_s is of a non-acyclic type, then no invalid cycle can be formed (recall that an invalid cycle involves acyclic objects).

Fig. 7 shows an example of a store instruction that introduces a cycle. The important thing to observe is that this can happen if and only if o_2 reaches o_1. This motivates rule STORE^a shown in Fig. 8, which replaces rule STORE when both the target and source of the store are references to acyclic classes. The only difference with rule STORE is the additional pre-condition $\sigma_e(v_t) \notin reg_\sigma(v_s)$.

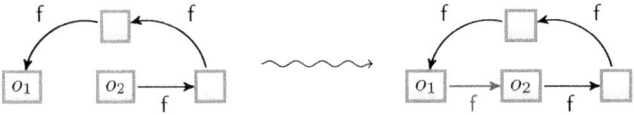

Fig. 7. Execution of $o_1.f = o_.2$ introduces an invalid cycle

$$\text{STORE}^a \quad \frac{\models Pre^{\mathbf{P}}(\sigma(\mathsf{v}_t) \neq \sigma(\emptyset), \sigma_e(\mathsf{v}_t) \notin reg_\sigma(\mathsf{v}_s))}{\models Pst_{\mathbf{Q}}^{\mathbf{P}}(\sigma \triangleright_{\{\sigma_e(\mathsf{v}_t)\}} \tau, \tau(\mathsf{v}_t) \approx \sigma(\mathsf{v}_t)[\mathsf{f} \mapsto \sigma(\mathsf{v}_s)], reg_\tau(\mathsf{v}_t) = reg_\sigma(\mathsf{v}_t) \cup reg_\sigma(\mathsf{v}_s))}{\{\mathbf{P}\} \; \mathsf{v}_t.\mathsf{f} := \mathsf{v}_s \; \{\mathbf{Q}\}}$$

Fig. 8. STORE^a rule applies whenever v_t and v_s are references of acyclic classes

Interestingly, the introduction of the STORE^a rule is the only modification we need to do to the semantics. To see this, consider a method invocation $\mathsf{v}_t = \mathsf{v}_s.\mathsf{m}(\mathsf{v}_1, \ldots, \mathsf{v}_n)$. According to rule CALL, it must be the case that the precondition of method m, $\alpha(\mathsf{v}_s, \mathsf{v}_1, \ldots, \mathsf{v}_n, \sigma)$, holds before the invocation. But this precondition must have been strong enough to verify that the body of method m introduces no invalid cycles (cf. (7) on p.70).

4   Trees and Similar Data Structures

Acyclicity is not the only common invariant that is impossible to express using classical logic. In this section we will discuss that of *being a tree*, which requires not only having no cycles but also having a *unique path* to every reachable node. We will show that by incorporating a second region to objects, we can handle an additional class qualifier "**tree**", whose precise meaning we will give below.

Before going into the details, it is worth observing that the dynamic frames methodology does not preclude the inclusion of more than one region per object. We will be exploiting this possibility also on the following sections.

For succinctness, we will say that an object o *is a tree* if it is an instance of a class tagged as **tree**. We will also assume that the **tree** qualifier implies the **acyclic** qualifier.

Just like in the previous section, we will consider a notion of "being a tree" that constitutes a compromise between what can be expressed and what can be enforced. Therefore, we want the following invariant for a *tree* o: i) o satisfies the invariant for acyclic objects (see §3), ii) if o reaches an object o' that is a tree, then there is only one path between o and o' where every object in the path is also a tree.

Again, in order to see how this invariant can be preserved, we need to look only at store instructions $\mathsf{v}_t.\mathsf{f} := \mathsf{v}_s$, where both v_t and v_s are trees. One can then see that there are essentially two ways in which the invariant can get broken. The first one, illustrated in Fig. 9 corresponds to the case when both v_t and v_s can reach a common (tree) object o: there is already a unique path from each to o, but executing the store would introduce an additional path from the target to o. This can be avoided by adding the following additional precondition:

$$reg_\sigma(\mathsf{v}_t) \parallel reg_\sigma(\mathsf{v}_s) \tag{8}$$

This clause implies the condition $\sigma_e(\mathsf{v}_t) \notin reg_\sigma(\mathsf{v}_s)$ required to enforce acyclicity.

It is not hard to verify that if v_t and v_s satisfy (8) then the tree invariant must hold on every node reachable from either v_t or v_s. But what can we say about

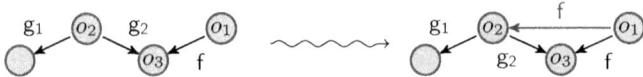

Fig. 9. Execution of $o_1.f = o._2$ introduces an extra path from o_1 to o_3

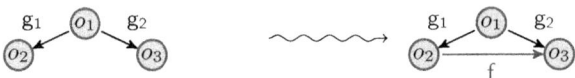

Fig. 10. Execution of $o_2.f = o._3$ introduces an extra path from o_1 to o_3

the invariant at tree nodes that reach either v_t or v_s? This question leads us to the second case, illustrated in Fig. 10. If v_t and v_s share a common ancestor, then the store will necessarily break the invariant of this ancestor.

Of course, it is not possible to express this condition using a first-order formula and the *reg* predicate. But using analogous ideas and techniques, one can associate to each object o an additional region $ger(o)$ that over-approximates the set of references that reach o and specify its evolution in semantic rules and method contracts. Using this, we can express the missing pre-condition for the store (when v_t and v_s are trees) as:

$$ger_\sigma(v_t) \parallel ger_\sigma(v_s) \tag{9}$$

To make everything fit well, we add to invariant $I_s(\sigma)$ (cf. p.68) the requirements:

$$ger_\sigma(\emptyset) = \emptyset \tag{10}$$

$$\forall \iota \in \operatorname{dm}\sigma_h \cdot \iota \neq \emptyset \Rightarrow \iota \in ger_\sigma(\iota) \wedge \forall f \cdot ger_\sigma(\iota) \subseteq ger_\sigma(\iota.f) \subseteq \operatorname{dm}\sigma_h \tag{11}$$

Similarly, the $\rhd_\rho^\nu$ predicate must be extended with the following clause:

$$\forall \iota \in \operatorname{dm}\sigma_h \cdot ger_\tau(\iota) = ger_\sigma(\iota) \cup gers_\tau(\rho \cap ger_\sigma(\iota)) \tag{12}$$

where $gers_\sigma(\rho) = \bigcup_{\kappa \in \rho} ger_\sigma(\kappa)$. Fig. 11 finally shows the formal semantics of the store rule for trees. It states that $ger(v_t)$ remains unchanged but, of course, $ger(v_s)$ is expanded. The latter implies that v_s is modified by the operation and therefore it must be included in the argument of the $\rhd$ predicate and the fact that every other field remains unchanged must be explicitly stated.

$$\text{STORE}^t \quad \frac{\begin{array}{l} \models Pre^{\mathbf{P}}(\sigma(v_t) \neq \sigma(\emptyset), reg_\sigma(v_t) \parallel reg_\sigma(v_s), ger_\sigma(v_t) \parallel ger_\sigma(v_s)) \\ \models Pst_{\mathbf{Q}}^{\mathbf{P}} \begin{pmatrix} \sigma \rhd_{\{\sigma_e(v_t), \sigma_e(v_s)\}} \tau, \tau(v_s) \approx \sigma(v_s), \tau(v_t) \approx \sigma(v_t)[f \mapsto \sigma(v_s)], \\ reg_\tau(v_t) = reg_\sigma(v_t) \cup reg_\sigma(v_s), ger_\tau(v_t) = ger_\sigma(v_t) \\ ger_\tau(v_s) = ger_\sigma(v_s) \cup ger_\sigma(v_t), reg_\tau(v_s) = reg_\sigma(v_s) \end{pmatrix} \end{array}}{\{\mathbf{P}\}\ v_t.f := v_s\ \{\mathbf{Q}\}}$$

Fig. 11. STORE^t rule applies whenever v_t and v_s are references to trees

The rules in Fig. 3 and 8 need to be modified to accommodate predicate ger but this is straightforward so we leave the details for the reader. It is not difficult to see that, in the resulting system, $ger(o)$ represents the set of objects that reach o and that every object of a class tagged as **tree** verifies its invariant.

5   Improving Precision

We have shown thus far that representation regions, used in principle for framing, can be also used to enforce complex structural invariants (e.g., acyclicity, etc.). In this section we will see examples of code that would be rejected by our proposed rules, although the invariants are clearly preserved.

Let us start considering the example in Fig. 12, where $LLNode$ is the acyclic class defined in Fig. 6 and T is not tagged as acyclic. Call σ the state before $b.next:=a$; then $reg_\sigma(a) = \{\sigma_e(a), \sigma_e(c), \sigma_e(b)\}$ which means that the precondition of the STORE^a rule does not hold. That is, it is detected that executing this instruction may lead to an invalid cycle involving a and b and the code is therefore rejected. But as Fig. 12 shows graphically, this would be indeed a valid cycle, since it passes through c that is not an instance of an acyclic class.

```
method rejected1() returns a {
    a := new LLNode;
    b := new LLNode;
    c := new T;
    a.payload := c;
    c.f := b;
    b.next := a;
}
```

Fig. 12. Rejected code snippet and the shape of a after $b.next:=a$

The technique is imprecise in this case, and the imprecision comes from the fact that $reg(o)$ contains the location of every reachable object, and not just of those that are reachable using only acyclic objects.

In §4 we already explored the possibility of incorporating additional regions in the context of handling the **tree** qualifier; we can take a similar approach here to improve the precision of the methodology. We propose, therefore, adding to each o a region $reg^a(o)$ that contains every acyclic object reachable from o passing only through acyclic objects. Again, this requires adding additional clauses to the state invariant $I_s(\sigma)$ and the $\triangleright$ predicate and extending the semantic rules. This changes are straightforward and were already mentioned in more detail in §4, so we will skip the details. The important part is replacing reg by reg^a in the pre-condition of rule STORE^a.

Of course, the same approach can be used to improve the precision on code handling trees: we can add regions reg^t and ger^t that only hold references to tree objects based on reachability via paths that contain only trees.

Let us assume this was indeed done and consider now the scenario in Fig. 13, where $BTNode$ is a class qualified as **tree**. The code in question "moves to

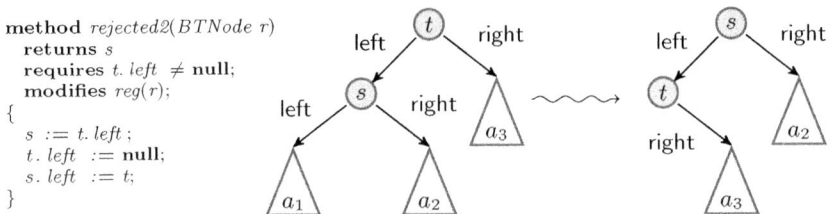

Fig. 13. Rejected code snippet, it is due to the *monotonic* nature of regions

the root" a node in a binary tree and it is not hard to see that the resulting structure would satisfy the required invariant. The problem is that before the method execution we have $r.\,left \in reg^t(r)$ and, therefore, $reg^t(s) \not\Vdash reg^t(r)$ holds after executing $s := t.\,left$. Since regions are monotonic (cf. §2), this is also true before executing $s.\,right := t$ and therefore, the requirements of STORE^t are not satisfied and this program is rejected.

More precisely, the problem originates after the execution of instruction $t.\,left := \textbf{null}$: s is no longer reachable from t although this is not reflected in $reg^t(t)$ nor in $ger^t(s)$. But since s and t are trees, there is only one path from s to any tree reachable from t; hence, if σ and τ are the pre and post-states of this instruction, it is safe to assume $reg^t_\tau(t) = reg^t_\sigma(t) \setminus reg^t_\sigma(s)$ and $ger^t_\tau(s) = ger^t_\sigma(s) \setminus ger^t_\sigma(t)$ (although it is not necessarily the case, for instance, that $reg_\tau(t) = reg_\sigma(t) \setminus reg_\sigma(s)$). Hence, we can improve rule STORE^t even further, by including in $Pst^{\textbf{P}}_{\textbf{Q}}$ the following clauses:

$$reg^t_\tau(\mathsf{v}_t) = (reg^t_\sigma(\mathsf{v}_t) \setminus reg^t_\sigma(\mathsf{v}_t.\mathsf{f})) \cup reg^t_\sigma(\mathsf{v}_s) \tag{13}$$

$$ger^t_\tau(\sigma_e(\mathsf{v}_t.\mathsf{f}))) = ger^t_\sigma(\mathsf{v}_t.\mathsf{f}) \setminus ger^t_\sigma(\mathsf{v}_t) \tag{14}$$

Of course, we need also add to the definition of $\rhd^\nu_\rho$ clauses:

$$\forall \iota \in \text{dm}\,\sigma_h \cdot reg^t_\tau(\iota) = (reg^t_\sigma(\iota) \setminus regs^t_\sigma(\rho \cap reg^t_\sigma(\iota))) \cup regs^t_\tau(\rho \cap reg^t_\sigma(\iota)) \tag{15}$$

$$\forall \iota \in \text{dm}\,\sigma_h \cdot ger^t_\tau(\iota) = (ger^t_\sigma(\iota) \setminus gers^t_\sigma(\rho \cap ger^t_\sigma(\iota))) \cup gers^t_\tau(\rho \cap ger^t_\sigma(\iota)) \tag{16}$$

where $regs^t_\sigma(\rho) = \bigcup_{\kappa \in \rho} reg^t_\sigma(\kappa)$ and $gers^t_\sigma(\rho) = \bigcup_{\kappa \in \rho} ger^t_\sigma(\kappa)$.

Using these new rules, it is not hard to see that the example in Fig. 13 is not rejected. Now, we must insist that the soundness of these rules depends on the fact that there is at most one path (passing only through tree nodes) that connects any two tree nodes. To illustrate this point, assume a class *ANode* is tagged as **acyclic** and consider Fig. 14. If after instruction $a.f1 := \textbf{null}$, b is removed from $reg^a(a)$, then the precondition of the last instruction will hold, which would permit the introduction of a cycle.

In a way, the problem in this case is that the set of reachable regions does not give us enough information to decide if we no longer reach an object. It might be interesting to consider, as future work, the possibility of turning $reg^a(o)$ into a *multiset* of locations, i.e., a total function from *Ref* to $\mathbb{N}$. The intended semantics for $reg^a(o)$ would be that it counts the total *number of paths* going only through

```
method invalid() {
    a := new ANode;
    b := new ANode;
    a.f1 := b;
    b.f2 := b;
    a.f1 := null;
    b.f1 := a;
}
```

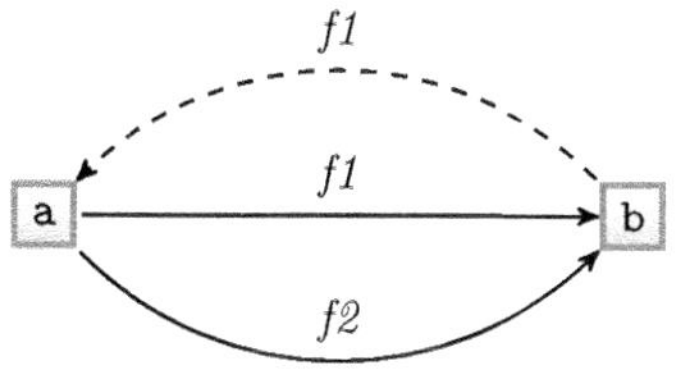

Fig. 14. If *ANode* is tagged **acyclic** this method should be rejected

acyclic nodes between o and any location ι and the tricky part would be to actually maintain this invariant.

6   Evaluation

We developed a prototype implementation[5] based on Dafny's language tool chain. Dafny is an experimental language that explores the use of dynamic frames in object-based sequential programs by enabling the use of ghost fields in contracts and statements [7]. We extended the language with automatic region inference and support for acyclic class qualification. Details can be found in [12].

We then implemented various basic abstract data types using acyclic structures. STACK, SEQUENCE and QUEUE use a linked list (the last two keep references to both head and tail), DICTIONARY uses a non-balanced binary search tree (BST). We evaluated the cost of enforcing acyclicity in terms of the number of atoms in region-related annotations, both in specifications (S.Ann) and inside method bodies such as loop invariants (B.Ann), and the verification time (in seconds). The following table summarizes the results obtained.[6]

Module	LOC	#classes	#methods	S.Ann	B.Ann	Verif. Time
Sequence	136	2	4	6	2	2.7s
Stack	48	2	4	5	0	2.3s
Queue	75	2	4	7	1	2.4s
Dictionary	140	2	5	11	4	4.0s

For most cases we were able to automatically check that a method preserves acyclicity. We could not do it, for instance, in the *remove* method of DICTIONARY because of monotonicity of the regions. A variation of the standard deletion algorithm for BSTs in which values are swapped instead of nodes should be more amenable to verification. It is worth observing that we did not use manual update of ghost fields in method bodies, only loop invariants were provided.

As a second experiment, we tried to assess if the inferred regions together with the **acyclic** qualifier could be used to significantly reduce the overall number of

---

[5] Tool and experiments available at `http://lafhis.dc.uba.ar/dynframes`
[6] Times measured on a 3GHz INTEL® CORE™2 DUO based desktop with 4GB of RAM, running MICROSOFT WINDOWS VISTA (32 bits), under regular load.

annotations on *idiomatic* Dafny programs. For this, we took some examples from the Dafny distribution (a linked list, an unbounded stack, a queue, and BST based dictionary) and tried to simplify them by tagging the relevant classes as **acyclic**, removing the ghost field used for the representation region (together with their manual updates) and pruning the contracts and invariants accordingly.

We were able to remove about 25%-30% of the annotations in contracts (plus the manual updates of ghost fields that were also removed). The only method we failed to verify was list reversal, which relies on temporarily breaking acyclicity.

It is worth noticing that our implementation cannot currently handle some constructs available in the latest version of the Dafny language; in particular, universal quantification restricted to objects of a given type. Once this is corrected, we might be able to reduce even further the number of annotations in some examples (most notably, in linked lists, which require a ghost field which represents the *spine* of the list, that is, the set of nodes that form the list).

7   Related Work

Dafny is a programming language with a verifying compiler and support for dynamic frames. It is possible to verify the correctness of Dafny programs (that is, without our extensions) that contain class invariants that entail acyclicity or tree-shapedness. It is therefore important to discuss the differences with our approach.

In Dafny, regions are explicitly declared as ghost fields of a class and the programmer is responsible for maintaining them with explicit update instructions. These are, arguably, a form of annotation. With our extensions there are also pre-defined regions handled by the tool. While this scheme is perhaps less flexible, it demands no region update annotations in program text.

Acyclicity can be enforced by way of user-provided class invariants that rely on the fact that certain ghost fields over-approximate reachability (even without manually maintaining regions, a similar effect could be achieved with the incomplete encoding of transitive closure in first-order logic given in [10]). These invariants can be temporally violated and must be provably restored at a later stage. This differs from our approach in two important ways. Firstly, it is the user, but not the compiler, who is aware of these properties. Therefore this knowledge cannot be transparently exploited by the compiler, for instance, for better memory-management (e.g., switching to a reference counting scheme for certain types) or loop-termination analysis. The second difference is that in our approach the properties cannot be temporarily broken and later restored but are preserved throughout the execution of the methods. This has the advantage of being enforceable by relatively simple checks. The price to pay, in the case of trees, is the introduction of a second region (cf. Fig. 10).

One can argue that class invariants expressing acyclicity constraints on complex types such as abstract syntax trees with subformula sharing could easily become unwiedly; on the other hand, they would be trivial to express with our approach. However, a systematic comparison of the annotation burden, both qualitative and quantitative, in Dafny with and without our extensions is needed.

Instead of using the representation regions of the dynamic frames methodology, one could instead enforce acyclicity using the *representation containment* concept of Ownership Types (see, e.g. [2]). In fact, the object graph structure obtained using this approach is typically a tree. However, since every object is required to have at most one owner, it is very non-trivial (if at all possible, depending the setting) to enforce a DAG, like it would be the case, for instance, in a Queue implemented with pointers to the first and last nodes of a linked list.

The most common approach to this problem is via *shape analysis* techniques (e.g., [13]). These are used to determine shape invariants for programs that perform destructive updates on dynamically allocated storage. In [6] a method is presented that automatically verifies acyclic linked lists, by carefully defining axioms for modeling the standard list operations. In [11] the authors introduce a type system which controls acyclicity by defining the concept of regions in which cycles are only allowed inside a region and forcing a partial order within regions.

8   Conclusions

In this work we tried to demonstrate that one can take advantage of the machinery required for dynamic frames to enforce non-trivial structural invariants in abstract data types. In particular, it is possible to guarantee acyclicity practically without performing changes to code. We developed similar techniques to enforce tree-shapedness and, finally, we discussed the use of additional representation regions as means to reject fewer valid programs. We also reported on some preliminary results using a prototypic implementation of some of these ideas.

We believe that besides providing correctness guarantees, these kind of structural invariants can be exploited by the compiler; examples include sound heuristics for analyzing program termination, and enabling cheaper memory management schemes (e.g., reference counting).

As future work, we would like to look at the interplay between the introduced techniques and classical features of object-oriented languages such us inheritance and dynamic binding. Moreover, while acyclic data structures are ubiquitous, so are cyclic ones (e.g. doubly-linked lists, circular buffers, etc.). We believe that common *cyclicity patterns* exist that can be enforced using similar techniques.

References

1. Barnett, M., Leino, K.R.M., Schulte, W.: The spec# programming system: An overview. In: Barthe, G., Burdy, L., Huisman, M., Lanet, J.-L., Muntean, T. (eds.) CASSIS 2004. LNCS, vol. 3362, pp. 49–69. Springer, Heidelberg (2005)
2. Clarke, D., Potter, J., Noble, J.: Ownership types for flexible alias protection. In: Proceedings of the 13th ACM SIGPLAN Conference on Object-Oriented Programming, Systems, Languages, and Applications (OOPSLA 1998), pp. 48–64. ACM Press, New York (1998)
3. Huizing, K., Kuiper, R.: Verification of object oriented programs using class invariants. In: FASE 2000. LNCS, vol. 1783, pp. 208–221. Springer, Heidelberg (2000)

4. Jacobs, B., Smans, J., Piessens, F.: VeriFast: Imperative Programs as Proofs. In: VSTTE Workshop on Tools & Experiments (2010)
5. Kassios, I.T.: The dynamic frames theory. In: Formal Aspects of Computing (2010)
6. Lahiri, S., Qadeer, S.: Verifying properties of well-founded linked lists. In: Proceedings of the 33rd ACM SIGPLAN-SIGACT Symposium on Principles of Programming Languages (POPL 2006), pp. 115–126. ACM Press, New York (2006)
7. Leino, K.R.M.: Dafny: An automatic program verifier for functional correctness. In: Clarke, E.M., Voronkov, A. (eds.) LPAR-16 2010. LNCS, vol. 6355, pp. 348–370. Springer, Heidelberg (2010)
8. Leino, K.R.M., Müller, P.: A basis for verifying multi-threaded programs. In: Castagna, G. (ed.) ESOP 2009. LNCS, vol. 5502, pp. 378–393. Springer, Heidelberg (2009)
9. Leino, K.R.M., Nelson, G.: Data abstraction and information hiding. ACM TOPLAS 24(5), 491–553 (2002)
10. Lev-Ami, T., Immerman, N., Reps, T., Sagiv, M., Srivastava, S., Yorsh, G.: Simulating reachability using first-order logic with applications to verification of linked data structures. In: Nieuwenhuis, R. (ed.) CADE 2005. LNCS (LNAI), vol. 3632, pp. 99–115. Springer, Heidelberg (2005)
11. Lu, Y., Potter, J.: A type system for reachability and acyclicity. In: Gao, X.-X. (ed.) ECOOP 2005. LNCS, vol. 3586, pp. 479–503. Springer, Heidelberg (2005)
12. Neisen, A.: Automatic verification of acyclic data structures using theorem provers. Master's thesis, Universidad de Buenos Aires (2010)
13. Sagiv, M., Reps, T., Wilhelm, R.: Parametric shape analysis via 3-valued logic. ACM TOPLAS 24(3), 217–298 (2002)
14. Smans, J., Jacobs, B., Piessens, F.: VeriCool: An automatic verifier for a concurrent object-oriented language. In: Barthe, G., de Boer, F.S. (eds.) FMOODS 2008. LNCS, vol. 5051, pp. 220–239. Springer, Heidelberg (2008)

Loop Summarization and Termination Analysis[*]

Aliaksei Tsitovich[1], Natasha Sharygina[1],
Christoph M. Wintersteiger[2], and Daniel Kroening[2]

[1] Formal Verification and Security Group, University of Lugano. Switzerland
[2] Oxford University, Computing Laboratory, UK

Abstract. We present a technique for program termination analysis based on loop summarization. The algorithm relies on a library of abstract domains to discover well-founded transition invariants. In contrast to state-of-the-art methods it aims to construct a complete ranking argument for all paths through a loop at once, thus avoiding expensive enumeration of individual paths. Compositionality is used as a completeness criterion for the discovered transition invariants. The practical efficiency of the approach is evaluated using a set of Windows device drivers.

1 Introduction

The program termination problem has received increased interest in the recent past. In practice, termination analysis is at a point where industrial application of termination proving tools is feasible. This is possible through a series of improvements upon methods that prove program termination by constructing *well-founded ranking relations*.

Podelski and Rybalchenko propose *disjunctive* well-foundedness of transition invariants [1] as a means to improve the performance of termination proving, as well as to simplify synthesis of ranking relations. Based on their crucial discovery, the same authors together with Cook give an algorithm to verify program termination using iterative construction of transition invariants — the *Terminator* algorithm [2,3]. This algorithm exploits the relative simplicity of ranking relations for a single path of a program. It relies on a safety checker to find previously unranked paths of a program, computes a ranking relation for each of them individually, and disjunctively combines them to form a global (disjunctively well-founded) termination argument. This strategy shifts the complexity of the problem from ranking relation synthesis to safety checking, a problem for which many efficient solutions exist.

The Terminator algorithm was successfully implemented in tools (e.g., TERMINATOR [3], ARMC [4], SATABS [5]) and applied to verify industrial code, most notably Windows device drivers. However, it has subsequently become apparent that the safety check is a bottleneck of the algorithm, taking up to 99% of the

---

[*] Supported by the Swiss National Science Foundation under grant no. 200020-122077 and by a Microsoft Software Engineering Innovation Foundation (SEIF) Award. Christoph M. Wintersteiger is now with Microsoft Research, Cambridge, UK.

P.A. Abdulla and K.R.M. Leino (Eds.): TACAS 2011, LNCS 6605, pp. 81–95, 2011.
© Springer-Verlag Berlin Heidelberg 2011

runtime in practice [3,5]. The runtime required for ranking relation synthesis is negligible in comparison. A possible solution to this performance issue is *Compositional Termination Analysis (CTA)* [6]. This method limits path exploration to several iterations of each loop of the program. Transitivity (or *compositionality*) of the intermediate ranking arguments is used as a criterion to determine when to stop the loop unwinding. This allows for a reduction in runtime, but introduces incompleteness since a transitive termination argument may not be found for each loop of a program. However, an experimental evaluation on Windows device drivers indicates that this case is rare in practice.

The complexity of the termination problem together with the observation that most loops in practice have (relatively) simple termination arguments suggests the use of light-weight static analysis for this purpose. In this paper, we propose a new technique for termination analysis, which extends a known algorithm for loop summarization [7] based on abstract interpretation [8]. The crucial difference between the previous approach and our proposal is the use of (disjunctively well-founded) *transition invariants* instead of *state invariants* during summarization. Furthermore, fixpoint computation of abstract transformers is avoided (but required by other methods, e.g., [9,10]).

Our algorithm constructs summaries for loops, starting from the inner-most loop in the control flow graph of the program. In case of nested loops, inner loops are replaced with (loop-free) summaries during verification. At any point during the analysis, the problem is therefore reduced to the analysis of a single loop. During construction of the loop summaries, our algorithm relies on a library of templates for abstract domains. These are used to construct candidates for transition invariants, which subsequently are verified to be actual disjunctively well-founded transition invariants by means of a safety checker and a satisfiability decision procedure. Due to the fact that the safety checker is employed to analyze only a single unwinding of a loop at any point, we gain large speedups compared to algorithms like Terminator or CTA. At the same time, the false-positive rate of our algorithm is very low in practice, which we demonstrate using an experimental evaluation on a diverse suite of C programs.

This paper is organized as follows: Section 2 introduces the theoretical background, Section 3 presents our new methods. Section 4 proposes an optimization that simplifies the selection of candidates for transition invariants. In Section 5 we give experimental evidence of the practicality of our approach. Section 6 relates this approach to size-change termination principle and discusses the other related work. Finally, Section 7 suggests future work and concludes.

2   Background

We formalize programs as *transition systems*.

Definition 1 (Transition System). *A transition system (program) P is a three tuple $\langle S, I, R \rangle$, where*

- S is a (possibly infinite) set of states,
- $I \subseteq S$ is the set of initial states,
- $R \subseteq S \times S$ is the transition relation.

A *computation* of a transition system is a (maximal) sequence of states $s_0, s_1, \ldots$ such that $s_0 \in I$ and $(s_i, s_{i+1}) \in R$ for all $i \geq 0$.

The reflexive and non-reflexive transitive closures of R are denoted as R^* and R^+ respectively. The set of reachable states is $R^*(I)$. We also define the relational composition operator $\circ$ for two relations $R_1, R_2 : S \times S$ by

$$R_1 \circ R_2 := \{ (s, s') \mid \exists s''.(s, s'') \in R_1 \wedge (s'', s') \in R_2 \} .$$

Note that a relation R is transitive if it is closed under relational composition, i.e., when $R \circ R \subseteq R$.

2.1 Termination

A program is terminating if it does not allow infinite computations, which follows from well-foundedness of the transition relation (restricted to the reachable states). A *well-founded* relation is a relation that does not contain infinite descending chains or, more formally:

Definition 2 (Well-foundedness). *A relation R is* well-founded (wf.) *over S if for any non-empty subset of S there exists a minimal element (with respect to R), i.e. $\forall X \subseteq S . X \neq \emptyset \implies \exists m \in X, \forall s \in S(s, m) \notin R$.*

The same does not hold true for the weaker notion of *disjunctive well-foundedness*. However, Podelski and Rybalchenko show that disjunctive well-foundedness of a *transition invariant* is equivalent to program termination:

Definition 3 (Disjunctive Well-foundedness [1]). *A relation T is* disjunctively well-founded (d.wf.) *if it is a finite union $T = T_1 \cup \ldots \cup T_n$ of well-founded relations.*

Definition 4 (Transition Invariant [1]). *A transition invariant T for program $P = \langle S, I, R \rangle$ is a superset of the transitive closure of R restricted to the reachable state space, i.e., $R^+ \cap (R^*(I) \times R^*(I)) \subseteq T$.*

The crucial theorem is as follows:

Theorem 1 (Termination [1]). *A program P is terminating iff there exists a d.wf. transition invariant for P.*

The Terminator algorithm [3] automates the construction of d.wf. transition invariants. It starts with an empty termination condition $T = \emptyset$ and queries a safety checker for a counterexample — a computation that is not covered by the current termination condition T. Next, a ranking relation synthesis algorithm is used to obtain a termination argument T' covering the transitions in the counterexample. The termination argument is then updated to $T = T \cup T'$

and the algorithm continues to search for counterexamples. Finally, either a complete (d.wf.) transition invariant is constructed or there does not exist a ranking relation for some counterexample, in which case the program is reported as non-terminating.

To comply with the terminology in the existing literature, we define the notion of compositionality for transition invariants as follows:

Definition 5 (Compositional Transition Invariant [1,6]). *A d.wf. transition invariant T is called* compositional *if it is also transitive, or equivalently, closed under composition with itself, i.e., when $T \circ T \subseteq T$.*

Podelski and Rybalchenko made an interesting remark regarding the compositionality (transitivity) of transition invariants: If T is transitive, it is sufficient to show that $T \supseteq R$ instead of $T \supseteq R^+$ to conclude termination, because a compositional and d.wf. transition invariant is well-founded, since it is an inductive transition invariant for itself [1]. Therefore, compositionality of a d.wf. transition invariant implies program termination. This fact is exploited in *Compositional Termination Analysis* [6], which iteratively constructs a termination argument, similar to the Terminator algorithm. In contrast to Terminator however, the safety checker is not required to analyze complete loops. Instead, the algorithm checks an increasing number of unwindings of the loops in the program until a compositional transition invariant is established. This technique results in significant speed-ups in practice, but comes at a price: there is no guarantee that a compositional transition invariant can be found for every loop.

2.2 Loop Summarization

In the following section we present a method for static analysis based on a previously presented loop summarization algorithm [7]. This technique constructs sound program abstractions for the purpose of scalable static analysis. It replaces loops in a program by smaller loop-free program fragments that over-approximate the original behavior of the loop.

Algorithm 1 presents an outline of this procedure. The function SUMMARIZE traverses the control-flow graph of the program P and calls itself recursively for each block with nested loops. If a block contains a non-nested loop, it is summarized using the function SUMMARIZELOOP and the resulting summary replaces the original loop in P'. Consequently, any outer loop eventually becomes non-nested, which enables further progress.

The function SUMMARIZELOOP computes the summaries. A simple over-approximation can be obtained by replacing a loop by a program fragment that 'havocs' the state, i.e., by setting all variables which are (potentially) modified during loop execution to non-deterministic values. To improve the precision of these summaries, they are strengthened by (partial) loop invariants. SUMMARIZELOOP has two subroutines that are related to invariant discovery: 1) PICKINVARIANTCANDIDATES, which generates a set of 'invariant candidates' using a library of abstract domains, and 2) ISINVARIANT, which checks whether a candidate is an actual invariant for a given loop.

```
 1  SUMMARIZE(P)
 2  input: program P
 3  output: Program summary
 4  begin
 5      foreach Block B in CONTROLFLOWGRAPH(P) do
 6          if B has nested loops then
 7              B :=SUMMARIZE(B)
 8          else if B is a single loop then
 9              B :=SUMMARIZELOOP(B)
10      return P
11  end

12  SUMMARIZELOOP(L)
13  input: Single-loop program L (over variable set X)
14  output: Loop summary
15  begin
16      I := ⊤
17      foreach Candidate C(X) in PICKINVARIANTCANDIDATES(Loop) do
18          if ISINVARIANT(L, C) then
19              I := I ∧ C
20      return "X^{pre} := X; havoc(L); assume(I(X^{pre}) ⟹ I(X));"
21  end

22  ISINVARIANT(L, C)
23  input: Single-loop program L (over variable set X), invariant candidate C
24  output: TRUE if C is invariant for L; FALSE otherwise
25  begin
26      return UNSAT(L(X, X') ∧ C(X) ⇒ C(X'))
27  end
```

Algorithm 1. Basic routines of loop summarization

Note that this summarization algorithm does not preserve loop termination: the summaries computed by the algorithm are always terminating program fragments. This abstraction is a sound over-approximation, but it may be too coarse for programs that contain unreachable paths.

3   Loop Summarization with Transition Invariants

In this section, we introduce a method that allows *transition* invariants to be included as a strengthening of loop summaries. This increases the precision of loop summaries and enables construction of termination proofs over summaries.

According to Definition 4, a binary relation T is a transition invariant for a program P if it contains R^+ (restricted to the reachable states). However, the transitivity of T is also a sufficient condition when T is only a superset of R:

Theorem 2. *A binary relation T is a transition invariant for the program $\langle S, I, R \rangle$ if it is transitive and $R \subseteq T$.*

Proof. From transitivity of T it follows that $T^+ \subseteq T$. Since $R \subseteq T$ it follows that $R^+ \subseteq T$. □

This simple fact allows for an integration of transition invariants into the loop summarization framework by a few adjustments to the original algorithm. Consider line 16 of Algorithm 1, where candidate invariants are selected. Clearly, we need to allow selection of transition invariants here, i.e., invariant candidates now have the form $C(X, X')$, where X' is the post-state of L.

What follows is a check for invariance of C over $L(X, X')$, i.e., a single unwinding of the loop. Consider the temporary (sub-)program $\langle S, S, L \rangle$ to represent the execution of the loop from a non-deterministic entry state, as required by ISINVARIANT. A transition invariant for this program is required to cover L^+, which, according to Theorem 2, is implied by $L \subseteq C$ and transitivity of C. The original invariant check in ISINVARIANT establishes $L \subseteq C$, when the check for unsatisfiability receives the more general formula $L(X, X') \wedge C(X, X')$ as a parameter. The summarization procedure furthermore requires a slight change to to include a check for compositionality. The resulting procedure is Algorithm 2.

---

1 SUMMARIZELOOP-TI(L)
2 **input**: Single-loop program L with a set of variables X
3 **output**: Loop summary
4 **begin**
5     $T := \top$
6   **foreach** *Candidate* $C(X, X')$ *in* PICKINVARIANTCANDIDATES*(Loop)* **do**
7   **if** ISINVARIANT*(L, C)* $\wedge$ ISCOMPOSITIONAL*(C)* **then**
8             $T := T \wedge C$
9   **return** "$X^{pre} := X; \texttt{havoc}(L); \texttt{assume}(T(X^{pre}, X));$"
10 **end**

Algorithm 2. Loop summarization with transition invariants

The additional compositionality (transitivity) check at line 7 of Algorithm 2 corresponds to a check for satisfiability of

$$\exists s_i, s_j, s_k \in S \; . \; \neg (C(s_i, s_j) \wedge C(s_j, s_k) \Rightarrow C(s_i, s_k)) \; , \tag{1}$$

which may be decided by a suitable decision procedure, e.g., an SMT solver. Of course, this check may be omitted if the selected invariant candidates are compositional by construction.

Termination. The changes to the summarization algorithm allow for termination checks during summarization through application of Theorem 1, which requires a transition invariant to be disjunctively well-founded. This property may be established by allowing only disjunctively well-founded invariant candidates, or it may be checked by means of decision procedures (e.g., SMT solvers where applicable). According to Definition 3, d.wf.-ness of a candidate relation T requires establishing well-foundedness of each of its disjuncts. This can be done by an explicit encoding of the well-foundedness criteria of Definition 2. However, the resulting formula contains quantifiers, which severely limits the applicability of existing decision procedures.

4   Selection of Candidate Invariants

In this section, we propose a set of specialized candidate relations, which we find to be useful in practice. We focus on transition invariants for machine-level integers for programs implemented in low-level languages like ANSI-C.

In contrast to other work on termination proving with abstract domains (e.g., [10]), we do not aim at general domains like Octagons or Polyhedra, as they are not designed for termination and the required d.wf.-ness and compositionality checks can be costly. Instead we prefer domains that

- generate few, relatively simple candidate relations, and
- allow for efficient d.wf. and compositionality checks.

Note that very similar criteria are applied in termination provers based on the size-change termination principle. This connection is discussed in more detail in Sec. 6.1.

Arithmetic operations on machine-level integers usually allow overflows, e.g., the instruction $i = i + 1$ for a pre-state $i = 2^k - 1$ results in a post-state $i' = -2^{k-1}$ (when represented in two's-complement), complicating termination arguments. If termination of the loop depends only on machine-level integers, there is however a way to simplify the argument:

Observation 3. *If $T : K \times K$ is a strict order relation for a finite set $K \subseteq S$ and is a transition invariant for the program $\langle S, I, R \rangle$, then T is well-founded.*

Proof. T is a transition invariant, i.e., it holds for all pairs $(k_1, k_2) \in K \times K$. Thus it is total. Non-empty finite totally-ordered sets always have a least element and, therefore, T is well-founded. □

A total strict order relation is also transitive, which gives rise to a criterion weaker than Theorem 1:

Corollary 1. *A program terminates if it has a transition invariant T that is also a finite strict order relation.*

This corollary allows for a selection of invariant candidates that ensures (disjunctive) well-foundedness of transition invariants. An explicit check is therefore not required.

Note that strictly ordered and finite transition invariants exist for many programs in practice: machine-level integers or strings of fixed length have a finite number of possible distinct pairs and strict natural or lexicographical orders are defined for them as well.

Table 1. Templates used to generate transition invariant candidates

#	Constraint	Meaning
1	$i' < i$ $i' > i$	A numeric variable i is strictly decreasing (increasing).
2	$x' < x$ $x' > x$	Any loop variable x is strictly decreasing (increasing).
3	$sum(x', y') < sum(x, y)$ $sum(x', y') > sum(x, y)$	The sum of all numeric loop variables is strictly decreasing (increasing).
4	$max(x', y') < max(x, y)$ $max(x', y') > max(x, y)$ $min(x', y') < min(x, y)$ $min(x', y') > min(x, y)$	The maximum or minimum of all numeric loop variables is strictly decreasing (increasing).
5	$(x' < x \wedge y' = y) \vee$ $(x' > x \wedge y' = y) \vee$ $(y' < y \wedge x' = x) \vee$ $(y' > y \wedge x' = x)$	A combination of strict increasing or decreasing for one of loop variables while the remaining ones are not updated.

5 Evaluation

We have implemented the algorithm described in the previous section in a new version of the static analyzer LOOPFROG [11]. The tool operates on program models produced by the GOTO-CC model extractor; ANSI-C programs are considered as the primary target.

We implemented a number of domains based on strict orders, thus, following Corollary 1, additional checks for compositionality and d.wf.-ness of candidate relations are not required. The domains are listed in Table 1.

The full set of experimental results is available on-line. Here, we only report the results for the two most illustrative schemata:

- LOOPFROG 1: domain #3 in Table 1. Expresses the fact that a sum of all numeric variables of a loop is strictly decreasing (increasing). This is the fastest approach, because it generates very few (but large) invariant candidates per loop;
- LOOPFROG 2: domain #1 in Table 1. Expresses strict decreasing (increasing) for every numeric variable of a loop. This generates twice as many simple strict orders as there are variables in a loop;

As reference points we use termination provers built upon the CBMC/SatAbs framework [12]. This tool implements both Compositional Termination Analysis (CTA) [6] and the TERMINATOR algorithm [2] (referred to as SATABS+T in all tables). For both the default ranking function synthesis methods were enabled; for more details see [5].

We experimented with a large number of ANSI-C programs including:

- The SNU real-time benchmark suite that contains small C programs used for worst-case execution time analysis[1];
- The Powerstone benchmark suite as an example set of C programs for embedded systems [13];
- The Verisec 0.2 benchmark suite [14];
- Windows device drivers (from the Windows Device Driver Kit 6.0).

All experiments were run on an Ubuntu server equipped with a Dual-Core 2 GHz Opteron 2212 CPU and 4 GB of memory. The timeout was set to 120 minutes if an analysis is applied to all loops at once (LOOPFROG) or to 60 minutes per loop (CTA and SATABS+T).

The results for SNU and Power-Stone are presented in Tables 2 and 3. Each table reports the number of loops that were proven as terminating (T), potentially non-terminating (NT) and time-out (TO) for each of the compared techniques. The time column contains the wall-clock time spend for the analysis of completed loops; time-outs are not included in the total time.

The results for the Verisec 0.2 benchmark suite are given in the aggregated form in Table 4. The suite consists a large number of stripped C programs that correspond to known security bugs. Although each program has only very few loops, the benchmark set offers a large variety of loop types and is therefore interesting for termination analysis.

The aggregated data on experiments with Windows device drivers is provided in Table 5. The benchmarks are grouped according to the harness used upon extraction of a model with GOTO-CC.[2] We omit results of the TERMINATOR algorithm for this benchmark set, as a corresponding comparison was already reported previously [6].

Discussion. Note that direct comparison of the runtime of LOOPFROG with that of iterative techniques like CTA and TERMINATOR is not fair. The latter methods are complete at least for finite-state programs, relative to the completeness of the ranking synthesis method. Our loop summarization technique, on the other hand, is a static analysis that only aims at conservative abstractions. In particular, it does not try to prove unreachability of a loop or of preconditions that lead to non-termination[3].

The timing information provided here serves as a reference that allows to compare efforts of achieving the same result. In summary, the three techniques can be compared as follows:

- LOOPFROG spends time enumerating invariant candidates, provided by the chosen abstract domain, and has to check just one loop iteration. Compositionality and d.wf. checks are not required for the domains we use.

---

[1] http://archi.snu.ac.kr/realtime/benchmark/

[2] The groups in Table 5 have varying numbers of benchmarks/loops as we omit the benchmarks without loops.

[3] In future, we plan to use a loop-free stem to prove unreachability of certain loop preconditions.

Table 2. SNU real-time benchmark suite

Benchmark	Method	T	NT	TO	Time
adpcm-test 18 loops	LoopFrog 1	13	5	0	470.052
	LoopFrog 2	17	1	0	644.092
	CTA	13	3	2	260.982+
	SatAbs+T	12	2	4	165.673+
bs 1 loop	LoopFrog 1	0	1	0	0.05
	LoopFrog 2	0	1	0	0.118
	CTA	0	1	0	12.218
	SatAbs+T	0	1	0	18.469
crc 3 loops	LoopFrog 1	1	2	0	0.17
	LoopFrog 2	2	1	0	0.255
	CTA	1	1	1	0.206+
	SatAbs+T	2	1	0	13.878
fft1k 7 loops	LoopFrog 1	2	5	0	0.356
	LoopFrog 2	5	2	0	0.668
	CTA	5	2	0	141.176
	SatAbs+T	5	2	0	116.81
fft1 11 loops	LoopFrog 1	3	8	0	3.68
	LoopFrog 2	7	4	0	4.976
	CTA	7	4	0	441.937
	SatAbs+T	7	4	0	427.355
fibcall 1 loop	LoopFrog 1	0	1	0	0.04
	LoopFrog 2	0	1	0	0.016
	CTA	0	1	0	0.335
	SatAbs+T	0	1	0	0.309
fir 8 loops	LoopFrog 1	2	6	0	2.897
	LoopFrog 2	6	2	0	8.481
	CTA	6	2	0	2817.08
	SatAbs+T	6	1	1	236.702+
insertsort 2 loops	LoopFrog 1	0	2	0	0.054
	LoopFrog 2	1	1	0	0.063
	CTA	1	1	0	226.446
	SatAbs+T	1	1	0	209.12
jfdctint 3 loops	LoopFrog 1	0	3	0	5.612
	LoopFrog 2	3	0	0	0.05
	CTA	3	0	0	1.24
	SatAbs+T	3	0	0	0.975
lms 10 loops	LoopFrog 1	3	7	0	2.863
	LoopFrog 2	6	4	0	10.488
	CTA	6	4	0	2923.12
	SatAbs+T	6	3	1	251.031+
ludcmp 11 loops	LoopFrog 1	0	11	0	96.726
	LoopFrog 2	5	6	0	112.808
	CTA	3	5	3	3.256+
	SatAbs+T	3	8	0	94.657
matmul 5 loops	LoopFrog 1	0	5	0	0.148
	LoopFrog 2	5	0	0	0.086
	CTA	3	2	0	1.969
	SatAbs+T	3	2	0	2.152
minver 17 loops	LoopFrog 1	1	16	0	2.574
	LoopFrog 2	16	1	0	7.664
	CTA	14	1	2	105.26+
	SatAbs+T	14	1	2	87.088+
qsort-exam 6 loops	LoopFrog 1	0	6	0	0.671
	LoopFrog 2	0	6	0	3.96
	CTA	0	5	1	45.918+
	SatAbs+T	0	5	1	2530.58+
select 4 loops	LoopFrog 1	0	4	0	0.548
	LoopFrog 2	0	4	0	3.561
	CTA	0	3	1	32.599+
	SatAbs+T	0	3	1	28.12+

Table 3. PowerStone benchmark suite

Benchmark	Method	T	NT	TO	Time
adpcm 11 loops	LoopFrog 1	8	3	0	59.655
	LoopFrog 2	10	1	0	162.752
	CTA	8	3	0	101.301
	SatAbs+T	6	2	3	94.449+
bcnt 2 loops	LoopFrog 1	0	2	0	2.634
	LoopFrog 2	0	2	0	2.822
	CTA	0	2	0	0.79
	SatAbs+T	0	2	0	0.299
blit 4 loops	LoopFrog 1	0	4	0	0.155
	LoopFrog 2	3	1	0	0.047
	CTA	3	1	0	5.945
	SatAbs+T	3	1	0	3.672
compress 18 loops	LoopFrog 1	5	13	0	3.134
	LoopFrog 2	6	12	0	33.924
	CTA	5	12	1	698.996+
	SatAbs+T	7	10	1	474.361+
crc 3 loops	LoopFrog 1	1	2	0	0.152
	LoopFrog 2	2	1	0	0.208
	CTA	1	1	1	0.328+
	SatAbs+T	2	1	0	14.583
engine 6 loops	LoopFrog 1	0	6	0	2.397
	LoopFrog 2	2	4	0	9.875
	CTA	2	4	0	16.195
	SatAbs+T	2	4	0	4.877
fir 9 loops	LoopFrog 1	2	7	0	5.993
	LoopFrog 2	6	3	0	21.592
	CTA	6	3	0	2957.06
	SatAbs+T	6	2	1	193.911+
g3fax 7 loops	LoopFrog 1	1	6	0	1.565
	LoopFrog 2	1	6	0	6.047
	CTA	1	5	1	256.899+
	SatAbs+T	1	5	1	206.847+
huff 11 loops	LoopFrog 1	3	8	0	24.368
	LoopFrog 2	8	3	0	94.613
	CTA	7	3	1	16.353+
	SatAbs+T	7	4	0	52.323
jpeg 23 loops	LoopFrog 1	2	21	0	8.366
	LoopFrog 2	16	7	0	32.9
	CTA	15	8	0	2279.13
	SatAbs+T	15	8	0	2121.36
pocsag 12 loops	LoopFrog 1	3	9	0	2.07
	LoopFrog 2	9	3	0	6.906
	CTA	9	3	0	10.392
	SatAbs+T	7	3	2	1557.57+
ucbqsort 15 loops	LoopFrog 1	1	14	0	0.789
	LoopFrog 2	2	13	0	2.059
	CTA	2	12	1	71.729+
	SatAbs+T	2	12	1	51.084+
v42 12 loops	LoopFrog 1	0	12	0	82.836
	LoopFrog 2	0	12	0	2587.22
	CTA	0	12	0	73.565
	SatAbs+T	1	11	0	335.688

Table 4. Aggregated data on Verisec 0.2 suite

Benchmark group	Method	T	NT	TO	Time
244 loops in 160 programs	LoopFrog 1	33	211	0	11.381
	LoopFrog 2	44	200	0	22.494
	CTA	34	208	2	1207.62+
	SatAbs+T	40	204	0	4040.53

Columns 3 to 5 state the number of loops proven to terminate (T), possibly non-terminate (NT) and time-out (TO) for each benchmark. Time is computed only for T/NT loops; '+' is used to denote the resulting time for the cases where at least one time-outed loop was not considered.

Table 5. Aggregated data on Windows device drivers

Benchmark group	Method	T	NT	TO	Time
SDV FLAT DISPATCH HARNESS 557 loops in 30 benchmarks	LOOPFROG 1	135	389	33	1752.08
	LOOPFROG 2	215	201	141	10584.4
	CTA	166	160	231	25399.5
SDV FLAT DISPATCH STARTIO HARNESS 557 loops in 30 benchmarks	LOOPFROG 1	135	389	33	1396.01
	LOOPFROG 2	215	201	141	9265.81
	CTA	166	160	231	28033.3
SDV FLAT HARNESS 635 loops in 45 benchmarks	LOOPFROG 1	170	416	49	1323
	LOOPFROG 2	239	205	191	6816.37
	CTA	201	186	248	31003.2
SDV FLAT SIMPLE HARNESS 573 loops in 31 benchmarks	LOOPFROG 1	135	398	40	1510
	LOOPFROG 2	200	191	182	6813.99
	CTA	166	169	238	30292.7
SDV HARNESS DRIVER CREATE 9 loops in 5 benchmarks	LOOPFROG 1	1	8	0	0.135
	LOOPFROG 2	1	8	0	0.234
	CTA	1	8	0	151.846
SDV HARNESS PNP DEFERRED IO REQUESTS 177 loops in 31 benchmarks	LOOPFROG 1	22	98	57	47.934
	LOOPFROG 2	66	54	57	617.41
	CTA	80	94	3	44645
SDV HARNESS PNP IO REQUESTS 173 loops in 31 benchmarks	LOOPFROG 1	25	94	54	46.568
	LOOPFROG 2	68	51	54	568.705
	CTA	85	86	2	15673.9
SDV PNP HARNESS SMALL 618 loops in 44 benchmarks	LOOPFROG 1	172	417	29	8209.51
	LOOPFROG 2	261	231	126	12373.2
	CTA	200	177	241	26613.7
SDV PNP HARNESS 635 loops in 45 benchmarks	LOOPFROG 1	173	426	36	7402.23
	LOOPFROG 2	261	230	144	13500.2
	CTA	201	186	248	41566.6
SDV PNP HARNESS UNLOAD 506 loops in 41 benchmarks	LOOPFROG 1	128	355	23	8082.51
	LOOPFROG 2	189	188	129	13584.6
	CTA	137	130	239	20967.8
SDV WDF FLAT SIMPLE HARNESS 172 loops in 18 benchmarks	LOOPFROG 1	27	125	20	30.281
	LOOPFROG 2	61	91	20	201.96
	CTA	73	95	4	70663

- CTA spends time 1) unwinding loop iterations, 2) discovering a ranking function for each unwound program fragment and 3) checking compositionality of a discovered relation.
- TERMINATOR spends time 1) enumerating paths through the loop and 2) discovering a ranking function for each path.

The techniques can greatly vary in the time required for a particular loop or program. CTA and TERMINATOR give up on a loop once a they hit a path on which ranking synthesis fails. LOOPFROG gives up on a loop if it runs out of transition invariant candidates to try. Given a large number of candidates, this behavior results in an advantage for TERMINATOR on loops that cannot be shown to terminate (huff and engine in Table 3). However, we observe in almost every other test that the LOOPFROG technique is generally cheaper (often orders of magnitude) in computational effort required to discover a valid termination argument.

The comparison demonstrates some weak points of iterative analysis:

- Enumeration of paths through the loop can require many iterations or even can be non-terminating for infinite state systems (as are many realistic programs).

- Ranking procedures often fail to produce a ranking argument; the same time if successful, a simpler relation could be sufficient as well.
- CTA suffers from the fact that the search for a compositional transition invariant sometimes results in exponential growth of the loop unrolling depth.

LOOPFROG does not suffer from at least the first of these problems: the analysis of each loop requires a finite number of calls to a decision procedure. The second issue is leveraged by relative simplicity of adding new abstract domains over implementing complex ranking function methods. The third issue is transformed into the generation of suitable candidate invariants, which, in general, may result in a large number candidates, which slow the procedure down. However, as we can control the ordering of the candidates by prioritizing some domains over the others, simple ranking arguments can be expected to be discovered early.

The complete results of the experiments as well as the LOOPFROG tool are available at `www.verify.inf.usi.ch/loopfrog/termination`

6   Related Work

Although the field of program termination analysis is mature (the first results date back to Turing [15]), recent years have seen a tremendous increase in practical applications of termination proving. Two directions of research contributed to the efficacy of termination provers in practice:

- the size-change termination principle (SCT) presented by Lee, Jones and Ben-Amram [16], and
- transition invariants by Podelski and Rybalchenko [1],

where the former has its roots in previous research on termination of declarative programs. Until very recently, these two lines of research did not intersect much. The first systematic attempt to understand their connections is a recent publication by Heizmann et al. [17].

6.1   Relation to Size-Change Termination Principle

Termination analysis based on the SCT principle usually involves two steps:

1. construction of an abstract model of the original program in the form of *size-change graphs* (SC-graphs) and
2. analysis of the SC-graphs for termination.

SC-graphs contain abstract program values as nodes and use two types of edges, along which values of variables *must decrease*, or *decrease or stay the same*. No edge between nodes means that none of the relations can be ensured. Graphs G which are closed under composition with itself are called *idempotent*, i.e., $G; G = G$.[4]

---

[4] In this discussion we omit introducing the notation necessary for a formal description of SCT; see Lee et al. [16,17] for more detail.

Lee et al. [16] identify two termination criteria based on a size-change graph:

1. The SC-graph is well-founded, or
2. the idempotent components of an SC-graph are well-founded.

An SC-graph can be related to transition invariants as follows. Each sub-graph corresponds to a conjunction of relations, which constitutes a transition invariant. The whole graph forms a disjunction, resulting in a termination criterion very similar to that presented as Theorem 1: if an SC-graph is well-founded then there exists a d.wf. transition invariant. Indeed, Heizmann et al. identify the SCT termination criterion as strictly stronger than the argument via transition invariants [17]. An intuitive argument is that SC-graphs abstract from the reachability of states in a program, i.e., arguments based on SC-graphs require termination of all paths irrespectively of whether those paths are reachable or not. Transition invariants, on the other hand, require the computation of the reachable states of the program. In this respect, our light-weight analysis is closely related to SCT, as it havocs the input to individual loop iterations before checking a candidate transition invariant.

The domains of SC-graphs correspond to abstract domains in our approach. The initial inspiration for the domains we experimented with comes from a recent survey on ranking functions for SCT [18]. The domains #1–4 in Table 1 encode those graphs with only down-arcs. Domain #5 has down-arcs and edges that preserve the value. However, note that, in order to avoid costly well-foundedness checks, we omit domains that have mixed edge types.

Program abstraction using our loop summarization algorithm can be seen as construction of size-change graphs. The domains suggested in Sec. 4 result in SC-graphs that are idempotent and well-founded by construction.

Another similarity to SCT relates to the second SCT criterion based on idempotent SC-components. In [17], the relation of idempotency to analyses using transition invariants was stated as an open question. We remark that there is a close relation between the idempotent SC-components and compositional transition invariants (Definition 5) used here and in compositional termination analysis [6]. The d.wf. transition invariant constructed from idempotent graphs is also a compositional transition invariant.

6.2   Relation to Other Research Using Transition Invariants

The work in this paper is a continuation of the research on proving termination using transition invariants initiated by Podelski and Rybalchenko [1]. Methods developed on the basis of transition invariants rely on an iterative, abstraction refinement-like construction of d.wf. transition invariants [2,3,5,6]. Our approach differs in that it aims to construct a d.wf. transition invariant without refinement. Instead of applying ranking function discovery for every non-ranked path, we use abstract domains that express ranking arguments for all paths at the same time.

Chawdhary et al. [9] propose a termination analysis using a combination of fixpoint-based abstract interpretation and an abstract domain of disjunctively well-founded relations. The abstract domain they suggest is of the same form as domain #5 in Table 1. However, their method performs an iterative computation of the set of abstract values and has a fixpoint detection of the form

$T \subseteq R^+$, while in our approach it is sufficient to check $T \subseteq R$, combined with the compositionality criterion. This allows a richer set of abstract domains to be applied for summarization, as the resulting satisfiability problems are low-cost.

Dams et al. [19] present a set of heuristics for inferring candidate ranking relations from a program. These heuristics can be seen as abstract domains in our framework. Moreover, we also show how candidate relations can be checked effectively using SAT/SMT.

Cook et al. [20] use relational predicates to extend the framework of Reps et al. [21] to support termination properties during computation of inter-procedural program summaries. Our approach shares a similar motivation and adds termination support to loop summarization based on abstract domains. However, we concentrate on scalable non-iterating methods to construct the summary while Cook et al. [20] rely on a refinement-based approach. The same argument applies in the case of Balaban et al.'s framework [22] for procedure summarization with support for liveness properties.

Berdine et al. [10] use the Octagon and Polyhedra abstract domains to discover invariance constraints sufficient to ensure termination. Well-foundedness checks, which we identify as an expensive part of the analysis, are left to iterative verification by an external procedure as in the TERMINATOR algorithm [3] and CTA [6]. In contrast to these methods, our approach relies on abstract domains that yield well-founded relations by construction and therefore do not require explicit checks.

7   Conclusion and Future Work

In this paper, we present an extension to a loop summarization algorithm such that it correctly handles termination properties while constructing a loop-free program over-approximation. To that end, we employ abstract domains that encode transition invariants, i.e., relations over pre- and post-states of the summarized loop. Termination of loops may be established at the same time, by checking disjunctive well-foundedness of the discovered transition invariants. We demonstrate the practicality of our approach on a large set of benchmarks including open-source programs and Windows device drivers.

Further research includes an investigation of abstract domains that allow effective summarization with termination support. We are especially interested in encoding forms of Size-Change-Graphs into schemata for generating candidate invariants.

References

1. Podelski, A., Rybalchenko, A.: Transition invariants. In: LICS, pp. 32–41. IEEE Computer Society, Los Alamitos (2004)
2. Cook, B., Podelski, A., Rybalchenko, A.: Abstraction refinement for termination. In: Hankin, C., Siveroni, I. (eds.) SAS 2005. LNCS, vol. 3672, pp. 87–101. Springer, Heidelberg (2005)
3. Cook, B., Podelski, A., Rybalchenko, A.: Termination proofs for systems code. In: PLDI, pp. 415–426. ACM, New York (2006)

4. Podelski, A., Rybalchenko, A.: ARMC: The logical choice for software model checking with abstraction refinement. In: Hanus, M. (ed.) PADL 2007. LNCS, vol. 4354, pp. 245–259. Springer, Heidelberg (2006)
5. Cook, B., Kroening, D., Ruemmer, P., Wintersteiger, C.: Ranking function synthesis for bit-vector relations. In: Esparza, J., Majumdar, R. (eds.) TACAS 2010. LNCS, vol. 6015, pp. 236–250. Springer, Heidelberg (2010)
6. Kroening, D., Sharygina, N., Tsitovich, A., Wintersteiger, C.M.: Termination analysis with compositional transition invariants. In: Touili, T., Cook, B., Jackson, P. (eds.) CAV 2010. LNCS, vol. 6174, pp. 89–103. Springer, Heidelberg (2010)
7. Kroening, D., Sharygina, N., Tonetta, S., Tsitovich, A., Wintersteiger, C.M.: Loop summarization using abstract transformers. In: Cha, S(S.), Choi, J.-Y., Kim, M., Lee, I., Viswanathan, M. (eds.) ATVA 2008. LNCS, vol. 5311, pp. 111–125. Springer, Heidelberg (2008)
8. Cousot, P., Cousot, R.: Abstract interpretation: A unified lattice model for static analysis of programs by construction or approximation of fixpoints. In: POPL, pp. 238–252 (1977)
9. Chawdhary, A., Cook, B., Gulwani, S., Sagiv, M., Yang, H.: Ranking abstractions. In: Gairing, M. (ed.) ESOP 2008. LNCS, vol. 4960, pp. 148–162. Springer, Heidelberg (2008)
10. Berdine, J., Chawdhary, A., Cook, B., Distefano, D., O'Hearn, P.: Variance analyses from invariance analyses. SIGPLAN Not. 42(1), 211–224 (2007)
11. Kroening, D., Sharygina, N., Tonetta, S., Tsitovich, A., Wintersteiger, C.M.: Loopfrog: A static analyzer for ANSI-C programs. In: The 24th IEEE/ACM International Conference on Automated Software Engineering, pp. 668–670. IEEE Computer Society, Los Alamitos (2009)
12. Clarke, E., Kröning, D., Sharygina, N., Yorav, K.: SATABS: SAT-based predicate abstraction for ANSI-C. In: Halbwachs, N., Zuck, L.D. (eds.) TACAS 2005. LNCS, vol. 3440, pp. 570–574. Springer, Heidelberg (2005)
13. Scott, J., Lee, L.H., Arends, J., Moyer, B.: Designing the low-power M*CORE architecture. In: Proc. IEEE Power Driven Microarchitecture Workshop (1998)
14. Ku, K., Hart, T.E., Chechik, M., Lie, D.: A buffer overflow benchmark for software model checkers. In: ASE 2007, pp. 389–392. ACM Press, New York (2007)
15. Turing, A.: Checking a large routine. In: Report of a Conference on High Speed Automatic Calculating Machines, pp. 67–69. Univ. Math. Lab., Cambridge (1949)
16. Lee, C.S., Jones, N.D., Ben-Amram, A.M.: The size-change principle for program termination. In: POPL, pp. 81–92. ACM, New York (2001)
17. Heizmann, M., Jones, N., Podelski, A.: Size-change termination and transition invariants. In: Cousot, R., Martel, M. (eds.) SAS 2010. LNCS, vol. 6337, pp. 22–50. Springer, Heidelberg (2010)
18. Ben-Amram, A.M., Lee, C.S.: Ranking functions for size-change termination II. Logical Methods in Computer Science 5(2) (2009)
19. Dams, D., Gerth, R., Grumberg, O.: A heuristic for the automatic generation of ranking functions. In: Workshop on Advances in Verification, pp. 1–8 (2000)
20. Cook, B., Podelski, A., Rybalchenko, A.: Summarization for termination: no return! Formal Methods in System Design 35(3), 369–387 (2009)
21. Reps, T., Horwitz, S., Sagiv, M.: Precise interprocedural dataflow analysis via graph reachability. In: Symposium on Principles of Programming Languages (POPL), pp. 49–61. ACM, New York (1995)
22. Balaban, I., Cohen, A., Pnueli, A.: Ranking abstraction of recursive programs. In: Emerson, E.A., Namjoshi, K.S. (eds.) VMCAI 2006. LNCS, vol. 3855, pp. 267–281. Springer, Heidelberg (2005)

Off-Line Test Selection with Test Purposes for Non-deterministic Timed Automata*

Nathalie Bertrand[1], Thierry Jéron[1], Amélie Stainer[1], and Moez Krichen[2]

[1] INRIA Rennes - Bretagne Atlantique, Rennes, France
[2] Institute of Computer Science and Multimedia, Sfax, Tunisia

Abstract. This paper proposes novel off-line test generation techniques for non-deterministic timed automata with inputs and outputs (TAIOs) in the formal framework of the **tioco** conformance theory. In this context, a first problem is the determinization of TAIOs, which is necessary to foresee next enabled actions, but is in general impossible. This problem is solved here thanks to an approximate determinization using a game approach, which preserves **tioco** and guarantees the soundness of generated test cases. A second problem is test selection for which a precise description of timed behaviors to be tested is carried out by expressive test purposes modeled by a generalization of TAIOs. Finally, using a symbolic co-reachability analysis guided by the test purpose, test cases are generated in the form of TAIOs equipped with verdicts.

Keywords: Conformance testing, timed automata, partial observability, urgency, approximate determinization, game, test purpose.

1 Introduction

Conformance testing is the process of testing whether an implementation behaves correctly with respect to a specification. Implementations are considered as *black boxes*, *i.e.* the source code is unknown, only their interface with the environment is known and used to interact with the tester. In *formal model-based conformance testing* models are used to describe testing artifacts (specifications, implementations, test cases, ...), conformance is formally defined and test cases with verdicts are generated automatically. Then, the quality of testing may be characterized by properties of test cases which relate the verdicts of their executions with conformance (*e.g.* soundness). For timed models, model-based conformance testing has already been explored in the last decade, with different models and conformance relations (see *e.g.* [16] for a survey), and test generation algorithms (e.g. [6,14,15]). In this context, a very popular model is *timed automata with inputs and outputs* (TAIOs), a variant of *timed automata* (TAs) [1], in which observable actions are partitioned into inputs and outputs. We consider here partially observable and non-deterministic TAIOs with invariants for the modeling of urgency.

---

* This work was partly funded by the french ANR project Testec. An extended version of the paper with full proofs is available as a technical report[4].

P.A. Abdulla and K.R.M. Leino (Eds.): TACAS 2011, LNCS 6605, pp. 96–111, 2011.

One of the main difficulties encountered in test generation for TAIOs is determinization, which is impossible in general, as for TAs [1], but is required in order to foresee the next enabled actions during execution and to emit a correct verdict. Two different approaches have been taken for test generation from timed models, which induce different treatments of non-determinism. In *off-line test generation* test cases are first generated as TAs (or timed sequences, trees, or timed transition systems) and subsequently executed on the implementation. Test cases can then be stored and further used e.g. for regression testing and documentation. However, due to the non-determinizability of TAIOs, the approach has often been limited to deterministic or determinizable TAIOs (see e.g. [12,15]), except in [14] where the problem is solved by the use of an over-approximate determinization with fixed resources, or [8] where winning strategies of timed games are used as test cases. In *on-line test generation*, test cases are generated during their execution, thus can be applied to any TAIO as only possible observable actions are computed along the current finite execution, thus avoiding a complete determinization. This is of particular interest to rapidly discover errors, but may sometimes be impracticable due to a lack of reactivity (the time needed to compute successor states on-line may sometimes be incompatible with delays).

In this paper, we propose to generate test cases off-line for non-deterministic TAIOs, in the formal context of the **tioco** conformance theory. The determinization problem is tackled thanks to an approximate determinization with fixed resources in the spirit of [14], using a game approach [5]. Determinization is exact for known classes of determinizable TAIOs (e.g. event-clock TAs, TAs with integer resets, strongly non-Zeno TAs) if resources are sufficient. In the general case, approximate determinization guarantees soundness of generated test cases by producing a deterministic *io-abstraction* of the TAIO for a particular *io-refinement* relation, generalizing the io-refinement of [7]. Our method is more precise than [14] (see [5] for details) and preserves the richness of our model by dealing with partial observability and urgency. Behaviors of specifications to be tested are identified by means of test purposes defined as *open timed automata with inputs and outputs* (OTAIOs), a model generalizing TAIOs, allowing to precisely describe behaviors according to actions and clocks of the specification as well as proper clocks. Then, in the same spirit as for the TGV tool in the untimed case [11], test selection is performed by a co-reachability analysis, producing a test case in the form of a TAIO. To our knowledge, this work constitutes the most general and advanced off-line test selection approach for TAIOs.

The paper is structured as follows. In the next section we introduce the model of OTAIOs, its semantics, some notations and operations. Section 3 recalls the **tioco** conformance theory including expected properties relating conformance and verdicts, and an io-refinement relation preserving **tioco**. Section 4 presents our game approach for the approximate determinization compatible with the io-refinement. In Section 5 we detail the test selection mechanism using test purposes and prove some properties on generated test cases.

2   A Model of Open Timed Automata with Inputs/Outputs

Timed automata (TAs) [1] is a usual model for time constrained systems. In the context of model-based testing, TAs have been extended to timed automata with inputs and outputs (TAIOs) whose sets of actions are partitioned into inputs, outputs and unobservable actions. In this section, we further extend TAIOs into the model of *open timed automata with inputs/outputs* (OTAIOs for short), by partitioning the set of clocks into proper and observed clocks. While the submodel of TAIOs (with only proper clocks) is sufficient for most testing artifacts, observed clocks of OTAIOs will be useful to express test purposes whose aim is to focus on the timed behavior of the specification. Like in [1] for TAs, we consider OTAIOs and TAIOs with location invariants to model urgency.

2.1   Open Timed Automata with Inputs/Outputs

We start by introducing notations and definitions concerning TAIOs and OTAIOs.

Given X a finite set of *clocks*, and $\mathbb{R}_{\geq 0}$ the set of non-negative real numbers, a *clock valuation* is a mapping $v : X \to \mathbb{R}_{\geq 0}$. If v is a valuation over X and $t \in \mathbb{R}$, then $v + t$ denotes the valuation which assigns to every clock $x \in X$ the value $v(x) + t$. For $X' \subseteq X$ we write $v_{[X' \leftarrow 0]}$ for the valuation equal to v on $X \setminus X'$ and assigning 0 to all clocks of X'.

Given M a non-negative integer, an *M-bounded guard* (or simply guard) over X is a finite conjunction of constraints of the form $x \sim c$ where $x \in X$, $c \in [0, M] \cap \mathbb{N}$ and $\sim \in \{<, \leq, =, \geq, >\}$. Given g a guard and v a valuation, we write $v \models g$ if v satisfies g. We abuse notations and write g for the set of valuations satisfying g. *Invariants* are restricted cases of guards: given $M \in \mathbb{N}$, an *M-bounded invariant* over X is a finite conjunction of constraints of the form $x \triangleleft c$ where $x \in X, c \in [0, M] \cap \mathbb{N}$ and $\triangleleft \in \{<, \leq\}$. We denote by $G_M(X)$ (resp. $I_M(X)$) the set of M-bounded guards (resp. invariants) over X.

Definition 1 (OTAIO). *An* open timed automaton with inputs and outputs *(OTAIO) is a tuple $\mathcal{A} = (L^{\mathcal{A}}, \ell_0^{\mathcal{A}}, \Sigma_?^{\mathcal{A}}, \Sigma_!^{\mathcal{A}}, \Sigma_\tau^{\mathcal{A}}, X_p^{\mathcal{A}}, X_o^{\mathcal{A}}, M^{\mathcal{A}}, I^{\mathcal{A}}, E^{\mathcal{A}})$ such that:*

- $L^{\mathcal{A}}$ *is a finite set of* locations, *with $\ell_0^{\mathcal{A}} \in L^{\mathcal{A}}$ the* initial location,
- $\Sigma_?^{\mathcal{A}}$, $\Sigma_!^{\mathcal{A}}$ *and $\Sigma_\tau^{\mathcal{A}}$ are disjoint finite alphabets of* input actions *(noted $a?, b?, \ldots$),* output actions *(noted $a!, b!, \ldots$), and* internal actions *(noted $\tau_1, \tau_2, \ldots$). We note $\Sigma_{obs}^{\mathcal{A}} = \Sigma_?^{\mathcal{A}} \sqcup \Sigma_!^{\mathcal{A}}$ (where $\sqcup$ denotes the disjoint union) for the alphabet of observable actions, and $\Sigma^{\mathcal{A}} = \Sigma_?^{\mathcal{A}} \sqcup \Sigma_!^{\mathcal{A}} \sqcup \Sigma_\tau^{\mathcal{A}}$ for the whole set of actions.*
- $X_p^{\mathcal{A}}$ *and $X_o^{\mathcal{A}}$ are disjoint finite sets of* proper clocks *and* observed clocks, *respectively. We note $X^{\mathcal{A}} = X_p^{\mathcal{A}} \sqcup X_o^{\mathcal{A}}$ for the whole set of clocks.*
- $M^{\mathcal{A}} \in \mathbb{N}$ *is the* maximal constant *of $\mathcal{A}$, and we will refer to $(|X^{\mathcal{A}}|, M^{\mathcal{A}})$ as the* resources *of $\mathcal{A}$,*

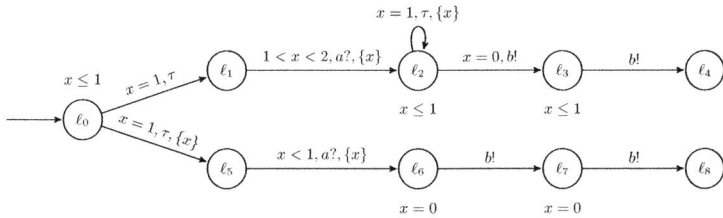

Fig. 1. Specification $\mathcal{A}$

- $I^{\mathcal{A}} : L^{\mathcal{A}} \to I_{M^{\mathcal{A}}}(X^{\mathcal{A}})$ *is a mapping labeling each location with an* invariant,
- $E^{\mathcal{A}} \subseteq L^{\mathcal{A}} \times G_{M^{\mathcal{A}}}(X^{\mathcal{A}}) \times \Sigma^{\mathcal{A}} \times 2^{X_p^{\mathcal{A}}} \times L^{\mathcal{A}}$ *is a finite set of* edges *where* guards *are defined on* $X^{\mathcal{A}}$, *but* resets *are restricted to proper clocks in* $X_p^{\mathcal{A}}$.

The reason for introducing the OTAIO model is to have a unique model (syntax and semantics) that will be next specialized for particular testing artifacts. In particular, an OTAIO with an empty set of observed clocks $X_o^{\mathcal{A}}$ is a classical TAIO, and will be the model for specifications, implementations and test cases. For example, Fig. 1 represents such a TAIO for a specification $\mathcal{A}$ with clock x, input a, output b and internal action τ. The partition of actions reflects their roles in the testing context: the environment cannot observe internal actions, but controls inputs and observes outputs (and delays). The set of clocks is also partitioned into *proper clocks, i.e.* usual clocks controlled by $\mathcal{A}$, and *observed clocks* referring to proper clocks of another OTAIO. These cannot be reset to avoid intrusiveness, but synchronization with them in guards and invariants is allowed. In particular, test purposes have observed clocks which observe proper clocks of specifications in order to describe time constrained behaviors to be tested.

2.2   The Semantics of OTAIOs

The semantics of an OTAIO $\mathcal{A} = (L^{\mathcal{A}}, \ell_0^{\mathcal{A}}, \Sigma_?^{\mathcal{A}}, \Sigma_!^{\mathcal{A}}, \Sigma_{\tau}^{\mathcal{A}}, X_p^{\mathcal{A}}, X_o^{\mathcal{A}}, M^{\mathcal{A}}, I^{\mathcal{A}}, E^{\mathcal{A}})$ is a timed transition system $\mathcal{T}^{\mathcal{A}} = (S^{\mathcal{A}}, s_0^{\mathcal{A}}, \Gamma^{\mathcal{A}}, \to_{\mathcal{A}})$ where $S^{\mathcal{A}} = L^{\mathcal{A}} \times \mathbb{R}_{\geq 0}^{X^{\mathcal{A}}}$ is the set of *states i.e.* pairs (ℓ, v) consisting in a location and a valuation of clocks; $s_0^{\mathcal{A}} = (\ell_0^{\mathcal{A}}, \bar{0}) \in S^{\mathcal{A}}$ is the *initial state*; $\Gamma^{\mathcal{A}} = \mathbb{R}_{\geq 0} \sqcup E^{\mathcal{A}} \times 2^{X_o^{\mathcal{A}}}$ is the set of transition *labels* consisting in either a delay δ or a pair (e, X_o') formed by an edge and a set of observed clocks; the transition relation $\to_{\mathcal{A}} \subseteq S^{\mathcal{A}} \times \Gamma^{\mathcal{A}} \times S^{\mathcal{A}}$ is the smallest set of the following moves:

- *Discrete moves:* $(\ell, v) \xrightarrow{(e, X_o')}_{\mathcal{A}} (\ell', v')$ whenever there exists $e = (\ell, g, a, X_p', \ell') \in E^{\mathcal{A}}$ such that $v \models g \wedge I^{\mathcal{A}}(\ell)$, $X_o' \subseteq X_o^{\mathcal{A}}$ is an arbitrary subset of observed clocks, $v' = v_{[X_p' \sqcup X_o' \leftarrow 0]}$ and $v' \models I^{\mathcal{A}}(\ell')$. Note that X_o' is unconstrained as observed clocks are controlled by another OTAIO.
- *Time elapse:* $(\ell, v) \xrightarrow{\delta}_{\mathcal{A}} (\ell, v + \delta)$ for $\delta \in \mathbb{R}_{\geq 0}$ if $v + \delta \models I^{\mathcal{A}}(\ell)$.

A *partial run* of $\mathcal{A}$ is a finite sequence of subsequent moves in $(S^{\mathcal{A}} \times \Gamma^{\mathcal{A}})^* . S^{\mathcal{A}}$.

For example $\rho = s_0 \xrightarrow{\delta_1}_\mathcal{A} s_0' \xrightarrow{(e_1, X_o^1)}_\mathcal{A} s_1 \cdots s_{k-1} \xrightarrow{\delta_k}_\mathcal{A} s_{k-1}' \xrightarrow{(e_k, X_o^k)}_\mathcal{A} s_k$. The sum of delays in ρ is noted $time(\rho)$. A *run* is a partial run starting in $s_0^\mathcal{A}$. Run($\mathcal{A}$) and pRun($\mathcal{A}$) denote respectively runs and partial runs of $\mathcal{A}$. A state s is *reachable* if there exists a run leading to s. A state s is *co-reachable* from a set $S' \subseteq S^\mathcal{A}$ if there is a partial run from s to a state in S'. We note reach($\mathcal{A}$) the set of reachable states and coreach($\mathcal{A}, S'$) the set of states co-reachable from S'.

A (partial) *sequence* is a projection of a (partial) run where states are forgotten, and discrete transitions are abstracted to actions and proper resets which are grouped with observed resets. The sequence corresponding to a run $\rho = s_0 \xrightarrow{\delta_1}_\mathcal{A}$ $s_0' \xrightarrow{(e_1, X_o^1)}_\mathcal{A} s_1 \cdots s_{k-1} \xrightarrow{\delta_k}_\mathcal{A} s_{k-1}' \xrightarrow{(e_k, X_o^k)}_\mathcal{A} s_k$ is $\mu = \delta_1.(a_1, X_p^1 \sqcup X_o^1) \cdots \delta_k.(a_k, X_p^k \sqcup X_o^k)$ where $\forall i \in [1, k], e_i = (\ell_i, g_i, a_i, X_p^i, \ell_i')$. We then note $s_0^\mathcal{A} \xrightarrow{\mu}_\mathcal{A} s_k$. We write $s_0^\mathcal{A} \xrightarrow{\mu}_\mathcal{A}$ for $\exists s_k, s_0^\mathcal{A} \xrightarrow{\mu}_\mathcal{A} s_k$. We note Seq($\mathcal{A}$) $\subseteq (\mathbb{R}_{\geq 0} \sqcup (\Sigma^\mathcal{A} \times 2^{X^\mathcal{A}}))^*$ (respectively pSeq($\mathcal{A}$)) the set of sequences (resp. partial sequences) of $\mathcal{A}$. For a sequence μ, $time(\mu)$ denotes the sum of delays in μ. For $\mu \in$ pSeq($\mathcal{A}$), $Trace(\mu) \in (\mathbb{R}_{\geq 0} \sqcup \Sigma_{obs}^\mathcal{A})^*.\mathbb{R}_{\geq 0}$ denotes the observable timed word obtained by erasing internal actions and summing delays between observable ones. It is defined inductively as follows: $Trace(\varepsilon) = 0$, $Trace((\tau, X).\mu) = Trace(\mu)$, $Trace(\delta_1 \ldots \delta_k) = \Sigma_{i=1}^k \delta_i$ and $Trace(\delta_1 \ldots \delta_k.(a, X').\mu) = (\Sigma_{i=1}^k \delta_i).a.Trace(\mu)$ if $a \in \Sigma_{obs}^\mathcal{A}$. For example $Trace(1.(\tau, X^1).2.(a, X^2).2.(\tau, X^3)) = (3, a).2$ and $Trace(1.(\tau, X^1).2.(a, X^2)) = (3, a).0$. For a run ρ projecting onto a sequence μ, we write $Trace(\rho)$ for $Trace(\mu)$. The set of traces of runs of $\mathcal{A}$ is denoted by Traces($\mathcal{A}$) $\subseteq (\mathbb{R}_{\geq 0} \sqcup \Sigma_{obs}^\mathcal{A})^*.\mathbb{R}_{\geq 0}$. Two OTAIOs with same sets of traces are said *equivalent*.

Let $\sigma \in (\mathbb{R}_{\geq 0} \sqcup \Sigma_{obs}^\mathcal{A})^*.\mathbb{R}_{\geq 0}$ be an observable timed word, and $s \in S^\mathcal{A}$ a state, $\mathcal{A}$ after $\sigma = \{s \in S^\mathcal{A} \mid \exists \mu \in \text{Seq}(\mathcal{A}), s_0^\mathcal{A} \xrightarrow{\mu}_\mathcal{A} s \wedge trace(\mu) = \sigma\}$ denotes the set of states where $\mathcal{A}$ can stay after observing the trace σ. We note $elapse(s) = \{t \in \mathbb{R}_{\geq 0} \mid s \xrightarrow{t}_\mathcal{A}\}$ the set of possible delays in s, and $out(s) = \{a \in \Sigma_!^\mathcal{A} \mid \exists X \subseteq X^\mathcal{A}, s \xrightarrow{(a, X)}_\mathcal{A}\} \sqcup elapse(s)$ (and $in(s) = \{a \in \Sigma_?^\mathcal{A} \mid s \xrightarrow{(a, X)}_\mathcal{A}\}$) for the set of outputs and delays (respectively inputs) that can be observed from s. For $S' \subset S^\mathcal{A}$, $out(S') = \bigcup_{s \in S'} out(s)$ and $in(S') = \bigcup_{s \in S'} in(s)$.

2.3   Properties and Operations

An TAIO $\mathcal{A}$ is *deterministic* (and called a DTAIO) whenever for any $\sigma \in$ Traces($\mathcal{A}$), $s_0^\mathcal{A}$ after σ is a singleton[1]. A TAIO $\mathcal{A}$ is *determinizable* if there exists an equivalent DTAIO. It is well-known that some TAs are not determinizable [1]; moreover, the determinizability of TAs is an undecidable problem, even with fixed resources [18,10].

An OTAIO $\mathcal{A}$ is *complete* if in any location ℓ, $I^\mathcal{A}(\ell) = $ true, for any action $a \in \Sigma^\mathcal{A}$, the disjunction of guards of transitions leaving ℓ and labeled by a is true.

---

[1] The notion of determinism is needed here and defined only for TAIOs. For OTAIOs the right definition would consider the projection of $s_0^\mathcal{A}$ after σ which forgets values of observed clocks, as these introduce "environmental" non-determinism.

This entails $\mathsf{Traces}(\mathcal{A}) = (\Sigma^{\mathcal{A}})^*$ (the universal language). $\mathcal{A}$ is *input-complete* in $s \in reach(\mathcal{A})$, if $\forall a \in \Sigma^{\mathcal{A}}_?$, $s \xrightarrow{a}$. $\mathcal{A}$ is *non-blocking* if $\forall s \in reach(\mathcal{A}), \forall t \in \mathbb{R}_{\geq 0}, \exists \mu \in \mathsf{pSeq}(\mathcal{A}) \cap (\mathbb{R}_{\geq 0} \sqcup (\Sigma^{\mathcal{A}}_! \sqcup \Sigma^{\mathcal{A}}_\tau) \times 2^{X^{\mathcal{A}}}))^*, time(\mu) = t \wedge s \xrightarrow{\mu}$.

We now define a product operation on OTAIOs which extends the classical product of TAs, with a particular attention to observed clocks:

Definition 2 (Product). *The* product *of two OTAIOs with same alphabets* $\mathcal{A}^i = (L^i, \ell^i_0, \Sigma_?, \Sigma_!, \Sigma_\tau, X^i_p, X^i_o, M^i, I^i, E^i)$, $i = 1, 2$, *and disjoint sets of proper clocks* $(X^1_p \cap X^2_p = \emptyset)$ *is the OTAIO* $\mathcal{A}^1 \times \mathcal{A}^2 = (L, \ell_0, \Sigma_?, \Sigma_!, \Sigma_\tau, X_p, X_o, M, I, E)$ *where:* $L = L^1 \times L^2$; $\ell_0 = (\ell^1_0, \ell^2_0)$; $X_p = X^1_p \sqcup X^2_p$, $X_o = (X^1_o \cup X^2_o) \setminus X_p$; $M = max(M^1, M^2)$; $\forall (\ell^1, \ell^2) \in L, I((\ell^1, \ell^2)) = I^1(\ell^1) \wedge I^2(\ell^2)$; *and* $((\ell^1, \ell^2), g^1 \wedge g^2, a, X'^1_p \sqcup X'^2_p, (\ell'^1, \ell'^2)) \in E$ *if* $(\ell^i, g^i, a, X'^i_p, \ell'^i) \in E^i$, *i=1,2.*

Intuitively, $\mathcal{A}^1$ and $\mathcal{A}^2$ synchronize on both time and common actions (including internal ones). $\mathcal{A}^2$ may observe proper clocks of $\mathcal{A}^1$ with its observed clocks $X^1_o \cap X^2_o$, and vice versa. The set of proper clocks of $\mathcal{A}^1 \times A^2$ is the union of proper clocks of $\mathcal{A}^1$ and A^2, and observed clocks are those observed clocks of one OTAIO that are not proper. For example, the OTAIO in Fig. 3 represents the product of the TAIO $\mathcal{A}$ in Fig. 1 and the OTAIO $\mathcal{TP}$ of Fig. 2.

The product is the right operation for intersecting sets of sequences. In fact, let $\mathcal{A}^1 \uparrow^{(X^2_p, X^2_o)}$ (respectively $\mathcal{A}^2 \uparrow^{(X^1_p, X^1_o)}$) denote the same TAIO $\mathcal{A}^1$ (resp. $\mathcal{A}^2$) defined on $(X^1_p, X^2_p \cup X^2_o \cup X^1_o \setminus X^1_p)$ (resp. on $(X^2_p, X^1_p \cup X^2_o \cup X^2_o \setminus X^2_p)$). Then we get: $\mathsf{Seq}(\mathcal{A}^1 \times \mathcal{A}^2) = \mathsf{Seq}(\mathcal{A}^1 \uparrow^{(X^2_p, X^2_o)}) \cap \mathsf{Seq}(\mathcal{A}^2 \uparrow^{(X^1_p, X^1_o)})$.

An OTAIO equipped with a set of states $F \subseteq S^{\mathcal{A}}$ can play the role of an acceptor. $\mathsf{Run}_F(\mathcal{A})$ denotes the set of runs *accepted* in F, those runs ending in F, $\mathsf{Seq}_F(\mathcal{A})$ denotes the set of sequences of accepted runs and $\mathsf{Traces}_F(\mathcal{A})$ the set of their traces. By abuse of notation, if L is a subset of locations $L^{\mathcal{A}}$, we write $\mathsf{Run}_L(\mathcal{A})$ for $\mathsf{Run}_{L \times \mathbb{R}^{X^{\mathcal{A}}}_{\geq 0}}(\mathcal{A})$ and similarly for $\mathsf{Seq}_L(\mathcal{A})$ and $\mathsf{Traces}_L(\mathcal{A})$. Note that for the product $\mathcal{A}^1 \times \mathcal{A}^2$, if F^1 and F^2 are subsets of states of $\mathcal{A}^1$ and $\mathcal{A}^2$ respectively, we get: $\mathsf{Seq}_{F^1 \times F^2}(\mathcal{A}^1 \times \mathcal{A}^2) = \mathsf{Seq}_{F^1}(\mathcal{A}^1 \uparrow^{X^2_p, X^2_o}) \cap \mathsf{Seq}_{F^2}(\mathcal{A}^2 \uparrow^{X^1_p, X^1_o})$.

3   Conformance Testing Theory

In this section, we recall the conformance relation **tioco** [14], that formally defines the set of correct implementations of a given TAIO specification. We then define test cases, formalize their executions, verdicts and expected properties. Finally, we introduce a refinement relation between TAIOs that preserves **tioco**.

3.1   The tioco Conformance Theory

We consider that the specification is given as a (possibly non-deterministic) TAIO $\mathcal{A} = (L^{\mathcal{A}}, \ell^{\mathcal{A}}_0, \Sigma_?, \Sigma_!, \Sigma_\tau, X^{\mathcal{A}}_p, \emptyset, M^{\mathcal{A}}, I^{\mathcal{A}}, E^{\mathcal{A}})$. The implementation is a black box, unknown except for its alphabet of observable actions, which is the same as the one of $\mathcal{A}$. As usual, in order to formally reason about conformance, we assume that the implementation can be modeled by an (unknown) TAIO $\mathcal{I} = (L^{\mathcal{I}}, \ell^{\mathcal{I}}_0, \Sigma_?, \Sigma_!, \Sigma^{\mathcal{I}}_\tau, X^{\mathcal{I}}_p, \emptyset, M^{\mathcal{I}}, I^{\mathcal{I}}, E^{\mathcal{I}})$ with same observable alphabet as $\mathcal{A}$, and require that it is input-complete and non-blocking. The set of such possible implementations of $\mathcal{A}$ is denoted by $\mathcal{I}(\mathcal{A})$. Among these, the conformance relation **tioco** [14] formally defines which ones conform to $\mathcal{A}$, naturaly extending the **ioco** relation of Tretmans [17] to timed systems:

Definition 3 (Conformance relation). *Let $\mathcal{A}$ be a TAIO and $\mathcal{I} \in \mathcal{I}(\mathcal{A})$, $\mathcal{I}$ **tioco** $\mathcal{A}$ if $\forall \sigma \in \mathsf{Traces}(\mathcal{A}), out(\mathcal{I} \ \mathtt{after} \ \sigma) \subseteq out(\mathcal{A} \ \mathtt{after} \ \sigma)$.*

Intuitively, $\mathcal{I}$ conforms to $\mathcal{A}$ ($\mathcal{I}$ **tioco** $\mathcal{A}$) if after any timed trace enabled in $\mathcal{A}$, every output or delay of $\mathcal{I}$ is specified in $\mathcal{A}$. In practice, conformance is checked by test cases run on implementations. In our setting, we define test cases as deterministic TAIOs equipped with verdicts defined by a partition of states.

Definition 4 (Test case, test suite). *Given a specification TAIO $\mathcal{A}$, a test case for $\mathcal{A}$ is a pair $(\mathcal{TC}, \textbf{Verdicts})$ consisting of a deterministic TAIO (DTAIO) $\mathcal{TC} = (L^{TC}, \ell_0^{TC}, \Sigma_?^{TC}, \Sigma_!^{TC}, \Sigma_\tau^{TC}, X_p^{TC}, \emptyset, M^{TC}, I^{TC}, E^{TC})$ together with a partition $\textbf{Verdicts}$ of the set of states $S^{TC} = \textbf{None} \sqcup \textbf{Inconc} \sqcup \textbf{Pass} \sqcup \textbf{Fail}$. States outside $\textbf{None}$ are called verdict states. We require that $\Sigma_?^{TC} = \Sigma_!^{\mathcal{A}}$ and $\Sigma_!^{TC} = \Sigma_?^{\mathcal{A}}$, $I^{TC}(\ell) = \mathtt{true}$ for all $\ell \in L^{TC}$, and $\mathcal{TC}$ is input-complete in all $\textbf{None}$ states, meaning that it is ready to receive any input from the implementation before reaching a verdict. A test suite is a set of test cases.*

The *verdict* of an execution $\sigma \in \mathsf{Traces}(\mathcal{TC})$, noted $\mathsf{Verdict}(\sigma, \mathcal{TC})$, is **Pass**, **Fail**, **Inconc** or **None** if $\mathcal{TC} \ \mathtt{after} \ \sigma$ is included in the corresponding states set. We note $\mathcal{I} \ \mathtt{fails} \ TC$ if some execution σ of $\mathcal{TC} \| \mathcal{I}$ leads $\mathcal{TC}$ to a **Fail** state, *i.e.* when $\mathsf{Traces}_{\textbf{Fail}}(\mathcal{TC}) \cap \mathsf{Traces}(\mathcal{I}) \neq \emptyset$ [2]. Notice that this is only a possibility to reach the **Fail** verdict among the infinite set of executions.

We now introduce soundness, a crucial property ensured by our test generation method and strictness that will be ensured when determinization is exact.

Definition 5 (Test case properties). *A test suite $\mathcal{TS}$ for $\mathcal{A}$ is sound if no conformant implementation is rejected by the test suite i.e. $\forall \mathcal{I} \in \mathcal{I}(\mathcal{A})$, $\forall \mathcal{TC} \in \mathcal{TS}$, $\mathcal{I} \ \mathtt{fails} \ \mathcal{TC} \Rightarrow \neg(\mathcal{I} \ \textbf{tioco} \ \mathcal{A})$. It is strict if non-conformance is detected as soon as it occurs i.e. $\forall \mathcal{I} \in \mathcal{I}(A), \forall \mathcal{TC} \in \mathcal{TS}, \neg(\mathcal{I} \| \mathcal{TC} \ \textbf{tioco} \ \mathcal{A}) \Rightarrow \mathcal{I} \ \mathtt{fails} \ \mathcal{TC}$.*

3.2 Refinement Preserving tioco

We introduce an io-refinement relation between TAIOs, a generalization to non-deterministic TAIOs of the io-refinement between DTAIOs introduced in [7], itself a generalization of alternating simulation [2]. We prove that io-abstraction (the inverse relation) preserves **tioco**: if $\mathcal{I}$ conforms to $\mathcal{A}$, it also conforms to any io-abstraction $\mathcal{B}$ of $\mathcal{A}$. This will ensure that soundness of test cases is preserved by the approximate determinization defined in Section 4.

---

[2] The execution of a test case $\mathcal{TC}$ on an implementation $\mathcal{I}$ is usually modeled by the standard parallel composition $\mathcal{TC} \| \mathcal{I}$. Due to space limitations, $\|$ is not defined here, but we use its trace properties: $\mathsf{Traces}(\mathcal{I} \| \mathcal{TC}) = \mathsf{Traces}(\mathcal{I}) \cap \mathsf{Traces}(\mathcal{TC})$.

Definition 6. *Let $\mathcal{A}$ and $\mathcal{B}$ be two TAIOs with same input and output alphabets, we say that $\mathcal{A}$ io-refines $\mathcal{B}$ (or $\mathcal{B}$ io-abstracts $\mathcal{A}$) and note $\mathcal{A} \preceq \mathcal{B}$ if*

(i) $\forall \sigma \in \mathsf{Traces}(\mathcal{B})$, $out(\mathcal{A} \; \mathtt{after} \; \sigma) \subseteq out(\mathcal{B} \; \mathtt{after} \; \sigma)$ and

(ii) $\forall \sigma \in \mathsf{Traces}(\mathcal{A})$, $in(\mathcal{B} \; \mathtt{after} \; \sigma) \subseteq in(\mathcal{A} \; \mathtt{after} \; \sigma)$.

It can be proved that $\preceq$ is a preorder relation. Moreover, as (ii) is always satisfied if $\mathcal{A}$ is input-complete, for $\mathcal{I} \in \mathcal{I}(\mathcal{A})$, $\mathcal{I}$ **tioco** $\mathcal{A}$ is equivalent to $\mathcal{I} \preceq \mathcal{A}$. By transitivity of $\preceq$, Proposition 1 states that io-refinement preserves conformance. Its Corollary 1 says that io-abstraction preserves soundness of test suites and will later justify that if a TAIO $\mathcal{B}$ io-abstracting $\mathcal{A}$ is obtained by approximate determinization, a sound test suite generated from $\mathcal{B}$ is still sound for $\mathcal{A}$.

Proposition 1. *If $\mathcal{A} \preceq \mathcal{B}$ then $\forall \mathcal{I} \in \mathcal{I}(\mathcal{A})$ $(= \mathcal{I}(\mathcal{B}))$, $\mathcal{I}$ tioco $\mathcal{A} \Rightarrow \mathcal{I}$ tioco $\mathcal{B}$.*

Corollary 1. *If $A \preceq B$ then any sound test suite for $\mathcal{B}$ is also sound for $\mathcal{A}$.*

4   Approximate Determinization Preserving tioco

We recently proposed a game approach to determinize or provide a deterministic over-approximation for TAs [5]. Determinization is exact on all known classes of determinizable TAIOs (*e.g.* event-clock TAs, TAs with integer resets, strongly non-Zeno TAs) if resources are sufficient. Provided a couple of extensions, this method can be adapted to the context of testing for building a deterministic io-abstraction of a given TAIO. Thanks to Proposition 1, the construction preserves **tioco**, and Corollary 1 guarantees the soundness of generated test cases.

The approximate determinization uses the classical region construction [1]. As for classical TAs, the regions form a partition of valuations over a given set of clocks which allows to make abstractions and decide properties like the reachability of a location. We note $\mathsf{Reg}_{(X,M)}$ the set of regions over clocks X with maximal constant M. A region r' is a *time-successor* of a region r if $\exists v \in r, \exists t \in \mathbb{R}_{\geq 0}, v + t \in r'$. Given X and Y two finite sets of clocks, a *relation* between clocks of X and Y is a finite conjunction C of atomic constraints of the form $x - y \sim c$ where $x \in X$, $y \in Y$, $\sim \in \{<, =, >\}$ and $c \in \mathbb{N}$. When $c \in [-M', M]$, for $M, M' \in \mathbb{N}$, $\mathsf{Rel}_{M,M'}(X, Y)$ we denote the set of relations between X and Y.

4.1   A Game Approach to Determinize Timed Automata

The technique presented in [5] applies first to TAs, *i.e.* the alphabet only consists of one kind of actions (output actions), and the invariants are all trivial. Given such a TA $\mathcal{A}$ over the set of clocks $X^{\mathcal{A}}$, the goal is to build a deterministic TA $\mathcal{B}$ with $\mathsf{Traces}(\mathcal{A}) = \mathsf{Traces}(\mathcal{B})$ as often as possible, or $\mathsf{Traces}(\mathcal{A}) \subseteq \mathsf{Traces}(\mathcal{B})$. In order to do so, resources of $\mathcal{B}$ (number of clocks k and maximal constant $M^{\mathcal{B}}$) are fixed, and a finite 2-player turn-based safety game $\mathcal{G}_{\mathcal{A},(k,M^{\mathcal{B}})}$ is built. The two players, Spoiler and Determinizator, alternate moves, the objective of player Determinizator being to remain in a set of safe states where intuitively, for sure no over-approximation has been performed. Every strategy for Determinizator yields a deterministic automaton $\mathcal{B}$ with $\mathsf{Traces}(\mathcal{A}) \subseteq \mathsf{Traces}(\mathcal{B})$, and every winning strategy induces a deterministic TA $\mathcal{B}$ equivalent to $\mathcal{A}$. It is well known that for this kind of games, winning strategies can be chosen positional and computed in linear time in the size of the arena.

Let us now give more details on the definition of the game. Let $X^{\mathcal{B}}$ be a set of clocks of cardinality k. The initial state of the game is a state of Spoiler consisting of the initial location of $\mathcal{A}$, the simplest relation between $X^{\mathcal{A}}$ and $X^{\mathcal{B}}$: $\forall x \in X^{\mathcal{A}}, \forall y \in X^{\mathcal{B}}, x - y = 0$, a marking $\top$ indicating that no over-approximation was done so far, together with the null region over $X^{\mathcal{B}}$. In each of his states, Spoiler challenges Determinizator by proposing a region $r \in \mathsf{Reg}_{(X^{\mathcal{B}}, M^{\mathcal{B}})}$, and an action $a \in \Sigma$. Determinizator answers by deciding the subset of clocks $Y' \subseteq X^{\mathcal{B}}$ he wishes to reset. The next state of Spoiler contains a region over $X^{\mathcal{B}}$ ($r' = r_{[Y' \leftarrow 0]}$), and a finite set of configurations: triples formed of a location of $\mathcal{A}$, a relation between clocks in $X^{\mathcal{A}}$ and clocks in $X^{\mathcal{B}}$, and a boolean marking ($\top$ or $\bot$). A state of Spoiler thus constitutes a states estimate of $\mathcal{A}$, and the role of the markings is to indicate whether over-approximations possibly happened. Bad states Determinizator wants to avoid are states where all configurations are marked $\bot$, *i.e.* configurations where an approximation possibly happened.

A strategy for Determinizator thus assigns to each state of Determinizator a set $Y' \subseteq X^{\mathcal{B}}$ of clocks to be reset. With every strategy for Determinizator Π we associate the TA $\mathcal{B} = \mathsf{Aut}(\Pi)$ obtained by merging a transition of Spoiler with the transition chosen by Determinizator just after. The following theorem links strategies of Determinizator with deterministic over-approximations of the original traces language and enlightens the interest of the game:

Theorem 1 ([5]). *Let $\mathcal{A}$ be a TA, $k, M^{\mathcal{B}} \in \mathbb{N}$. For any strategy Π of Determinizator in $\mathcal{G}_{\mathcal{A}, (k, M^{\mathcal{B}})}$, $\mathcal{B} = \mathsf{Aut}(\Pi)$ is a deterministic TA over resources $(k, M^{\mathcal{B}})$ with $\mathsf{Traces}(\mathcal{A}) \subseteq \mathsf{Traces}(\mathcal{B})$. Moreover, if Π is winning, $\mathsf{Traces}(\mathcal{A}) = \mathsf{Traces}(\mathcal{B})$.*

4.2 Extensions to TAIOs and Adaptation to tioco

In the context of model-based testing, the above-mentioned determinization technique must be adapted to TAIOs, as detailed in [5], and summarized below. First the model of TAIOs is more expressive than TAs, incorporating internal actions and invariants. Second, inputs and outputs must be treated differently in order to build from a TAIO $\mathcal{A}$ a DTAIO $\mathcal{B}$ such that $\mathcal{A} \preceq \mathcal{B}$ and then preserve **tioco**.

Internal actions: Specifications naturally include internal actions that cannot be observed during test executions, and should thus be removed during determinization. In order to do so, a closure by internal actions is performed for each state during the construction of the game. To this attempt, states of the game have to be extended since internal actions might be enabled only from some time-successor of the region associated with the state. Therefore, each configuration is associated with a proper region which is a time-successor of the initial region of the state. The closure by silent transitions is effectively computed the same way as successors in the original construction when Determinizator does not reset any clock, computations thus terminate for the same reasons. It is well known that TAs with silent transitions are strictly more expressive than standard TAs [3]. Therefore, our approximation can be coarse, but it performs as well as possible with its available clock information.

Invariants: Modeling urgency is quite important and using invariants to this aim is classical. Without the ability to express urgency, for instance, any inactive system would conform to all specifications. Ignoring all invariants in the approximation surely yields an io-abstraction: delays (considered as outputs) are over-approximated. In order to be more precise while preserving $\preceq$, with each state of the game is associated the most restrictive invariant containing invariants of all the configurations in the state. In the computation of the successors, invariants are treated as guards and their validity is verified at both extremities of the transition. A state whose invariant is strictly over-approximated is unsafe.

io-abstraction vs. *over-approximation:* Rather than over-approximating a given TAIO $\mathcal{A}$, we aim here at building a DTAIO $\mathcal{B}$ io-abstracting $\mathcal{A}$ ($\mathcal{A} \preceq \mathcal{B}$). Successors by output are over-approximated as in the original game, while successors by inputs must be under-approximated. The over-approximated closure by silent transitions is not suitable to under-approximation. Therefore, states of the game are extended to contain both over- and under-approximated closures. Thus, the unsafe successors by an input are not built.

All in all, these modifications allow to deal with the full TAIO model with invariants, silent transitions and inputs/outputs, consistently with the io-abstraction. Fig.4 represents a part of this game for the TAIO of Fig.3. The new game then enjoys the following nice property:

Proposition 2 ([5]³). *Let $\mathcal{A}$ be a TAIO, $k, M^{\mathcal{B}} \in \mathbb{N}$. For any strategy Π of Determinizator in $\mathcal{G}_{\mathcal{A},(k,M^{\mathcal{B}})}$, $\mathcal{B} = \mathsf{Aut}(\Pi)$ is a DTAIO over resources $(k, M^{\mathcal{B}})$ with $\mathcal{A} \preceq \mathcal{B}$. Moreover, if Π is winning, $\mathsf{Traces}(\mathcal{A}) = \mathsf{Traces}(\mathcal{B})$.*

In other words, the approximations produced by our method are deterministic io-abstractions of the initial specification, hence our approach preserves **tioco** (Proposition 1) and soundness of test cases (Corollary 1). In comparison, the algorithm proposed in [14] is an over-approximation, thus preserves **tioco** only if the specification is input-complete. Moreover it does not preserve urgency.

5   Off-Line Test Case Generation

In this section we first define test purposes and then give the principles for off-line test selection with test purposes and properties of generated test cases.

5.1   Test Purposes

Test purposes are practical means to select behaviors to be tested, either focusing on usual behaviors, or on suspected errors in implementations. In this work we choose the following definition, and discuss alternatives in the conclusion.

---

³ Note that the proof of this proposition in [5] considers a stronger refinement relation, thus implies the same result for the present refinement relation.

Definition 7 (Test purpose). *For a specification TAIO $\mathcal{A}$, a test purpose is a pair $(\mathcal{TP}, \mathsf{Accept})$ where $\mathcal{TP} = (L^{TP}, \ell_0^{TP}, \Sigma_?, \Sigma_!, \Sigma_\tau, X_p^{TP}, X_o^{TP}, M^{TP}, I^{TP}, E^{TP})$ is a complete OTAIO (in particular $\forall \ell \in L^{TP}, I^{TP}(\ell) = \mathbf{true}$) with $X_o^{TP} = X_p^{\mathcal{A}}$ ($\mathcal{TP}$ observes proper clocks of $\mathcal{A}$), and $\mathsf{Accept} \subseteq L^{TP}$ is a subset of trap locations.*

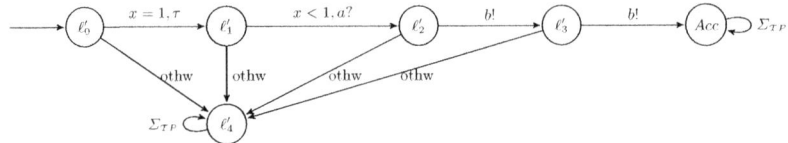

Fig. 2. Test purpose $\mathcal{TP}$

Fig. 2 represents a test purpose for the specification $\mathcal{A}$ of Fig. 1. It has no proper clock and observes the unique clock x of $\mathcal{A}$. It accepts sequences where τ occurs at $x = 1$, followed by an input a? at $x < 1$ (thus focusing on the lower branch of $\mathcal{A}$ where x is reset), and two subsequent b!'s. The label *othw* (for otherwise) is an abbreviation for the complement of specified transitions.

5.2   Principle of Test Generation

Given a specification TAIO $\mathcal{A}$ and a test purpose $(\mathcal{TP}, \mathsf{Accept}^{TP})$, the aim is to build a sound and, if possible strict test case $(\mathcal{TC}, \mathbf{Verdicts})$. It should also deliver **Pass** verdicts on traces of sequences of $\mathcal{A}$ accepted by $\mathcal{TP}$, as formalized by the following property:

Definition 8. *A test suite $\mathcal{TS}$ for $\mathcal{A}$ and $\mathcal{TP}$ is precise if $\forall \mathcal{TC} \in \mathcal{TS}, \forall \sigma \in (\Sigma_{obs}^{\mathcal{A}})^*, \mathsf{Verdict}(\sigma, \mathcal{TC}) = \mathbf{Pass} \iff \sigma \in \mathsf{Traces}(\mathsf{Seq}_{\mathsf{Accept}}^{TP}(\mathcal{TP}) \cap \mathsf{Seq}(\mathcal{A}))$.*

The different steps of test generation are described in the following paragraphs.

Product: we first build the TAIO $\mathcal{P} = \mathcal{A} \times \mathcal{TP}$ associated with the set of marked locations $\mathsf{Accept}^{\mathcal{P}} = L^{\mathcal{A}} \times \mathsf{Accept}^{TP}$. Fig. 3 represents this product $\mathcal{P}$ for the specification $\mathcal{A}$ in Fig. 1 and the test purpose $\mathcal{TP}$ in Fig. 2. The effect of the product is to unfold $\mathcal{A}$ and to mark those sequences of $\mathcal{A}$ accepted by $\mathcal{TP}$ in locations Accept^{TP}. $\mathcal{TP}$ is complete, thus $\mathsf{Seq}(\mathcal{P}) = \mathsf{Seq}(\mathcal{A} \uparrow^{X_p^{TP}, X_o^{TP}})$ (sequences of the product are sequences of $\mathcal{A}$ lifted to X^{TP}), and then $\mathsf{Traces}(\mathcal{P}) = \mathsf{Traces}(\mathcal{A})$, which implies that $\mathcal{P}$ and $\mathcal{A}$ define the same sets of conformant implementations. We also have $\mathsf{Seq}_{\mathsf{Accept}^{\mathcal{P}}}(\mathcal{P}) = \mathsf{Seq}(\mathcal{A} \uparrow^{X_p^{TP}; X_o^{TP}}) \cap \mathsf{Seq}_{\mathsf{Accept}^{TP}}(\mathcal{TP})$ which induces $\mathsf{Traces}_{\mathsf{Accept}^{\mathcal{P}}}(\mathcal{P}) = \mathsf{Traces}(\mathsf{Seq}(\mathcal{A}) \cap \mathsf{Seq}_{\mathsf{Accept}^{TP}}(\mathcal{TP}))$.

Let $\mathsf{ATraces}(\mathcal{A}, \mathcal{TP}) = \mathsf{Traces}_{\mathsf{Accept}^{\mathcal{P}}}(\mathcal{P})$ and $\mathsf{RTraces}(\mathcal{A}, \mathcal{TP}) = \mathsf{Traces}(\mathcal{A}) \setminus pref(\mathsf{ATraces}(\mathcal{A}, \mathcal{TP}))$ where, for a set of traces T, $pref(T)$ denotes the set of prefixes of traces in T. The principle is to select traces in $\mathsf{ATraces}(\mathcal{A}, \mathcal{TP})$ and try to avoid or at least detect those in $\mathsf{RTraces}(\mathcal{A}, \mathcal{TP})$ as these traces cannot be prefixes of traces of sequences satisfying the test purpose.

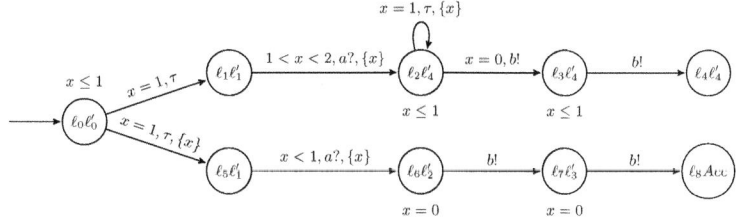

Fig. 3. Product $\mathcal{P} = \mathcal{A} \times \mathcal{TP}$

Approximate determinization of $\mathcal{P}$ into $\mathcal{DP}$: If $\mathcal{P}$ is already deterministic, we simply take $\mathcal{DP} = \mathcal{P}$. Otherwise, with the approximate determinization of Section 4, we can build a deterministic io-abstraction $\mathcal{DP}$ of $\mathcal{P}$ with resources $(k, M^{\mathcal{DP}})$ fixed by the user, thus $\mathcal{P} \preceq \mathcal{DP}$. $\mathcal{DP}$ is equipped with the set of marked locations $\mathsf{Accept}^{\mathcal{DP}}$ consisting of locations in $L^{\mathcal{DP}}$ containing some configuration whose location is in $\mathsf{Accept}^{\mathcal{P}}$. If the determinization is exact, we get $\mathsf{Traces}(\mathcal{DP}) = \mathsf{Traces}(\mathcal{P})$ and $\mathsf{Traces}_{\mathsf{Accept}^{\mathcal{DP}}}(\mathcal{DP}) = \mathsf{ATraces}(\mathcal{A}, \mathcal{TP})$. Fig. 4 partially represents the game $\mathcal{G}_{\mathcal{P},(1,2)}$ for the TAIO $\mathcal{P}$ of Fig. 3 where, for readability reasons, some behaviors not co-reachable from $\mathsf{Accept}^{\mathcal{DP}}$ are omitted. $\mathcal{DP}$ is simply obtained from $\mathcal{G}_{\mathcal{P},(1,2)}$ by merging transitions of Spoiler and Determinizator.

Generating $\mathcal{TC}$ from $\mathcal{DP}$: The next step consists in building $(\mathcal{TC}, \mathbf{Verdicts})$ from $\mathcal{DP}$, using an analysis of the co-reachability to locations $\mathsf{Accept}^{\mathcal{DP}}$ in $\mathcal{DP}$.

The test case built from $\mathcal{DP} = (L^{\mathcal{DP}}, \ell_0^{\mathcal{DP}}, \Sigma_?^{\mathcal{DP}}, \Sigma_!^{\mathcal{DP}}, X_p^{\mathcal{DP}}, \emptyset, M^{\mathcal{DP}}, I^{\mathcal{DP}}, E^{\mathcal{DP}})$ and $\mathsf{Accept}^{\mathcal{DP}}$ is the TAIO $\mathcal{TC} = (L^{\mathcal{TC}}, \ell_0^{\mathcal{TC}}, \Sigma_?^{\mathcal{TC}}, \Sigma_!^{\mathcal{TC}}, X_p^{\mathcal{TC}}, \emptyset, M^{\mathcal{TC}}, I^{\mathcal{TC}}, E^{\mathcal{TC}})$

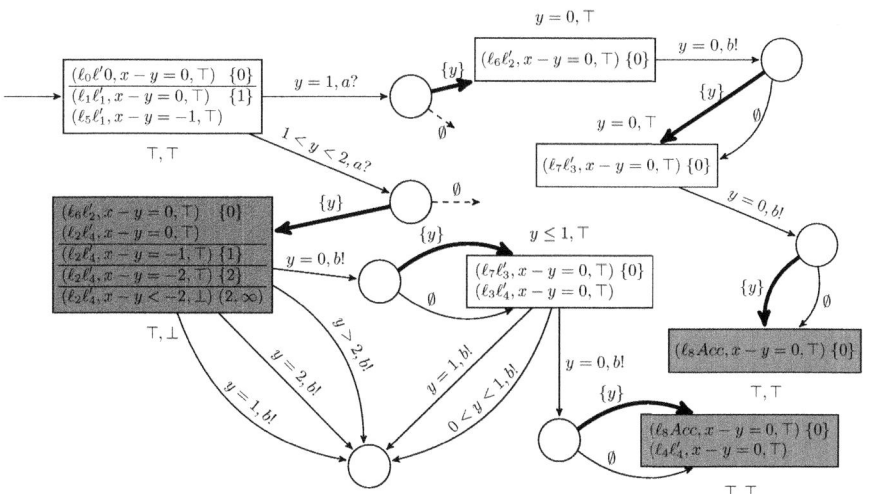

Fig. 4. Game $\mathcal{G}_{\mathcal{P},(1,2)}$

such that $L^{TC} = L^{DP} \sqcup \{\ell_{\mathbf{Fail}}\}$ where $\ell_{\mathbf{Fail}}$ is a new location; $\ell_0^{TC} = \ell_0^{DP}$; $\Sigma_?^{TC} = \Sigma_!^{DP} = \Sigma_!^{A}$ and $\Sigma_!^{TC} = \Sigma_?^{DP} = \Sigma_?^{A}$, *i.e.* input/output alphabets are mirrored in order to reflect the opposite role of actions in the synchronization of TC and I; $X_p^{TC} = X_p^{DP}$ and $X_o^{TC} = \emptyset$; $M^{TC} = M^{DP}$; **Verdicts** is the partition of S^{TC} with $\mathbf{Pass} = \bigcup_{\ell \in \mathbf{Accept}^{DP}} \{\ell\} \times I^{DP}(\ell)$, $\mathbf{None} = \mathtt{coreach}(DP, \mathbf{Pass}) \setminus \mathbf{Pass}$, $\mathbf{Inconc} = S^{DP} \setminus \mathtt{coreach}(DP, \mathbf{Pass})$, and $\mathbf{Fail} = \{\ell_{\mathbf{Fail}}\} \times \mathbb{R}_+^{X^{TC}} \sqcup \{(\ell, \neg I^{DP}(\ell)) | \ell \in L^{DP}\}$; $I^{TC}(\ell) = \mathtt{true}$ for any $\ell \in L^{TC}$; $E^{TC} = E_I^{DP} \sqcup E_{\ell_{\mathbf{Fail}}}$ where $E_I^{DP} = \{(\ell, g \wedge I^{DP}(\ell), a, X, \ell') \mid (\ell, g, a, X, \ell') \in E^{DP}\}$ and $E_{\ell_{\mathbf{Fail}}} = \{(\ell, \bar{g}, a, X_p^{TC}, \ell_{\mathbf{Fail}}) \mid \ell \in L^{DP}, a \in \Sigma_?^{DP}, \bar{g} = \neg \bigvee_{(\ell,g,a,X,\ell') \in E^{DP}} g\}$.

The important points to understand in the construction of TC are the completion to **Fail** and the computation of **Inconc**. For the completion, the idea is to detect unspecified outputs and delays of DP. Outputs of DP being inputs of TC, in any location ℓ, for each input $a \in \Sigma_?^{TC} = \Sigma_!^{DP}$, a transition leading to $\ell_{\mathbf{Fail}}$ is added, labeled with a, and whose guard is the negation of the disjunction of all guards of transitions labeled by a and leaving ℓ (thus $\mathtt{true}$ if no a-action leaves ℓ). Authorized delays in DP being defined by invariants, all states in $(\ell, \neg I^{DP}(\ell)), \ell \in L^{DP}$, *i.e.* states where the invariant runs out, are put into **Fail**. Moreover, in each location ℓ, the invariant $I^{DP}(\ell)$ in DP is removed and shifted to guards of all transitions leaving ℓ in TC.

The computation of **Inconc** is based on an analysis of the co-reachability to **Pass**. **Inconc** contains all states not co-reachable from locations in **Pass**. Notice that $\mathtt{coreach}(DP, \mathbf{Pass})$, and thus **Inconc**, can be computed symbolically in the region graph of DP. Fig.5 represents the test case obtained from A and TP.

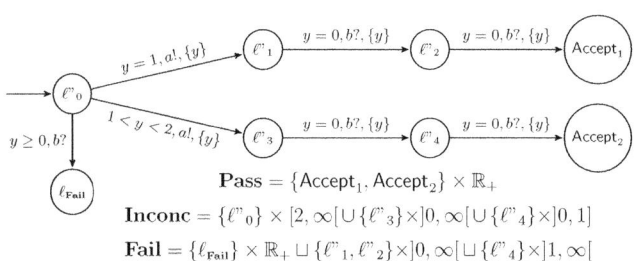

$$\mathbf{Pass} = \{\mathrm{Accept}_1, \mathrm{Accept}_2\} \times \mathbb{R}_+$$
$$\mathbf{Inconc} = \{\ell''_0\} \times [2, \infty[\sqcup \{\ell''_3\} \times]0, \infty[\sqcup \{\ell''_4\} \times]0, 1]$$
$$\mathbf{Fail} = \{\ell_{\mathbf{Fail}}\} \times \mathbb{R}_+ \sqcup \{\ell''_1, \ell''_2\} \times]0, \infty[\sqcup \{\ell''_4\} \times]1, \infty[$$

Fig. 5. Test case TC

Test selection: So far, the construction of TC determines **Verdicts**, but does not perform any selection of behaviors. A last step consists in trying to control the behavior of TC in order to avoid **Inconc** states (thus stay in $pref(\mathsf{ATraces}(A, TP)))$, or produce an **Inconc** verdict when this is impossible. To this aim, guards of transitions are refined in two complementary ways. First, transitions leaving a verdict state are useless, thus for each transition, the guard is intersected with the set of valuations associated with **None** in the source location. Second, transitions arriving in **Inconc** states and carrying inputs are also useless, thus for any transition labeled by an input, the guard is intersected with the set of valuations associated with $\mathtt{coreach}(DP, \mathbf{Pass})$ in the target location.

For example in $\mathcal{TC}$ (Fig. 5), the bottom-left state of the game in Fig. 4 has been removed.

After these steps, generated test cases exhibit the following properties:

Theorem 2. *Any test case $\mathcal{TC}$ built by the procedure is sound for $\mathcal{A}$. If $\mathcal{DP}$ is an exact approximation of $\mathcal{P}$, $\mathcal{TC}$ is also strict and precise for $\mathcal{A}$ and $\mathcal{TP}$.*

The proof is given in the technical report[4]. Soundness comes from the construction of $E_{\mathbf{Fail}}$ in $\mathcal{TC}$ and preservation of soundness by the approximate determinization $\mathcal{DP}$ of $\mathcal{P}$ given by Corollary 1. When $\mathcal{DP}$ is an exact determinization of $\mathcal{P}$, $\mathsf{Traces}(\mathcal{DP}) = \mathsf{Traces}(\mathcal{P}) = \mathsf{Traces}(\mathcal{A})$. Strictness then comes from the fact that $\mathcal{DP}$ and $\mathcal{A}$ have the same non-conformant traces and from the definition of $E_{\mathbf{Fail}}$ in $\mathcal{TC}$. Precision comes from $\mathsf{Traces}_{\mathsf{Accept}^{\mathcal{DP}}}(\mathcal{DP}) = \mathsf{ATraces}(\mathcal{A}, \mathcal{TP})$ and from the definition of **Pass**. When $\mathcal{DP}$ is not exact however, there is a risk that some behaviors allowed in $\mathcal{DP}$ are not in $\mathcal{P}$, thus some non-conformant behaviors are not detected, even if they are executed by $\mathcal{TC}$. Similarly, some **Pass** verdicts may be produced for non-accepted or non-conformant behaviors.

Test execution. After test selection, it remains to execute test cases on a real implementation. As the test case is a TAIO, a number of decisions still need to be made at each node of the test case: (1) whether to wait for a certain delay, to receive an input or emit an output (2) which output to send, in case there is a choice. Some of these choices can be made either randomly, or according to user-defined strategies, for example by applying a technique similar to the control approach of [8] whose goal is to avoid $\mathsf{RTraces}(\mathcal{A}, \mathcal{TP})$.

6   Conclusion

In this paper, we presented a complete formalization and operations for the automatic off-line generation of test cases from non-deterministic timed automata with inputs and outputs (TAIOs). The model of TAIOs is general enough to take into account non-determinism, partial observation and urgency. One main contribution is the ability to tackle any TAIO, thanks to an original approximate determinization procedure. Another main contribution is the selection of test cases with expressive OTAIOs test purposes, able to precisely select behaviors based on clocks and actions of the specification as well as proper clocks. Test cases are generated as TAIOs using a symbolic co-reachability analysis of the observable behaviors of the specification guided by the test purpose.

Related work and discussion: As mentioned in the introduction, off-line test selection is in general limited to deterministic or determinizable timed automata, except in [14] which relies on an approximate determinization. Compared to this work, our approximate determinization is more precise (it is exact in more cases) and preserves urgency in test cases as much as possible.

In several other works [13,9], test purposes are used for test case selection from TAIOs. In all these works, test purposes only have proper clocks, thus cannot observe clocks of the specification. The advantage of our definition is its generality and a fine tuning of selection. One could argue that the cost of producing a test suite can be heavy, as for each test purpose, the whole sequence of operations, including determinization, must be done. In order to avoid this, an alternative would be to define test purposes recognizing timed traces and perform selection on the approximate determinization $\mathcal{B}$ of $\mathcal{A}$. But then, the test purpose should not use $\mathcal{A}$'s clocks as these are lost by determinization. Then, test purposes are either defined after determinization and observe $\mathcal{B}$'s clocks, or their expressive power is further restricted by using only proper clocks in order not to depend on $\mathcal{B}$.

Concerning test selection, in [8], the authors propose a game approach which effect can be understood as a way to completely avoid RTraces$(\mathcal{A}, \mathcal{TP})$, with the possible risk to miss some or even all traces in $pref(\mathsf{ATraces}(\mathcal{A}, \mathcal{TP}))$. Our selection, which allows to lose the game and produce an **Inconc** verdict when this happens, is both more liberal and closer to usual practice.

It should be noticed that selection by test purposes can be used for test selection with respect to coverage criteria. Those coverage criteria define a set of elements (generally syntactic ones) to be covered (e.g. locations, transitions, branches, etc). Each element can then be translated into a test purpose, the produced test suite covering the given criteria.

Acknowledgements. We wish to thank the reviewers for their helpful comments.

References

1. Alur, R., Dill, D.L.: A theory of timed automata. Theoretical Computer Science 126(2), 183–235 (1994)
2. Alur, R., Henzinger, T.A., Kupferman, O., Vardi, M.Y.: Alternating refinement relations. In: Sangiorgi, D., de Simone, R. (eds.) CONCUR 1998. LNCS, vol. 1466, pp. 163–178. Springer, Heidelberg (1998)
3. Bérard, B., Gastin, P., Petit, A.: On the power of non-observable actions in timed automata. In: Puech, C., Reischuk, R. (eds.) STACS 1996. LNCS, vol. 1046, pp. 255–268. Springer, Heidelberg (1996)
4. Bertrand, N., Jéron, T., Stainer, A., Krichen, M.: Off-line test selection with test purposes for non-deterministic timed automata. Technical Report 7501, INRIA (January 2011), http://hal.inria.fr/inria-00550923
5. Bertrand, N., Stainer, A., Jéron, T., Krichen, M.: A game approach to determinize timed automata. In: FOSSACS 2011 (to appear, 2011); Extended version as INRIA report 7381, http://hal.inria.fr/inria-00524830
6. Briones, L.B., Brinksma, E.: A test generation framework for *quiescent* real-time systems. In: Grabowski, J., Nielsen, B. (eds.) FATES 2004. LNCS, vol. 3395, pp. 64–78. Springer, Heidelberg (2005)
7. David, A., Larsen, K.G., Legay, A., Nyman, U., Wasowski, A.: Timed I/O automata: a complete specification theory for real-time systems. In: HSCC 2010, pp. 91–100. ACM Press, New York (2010)
8. David, A., Larsen, K.G., Li, S., Nielsen, B.: Timed testing under partial observability. In: ICST 2009, pp. 61–70. IEEE Computer Society, Los Alamitos (2009)

9. En-Nouaary, A., Dssouli, R.: A guided method for testing timed input output automata. In: Hogrefe, D., Wiles, A. (eds.) TestCom 2003. LNCS, vol. 2644, pp. 211–225. Springer, Heidelberg (2003)

10. Finkel, O.: Undecidable problems about timed automata. In: Asarin, E., Bouyer, P. (eds.) FORMATS 2006. LNCS, vol. 4202, pp. 187–199. Springer, Heidelberg (2006)

11. Jard, C., Jéron, T.: TGV: theory, principles and algorithms. Software Tools for Technology Transfer 7(4), 297–315 (2005)

12. Khoumsi, A., Jéron, T., Marchand, H.: Test cases generation for nondeterministic real-time systems. In: Petrenko, A., Ulrich, A. (eds.) FATES 2003. LNCS, vol. 2931, pp. 131–145. Springer, Heidelberg (2004)

13. Koné, O., Castanet, R., Laurencot, P.: On the fly test generation for real time protocols. In: ICCCN 1998, pp. 378–387. IEEE, Los Alamitos (1998)

14. Krichen, M., Tripakis, S.: Conformance testing for real-time systems. Formal Methods in System Design 34(3), 238–304 (2009)

15. Nielsen, B., Skou, A.: Automated test generation from timed automata. Software Tools for Technology Transfer 5(1), 59–77 (2003)

16. Schmaltz, J., Tretmans, J.: On conformance testing for timed systems. In: Cassez, F., Jard, C. (eds.) FORMATS 2008. LNCS, vol. 5215, pp. 250–264. Springer, Heidelberg (2008)

17. Tretmans, J.: Test generation with inputs, outputs and repetitive quiescence. Software - Concepts and Tools 3, 103–120 (1996)

18. Tripakis, S.: Folk theorems on the determinization and minimization of timed automata. Information Processing Letters 99(6), 222–226 (2006)

Quantitative Multi-objective Verification for Probabilistic Systems

Vojtěch Forejt[1], Marta Kwiatkowska[1],
Gethin Norman[2], David Parker[1], and Hongyang Qu[1]

[1] Oxford University Computing Laboratory, Parks Road, Oxford, OX1 3QD, UK
[2] School of Computing Science, University of Glasgow, Glasgow, G12 8RZ, UK

Abstract. We present a verification framework for analysing multiple quantitative objectives of systems that exhibit both nondeterministic and stochastic behaviour. These systems are modelled as probabilistic automata, enriched with cost or reward structures that capture, for example, energy usage or performance metrics. Quantitative properties of these models are expressed in a specification language that incorporates probabilistic safety and liveness properties, expected total cost or reward, and supports multiple objectives of these types. We propose and implement an efficient verification framework for such properties and then present two distinct applications of it: firstly, *controller synthesis* subject to multiple quantitative objectives; and, secondly, quantitative *compositional verification*. The practical applicability of both approaches is illustrated with experimental results from several large case studies.

1 Introduction

Automated formal verification techniques such as model checking have proved to be an effective way of establishing rigorous guarantees about the correctness of real-life systems. In many instances, though, it is important to also take stochastic behaviour of these systems into account. This might be because of the presence of components that are prone to failure, because of unpredictable behaviour, e.g. of lossy communication media, or due to the use of randomisation, e.g. in distributed communication protocols such as Bluetooth.

Probabilistic verification offers techniques to automatically check quantitative properties of such systems. Models, typically labelled transition systems augmented with probabilistic information, are verified against properties specified in probabilistic extensions of temporal logics. Examples of such properties include "the probability of both devices failing within 24 hours is less than 0.001" or "with probability at least 0.99, all message packets are sent successfully".

In this paper, we focus on verification techniques for *probabilistic automata* (PAs) [23], which model both nondeterministic and probabilistic behaviour. We augment these models with one or more *reward* structures that assign real values to certain transitions of the model. In fact, these can associate a notion of either *cost* or *reward* with the executions of the model and capture a wide range of quantitive measures of system behaviour, for example "number of time steps", "energy usage" or "number of messages successfully sent".

P.A. Abdulla and K.R.M. Leino (Eds.): TACAS 2011, LNCS 6605, pp. 112–127, 2011.
© Springer-Verlag Berlin Heidelberg 2011

Properties of PAs can be specified using well-known temporal logics such as PCTL, LTL or PCTL* [6] and extensions for reward-based properties [1]. The corresponding verification problems can be executed reasonably efficiently and are implemented in tools such as PRISM, LiQuor and RAPTURE.

A natural extension of these techniques is to consider *multiple objectives*. For example, rather than verifying two separate properties such as "message loss occurs with probability at most 0.001" and "the expected total energy consumption is below 50 units", we might ask whether it is possible to satisfy both properties *simultaneously*, or to investigate the *trade-off* between the two objectives as some parameters of the system are varied.

In this paper, we consider verification problems for probabilistic automata on properties with *multiple, quantitative* objectives. We define a language that expresses Boolean combinations of probabilistic ω-regular properties (which subsumes e.g. LTL) and expected total reward measures. We then present, for properties expressed in this language, techniques both to *verify* that a property holds for all adversaries (strategies) of a PA and to *synthesise* an adversary of a PA under which a property holds. We also consider *numerical* queries, which yield an optimal value for one objective, subject to constraints imposed on one or more other objectives. This is done via reduction to a linear programming problem, which can be solved efficiently. It takes time polynomial in the size of the model and doubly exponential in the size of the property (for LTL objectives), i.e. the same as for the single-objective case [14].

Multi-criteria optimisation for PAs or, equivalently, Markov decision processes (MDPs) is well studied in operations research [13]. More recently, the topic has also been considered from a probabilistic verification point of view [12,16,9]. In [16], ω-regular properties are considered, but not rewards which, as illustrated by the examples above, offer an additional range of useful properties. In, [12] *discounted* reward properties are used. In practice, though, a large class of properties, such as "expected total time for algorithm completion" are not accurately captured when using discounting. Finally, [9] handles a complementary class of long-run average reward properties. All of [12,16,9] present algorithms and complexity results for verifying properties and approximating Pareto curves; however, unlike this paper, they do not consider implementations.

We implement our multi-objective verification techniques and present two distinct applications. Firstly, we illustrate the feasibility of performing *controller synthesis*. Secondly, we develop *compositional verification* methods based on assume-guarantee reasoning and quantitative multi-objective properties.

Controller synthesis. Synthesis, which aims to build correct-by-construction systems from formal specifications of their intended behaviour, represents a long-standing and challenging goal in the field of formal methods. One area where progress has been made is *controller synthesis*, a classic problem in control engineering which devises a strategy to control a system such that it meets its specification. We demonstrate the application of our techniques to synthesising controllers under multiple quantitative objectives, illustrating this with experimental results from a realistic model of a disk driver controller.

Compositional verification. Perhaps the biggest challenge to the practical applicability of formal verification is scalability. *Compositional verification* offers a powerful means to address this challenge. It works by breaking a verification problem down into manageable sub-tasks, based on the structure of the system being analysed. One particularly successful approach is the *assume-guarantee* paradigm, in which properties (*guarantees*) of individual system components are verified under *assumptions* about their environment. Desired properties of the combined system, which is typically too large to verify, are then obtained by combining separate verification results using proof rules. Compositional analysis techniques are of particular importance for probabilistic systems because verification is often more expensive than for non-probabilistic models.

Recent work in [21] presents an assume-guarantee framework for probabilistic automata, based on a reduction to the multi-objective techniques of [16]. However, the assumptions and guarantees in this framework are restricted to probabilistic safety properties. This limits the range of properties that can be verified and, more importantly, can be too restrictive to express assumptions of the environment. We use our techniques to introduce an alternative framework where assumptions and guarantees are the *quantitative multi-objective properties* defined in this paper. This adds the ability to reason compositionally about, for example, probabilistic liveness or expected rewards. To facilitate this, we also incorporate a notion of *fairness* into the framework. We have implemented the techniques and present results from compositional verification of several large case studies, including instances where it is infeasible non-compositionally.

Related work. Existing research on *multi-objective* analysis of MDPs and its relationship with this work has been discussed above. On the topic of *controller synthesis*, the problem of synthesising MDP adversaries to satisfy a temporal logic specification has been addressed several times, e.g. [3,7]. Also relevant is [11], which synthesises non-probabilistic automata based on quantitative measures. In terms of *compositional verification*, the results in this paper significantly extend the recent work in [21]. Other related approaches include: [8,15], which present specification theories for compositional reasoning about probabilistic systems; and [10], which presents a theoretical framework for compositional verification of quantitative (but not probabilistic) properties. None of [8,15,10], however, consider practical implementations of their techniques.

Contributions. In summary, the contributions of this paper are as follows:

- novel *multi-objective* verification techniques for probabilistic automata (and MDPs) that include both ω-regular and *expected total reward* properties;
- a corresponding method to generate optimal adversaries, with direct applicability to the problem of *controller synthesis* for these models;
- new *compositional verification* techniques for probabilistic automata using expressive quantitative properties for assumptions and guarantees.

An extended version of this paper, with proofs, is available as [18].

2 Background

We use $Dist(S)$ for the set of all discrete probability distributions over a set S, η_s for the point distribution on $s \in S$, and $\mu_1 \times \mu_2$ for the product distribution of $\mu_1 \in Dist(S_1)$ and $\mu_2 \in Dist(S_2)$, defined by $\mu_1 \times \mu_2((s_1, s_2)) = \mu_1(s_1) \cdot \mu_2(s_2)$.

2.1 Probabilistic Automata (PAs)

Probabilistic automata [23] are a commonly used model for systems that exhibit both probabilistic and nondeterministic behaviour. PAs are very similar to Markov decision processes (MDPs).[1] For the purposes of verification (as in Section 3), they can often be treated identically; however, for compositional analysis (as in Section 4), the distinction becomes important.

Definition 1 (Probabilistic automata). *A probabilistic automaton (PA) is a tuple $\mathcal{M} = (S, \bar{s}, \alpha_{\mathcal{M}}, \delta_{\mathcal{M}})$ where S is a set of states, $\bar{s} \in S$ is an initial state, $\alpha_{\mathcal{M}}$ is an alphabet and $\delta_{\mathcal{M}} \subseteq S \times \alpha_{\mathcal{M}} \times Dist(S)$ is a probabilistic transition relation.*

In a state s of a PA $\mathcal{M}$, a *transition* $s \xrightarrow{a} \mu$, where a is an *action* and μ is a distribution over states, is available if $(s, a, \mu) \in \delta$. The selection of an available transition is nondeterministic and the subsequent choice of successor state is probabilistic, according to the distribution of the chosen transition.

A *path* is a sequence $\omega = s_0 \xrightarrow{a_0, \mu_0} s_1 \xrightarrow{a_1, \mu_1} \cdots$ where $s_0 = \bar{s}$, $s_i \xrightarrow{a_i} \mu_i$ is an available transition and $\mu_i(s_{i+1}) > 0$ for all $i \in \mathbb{N}$. We denote by *IPaths* (*FPaths*) the set of all infinite (finite) paths. If ω is finite, $|\omega|$ denotes its length and $last(\omega)$ its last state. The trace, $tr(\omega)$, of ω is the sequence of actions $a_0 a_1 \ldots$ and we use $tr(\omega)\!\upharpoonright_\alpha$ to indicate the projection of such a trace onto an alphabet $\alpha \subseteq \alpha_{\mathcal{M}}$.

A *reward structure* for $\mathcal{M}$ is a mapping $\rho : \alpha_\rho \to \mathbb{R}_{>0}$ from some alphabet $\alpha_\rho \subseteq \alpha_{\mathcal{M}}$ to the positive reals. We sometimes write $\rho(a) = 0$ to indicate that $a \notin \alpha_\rho$. For an infinite path $\omega = s_0 \xrightarrow{a_0, \mu_0} s_1 \xrightarrow{a_1, \mu_1} \cdots$, the *total reward* for ω over ρ is $\rho(\omega) = \sum_{i \in \mathbb{N}, a_i \in \alpha_\rho} \rho(a_i)$.

An *adversary* of $\mathcal{M}$ is a function $\sigma : FPaths \to Dist(\alpha_{\mathcal{M}} \times Dist(S))$ such that, for a finite path ω, $\sigma(\omega)$ only assigns non-zero probabilities to action-distribution pairs (a, μ) for which $(last(\omega), a, \mu) \in \delta$. Employing standard techniques [20], an adversary σ induces a probability measure $Pr^\sigma_{\mathcal{M}}$ over *IPaths*. An adversary σ is *deterministic* if $\sigma(\omega)$ is a point distribution for all ω, *memoryless* if $\sigma(\omega)$ depends only on $last(\omega)$, and *finite-memory* if there are a finite number of memory configurations such that $\sigma(\omega)$ depends only on $last(\omega)$ and the current memory configuration, which is updated (possibly stochastically) when an action is performed. We let $Adv_{\mathcal{M}}$ denote the set of all adversaries for $\mathcal{M}$.

If $\mathcal{M}_i = (S_i, \bar{s}_i, \alpha_{\mathcal{M}_i}, \delta_{\mathcal{M}_i})$ for $i = 1, 2$, then their *parallel composition*, denoted $\mathcal{M}_1 \| \mathcal{M}_2$, is given by the PA $(S_1 \times S_2, (\bar{s}_1, \bar{s}_2), \alpha_{\mathcal{M}_1} \cup \alpha_{\mathcal{M}_2}, \delta_{\mathcal{M}_1 \| \mathcal{M}_2})$ where $\delta_{\mathcal{M}_1 \| \mathcal{M}_2}$ is defined such that $(s_1, s_2) \xrightarrow{a} \mu_1 \times \mu_2$ if and only if one of the following holds: (i) $s_1 \xrightarrow{a} \mu_1$, $s_2 \xrightarrow{a} \mu_2$ and $a \in \alpha_{\mathcal{M}_1} \cap \alpha_{\mathcal{M}_2}$; (ii) $s_1 \xrightarrow{a} \mu_1$, $\mu_2 = \eta_{s_2}$ and $a \in \alpha_{\mathcal{M}_1} \setminus \alpha_{\mathcal{M}_2}$; or (iii) $s_2 \xrightarrow{a} \mu_2$, $\mu_1 = \eta_{s_1}$ and $a \in \alpha_{\mathcal{M}_2} \setminus \alpha_{\mathcal{M}_1}$.

---

[1] For MDPs, $\delta_{\mathcal{M}}$ in Definition 1 becomes a partial function $S \times \alpha_{\mathcal{M}} \to Dist(S)$.

When verifying systems of PAs composed in parallel, it is often essential to consider *fairness*. In this paper, we use a simple but effective notion of fairness called *unconditional fairness*, in which it is required that each process makes a transition infinitely often. For probabilistic automata, a natural approach to incorporating fairness (as taken in, e.g., [4,2]) is to restrict analysis of the system to a class of adversaries in which fair behaviour occurs with probability 1.

If $\mathcal{M} = \mathcal{M}_1 \| \ldots \| \mathcal{M}_n$ is a PA comprising n components, then an (unconditionally) *fair path* of $\mathcal{M}$ is an infinite path $\omega \in IPaths$ in which, for each component M_i, there exists an action $a \in \alpha_{M_i}$ that appears infinitely often. A *fair adversary* σ of $\mathcal{M}$ is an adversary for which $Pr^{\sigma}_{\mathcal{M}}\{\omega \in IPaths \mid \omega \text{ is fair}\} = 1$. We let $Adv^{\text{fair}}_{\mathcal{M}}$ denote the set of fair adversaries of $\mathcal{M}$.

2.2   Verification of PAs

Throughout this section, let $\mathcal{M} = (S, \bar{s}, \alpha_{\mathcal{M}}, \delta_{\mathcal{M}})$ be a PA.

Definition 2 (Probabilistic predicates). *A probabilistic predicate* $[\phi]_{\sim p}$ *comprises an ω-regular property $\phi \subseteq (\alpha_\phi)^\omega$ over some alphabet $\alpha_\phi \subseteq \alpha_{\mathcal{M}}$, a relational operator $\sim \in \{<, \leqslant, >, \geqslant\}$ and a rational probability bound p. Satisfaction of $[\phi]_{\sim p}$ by $\mathcal{M}$, under adversary σ, denoted $\mathcal{M}, \sigma \models [\phi]_{\sim p}$, is defined as follows:*

$$\mathcal{M}, \sigma \models [\phi]_{\sim p} \Leftrightarrow Pr^{\sigma}_{\mathcal{M}}(\phi) \sim p \text{ where } Pr^{\sigma}_{\mathcal{M}}(\phi) \stackrel{\text{def}}{=} Pr^{\sigma}_{\mathcal{M}}(\{\omega \in IPaths \mid tr(\omega)\restriction_{\alpha_\phi} \in \phi\}).$$

Definition 3 (Reward predicates). *A reward predicate* $[\rho]_{\sim r}$ *comprises a reward structure $\rho : \alpha_\rho \rightarrow \mathbb{R}_{>0}$ over some alphabet $\alpha_\rho \subseteq \alpha_{\mathcal{M}}$, a relational operator $\sim \in \{<, \leqslant, >, \geqslant\}$ and a rational reward bound r. Satisfaction of $[\rho]_{\sim r}$ by $\mathcal{M}$, under adversary σ, denoted $\mathcal{M}, \sigma \models [\rho]_{\sim r}$, is defined as follows:*

$$\mathcal{M}, \sigma \models [\rho]_{\sim r} \Leftrightarrow ExpTot^{\sigma}_{\mathcal{M}}(\rho) \sim r \text{ where } ExpTot^{\sigma}_{\mathcal{M}}(\rho) \stackrel{\text{def}}{=} \int_\omega \rho(\omega) \, dPr^{\sigma}_{\mathcal{M}}.$$

Verification of PAs is based on quantifying over all adversaries. For example, we define satisfaction of probabilistic predicate $[\phi]_{\sim p}$ by $\mathcal{M}$, denoted $\mathcal{M} \models [\phi]_{\sim p}$, as:

$$\mathcal{M} \models [\phi]_{\sim p} \Leftrightarrow \forall \sigma \in Adv_{\mathcal{M}} . \mathcal{M}, \sigma \models [\phi]_{\sim p}.$$

In similar fashion, we can verify a multi-component PA $\mathcal{M}_1 \| \ldots \| \mathcal{M}_n$ *under fairness* by quantifying only over fair adversaries:

$$\mathcal{M}_1 \| \ldots \| \mathcal{M}_n \models_{\text{fair}} [\phi]_{\sim p} \Leftrightarrow \forall \sigma \in Adv^{\text{fair}}_{\mathcal{M}_1 \| \ldots \| \mathcal{M}_n} . \mathcal{M}_1 \| \ldots \| \mathcal{M}_n, \sigma \models [\phi]_{\sim p}.$$

Verifying whether $\mathcal{M}$ satisfies a probabilistic predicate $[\phi]_{\sim p}$ or reward predicate $[\rho]_{\sim r}$ can be done with, for example, the techniques in [14,1]. In the remainder of this section, we give further details of the case for ω-regular properties, since we need these later in the paper. We follow the approach of [1], which is based on the use of *deterministic Rabin automata* and *end components*.

An *end component* (EC) of $\mathcal{M}$ is a pair (S', δ') comprising a subset $S' \subseteq S$ of states and a probabilistic transition relation $\delta' \subseteq \delta$ that is strongly connected when restricted to S' and closed under probabilistic branching, i.e., $\{s \in S \mid \exists (s, a, \mu) \in$

$\delta'\} \subseteq S'$ and $\{s' \in S \mid \mu(s')>0$ for some $(s,a,\mu) \in \delta\} \subseteq S'$. An EC (S',δ') is *maximal* if there is no EC (S'',δ'') such that $\delta'\subsetneq\delta''$.

A *deterministic Rabin automaton* (DRA) is a tuple $\mathcal{A} = (Q,\overline{q},\alpha,\delta,Acc)$ of states Q, initial state $\overline{q}$, alphabet α, transition function $\delta : Q\times\alpha \rightarrow Q$, and acceptance condition $Acc = \{(L_i,K_i)\}_{i=1}^{k}$ with $L_i,K_i \subseteq Q$. Any infinite word $w \in (\alpha)^{\omega}$ has a unique corresponding run $\overline{q}\,q_1q_2\ldots$ through $\mathcal{A}$ and we say that $\mathcal{A}$ accepts w if the run contains, for some $1\leqslant i\leqslant k$, finitely many states from L_i and infinitely many from K_i. For any ω-regular property $\phi \subseteq (\alpha_\phi)^{\omega}$ we can construct a DRA, say $\mathcal{A}_\phi$, over α_ϕ that accepts precisely ϕ.

Verification of $[\phi]_{\sim p}$ on $\mathcal{M}$ is done by constructing the *product* of $\mathcal{M}$ and $\mathcal{A}_\phi$, and then identifying *accepting end components*. The *product* $\mathcal{M}\otimes\mathcal{A}_\phi$ of $\mathcal{M}$ and DRA $\mathcal{A}_\phi = (Q,\overline{q},\alpha_\mathcal{M},\delta,\{(L_i,K_i)\}_{i=1}^{k})$ is the PA $(S \times Q, (\overline{s},\overline{q}),\alpha_\mathcal{M},\delta'_\mathcal{M})$ where for all $(s,a,\mu) \in \delta_M$ there is $((s,q),a,\mu') \in \delta'_\mathcal{M}$ such that $\mu'(s',q') = \mu(s')$ for $q' = \delta(q,a)$ and all $s' \in S$. An *accepting EC* for ϕ in $\mathcal{M}\otimes\mathcal{A}_\phi$ is an EC (S',δ') for which there exists an $1\leqslant i\leqslant k$ such that the set of states S', when projected onto Q, contains some state from K_i, but no states from L_i. Verifying, for example, that $\mathcal{M} \models [\phi]_{\sim p}$, when $\sim \in \{<,\leqslant\}$, reduces to checking that $\mathcal{M}\otimes\mathcal{A}_\phi \models [\lozenge T]_{\sim p}$, where T is the union of states of accepting ECs for ϕ in $\mathcal{M}\otimes\mathcal{A}_\phi$.

Verification of such properties under fairness, e.g. checking $\mathcal{M} \models_{\text{fair}} [\phi]_{\sim p}$, can be done by further restricting the set of accepting ECs. For details, see [2], which describes verification of PAs under strong and weak fairness conditions, of which unconditional fairness is a special case.

3   Quantitative Multi-objective Verification

In this section, we define a language for expressing multiple quantitative objectives of a probabilistic automaton. We then describe, for properties expressed in this language, techniques both to *verify* that the property holds for all adversaries of a PA and to *synthesise* an adversary of a PA under which the property holds. We also consider *numerical* queries, which yield an optimal value for one objective, subject to constraints imposed on one or more other objectives.

Definition 4 (Quantitative multi-objective properties). *A quantitative multi-objective property (qmo-property) for a PA $\mathcal{M}$ is a Boolean combination of probabilistic and reward predicates, i.e. an expression produced by the grammar:*

$$\Psi ::= \texttt{true} \mid \Psi \wedge \Psi \mid \Psi \vee \Psi \mid \neg\Psi \mid [\phi]_{\sim p} \mid [\rho]_{\sim r}$$

where $[\phi]_{\sim p}$ and $[\rho]_{\sim r}$ are probabilistic and reward predicates for $\mathcal{M}$, respectively. A simple qmo-property comprises a single conjunction of predicates, i.e. is of the form $(\wedge_{i=1}^{n}[\phi_i]_{\sim_i p_i}) \wedge (\wedge_{j=1}^{m}[\rho_j]_{\sim_j r_j})$. We refer to the predicates occurring in a formula as objectives. For property Ψ_P, we use α_P to denote the set of actions used in Ψ_P, i.e. the union of α_ϕ and α_ρ over $[\phi]_{\sim p}$ and $[\rho]_{\sim r}$ occurring in Ψ_P.

A quantitative multi-objective property Ψ is evaluated over a PA $\mathcal{M}$ and an adversary σ of $\mathcal{M}$. We say that $\mathcal{M}$ satisfies Ψ under σ, denoted $\mathcal{M},\sigma\models\Psi$, if Ψ evaluates to $\texttt{true}$ when substituting each predicate x with the result of $\mathcal{M},\sigma \models x$. Verification of Ψ over a PA $\mathcal{M}$ is defined as follows.

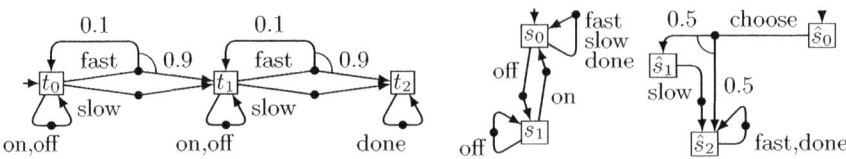

Fig. 1. PAs for a machine $\mathcal{M}_m$ (left) and two controllers, $\mathcal{M}_{c_1}$ (centre) and $\mathcal{M}_{c_2}$ (right)

Definition 5 (Verification queries). *For a PA $\mathcal{M}$ and a qmo-property Ψ, a verification query* asks whether Ψ is satisfied under all adversaries of $\mathcal{M}$:

$$\mathcal{M} \models \Psi \; \Leftrightarrow \; \forall \sigma \in Adv_{\mathcal{M}} \,.\, \mathcal{M}, \sigma \models \Psi.$$

For a *simple* qmo-property Ψ, we can verify whether $\mathcal{M} \models \Psi$ using standard techniques [14,1] (since each conjunct can be verified separately). To treat the general case, we will use multi-objective model checking, proceeding via a reduction to the dual notion of *achievability queries*.

Definition 6 (Achievability queries). *For a PA $\mathcal{M}$ and qmo-property Ψ, an achievability query* asks if there exists a satisfying adversary of $\mathcal{M}$, i.e. whether there exists $\sigma \in Adv_{\mathcal{M}}$ such that $\mathcal{M}, \sigma \models \Psi$.

Remark. Since qmo-properties are closed under negation, we can convert any *verification* query into an equivalent (negated) *achievability* query. Furthermore, any qmo-property can be translated to an equivalent disjunction of *simple* qmo-properties (obtained by converting to disjunctive normal form and pushing negation into predicates, e.g. $\neg([\phi]_{>p}) \equiv [\phi]_{\leqslant p}$).

In practice, it is also often useful to obtain the minimum/maximum *value* of an objective, subject to constraints on others. For this, we use *numerical queries*.

Definition 7 (Numerical queries). *For a PA $\mathcal{M}$, qmo-property Ψ and ω-regular property ϕ or reward structure ρ, a (maximising) numerical query* is:

$$Pr_{\mathcal{M}}^{\max}(\phi \,|\, \Psi) \stackrel{\text{def}}{=} \sup\{Pr_{\mathcal{M}}^{\sigma}(\phi) \mid \sigma \in Adv_{\mathcal{M}} \wedge \mathcal{M}, \sigma \models \Psi\},$$
$$or \; ExpTot_{\mathcal{M}}^{\max}(\rho \,|\, \Psi) \stackrel{\text{def}}{=} \sup\{ExpTot_{\mathcal{M}}^{\sigma}(\rho) \mid \sigma \in Adv_{\mathcal{M}} \wedge \mathcal{M}, \sigma \models \Psi\}.$$

If the property Ψ is not satisfied by any adversary of $\mathcal{M}$, these queries return $\bot$. A minimising numerical query is defined similarly.

Example 1. Figure 1 shows the PAs we use as a running example. A machine, $\mathcal{M}_m$, executes 2 consecutive jobs, each in 1 of 2 ways: *fast*, which requires 1 time unit and 20 units of energy, but fails with probability 0.1; or *slow*, which requires 3 time units and 10 units of energy, and never fails. The reward structures $\rho_{time} = \{fast \mapsto 1, slow \mapsto 3\}$ and $\rho_{pow} = \{fast \mapsto 20, slow \mapsto 10\}$ capture the time elapse and power consumption of the system. The controllers, $\mathcal{M}_{c_1}$ and $\mathcal{M}_{c_2}$, can (each) control the machine, when composed in parallel with $\mathcal{M}_m$. Using qmo-property $\Psi = [\lozenge done]_{\geq 1} \wedge [\rho_{pow}]_{\leq 20}$, we can write a verification query $\mathcal{M}_m \models \Psi$

(which is false) or a numerical query $ExpTot^{\min}_{\mathcal{M}_m}(\rho_{time} \,|\, \Psi)$ (which yields 6). Before describing our techniques to check verification, achievability and numerical queries, we first need to discuss some *assumptions* made about PAs. One of the main complications when introducing rewards into multi-objective queries is the possibility of *infinite* expected total rewards. For the classical, single-objective case (see e.g. [1]), it is usual to impose assumptions so that such behaviour does not occur. For the multi-objective case, the situation is more subtle, and requires careful treatment. We now outline what assumptions should be imposed; later we describe how they can be checked algorithmically.

A key observation is that, if we allow arbitrary reward structures, situations may occur where extremely improbable (but non-zero probability) behaviour still yields infinite expected reward. Consider e.g. the PA $(\{s_0, s_1, s_2\}, s_0, \{a, b\}, \delta)$ with $\delta = \{(s_0, b, \eta_{s_1}), (s_0, b, \eta_{s_2}), (s_1, a, \eta_{s_1}), (s_2, b, \eta_{s_2})\}$, reward structure $\rho = \{a \mapsto 1\}$, and the qmo-property $\Psi = [\Box b]_{\geq p} \wedge [\rho]_{\geq r}$. For any p, including values arbitrarily close to 1, there is an adversary satisfying Ψ for any $r \in \mathbb{R}_{\geq 0}$, because it suffices to take action a with non-zero probability. This rather unnatural behaviour would lead to misleading verification results, masking possible errors in the model design.

Motivated by such problems, we enforce the restriction below on multi-objective queries. To match the contents of the next section, we state this for a maximising numerical query on rewards. We describe *how* to check the restriction holds in the next section.

Assumption 1. *Let $ExpTot^{\max}_{\mathcal{M}}(\rho \,|\, \Psi)$ be a numerical query for a PA $\mathcal{M}$ and qmo-property Ψ which is a disjunction[2] of simple qmo-properties $\Psi_1, \ldots, \Psi_l$. For each $\Psi_k = (\wedge^n_{i=1} [\phi_i]_{\sim_i p_i}) \wedge (\wedge^m_{j=1} [\rho_j]_{\sim_j r_j})$, we require that:*

$$\sup\{ExpTot^{\sigma}_{\mathcal{M}}(\zeta) \mid \mathcal{M}, \sigma \models \wedge^n_{i=1} [\phi_i]_{\sim_i p_i}\} < \infty$$

for all $\zeta \in \{\rho\} \cup \{\rho_j \,|\, 1 \leqslant j \leqslant m \,\wedge\, \sim_j \in \{>, \geqslant\}\}$.

3.1 Checking Multi-objective Queries

We now describe techniques for checking the multi-objective queries described previously. For presentational purposes, we focus on *numerical* queries. It is straightforward to adapt this to *achievability* queries by introducing, and then ignoring, a dummy property to maximise (with no loss in complexity). As mentioned earlier, *verification* queries are directly reducible to achievability queries.

Let $\mathcal{M}$ be a PA and $ExpTot^{\max}_{\mathcal{M}}(\rho \,|\, \Psi)$ be a maximising numerical query for reward structure ρ (the cases for minimising queries and ω-regular properties are analogous). As discussed earlier, we can convert Ψ to a disjunction of simple qmo-properties. Clearly, we can treat each element of the disjunction separately and then take the maximum. So, without loss of generality, we assume that Ψ is simple, i.e. $\Psi = (\wedge^n_{i=1} [\phi_i]_{\sim_i p_i}) \wedge (\wedge^m_{j=1} [\rho_j]_{\sim_j r_j})$. Furthermore, we assume that each $\sim_i$ is $\geqslant$ or $>$ (which we can do by changing e.g. $[\phi]_{<p}$ to $[\neg\phi]_{>1-p}$).

---

[2] This assumption extends to arbitrary properties Ψ by, as described earlier, first reducing to disjunctive normal form.

Our technique to compute $ExpTot^{\max}_{\mathcal{M}}(\rho\,|\,\Psi)$ proceeds via a sequence of modifications to $\mathcal{M}$, producing a PA $\hat{\mathcal{M}}$. From this, we construct a linear program $L(\hat{\mathcal{M}})$, whose solution yields both the desired numerical result and a corresponding adversary $\hat{\sigma}$ of $\hat{\mathcal{M}}$. Crucially, $\hat{\sigma}$ is *memoryless* and can thus be mapped to a matching *finite-memory* adversary of $\mathcal{M}$. The structure of $L(\hat{\mathcal{M}})$ is very similar to the one used in [16], but many of the steps to construct $\hat{\mathcal{M}}$ and the techniques to establish a memoryless adversary are substantially different. We also remark that, although not discussed here, $L(\hat{\mathcal{M}})$ can be adapted to a multi-objective linear program, or used to approximate the Pareto curve between objectives.

In the remainder of this section, we describe the process in detail, which comprises 4 steps: **1.** checking Assumption 1; **2.** building a PA $\bar{\mathcal{M}}$ in which unneccessary actions are removed; **3.** converting $\bar{\mathcal{M}}$ to a PA $\hat{\mathcal{M}}$; **4.** building and solving the linear program $L(\hat{\mathcal{M}})$. The correctness of the procedure is formalised with a corresponding sequence of propositions (see [18] for proofs).

Step 1. We start by constructing a PA $\mathcal{M}^{\phi} = \mathcal{M} \otimes \mathcal{A}_{\phi_1} \otimes \cdots \otimes \mathcal{A}_{\phi_n}$ which is the product of $\mathcal{M}$ and a DRA $\mathcal{A}_{\phi_i}$ for each ω-regular property ϕ_i appearing in Ψ. We check Assumption 1 by analysing $\mathcal{M}^{\phi}$: for each *maximising* reward structure ζ (i.e. letting $\zeta{=}\rho$ or $\zeta{=}\rho_j$ when $\sim_j \in \{>, \geq\}$) we use the proposition below. This requires a simpler multi-objective achievability query on probabilistic predicates only. In fact, this can be done with the techniques of [16].

Proposition 1. *We have* $\sup\{ExpTot^{\sigma}_{\mathcal{M}}(\zeta) \mid \mathcal{M}, \sigma \models \bigwedge_{i=1}^{n} [\phi_i]_{\sim_i p_i}\} = \infty$ *for a reward structure ζ of $\mathcal{M}$ iff there is an adversary σ of $\mathcal{M}^{\phi}$ such that $\mathcal{M}^{\phi}, \sigma \models [\lozenge pos]_{>0} \wedge \bigwedge_{i=1}^{n} [\phi_i]_{\sim_i p_i}$ where "pos" labels any transition (s, a, μ) that satisfies $\zeta(a){>}0$ and is contained in an EC.*

Step 2. Next, we build the PA $\bar{\mathcal{M}}$ from $\mathcal{M}^{\phi}$ by removing actions that, thanks to Assumption 1, will not be used by any adversary which satisfies Ψ and maximises the expected value for the reward ρ. Let $\mathcal{M}^{\phi} = (S^{\phi}, \bar{s}, \alpha_{\mathcal{M}}, \delta^{\phi}_{\mathcal{M}})$. Then $\bar{\mathcal{M}} = (\bar{S}, \bar{s}, \alpha_{\mathcal{M}}, \bar{\delta}_{\mathcal{M}})$ is the PA obtained from $\mathcal{M}^{\phi}$ as follows. First, we remove (s, a, μ) from $\delta^{\phi}_{\mathcal{M}}$ if it is contained in an EC and $\zeta(a){>}0$ for some *maximising* reward structure ζ. Second, we repeatedly remove states with no outgoing transitions and transitions that lead to non-existent states, until a fixpoint is reached. The following proposition holds whenever Assumption 1 is satisfied.

Proposition 2. *There is an adversary σ of $\mathcal{M}^{\phi}$ where $ExpTot^{\sigma}_{\mathcal{M}^{\phi}}(\rho){=}x$ and $\mathcal{M}^{\phi}, \sigma \models \Psi$ iff there is an adversary $\bar{\sigma}$ of $\bar{\mathcal{M}}$ where $ExpTot^{\bar{\sigma}}_{\bar{\mathcal{M}}}(\rho){=}x$ and $\bar{\mathcal{M}}, \bar{\sigma} \models \Psi$.*

Step 3. Then, we construct PA $\hat{\mathcal{M}}$ from $\bar{\mathcal{M}}$, by converting the n probabilistic predicates $[\phi_i]_{\sim_i p_i}$ into n reward predicates $[\lambda_i]_{\sim_i p_i}$. For each $R \subseteq \{1, \ldots, n\}$, we let S_R denote the set of states that are contained in an EC (S', δ') that: (i) is accepting for all $\{\phi_i \mid i \in R\}$; (ii) satisfies $\rho_j(a) = 0$ for all $1 \leq j \leq m$ and $(s, a, \mu) \in \delta'$. Thus, in each S_R, no reward is gained and almost all paths satisfy the ω-regular properties ϕ_i for $i \in R$. Note that identifying the sets S_R can be done in time polynomial in the size of $\bar{\mathcal{M}}$ (see [18] for clarification).

We then construct $\hat{\mathcal{M}}$ by adding a new terminal state s_{dead} and adding transitions from states in each S_R to s_{dead}, labelled with a new action a^R. Intuitively,

$$Maximise \ \sum_{(s,a,\mu)\in\hat\delta_{\mathcal{M}},\, s\neq s_{dead}} \rho(a)\cdot y_{(s,a,\mu)} \ subject\ to:$$

$$\sum_{(s,a^R,\mu)\in\hat\delta_{\mathcal{M}},\, s\neq s_{dead}} y_{(s,a^R,\mu)} = 1$$

$$\sum_{(s,a,\mu)\in\hat\delta_{\mathcal{M}},\, s\neq s_{dead}} \lambda_i(a)\cdot y_{(s,a,\mu)} \sim_i p_i \qquad for\ all\ 1\le i\le n$$

$$\sum_{(s,a,\mu)\in\hat\delta_{\mathcal{M}},\, s\neq s_{dead}} \rho_j(a)\cdot y_{(s,a,\mu)} \sim_j r_j \qquad for\ all\ 1\le j\le m$$

$$\sum_{(s,a,\mu)\in\hat\delta_{\mathcal{M}}} y_{(s,a,\mu)} - \sum_{(\hat s,\hat a,\hat\mu)\in\hat\delta_{\mathcal{M}}} \mu'(s)\cdot y_{(\hat s,\hat a,\hat\mu)} = init(s) \quad for\ all\ s\in\hat S\backslash\{s_{dead}\}$$

$$y_{(s,a,\mu)} \ge 0 \qquad for\ all\ (s,a,\mu)\in\hat\delta_{\mathcal{M}}$$

$where\ init(s)\ is\ 1\ if\ s = \bar s\ and\ 0\ otherwise.$

Fig. 2. The linear program $L(\hat{\mathcal{M}})$

taking an action a^R in $\hat{\mathcal{M}}$ corresponds to electing to remain forever in the corresponding EC of $\bar{\mathcal{M}}$. Formally, $\hat{\mathcal{M}} = (\hat S, \bar s, \hat\alpha_{\mathcal{M}}, \hat\delta_{\mathcal{M}})$ where $\hat S = \bar S \cup \{s_{dead}\}$, $\hat\alpha_{\mathcal{M}} = \alpha_{\mathcal{M}} \cup \{a^R \mid R \subseteq \{1,\ldots,n\}\}$, and $\hat\delta_{\mathcal{M}} = \bar\delta_{\mathcal{M}} \cup \{(s, a^R, \eta_{s_{dead}}) \mid s \in S_R\}$. Finally, we create, for each $1 \le i \le n$, a reward structure $\lambda_i : \{a^R \mid i \in R\} \to \mathbb{R}_{>0}$ with $\lambda_i(a^R) = 1$ whenever λ_i is defined.

Proposition 3. *There is an adversary $\bar\sigma$ of $\bar{\mathcal{M}}$ such that $ExpTot^{\bar\sigma}_{\bar{\mathcal{M}}}(\rho)=x$ and $\bar{\mathcal{M}}, \bar\sigma \models \Psi$ iff there is a memoryless adversary $\hat\sigma$ of $\hat{\mathcal{M}}$ such that $ExpTot^{\hat\sigma}_{\hat{\mathcal{M}}}(\rho)=x$ and $\hat{\mathcal{M}}, \hat\sigma \models (\wedge_{i=1}^n [\lambda_i]_{\sim_i p_i}) \wedge (\wedge_{j=1}^m [\rho_j]_{\sim_j r_j}) \wedge ([\Diamond s_{dead}]_{\geqslant 1}).$*

Step 4. Finally, we create a linear program $L(\hat{\mathcal{M}})$, given in Figure 2, which encodes the structure of $\hat{\mathcal{M}}$ as well as the objectives from Ψ. Intuitively, in a solution of $L(\hat{\mathcal{M}})$, the variables $y_{(s,a,\mu)}$ express the expected number of times that state s is visited and transition $s \xrightarrow{a} \mu$ is taken subsequently. The expected total reward w.r.t. ρ_i is then captured by $\sum_{(s,a,\mu)\in\hat\delta_{\mathcal{M}},\, s\neq s_{dead}} \rho_i(a)\cdot y_{(s,a,\mu)}$. The result of $L(\hat{\mathcal{M}})$ yields the desired value for our numerical query.

Proposition 4. *For $x \in \mathbb{R}_{\geqslant 0}$, there is a memoryless adversary $\hat\sigma$ of $\hat{\mathcal{M}}$ where $ExpTot^{\hat\sigma}_{\hat{\mathcal{M}}}(\rho)=x$ and $\hat{\mathcal{M}}, \hat\sigma \models (\wedge_{i=1}^n [\lambda_i]_{\sim_i p_i}) \wedge (\wedge_{j=1}^m [\rho_j]_{\sim_j r_j}) \wedge ([\Diamond s_{dead}]_{\geqslant 1})$ iff there is a feasible solution $(y^\star_{(s,a,\mu)})_{(s,a,\mu)\in\hat\delta_{\mathcal{M}}}$ of the linear program $L(\hat{\mathcal{M}})$ such that $\sum_{(s,a,\mu)\in\hat\delta_{\mathcal{M}},\, s\neq s_{dead}} \rho_i(a)\cdot y^\star_{(s,a,\mu)} = x$.*

In addition, a solution to $L(\hat{\mathcal{M}})$ gives a memoryless adversary σ_{prod} defined by $\sigma_{prod}(s)(a,\mu) = \frac{y_{(s,a,\mu)}}{\sum_{a',\mu'} y_{(s,a',\mu')}}$ if the denominator is nonzero (and defined arbitrarily otherwise). This can be converted into a finite memory adversary σ' for $\mathcal{M}^\phi$ by combining decisions of σ on actions in $\alpha_{\mathcal{M}}$ and, instead of taking actions a^R, mimicking adversaries witnessing that the state which precedes a^R in the history is in S_R. Adversary σ' can be translated into an adversary σ of $\mathcal{M}$ in standard fashion using the fact that every finite path in $\mathcal{M}^\phi$ has a counterpart in $\mathcal{M}$ given by projecting states of $\mathcal{M}^\phi$ to their first components.
The following is then a direct consequence of Propositions 2, 3 and 4.

Theorem 1. *Given a PA $\mathcal{M}$ and numerical query $ExpTot^{\max}_{\mathcal{M}}(\rho \mid \Psi)$ satisfying Assumption 1, the result of the query is equal to the solution of the linear program*

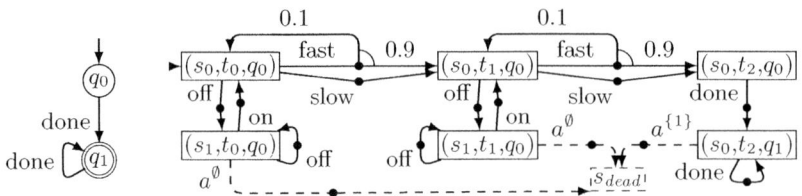

Fig. 3. A DRA $\mathcal{A}_\phi$ for the property $\phi = \Diamond done$ and the PA $\hat{\mathcal{M}}$ for $\mathcal{M} = \mathcal{M}_m \| \mathcal{M}_{c_1}$

$L(\hat{\mathcal{M}})$ *(see Figure 2). Furthermore, this requires time polynomial in the size of* $\mathcal{M}$ *and doubly exponential in the size of the property (for LTL objectives).*

An analogous result holds for numerical queries of the form $ExpTot_{\mathcal{M}}^{\min}(\rho\,|\,\Psi)$, $Pr_{\mathcal{M}}^{\max}(\phi\,|\,\Psi)$ or $Pr_{\mathcal{M}}^{\min}(\phi\,|\,\Psi)$. As discussed previously, this also yields a technique to solve both achievability and verification queries in the same manner.

3.2 Controller Synthesis

The achievability and numerical queries presented in the previous section are directly applicable to the problem of *controller synthesis*. We first illustrate these ideas on our simple example, and then apply them to a large case study.

Example 2. Consider the composition $\mathcal{M} = \mathcal{M}_{c_1} \| \mathcal{M}_m$ of PAs from Figure 1; $\mathcal{M}_{c_1}$ can be seen as a template for a controller of $\mathcal{M}_m$. We synthesise an adversary for $\mathcal{M}$ that minimises the expected execution time under the constraints that the machine completes both jobs and the expected power consumption is below some bound r. Thus, we use the minimising numerical query $ExpTot_{\mathcal{M}}^{\min}(\rho_{time}\,|\,[\rho_{power}]_{\leqslant r} \wedge [\Diamond done]_{\geqslant 1})$. Figure 3 shows the corresponding PA $\hat{\mathcal{M}}$, dashed lines indicating additions to construct $\hat{\mathcal{M}}$ from $\bar{\mathcal{M}}$. Solving the LP problem $L(\hat{\mathcal{M}})$ yields the minimum expected time under these constraints. If $r=30$, for example, the result is $\frac{49}{11}$. Examining the choices made in the corresponding (memoryless) adversary, we find that, to obtain this time, a controller could schedule the first job *fast* with probability $\frac{5}{6}$ and *slow* with $\frac{1}{6}$, and the second job *slow*. Figure 4(a) shows how the result changes as we vary the bound r and use different values for the failure probability of *fast* (0.1 in Figure 1).

Case study. We have implemented the techniques of Section 3 as an extension of PRISM [19] and using the ECLiPSe LP solver. We applied them to perform controller synthesis on a realistic case study: we build a *power manager* for an IBM TravelStar VP disk-drive [5]. Specific (randomised) power management policies can already be analysed in PRISM [22]; here, we *synthesise* such policies, subject to constraints specified as qmo-properties. More precisely, we minimise the expected power consumption under restrictions on, for example, the expected job-queue size, expected number of lost jobs, probability that a request waits more than K steps, or probability that N requests are lost. Further details are available from [24]. As an illustration, Figure 4(b) plots the minimal power consumption under restrictions on both the expected queue size and number

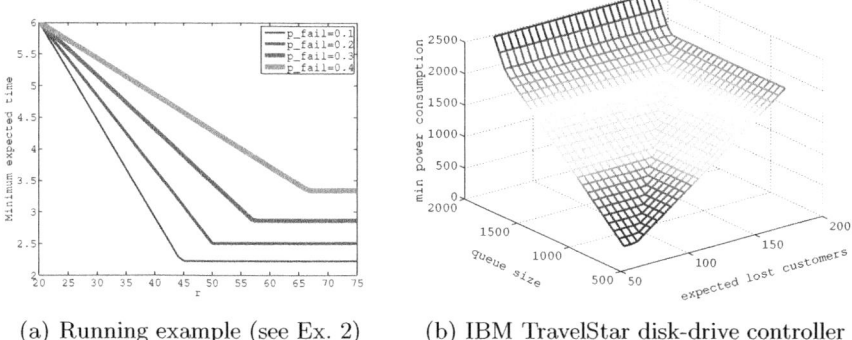

(a) Running example (see Ex. 2)   (b) IBM TravelStar disk-drive controller

Fig. 4. Experimental results illustrating controller synthesis

of lost customers. This shows the familiar power-versus-performance trade-off: policies can offer improved performance, but at the expense of using more power.

4   Quantitative Assume-Guarantee Verification

We now present novel compositional verification techniques for probabilistic automata, based on the quantitative multi-objective properties defined in Section 3. The key ingredient of this approach is the *assume-guarantee triple*, whose definition, like in [21], is based on quantification over adversaries. However, whereas [21] uses a single *probabilistic safety property* as an assumption or guarantee, we permit *quantitative multi-objective* properties. Another key factor is the incorporation of *fairness*.

Definition 8 (Assume-guarantee triples). *If $\mathcal{M} = (S, \bar{s}, \alpha_{\mathcal{M}}, \delta_{\mathcal{M}})$ is a PA and Ψ_A, Ψ_G are qmo-properties such that $\alpha_G \subseteq \alpha_A \cup \alpha_{\mathcal{M}}$, then $\langle \Psi_A \rangle \mathcal{M} \langle \Psi_G \rangle$ is an* assume-guarantee triple *with the following semantics:*

$$\langle \Psi_A \rangle \mathcal{M} \langle \Psi_G \rangle \;\Leftrightarrow\; \forall \sigma \in Adv_{\mathcal{M}[\alpha_A]} \,.\, (\mathcal{M}, \sigma \models \Psi_A \rightarrow \mathcal{M}, \sigma \models \Psi_G) \,.$$

where $\mathcal{M}[\alpha_A]$ denotes the alphabet extension *[21] of $\mathcal{M}$, which adds a-labelled self-loops to all states of $\mathcal{M}$ for each $a \in \alpha_A \backslash \alpha_{\mathcal{M}}$.*

Informally, an assume-guarantee triple $\langle \Psi_A \rangle \mathcal{M} \langle \Psi_G \rangle$, means "if $\mathcal{M}$ is a component of a system such that the environment of $\mathcal{M}$ satisfies Ψ_A, then the combined system (under fairness) satisfies Ψ_G".

Verification of an assume guarantee triple, i.e. checking whether $\langle \Psi_A \rangle \mathcal{M} \langle \Psi_G \rangle$ holds, reduces directly to the verification of a qmo-property since:

$$(\mathcal{M}, \sigma \models \Psi_A \rightarrow \mathcal{M}, \sigma \models \Psi_G) \;\Leftrightarrow\; \mathcal{M}, \sigma \models (\neg \Psi_A \vee \Psi_G) \,.$$

Thus, using the techniques of Section 3, we can reduce this to an achievability query, solvable via linear programming. Using these assume-guarantee triples as a basis, we can now formulate several proof rules that permit compositional verification of probabilistic automata. We first state two such rules, then explain their usage, illustrating with an example.

Theorem 2. *If $\mathcal{M}_1$ and $\mathcal{M}_2$ are PAs, and Ψ_{A_1}, Ψ_{A_2} and Ψ_G are quantitative multi-objective properties, then the following proof rules hold:*

$$\frac{\begin{array}{c} \mathcal{M}_1 \models_{\text{fair}} \Psi_{A_1} \\ \langle \Psi_{A_1} \rangle \, \mathcal{M}_2 \, \langle \Psi_G \rangle \end{array}}{\mathcal{M}_1 \| \mathcal{M}_2 \models_{\text{fair}} \Psi_G} \quad (\text{ASYM})$$

$$\frac{\begin{array}{c} \mathcal{M}_2 \models_{\text{fair}} \Psi_{A_2} \\ \langle \Psi_{A_2} \rangle \, \mathcal{M}_1 \, \langle \Psi_{A_1} \rangle \\ \langle \Psi_{A_1} \rangle \, \mathcal{M}_2 \, \langle \Psi_G \rangle \end{array}}{\mathcal{M}_1 \| \mathcal{M}_2 \models_{\text{fair}} \Psi_G} \quad (\text{CIRC})$$

where, for well-formedness, we assume that, if a rule contains an occurrence of the triple $\langle \Psi_A \rangle \, \mathcal{M} \, \langle \Psi_G \rangle$ in a premise, then $\alpha_G \subseteq \alpha_A \cup \alpha_{\mathcal{M}}$; similarly, for a premise that checks Ψ_A against $\mathcal{M}$, we assume that $\alpha_A \subseteq \alpha_{\mathcal{M}}$

Theorem 2 presents two assume-guarantee rules. The simpler, (ASYM), uses a single assumption Ψ_{A_1} about $\mathcal{M}_1$ to prove a property Ψ_G on $\mathcal{M}_1 \| \mathcal{M}_2$. This is done compositionally, in two steps. First, we verify $\mathcal{M}_1 \models_{\text{fair}} \Psi_{A_1}$. If $\mathcal{M}_1$ comprises just a single PA, the stronger (but easier) check $\mathcal{M}_1 \models \Psi_{A_1}$ suffices; the use of fairness in the first premise is to permit recursive application of the rule. Second, we check that $\langle \Psi_{A_1} \rangle \, \mathcal{M}_2 \, \langle \Psi_G \rangle$ holds. Again, optionally, we can consider fairness here.[3] In total, these two steps have the potential to be significantly cheaper than verifying $\mathcal{M}_1 \| \mathcal{M}_2$. The other rule, (CIRC), operates similarly, but using assumptions about both $\mathcal{M}_1$ and $\mathcal{M}_2$.

Example 3. We illustrate assume-guarantee verification using the PAs $\mathcal{M}_m$ and $\mathcal{M}_{c_2}$ from Figure 1. Our aim is to verify that $\mathcal{M}_{c_2} \| \mathcal{M}_m \models_{\text{fair}} [\rho_{time}]_{\leqslant \frac{19}{6}}$, which does indeed hold. We do so using the proof rule (ASYM) of Theorem 2, with $\mathcal{M}_1 = \mathcal{M}_{c_2}$ and $\mathcal{M}_2 = \mathcal{M}_m$. We use the assumption $\mathcal{A}_1 = [\rho_{slow}]_{\leqslant \frac{1}{2}}$ where $\rho_{slow} = \{slow \mapsto 1\}$, i.e. we assume the expected number of *slow* jobs requested is at most 0.5. We verify $\mathcal{M}_{c_2} \models [\rho_{slow}]_{\leqslant \frac{1}{2}}$ and the triple $\langle [\rho_{slow}]_{\leqslant \frac{1}{2}} \rangle \, \mathcal{M}_m \, \langle [\rho_{time}]_{\leqslant \frac{19}{6}} \rangle$. The triple is checked by verifying $\mathcal{M}_m \models \neg [\rho_{slow}]_{\leqslant \frac{1}{2}} \vee [\rho_{time}]_{\leqslant \frac{19}{6}}$ or, equivalently, that no adversary of $\mathcal{M}_m$ satisfies $[\rho_{slow}]_{\leqslant \frac{1}{2}} \wedge [\rho_{time}]_{> \frac{19}{6}}$.

Experimental Results. Using our implementation of the techniques in Section 3, we now demonstrate the application of our quantitative assume-guarantee verification framework to two large case studies: Aspnes & Herlihy's randomised *consensus* algorithm and the *Zeroconf* network configuration protocol. For *consensus*, we check the maximum expected number of steps required in the first R rounds; for *zeroconf*, we verify that the protocol terminates with probability 1 and the minimum/maximum expected time to do so. In each case, we use the (CIRC) rule, with a combination of probabilistic safety and liveness properties for assumptions. All models and properties are available from [24]. In fact, we execute numerical queries to obtain lower/upper bounds for system properties, rather than just verifying a specific bound.

---

[3] Adding fairness to checks of both qmo-properties and assume-guarantee triples is achieved by encoding the unconditional fairness constraint as additional objectives.

Table 1. Experimental results for compositional verification

Case study [parameters]			Non-compositional			Compositional		
			States	Time (s)	Result	LP size	Time (s)	Result
consensus (2 processes) (max. steps) [R K]	3	2	1,806	0.4	89.00	1,565	1.8	89.65
	3	20	11,598	27.8	5,057	6,749	10.8	5,057
	4	2	7,478	1.3	89.00	5,368	3.9	98.42
	4	20	51,830	155.0	5,057	15,160	16.2	5,120
	5	2	30,166	3.1	89.00	10,327	6.5	100.1
	5	20	212,758	552.8	5,057	24,727	21.9	5,121
consensus (3 processes) (max. steps) [R K]	3	2	114,559	20.5	212.0	43,712	12.1	214.3
	3	12	507,919	1,361.6	4,352	92,672	284.9	4,352
	3	20	822,607	time-out	-	131,840	901.8	11,552
	4	2	3,669,649	728.1	212.0	260,254	118.9	260.3
	4	12	29,797,249	mem-out	-	351,694	642.2	4,533
	4	20	65,629,249	mem-out	-	424,846	1,697.0	11,840
zeroconf (termination) [K]	4		57,960	8.7	1.0	155,458	23.8	1.0
	6		125,697	16.6	1.0	156,690	24.5	1.0
	8		163,229	19.4	1.0	157,922	25.5	1.0
zeroconf (min. time) [K]	4		57,960	6.7	13.49	155,600	23.0	16.90
	6		125,697	15.7	17.49	154,632	23.1	12.90
	8		163,229	22.2	21.49	156,568	23.9	20.90
zeroconf (max. time) [K]	4		57,960	5.8	14.28	154,632	23.7	17.33
	6		125,697	13.3	18.28	155,600	24.2	22.67
	8		163,229	18.9	22.28	156,568	25.1	28.00

Table 1 summarises the experiments on these case studies, which were run on a 2.66GHz PC with 8GB of RAM, using a time-out of 1 hour. The table shows the (numerical) result obtained and the time taken for verification done both compositionally and non-compositionally (with PRISM). As an indication of problem size, we give the size of the (non-compositional) PA, and the number of variables in the linear programs for multi-objective model checking.

Compositional verification performs very well. For the *consensus* models, it is almost always faster than the non-compositional case, often significantly so, and is able to scale up to larger models. For *zeroconf*, times are similar. Encouragingly, though, times for compositional verification grow much more slowly with model size. We therefore anticipate better scalability through improvements to the underlying LP solver. Finally, we note that the numerical results obtained compositionally are very close to the true results (where obtainable).

5 Conclusions

We have presented techniques for studying multi-objective properties of PAs, using a language that combines ω-regular properties, expected total reward and multiple objectives. We described how to verify a property over all adversaries of a PA, synthesise an adversary that satisfies and/or optimises objectives, and compute the minimum or maximum value of an objective, subject to constraints. We demonstrated direct applicability to controller synthesis, illustrated with a realistic disk-drive controller case study. Finally, we proposed an assume-guarantee framework for PAs that significantly improves existing ones [21], and demonstrated successful compositional verification on several large case studies.

Possible directions for future work include extending our compositional verification approach with learning-based assumption generation, as [17] does for the simpler framework of [21], and investigation of continuous-time models.

Acknowledgments. The authors are part supported by ERC Advanced Grant VERIWARE, EU FP7 project CONNECT and EPSRC grant EP/D07956X. Vojtěch Forejt is also supported by a Royal Society Newton Fellowship and the Institute for Theoretical Computer Science, project no. 1M0545.

References

1. de Alfaro, L.: Formal Verification of Probabilistic Systems. Ph.D. thesis, Stanford University (1997)
2. Baier, C., Größer, M., Ciesinski, F.: Quantitative analysis under fairness constraints. In: Liu, Z., Ravn, A.P. (eds.) ATVA 2009. LNCS, vol. 5799, pp. 135–150. Springer, Heidelberg (2009)
3. Baier, C., Größer, M., Leucker, M., Bollig, B., Ciesinski, F.: Controller synthesis for probabilistic systems. In: Proc. IFIP TCS 2004, pp. 493–506 (2004)
4. Baier, C., Kwiatkowska, M.: Model checking for a probabilistic branching time logic with fairness. Distributed Computing 11(3), 125–155 (1998)
5. Benini, L., Bogliolo, A., Paleologo, G., De Micheli, G.: Policy optimization for dynamic power management. IEEE Transactions on Computer-Aided Design of Integrated Circuits and Systems 8(3), 299–316 (2000)
6. Bianco, A., de Alfaro, L.: Model checking of probabilistic and nondeterministic systems. In: Thiagarajan, P.S. (ed.) FSTTCS 1995. LNCS, vol. 1026, pp. 499–513. Springer, Heidelberg (1995)
7. Brázdil, T., Forejt, V., Kučera, A.: Controller synthesis and verification for Markov decision processes with qualitative branching time objectives. In: Aceto, L., Damgård, I., Goldberg, L.A., Halldórsson, M.M., Ingólfsdóttir, A., Walukiewicz, I. (eds.) ICALP 2008, Part II. LNCS, vol. 5126, pp. 148–159. Springer, Heidelberg (2008)
8. Caillaud, B., Delahaye, B., Larsen, K., Legay, A., Pedersen, M., Wasowski, A.: Compositional design methodology with constraint Markov chains. In: QEST (2010)
9. Chatterjee, K.: Markov decision processes with multiple long-run average objectives. In: Arvind, V., Prasad, S. (eds.) FSTTCS 2007. LNCS, vol. 4855, pp. 473–484. Springer, Heidelberg (2007)
10. Chatterjee, K., de Alfaro, L., Faella, M., Henzinger, T., Majumdar, R., Stoelinga, M.: Compositional quantitative reasoning. In: Proc. QEST 2006 (2006)
11. Chatterjee, K., Henzinger, T., Jobstmann, B., Singh, R.: Measuring and synthesizing systems in probabilistic environments. In: Touili, T., Cook, B., Jackson, P. (eds.) CAV 2010. LNCS, vol. 6174, pp. 380–395. Springer, Heidelberg (2010)
12. Chatterjee, K., Majumdar, R., Henzinger, T.: Markov decision processes with multiple objectives. In: Durand, B., Thomas, W. (eds.) STACS 2006. LNCS, vol. 3884, pp. 325–336. Springer, Heidelberg (2006)
13. Clímaco, J. (ed.): Multicriteria Analysis. Springer, Heidelberg (1997)
14. Courcoubetis, C., Yannakakis, M.: Markov decision processes and regular events. IEEE Transactions on Automatic Control 43(10), 1399–1418 (1998)

15. Delahaye, B., Caillaud, B., Legay, A.: Probabilistic contracts: A compositional reasoning methodology for the design of stochastic systems. In: ACSD 2010 (2010)
16. Etessami, K., Kwiatkowska, M., Vardi, M., Yannakakis, M.: Multi-objective model checking of Markov decision processes. LMCS 4(4), 1–21 (2008)
17. Feng, L., Kwiatkowska, M., Parker, D.: Compositional verification of probabilistic systems using learning. In: Proc. QEST 2010, pp. 133–142. IEEE, Los Alamitos (2010)
18. Forejt, V., Kwiatkowska, M., Norman, G., Parker, D., Qu, H.: Quantitative multi-objective verification for probabilistic systems. Tech. Rep. RR-10-26, Oxford University Computing Laboratory (2010)
19. Hinton, A., Kwiatkowska, M., Norman, G., Parker, D.: PRISM: A tool for automatic verification of probabilistic systems. In: Hermanns, H. (ed.) TACAS 2006. LNCS, vol. 3920, pp. 441–444. Springer, Heidelberg (2006)
20. Kemeny, J., Snell, J., Knapp, A.: Denumerable Markov Chains, 2nd edn. Springer, Heidelberg (1976)
21. Kwiatkowska, M., Norman, G., Parker, D., Qu, H.: Assume-guarantee verification for probabilistic systems. In: Esparza, J., Majumdar, R. (eds.) TACAS 2010. LNCS, vol. 6015, pp. 23–37. Springer, Heidelberg (2010)
22. Norman, G., Parker, D., Kwiatkowska, M., Shukla, S., Gupta, R.: Using probabilistic model checking for dynamic power management. FAC 17(2) (2005)
23. Segala, R.: Modelling and Verification of Randomized Distributed Real Time Systems. Ph.D. thesis, Massachusetts Institute of Technology (1995)
24. http://www.prismmodelchecker.org/files/tacas11/

Efficient CTMC Model Checking of Linear Real-Time Objectives

Benoît Barbot[2], Taolue Chen[3], Tingting Han[1],
Joost-Pieter Katoen[1,3], and Alexandru Mereacre[1]

[1]MOVES, RWTH Aachen University, Germany
[2] ENS Cachan, France
[3] FMT, University of Twente, The Netherlands

Abstract. This paper makes verifying continuous-time Markov chains (CTMCs) against deterministic timed automata (DTA) objectives practical. We show that verifying 1-clock DTA can be done by analyzing subgraphs of the product of CTMC $\mathcal{C}$ and the region graph of DTA $\mathcal{A}$. This improves upon earlier results and allows to only use standard analysis algorithms. Our graph decomposition approach naturally enables bisimulation minimization as well as parallelization. Experiments with various examples confirm that these optimizations lead to significant speed-ups. We also report on experiments with multiple-clock DTA objectives. The objectives and the size of the problem instances that can be checked with our prototypical tool go (far) beyond what could be checked so far.

1 Introduction

For more than a decade, the verification of continuous-time Markov chains, CTMCs for short, has received considerable attention, cf. the recent survey [7]. Due to unremitting improvements on algorithms, (symbolic) data structures and abstraction techniques, CTMC model checking has emerged into a valuable analysis technique which – supported by powerful software tools– has been adopted by various researchers for systems biology, queueing networks, and dependability.

The focus of CTMC model checking has primarily been on checking stochastic versions of the branching temporal logic CTL, such as CSL [6]. The verification of LTL objectives reduces to applying well-known algorithms [20] to embedded discrete-time Markov chains (DTMCs). Linear objectives equipped with timing constraints, have just recently been considered. This paper treats linear *real-time* specifications that are given as deterministic timed automata (DTA). These include, e.g., properties of the form: what is the probability to reach a given target state within the deadline, while avoiding "forbidden" states and not staying too long in any of the "dangerous" states on the way. Such properties can neither be expressed in CSL nor in dialects thereof [5,14]. Model checking DTA objectives amounts to determining the probability of the set of paths of CTMC $\mathcal{C}$ that are accepted by the DTA $\mathcal{A}$, i.e., *Prob* $(\mathcal{C} \models \mathcal{A})$. We recently showed in [12] that this equals the reachability probability in a finite *piecewise deterministic Markov process* (PDP) that is obtained by a region construction on the product $\mathcal{C} \otimes \mathcal{A}$. This paper reports on how to make this approach practical, i.e., how to efficiently realize CTMC model checking against DTA objectives.

P.A. Abdulla and K.R.M. Leino (Eds.): TACAS 2011, LNCS 6605, pp. 128–142, 2011.
© Springer-Verlag Berlin Heidelberg 2011

As a first step, we show that rather than taking the region graph of the product $\mathcal{C} \otimes \mathcal{A}$, which is a somewhat ad-hoc mixture of CTMCs and DTA, we can apply a standard region construction on DTA $\mathcal{A}$ *prior to* building the product. This enables applying a standard region construction for timed automata. The product of this region graph with CTMC $\mathcal{C}$ yields the PDP to be analyzed. Subsequently, we exploit that for 1-clock DTA, the resulting PDP can be decomposed into subgraphs—each of which is a CTMC [12]. In this case, $Prob(\mathcal{C} \models \mathcal{A})$ is the solution of a system of linear equations whose coefficients are transient probability distributions of the (slightly amended) subgraph CTMCs. We adapt the algorithm for lumping [13,19] on CTMCs to our setting and prove that this preserves reachability probabilities, i.e., keeps $Prob(\mathcal{C} \models \mathcal{A})$ invariant. As the graph decomposition naturally enables parallelization, our tool implementation also supports the distribution of computing transient probabilities over multiple multi-core computers. Finally, multi-clock DTA objectives –for which the graph decomposition does not apply– are supported by a discretization of the product PDP.

Three case studies from different application fields are used to show the feasibility of this approach. The first case study has been taken from [3] which considers 1-clock DTA as time constraints of until modalities. Although using a quite different approach, our verification results coincide with [3]. The running time of our implementation (without lumping and parallelization) is about three orders of magnitude faster than [3]. Other considered case studies are a randomly moving robot, and a real case study from systems biology [15]. Bisimulation quotienting (i.e., lumping) yields state space reduction of up to one order of magnitude, whereas parallelizing transient analysis yields speedups of up to a factor 13 on 20 cores, depending on the number of subgraphs in the decomposition.

The discretization approach for multi-clock DTA may give rise to large models: checking the robot example (up to 5,000 states) against a two-clock DTA yields a 40-million state DTMC for which simple reachability probabilities are to be determined.

Related work. The logic asCSL [5] extends CSL by (time-bounded) regular expressions over actions and state formulas as path formulas. CSL^{TA} [14] allows 1-clock DTA as time constraints of until modalities; this subsumes acCSL. The joint behavior of $\mathcal{C}$ and DTA $\mathcal{A}$ is interpreted as a Markov renewal process. A prototypical implementation of this approach has recently been reported in [3]. Our algorithmic approach for 1-clock DTA is different, yields the same results, and is –as shown in this paper– significantly faster. Moreover, lumping is not easily possible (if at all) in [3].

In addition, it naturally supports bisimulation minimization and parallelization. Bisimulation quotienting in CSL model checking has been addressed in [18]. The works [4,9] provide a quantitative interpretation to timed automata where delays of unbounded clocks are governed by exponential distributions. Brazdil et al. present an algorithmic approaches towards checking continuous-time stochastic games (supporting more general distributions) against DTA specifications [11]. To our knowledge, tool implementations of [4,9,11] do not exist.

Organization of the paper. Section 2 provides the preliminaries and summarizes the main results of [12]. Section 3 describes our bisimulation quotienting algorithm. Section 4 reports on the experiments for 1-clock DTA objectives. Section 5 describes the discretization approach and gives experimental results for multi-clock DTA. Finally, Section 6 concludes. (Proofs are omitted and can be found in the full version available at http://moves.rwth-aachen.de/i2/han.)

2 Preliminaries

2.1 Continuous-Time Markov Chains

Definition 1 (CTMC). *A (labeled) continuous-time Markov chain (CTMC) is a tuple* $\mathcal{C} = (S, \text{AP}, L, \alpha, \mathbf{P}, E)$ *where* S *is a* finite *set of* states; AP *is a finite set of* atomic propositions; $L : S \to 2^{\text{AP}}$ *is the* labeling function; $\alpha \in Distr(S)$ *is the* initial distribution; $\mathbf{P} : S \times S \to [0, 1]$ *is a stochastic matrix; and* $E : S \to \mathbb{R}_{\geqslant 0}$ *is the* exit rate function.

Here, $Distr(S)$ is the set of probability distributions on set S. The probability to exit state s in t time units is given by $\int_0^t E(s) \cdot e^{-E(s)\tau} d\tau$. The probability to take the transition $s \to s'$ in t time units is $\mathbf{P}(s, s') \cdot \int_0^t E(s) e^{-E(s) \cdot \tau} d\tau$. The embedded DTMC of $\mathcal{C}$ is $(S, \text{AP}, L, \alpha, \mathbf{P})$. A *(timed) path* in CTMC $\mathcal{C}$ is a sequence $\rho = s_0 \xrightarrow{t_0} s_1 \xrightarrow{t_1} s_2 \cdots$ such that $s_i \in S$ and $t_i \in \mathbb{R}_{>0}$. Let $Paths^{\mathcal{C}}$ be the set of paths in $\mathcal{C}$ and $\rho[n] := s_n$ be the n-th state of ρ. The definitions of a *Borel space* on paths through CTMCs, and the *probability measure* Pr follow [20,6].

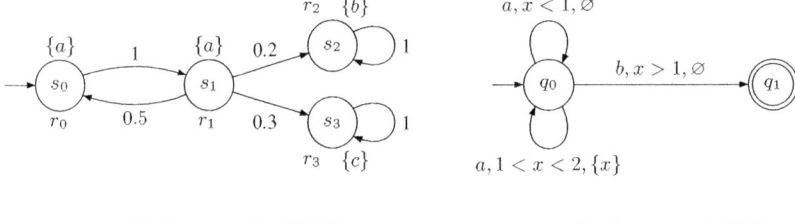

(a) An example CTMC   (b) An example DTA

Fig. 1. Example CTMC and DTA

Example 1. Fig. 1(a) shows an example CTMC with $\text{AP} = \{a, b, c\}$ and initial state s_0, i.e., $\alpha(s) = 1$ iff $s = s_0$. The exit rates are indicated at the states, whereas the transition probabilities are attached to the edges. An example path is $\rho = s_0 \xrightarrow{2.5} s_1 \xrightarrow{1.4} s_0 \xrightarrow{2} s_1 \xrightarrow{2\pi} s_2 \cdots$ with $\rho[2] = s_0$ and $\rho[3] = s_1$.

Definition 2 (Bisimulation). *Let* $\mathcal{C} = (S, \text{AP}, L, \alpha, \mathbf{P}, E)$ *be a CTMC. Equivalence relation* $\mathcal{R}$ *on* S *is a (strong) bisimulation on* $\mathcal{C}$ *if for* $s_1 \mathcal{R} s_2$:

$$L(s_1) = L(s_2) \text{ and } \mathbf{P}(s_1, C) = \mathbf{P}(s_2, C) \text{ for all } C \text{ in } S/\mathcal{R} \text{ and } E(s_1) = E(s_2).$$

Let $s_1 \sim s_2$ *if there exists a strong bisimulation* $\mathcal{R}$ *on* $\mathcal{C}$ *with* $s_1 \mathcal{R} s_2$.

The quotient CTMC under the coarsest bisimulation $\sim$ can be obtained by partition-refinement with time complexity $\mathcal{O}(m \log n)$ [13]. A simplified version with the same complexity is recently proposed in [19].

2.2 Deterministic Timed Automata

Clock variables and valuations. Let $\mathcal{X} = \{x_1, \ldots, x_n\}$ be a set of *clocks* in $\mathbb{R}_{\geqslant 0}$. An $\mathcal{X}$-valuation (valuation for short) is a function $\eta : \mathcal{X} \to \mathbb{R}_{\geqslant 0}$; $\mathbf{0}$ is the valuation that assigns 0 to all clocks. A *clock constraint* g on $\mathcal{X}$ has the form $x \bowtie c$, or $g' \wedge g''$, where $x \in \mathcal{X}$, $\bowtie \in \{<, \leqslant, >, \geqslant\}$ and $c \in \mathbb{N}$. Diagonal clock constraints like $x - y \bowtie c$ do not extend the expressiveness [8], and are not considered. Let $\mathcal{CC}(\mathcal{X})$ denote the set of clock constraints over $\mathcal{X}$. An $\mathcal{X}$-valuation η *satisfies* constraint g, denoted as $\eta \models g$ if $\eta(x) \bowtie c$ for g of the form $x \bowtie c$, or $\eta \models g' \wedge \eta \models g''$ for g of the form $g' \wedge g''$. The reset of η w.r.t. $X \subseteq \mathcal{X}$, denoted $\eta[X := 0]$, is the valuation η' with $\forall x \in X. \, \eta'(x) := 0$ and $\forall x \notin X. \, \eta'(x) := \eta(x)$. For $\delta \in \mathbb{R}_{\geqslant 0}$, $\eta + \delta$ is the valuation η'' such that $\forall x \in \mathcal{X}. \, \eta''(x) := \eta(x) + \delta$.

Definition 3 (DTA). *A deterministic timed automaton (DTA) is a tuple $\mathcal{A} = (\Sigma, \mathcal{X}, Q, q_0, Q_F, \to)$ where Σ is a finite alphabet; $\mathcal{X}$ is a finite set of clocks; Q is a nonempty finite set of locations; $q_0 \in Q$ is the initial location; $Q_F \subseteq Q$ is a set of accepting locations; and $\to \in (Q \setminus Q_F) \times \Sigma \times \mathcal{CC}(\mathcal{X}) \times 2^{\mathcal{X}} \times Q$ satisfies: $q \xrightarrow{a,g,X} q'$ and $q \xrightarrow{a,g',X'} q''$ with $g \neq g'$ implies $g \cap g' = \varnothing$.*

We refer to $q \xrightarrow{a,g,X} q'$ as an *edge*, where $a \in \Sigma$ is the input symbol, the *guard* g is a clock constraint on the clocks of $\mathcal{A}$, $X \subseteq \mathcal{X}$ is a set of clocks and q, q' are locations. The intuition is that the DTA $\mathcal{A}$ can move from location q to q' on reading the input symbol a if the guard g holds, while resetting the clocks in X on entering q'. By convention, we assume $q \in Q_F$ to be a sink, cf. Fig. 1(b).

Paths. A finite (timed) path in $\mathcal{A}$ is of the form $\theta = q_0 \xrightarrow{a_0, t_0} q_1 \cdots q_n \xrightarrow{a_n, t_n} q_{n+1}$ with $q_i \in Q$, $t_i > 0$ and $a_i \in \Sigma$ ($0 \leqslant i \leqslant n$). The path θ is *accepted* by $\mathcal{A}$ if $\theta[i] \in Q_F$ for some $0 \leqslant i \leqslant |\theta|$ and for all $0 \leqslant j < i$, it holds that $\eta_0 = \mathbf{0}$, $\eta_j + t_j \models g_j$ and $\eta_{j+1} = (\eta_j + t_j)[X_j := 0]$, where η_j is the clock valuation on entering q_j and g_j the guard of $q_i \to q_{i+1}$. The infinite path $\rho = s_0 \xrightarrow{t_0} s_1 \xrightarrow{t_1} \cdots$ in CTMC $\mathcal{C}$ is accepted by $\mathcal{A}$ if for some $n \in \mathbb{N}$, $s_0 \xrightarrow{t_0} s_1 \cdots s_{n-1} \xrightarrow{t_{n-1}} s_n$ induces the DTA path $\hat{\rho} = q_0 \xrightarrow{L(s_0), t_0} q_1 \cdots q_{n-1} \xrightarrow{L(s_{n-1}), t_{n-1}} q_n$, which is accepted by $\mathcal{A}$. The set of CTMC paths accepted by a DTA is measurable [12]; our measure of interest is given by:

$$Prob(\mathcal{C} \models \mathcal{A}) = \Pr\{\, \rho \in Paths^{\mathcal{C}} \mid \hat{\rho} \text{ is accepted by DTA } \mathcal{A} \,\}.$$

Regions. Regions are sets of valuations, often represented as constraints. Let $\mathcal{R}e(\mathcal{X})$ be the set of regions over $\mathcal{X}$. For regions $\Theta, \Theta' \in \mathcal{R}e(\mathcal{X})$, Θ' is the *successor region* of Θ if for all $\eta \models \Theta$ there exists $\delta \in \mathbb{R}_{>0}$ such that $\eta + \delta \models \Theta'$ and $\forall \delta' < \delta. \, \eta + \delta' \models \Theta \vee \Theta'$. The region Θ *satisfies* a guard g, denoted $\Theta \models g$, iff $\forall \eta \models \Theta. \, \eta \models g$. The *reset operation* on region Θ is defined as $\Theta[X := 0] := \{\eta[X := 0] \mid \eta \models \Theta\}$.

Definition 4 (Region graph [2]). *Let $\mathcal{A} = (\Sigma, \mathcal{X}, Q, q_0, Q_F, \to)$ be a DTA. The region graph $\mathcal{G}(\mathcal{A})$ of $\mathcal{A}$ is $(\Sigma, W, w_0, W_F, \dashrightarrow)$ with $W = Q \times \mathcal{R}e(\mathcal{X})$ the set of states; $w_0 = (q_0, \mathbf{0})$ the initial state; $W_F = Q_F \times \mathcal{R}e(\mathcal{X})$ the set of final states; and $\dashrightarrow \subset W \times ((\Sigma \times 2^{\mathcal{X}}) \uplus \{\delta\}) \times W$ the smallest relation such that:*

$(q, \Theta) \overset{\delta}{\dashrightarrow} (q, \Theta')$ *if* Θ' *is the successor region of* Θ; *and*

$(q, \Theta) \overset{a, X}{\dashrightarrow} (q', \Theta')$ *if* $\exists g \in CC(X)$ *s.t.* $q \xrightarrow{a, g, X} q'$ *with* $\Theta \models g$ *and* $\Theta[X := 0] = \Theta'$.

Product. Our aim is to determine $Prob(\mathcal{C} \models \mathcal{A})$. This can be accomplished by computing reachability probabilities in the region graph of $\mathcal{C} \otimes \mathcal{A}$ where $\otimes$ denotes a product construction between CTMC $\mathcal{C}$ and DTA $\mathcal{A}$ [12]. To ease the implementation, we consider the product $\mathcal{C} \otimes \mathcal{G}(\mathcal{A})$.

Definition 5 (Product). *The product of CTMC* $\mathcal{C} = (S, \mathrm{AP}, L, \alpha, \mathbf{P}, E)$ *and DTA region graph* $\mathcal{G}(\mathcal{A}) = (\Sigma, W, w_0, W_F, \dashrightarrow)$, *denoted* $\mathcal{C} \otimes \mathcal{G}(\mathcal{A})$, *is the tuple* $(V, \alpha', V_F, \Lambda, \hookrightarrow)$ *with* $V = S \times W, \alpha'(s, w_0) = \alpha(s), V_F = S \times W_F$, *and*

- $\hookrightarrow \subseteq V \times \left(\left([0,1] \times 2^X\right) \uplus \{\delta\} \right) \times V$ *is the smallest relation such that:*
  - $(s, w) \overset{\delta}{\hookrightarrow} (s, w')$ *iff* $w \overset{\delta}{\dashrightarrow} w'$; *and*
  - $(s, w) \overset{p, X}{\hookrightarrow} (s', w')$ *iff* $p = \mathbf{P}(s, s'), p > 0$, *and* $w \overset{L(s), X}{\dashrightarrow} w'$.
- $\Lambda : V \to \mathbb{R}_{\geqslant 0}$ *is the* exit rate function *where:*

$$\Lambda(s, w) = \begin{cases} E(s) & \text{if } (s, w) \overset{p, X}{\hookrightarrow} (s', w') \text{ for some } (s', w') \in V \\ 0 & \text{otherwise.} \end{cases}$$

The (reachable fragment of the) product of CTMC $\mathcal{C}$ in Fig. 1(a) and the region graph of DTA $\mathcal{A}$ in Fig. 1(b) is given in Fig. 2. It turns out that $\mathcal{C} \otimes \mathcal{G}(\mathcal{A})$ is identical to $\mathcal{G}(\mathcal{C} \otimes \mathcal{A})$, the region graph of $\mathcal{C} \otimes \mathcal{A}$ as defined in [12].

Theorem 1. *For any* CTMC $\mathcal{C}$ *and any* DTA $\mathcal{A}, \mathcal{C} \otimes \mathcal{G}(\mathcal{A}) = \mathcal{G}(\mathcal{C} \otimes \mathcal{A})$.

As a corollary [12], it follows that $Prob(\mathcal{C} \models \mathcal{A})$ equals the reachability probability of the accepting states in $\mathcal{C} \otimes \mathcal{G}(\mathcal{A})$. In addition, $\mathcal{C} \otimes \mathcal{G}(\mathcal{A})$ is a PDP.

2.3 Decomposition for 1-Clock DTA

Let $\mathcal{A}$ be a 1-clock DTA and $\{c_0, \ldots, c_m\} \subseteq \mathbb{N}$ with $0 = c_0 < c_1 < \cdots < c_m$ the constants appearing in its clock constraints. Let $\Delta c_i = c_{i+1} - c_i$ for $0 \leqslant i < m$. The product $\mathcal{C} \otimes \mathcal{G}(\mathcal{A})$ can be split into $m+1$ subgraphs, denoted $\mathcal{G}_i$ for $0 \leqslant i \leqslant m$, such that any state in $\mathcal{G}_i$ has a clock valuation in $[c_i, c_{i+1})$ for $0 \leqslant i < m$ and in $[c_m, \infty)$ for $i = m$. Each column in Fig. 2 constitutes such subgraph. Subgraph $\mathcal{G}_i$ thus captures the joint behavior of CTMC $\mathcal{C}$ and DTA $\mathcal{A}$ in the interval $[c_i, c_{i+1})$. All transitions within $\mathcal{G}_i$ are probabilistic; delay-transitions, i.e., δ-labeled transitions, yield a state in $\mathcal{G}_{i+1}$, whereas a clock reset (of the

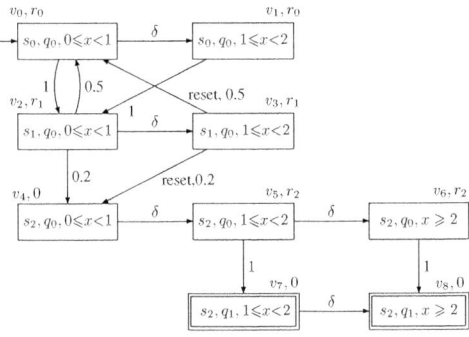

Fig. 2. Example $\mathcal{C} \otimes \mathcal{G}(\mathcal{A})$

only clock) yields a state in $\mathcal{G}_0$. In fact subgraph $\mathcal{G}_i$ is a CTMC. To take the effect of "reset" transitions into account, define the CTMC $\mathcal{C}_i$ with state space $V_i \cup V_0$, with all edges from V_i to V_0, all edges between V_i-vertices, but no outgoing edges from V_0-vertices.

Definition 6 (Augmented CTMC). *Let* $\mathcal{G} = \mathcal{C} \otimes \mathcal{G}(\mathcal{A}) = (V, \alpha, V_F, \Lambda, \hookrightarrow)$. *Subgraph* $\mathcal{G}_i$ *of* $\mathcal{G}$ *induces the CTMC* $\mathcal{C}_i = (V_i \cup V_0, \mathrm{AP}_i, L_i, \alpha_i, \mathbf{P}_i^a, E_i)$ *with:*

- $\mathrm{AP}_i = \begin{cases} \{\delta_v \mid v \in V_{i+1}\} \cup \{Rst_v \mid v \in V_0\} & \text{if } i < m \\ \{Rst_v \mid v \in V_0\} \cup \{F\} & \text{if } i = m \end{cases}$

- $L_i(v) = \{\delta_{v'}\}$ *if* $v \overset{\delta}{\hookrightarrow} v'$ *and* $L_i(v) = \{Rst_{v'}\}$ *if* $v \overset{p,X}{\hookrightarrow} v'$, *for* $i \leqslant m$; $L_m(v) = \{F\}$ *if* $v \in V_F \cap V_m$;

- $\alpha_0 = \alpha$, *and for* $0 \leqslant i < m$, α_{i+1} *is the transient distribution of* $\mathcal{C}_i$ *at* Δc_i;

- $\mathbf{P}_i^a(v, v') = \begin{cases} p & \text{if } v, v' \in V_i \wedge v \overset{p,\varnothing}{\hookrightarrow} v' \quad \text{or} \quad v \in V_i \wedge v' \in V_0 \wedge v \overset{p,X}{\hookrightarrow} v' \\ 1 & \text{if } v = v' \in V_0 \\ 0 & \text{otherwise.} \end{cases}$

- $E_i(s, w) = E(s)$ *where* E *is the exit-rate function of CTMC* $\mathcal{C}$.

State v in CTMC $\mathcal{C}_i$ is labeled with $Rst_{v'}$ if a clock reset in v yields v'; similarly it is labeled with $\delta_{v'}$ if a delay results in state v' in the successor CTMC $\mathcal{C}_{i+1}$. These labels are relevant for bisimulation quotienting as explained later. The proposition F indicates the "final" CTMC states in the CTMC $\mathcal{C}_m$.

The matrix $\mathbf{P}_i^a$ can be split into the matrices $\mathbf{P}_i$ and $\hat{\mathbf{P}}_i^a$ where $\mathbf{P}_i$ contains only (probabilistic) transitions inside V_i whereas $\hat{\mathbf{P}}_i^a$ contains transitions from V_i to V_0. The transient probability matrix for $\mathcal{C}_i$ is defined as the solution of the equation $\mathbf{\Pi}_i^a(x) = \int_0^x E_i \cdot e^{-E_i \tau} \cdot \mathbf{P}_i^a \cdot \mathbf{\Pi}_i^a(x-\tau) d\tau + e^{-E_i x}$. The matrices $\mathbf{\Pi}_i$ and $\hat{\mathbf{\Pi}}_i^a$ can be defined in a similar way as $\mathbf{P}$. Let $\mathbf{A}_i := \mathbf{\Pi}_0 \cdot \mathbf{\Pi}_1 \cdot \ldots \cdot \mathbf{\Pi}_i$ for $0 \leqslant i \leqslant m-1$ and $\mathbf{B}_i := \hat{\mathbf{\Pi}}_0^a$ if $i = 0$ and $\mathbf{B}_i := \mathbf{A}_{i-1} \cdot \hat{\mathbf{\Pi}}_i^a + \mathbf{B}_{i-1}$ if $1 \leqslant i < m$. We can now define the linear equation system $\mathbf{x} \cdot \mathbf{M} = \mathbf{f}$ with

$$\mathbf{M} = \left(\begin{array}{c|c} \mathbf{I}_{n_0} - \mathbf{B}_{m-1} & \mathbf{A}_{m-1} \\ \hline \hat{\mathbf{P}}_m^a & \mathbf{I}_{n_m} - \mathbf{P}_m \end{array} \right) \text{ and } \mathbf{f}(v) = \begin{cases} 1 & \text{if } v \in V_m \wedge F \in L_m(v) \\ 0 & \text{othewise} \end{cases}$$

Here $\mathbf{I}_n$ denotes the identity matrix of cardinality n. For details, consult [12].

Theorem 2 ([12]). *For CTMC* $\mathcal{C}$ *with initial distribution* α, *1-clock DTA* $\mathcal{A}$ *and linear equation system* $\mathbf{x} \cdot \mathbf{M} = \mathbf{f}$ *with solution* U, *we have that* $Prob(\mathcal{C} \models \mathcal{A}) = \alpha \cdot U$.

Algorithm 1 summarizes the main steps needed to model-check a CTMC against a 1-clock DTA, where as an (optional) optimization step, in line 7 all augmented CTMCs are lumped by the adapted lumping algorithm as explained below.

3   Lumping

It is known that bisimulation minimization is quite beneficial for CSL model checking as experimentally showed in [18]. Besides yielding a state space reduction, it may yield

---

Algorithm 1. Verifying a CTMC against a 1-clock DTA

Require: a CTMC $\mathcal{C}$ with initial distr. α, a 1-clock DTA $\mathcal{A}$ with constants $c_0, ..., c_m$
Ensure: $\Pr(\mathcal{C} \models \mathcal{A})$
1: $\mathcal{G}(\mathcal{A}) := \texttt{buildRegionGraph}(\mathcal{A})$;
2: $Product := \texttt{buildProduct}(\mathcal{C}, \mathcal{G}(\mathcal{A}))$; $\{\mathcal{C} \otimes \mathcal{G}(\mathcal{A})\}$
3: subGraphs $\{\mathcal{G}_i\}_{0 \leqslant i \leqslant m} := \texttt{partitionProduct}(Product)$;
4: **for** each subGraph $\mathcal{G}_i$ **do**
5:    $\mathcal{C}_i := \texttt{buildAugmentedCTMC}(\mathcal{G}_i)$; $\{$build augmented CTMC cf. Definition 6$\}$
6: **end for**
7: $\{\mathcal{C}_i'\}_{0 \leqslant i \leqslant m} := \texttt{lumpGroupCTMCs}(\{\mathcal{C}_i\}_{0 \leqslant i \leqslant m})$; $\{$lump a group of CTMCs, see Alg. 2$\}$
8: **for** each CTMC $\mathcal{C}_i'$ **do** $TransProb_i := \texttt{computeTransientProb}(\mathcal{C}_i', \Delta c_i)$; **end for**
9: $linearEqSystem := \texttt{buildLinearSystem}(\{TransProb_i\}_{0 \leqslant i \leqslant m})$; $\{$cf. Theorem 2$\}$
10: $probVector := \texttt{solveLinearSystem}(linearEqSystem)$;
11: **return** $\alpha \cdot probVector$;

---

substantial speed-ups. The decomposition of the product $\mathcal{C} \otimes \mathcal{G}(\mathcal{A})$ into the CTMCs $\mathcal{C}_i$ naturally facilitates a bisimulation reduction of each $\mathcal{C}_i$ prior to computing the transient probabilities in $\mathcal{C}_i$. In order to do so, an amendment of the standard lumping algorithm [13,19] is needed as the CTMCs to be lumped are connected by delay and reset transitions. Initial states in CTMC $\mathcal{C}_i$ might be the target states of edges whose source states are in a different CTMC, $\mathcal{C}_j$, say, with $i \neq j$. The partitioning of the target states in $\mathcal{C}_i$ will affect the partitioning of the source states in $\mathcal{C}_j$. For delay edges, $i = j + 1$ while for reset transition, $i = 0$. The intra-CTMC edges thus cause a "cyclic affection" between partitions among all sub-CTMCs. From the state labeling (cf. Def. 6), it follows that for any two states $v, v' \in \mathcal{C}_i$, if their respective successor states in $\mathcal{C}_{i+1}$ (or $\mathcal{C}_0$) can be lumped, then v and v' might be lumped. This implies that any refinement on the lumping blocks in $\mathcal{C}_{i+1}$ might affect the blocks in $\mathcal{C}_i$. Similarly, refining $\mathcal{C}_0$ might affect any $\mathcal{C}_i$, viz., the CTMCs that have a reset edge to $\mathcal{C}_0$.

We initiate the lumping algorithm (cf. Alg. 2) for CTMC $\mathcal{C}_i = (V_i \cup V_0, AP_i, L_i, \alpha_i, \mathbf{P}_i^a, E_i)$ by taking as initial partition the quotient induced by $\{(v_1, v_2) \in (V_i \cup V_0)^2 \mid L_i(v_1) = L_i(v_2)\}$. This initial partition is successively refined on each $\mathcal{C}_i$ by the standard approach [13,19], see lines 5–6. We then use the blocks in $\mathcal{C}_{i+1}$ to update AP_i in $\mathcal{C}_i$ and use blocks in $\mathcal{C}_0$ to update AP_i in all the affected $\mathcal{C}_i$'s, cf. lines 7–11. As a result, the new AP_i' may be coarser than the old AP_i:

$$AP_i' = \{Rst_{[v]_0} \mid Rst_v \in AP_i\} \cup \{\delta_{[v]_{i \oplus 1}} \mid \delta_v \in AP_i\}.$$

Here, $[v]_i$ is the equivalence class in CTMC $\mathcal{C}_i$ containing state v, and $i \oplus 1 = i + 1$ if $i < m$ and $m \oplus 1 = m$. With the new AP_i', this approach (cf. while-loop) is repeated until all CTMC partitions are stable.

Example 2. Let $v_1 \xrightarrow{\delta} v_1'$ and $v_2 \xrightarrow{\delta} v_2'$ be two delay transitions from CTMC $\mathcal{C}_i$ to $\mathcal{C}_{i+1}$. Then $L(v_1) = \delta_{v_1'}$ and $L(v_2) = \delta_{v_2'}$. Since v_1 and v_2 are labeled differently, they cannot be in one equivalence class. However, if in $\mathcal{C}_{i+1}$ it turns out that v_1' and v_2' are in one equivalence class, then we can update AP_i to AP_i' such that now $L(v_1) = L(v_2) = \delta_{v_1'}$. In this case, v_1 and v_2 can be lumped together.

Algorithm 2. LumpGroupCTMCs

Require: a set of CTMCs $\mathcal{C}_i$ with AP_i for $0 \leqslant i \leqslant m$
Ensure: a set of lumped CTMCs $\mathcal{C}_i'$ such that $\mathcal{C}_i \sim \mathcal{C}_i'$
1: *notStable* := *true*;
2: **while** *notStable* **do**
3:     *notStable* := *false*;
4:     **for** $i = m$ to 0 **do**
5:         *old*APSize := $|\mathrm{AP}_i|$;
6:         $(\mathcal{C}_i, \mathrm{AP}_i)$:= $\mathtt{lumpCTMC}(\mathcal{C}_i, \mathrm{AP}_i)$;   {lump $\mathcal{C}_i$ due to AP_i and update $\mathcal{C}_i, \mathrm{AP}_i$}
7:         **if** *old*APSize $> |\mathrm{AP}_i|$ **then**   {some states have been lumped in $\mathcal{C}_i$}
8:             *notStable* := *true*;
9:             **if** $i = 0$ **then**
10:                 **for** $j = 1$ to m **do** $\mathtt{updateResetEdge}(\mathrm{AP}_0, \mathrm{AP}_j)$;   {update AP_j}
11:             **else** $\mathtt{updateDelayEdge}(\mathrm{AP}_i, \mathrm{AP}_{i-1})$; {update AP_{i-1} according to AP_i}
12: **return** the new set of CTMCs lumped by the newest AP_i

---

If some states have been updated in $\mathcal{C}_i$ and AP_i has been updated (line 7), there are two cases: if $i{=}0$, then we update all AP_j, $j \neq 0$ that have a reset edge to $\mathcal{C}_0$ (line 9-10); otherwise, we update its directly predecessor AP_{i-1} which has a delay edge to $\mathcal{C}_i$ (line 11). This procedure is repeated until all AP_i's are stable.

Theorem 3. *The transient probability distribution in $\mathcal{C}_i$ and its quotient $\mathcal{C}_i'$, obtained by Alg. 2 are equal.*

As a corollary, it follows that the reachability probabilities of the accepting states in $\mathcal{C} \otimes \mathcal{G}(\mathcal{A})$, and its quotient obtained by applying Alg. 2 coincide.

4   Experimental Results

Implementation. We implemented our approach in approximately 4,000 lines of OCaml code. Transient probabilities are computed using uniformization, linear equation systems are solved using the Gauss-Seidel algorithm, and lumping has been realized by adapting [13] with the correction explained in [19]. Unreachable states (both forwards from the initial and backwards from the final states) are removed in $\mathcal{C} \otimes \mathcal{G}(\mathcal{A})$ prior to the analysis, and transient probabilities in $\mathcal{C}_i$ are only determined for its initial states, i.e., its entry points. The tool adopts the input format of the MRMC model checker [17]. Thanks to the output facility of PRISM [1], the verification of PRISM models is possible.

Case studies. We conducted extensive experiments with three case studies. The first case study has been taken from [3], and facilitates a comparison with the approach of [14]. The second case study, a random robot, is (to our taste) a nice example showing the need for DTA objectives. We use it for 1-clock as well as multi-clock objectives. The specifications of this example cannot be expressed using any other currently available techniques. The final case study originates from systems biology and is a more realistic case study. We first present experimental results using a sequential algorithm, with and without lumping. Section 4.4 presents the results when parallelizing the transient analysis (but not the lumping). All results (one and four cores) have been obtained on a 2×2.33GHz Intel Dual-Core computer with 32GB main memory. The experiments on 20 cores have been obtained using a cluster of five such computers with a GigaBit connection. All the results are obtained with precision 10^{-8}.

4.1 Cyclic Polling Server

This case study facilitates a comparison with [3]. The cyclic polling server (CPS) system [16] is a queuing system consisting of a server and N stations each equipped with a queue of capacity 1, cf. Fig. 3 for $N = 3$. Jobs arrive with rate λ and the server polls the stations in a round-robin order with rate γ. When the server is polling a station with a full queue, it can either serve the job in the queue or it can poll the next station (both with rate γ). Once the server decides to serve a job, it can successfully process the job with rate μ or it will fail with rate ρ. The 1-clock DTA objective (adapted from [3], see Fig. 4) requires that after consulting all queues for one round, the server should serve each queue one after the other within T time units. The label st_i indicates that the system is at station i; srv means that the system is serving the job in the current station and j_arr means a new job arrives at some station. The DTA starts from station 1 at q_0 and goes to q_1 when polling the next station (st_2). It stays at q_1 for not polling station 1 –implicitly it goes sequentially from station 2 to $N-$ until it sees station 1 again (and goes to state q_2). Note that the clock is reset before going to state q_2. From state q_2 to state q_5, it specifies serving stations $1, \ldots, N$ one by one within the deadline T. The dashed line indicates the intermediate transitions from station st_2 to station st_{N-1}.

Table 1 summarizes our results, where %transient and %lumping indicate the fraction of time to compute transient probabilities and to lump all CTMCs, respectively. The computed probabilities $Prob(\mathcal{C} \models \mathcal{A})$ are identical to [3] (that contains results up to $N=7$); our verification times are, however, three orders of magnitude faster. If lumping is not applied, then most of the time is spent on the transient analysis. Lumping can save approximately $\frac{2}{3}$ of the state space ($\frac{\#blocks}{\#product\ states}$), however, it has a major impact on the verification times.

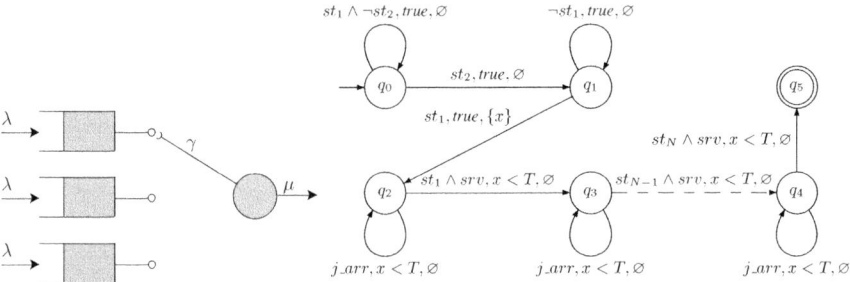

Fig. 3. Cyclic polling server ($\lambda=\frac{\mu}{N}, \mu=0.5, \gamma=10, \rho=1$)

Fig. 4. DTA for the polling server system ($T = 1$)

Table 1. Experimental results for polling server system (no parallelization)

#queues	#CTMC	No lumping			With lumping			
N	states	#product states	time(s)	%transient	#blocks	time(s)	%transient	%lumping
2	16	51	0	0%	21	0	0%	0%
3	48	143	0.01	0%	52	0.02	0%	60%
4	128	363	0.03	60%	126	0.08	13%	65%
5	320	875	0.13	84%	298	0.37	16%	79%
6	768	2043	0.57	88%	690	1.75	15%	82%
7	1792	4667	2.6	90%	1570	8.58	14%	84%
8	4096	10491	14.8	94%	3522	41.92	13%	85%
9	9216	23291	120	98%	7810	230	22%	77%
10	20480	51195	636	98%	17154	1381	25%	75%

4.2 Robot Navigation

A robot moves on a grid with $N \times N$ cells, see Fig. 5. It can move up, down, left and right with rate λ. The black squares from the grid represent walls, i.e., the robot is prohibited to pass through them. The robot is allowed to stay in consecutive C-cells for at most T_1 units of time, and in the D-cells for at most T_2 units of time. There is no time constraint on the residence times in the A-cells. The task is to compute the probability to reach the B-cell from the A-cell labeled with ↗. No time constraint is imposed on reaching the target. The DTA objective is shown in Fig. 6. Intuitively, q_A, q_B, q_C and q_D represent the states that the robot is in the respective cells. From q_A, q_C, and q_D, it is possible to go to the final state q_B. The outgoing edges from q_C and q_D have the guard $x < T_1$ or $x < T_2$; while their incoming edges reset the clock x.

Table 2 presents the results. Lumping is attractive as it reduces the state space by a factor two, and speeds-up the verification. As opposed to the polling system case, most time is spent on building the product and solving the linear equation system. The gray rows in Table 1 and 2 refer to similar product size whereas the verification times differ by two orders of magnitude (14.88 vs. 2433.31). This is due to the fact that there are two and three subgraphs, respectively. The resulting linear equation system has 2 and 3

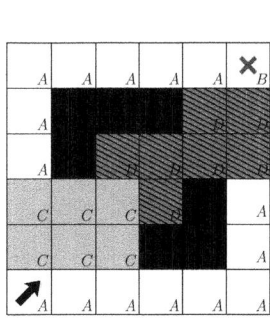
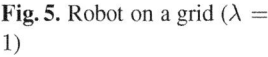
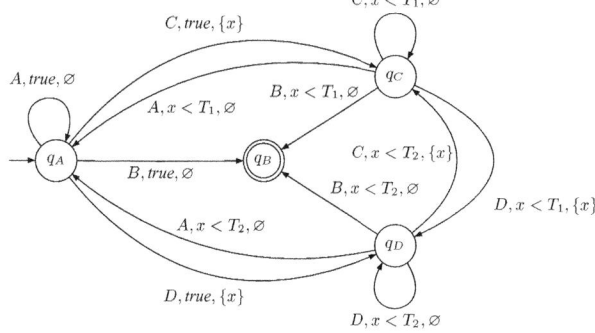

Fig. 5. Robot on a grid ($\lambda = 1$)

Fig. 6. DTA for robot case study ($T_1 = 3, T_2 = 5$)

Table 2. Experimental results for the robot example (without parallelization)

	#CTMC	No lumping			With lumping			
N	states	#product states	time(s)	%transient	#blocks	time(s)	%transient	%lumping
10	100	148	0.09	59%	78	0.09	43%	32%
20	400	702	6.7	18%	380	7.1	14%	7%
30	900	1248	32	17%	619	26	14%	6%
40	1600	2672	119	13%	1296	93	10%	5%
50	2500	4174	135	17%	2015	138	12%	7%
60	3600	4232	309	16%	1525	261	12%	7%
70	4900	8661	904	12%	4212	1130	7%	3%
80	6400	9529	1753	12%	4339	1429	14%	4%
90	8100	9812	2433	8%	2613	1922	6%	5%

variables accordingly and this influences the verification times. The number of blocks is not monotonically increasing, as the robot grid (how the walls and regions C and D are located) is randomly generated. The structure of the grid, e.g., whether it is symmetric or not, has a major influence on the lumping time and quotient size.

4.3 Systems Biology

The last case study stems from a real example [15] in systems biology. The goal is to generate activated messengers. M ligands can bind with a number of receptors, cf. Fig. 7). Initially each ligand binds with a free receptor R with rate k_{+1} and it forms a ligand-receptor (LR) B_0. The LR then undergoes a sequence of N modifications with a constant rate k_p and becomes $B_1, \ldots, B_N$. From every LR B_i ($0 \leqslant i \leqslant N$) the ligand can separate from the receptor with rate k_{-1}. The LR B_N can link with an inactive messenger with rate k_{+x} and then forms a new component ligand-receptor-messenger (LRM) (the last B_N in Fig. 7). The LRM can decompose into three separate components in their initial forms with rate k_{-1}, or the messenger can separate from LRM into an inactive (resp. active) messenger with rate k_{-x} (resp. k_{cat}).

The 1-clock DTA objective is given in Fig. 8. Intuitively, it requires a transformation from R to LR B_N directly without jumping back to R inbetween and manage to activate

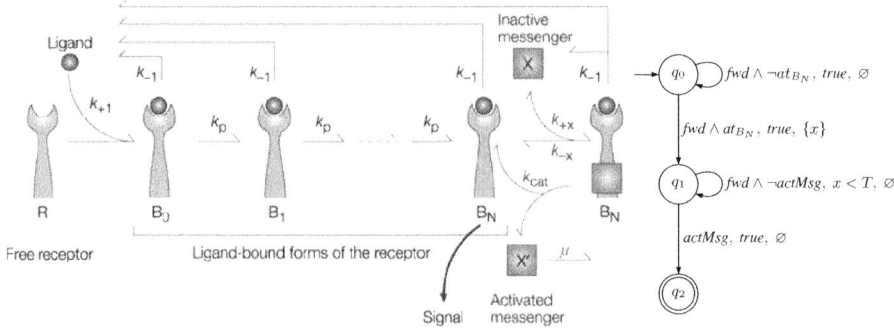

Fig. 7. Kinetik proof reading with a messenger [15] (#$R = 900$, #Inact. msg $= 10000$, $N = 6$, $k_{+1} = 6, 7 \times 10^{-3}$, $k_{-1} = 0, 5$, $k_p = 0, 25$, $k_{+x} = 1, 2 \times 10^{-3}$, $k_{-x} = 0, 01$, $k_{cat} = 100$)

Fig. 8. DTA for the biology case study ($T = 1$)

Table 3. Experimental results for the biology example (no parallelization)

	#CTMC	No lumping			With lumping			
M	states	#product states	time(s)	%transient	#blocks	time(s)	%transient	%lumping
1	18	31	0	0%	13	0	0%	0%
2	150	203	0.06	93%	56	0.05	58%	39%
3	774	837	1.36	94%	187	0.84	64%	30%
4	3024	2731	17.29	97%	512	9.19	73%	24%
5	9756	7579	152.54	97%	1213	73.4	76%	21%
6	27312	18643	1547.45	98%	2579	457.35	78%	20%
7	68496	41743	11426.46	99%	5038	3185.6	85%	14%
8	157299	86656	23356.5	99%	9200	11950.8	81%	18%
9	336049	169024	71079.15	99%	15906	38637.28	76%	22%
10	675817	312882	205552.36	99%	26256	116314.41	71%	26%

a messenger within T time units after reaching the LR B_N. In the DTA, *fwd* means that the last transformation is moving forward, i.e., not jumping back to R; at_{B_N} means that the process reaches LR B_N and *actMsg* means the active messenger is generated. In q_0, the process is on the way to reach LR B_N. When it reaches LR B_N, the DTA goes from q_0 to q_1 and resets the clock x. In q_1, there is no active messenger generated yet. It will go to q_2 when an active messenger is generated. Note that the time constraint $x < T$ is checked on the self-loop on q_1. As Table 3 indicates, lumping works very well on this example: it reduces both time and space by almost one order of magnitude.

4.4   Parallelization

Model checking a CTMC against a 1-clock DTA can be parallelized in a natural manner. In this section, we present the results when parallelization is applied on the above three case studies. We experimented on distributing the tasks on 1 machine with 4 cores as well as 5 machines with 4 cores each (20 cores in total). We focused on parallelizing the transient analysis; as lumping is not parallelized, we determine the speedup without lumping. For each CTMC $\mathcal{C}_i$, we need to compute for each state its transient probability vector which corresponds to a column in the transient probability matrix. We distribute the computation of different columns to different cores (which might be on different machines). To do so, we launch N different processes and send them the rate matrix and a list of initial states, and each process returns the transient probability vector for each initial state. The speedup factor is computed by $\frac{\text{time without para.}}{\text{time with para.}}$. From Table 4, it follows that parallelization mostly works well for larger models. For small models, it usually does not pay off to distribute the computation tasks, due to overhead. For the polling server system, as the number of stations N increases, the value of $\frac{\text{speedup for 20 cores}}{\text{speedup for 4 cores}}$ approximates 20/4=5. The same applies to the biology example. The parallelization does not work well on the robot example. The performance on 20 cores might even be worse than on 4 cores. This is due to the fact that only the transient analysis is parallelized. From Table 1 and 2, we can see that most of the computation time is spent on transient analysis for the polling and biology examples. This explains why parallelization works well here. In the robot example, however, the transient analysis does not dominate the computation time. This yields moderate speedups. An interesting future work is to apply parallel lumping as in [10] to our setting.

Table 4. Parallel verification of polling (left), robot (mid), and biology example (right)

	4 Cores		20 Cores	
N	time(s)	speedup	time(s)	speedup
4	0.03	1.21	0.18	0.18
5	0.08	1.70	0.22	0.59
6	0.32	1.77	1.54	0.37
7	1.04	2.58	2.08	1.29
8	7.35	2.02	4.04	3.68
9	40.28	2.98	13.76	8.73
10	186.02	3.42	54.97	11.58
11	863.3	3.35	233.99	12.35
12	3940.42	3.65	1089	13.22

	4 Cores		20 Cores	
N	time(s)	speedup	time(s)	speedup
30	23.59	1.37	27.18	1.19
40	84.05	1.42	81.64	1.47
50	122.01	1.11	117.17	1.16
60	266.67	1.16	265.48	1.17
70	793.48	1.14	778.69	1.16
80	1474.88	1.19	1441.99	1.22
90	2498.34	0.97	1917.1	1.27
100	1667.78	1.14	1342.26	1.41
110	4614.92	1.32	5165.7	1.18

	4 Cores		20 Cores	
N	time(s)	speedup	time(s)	speedup
3	0.45	3.03	0.42	3.22
4	5.3	3.26	3.44	5.02
5	44.73	3.41	15.87	9.61
6	620.16	2.50	160.58	9.64
7	4142.19	2.76	949.32	12.04
8	8168.62	2.86	1722.63	13.56
9	23865.17	2.98	5457.01	13.03
10	70623.46	2.91	16699.22	12.31

5 Multi-clock DTA Objectives

The graph decomposition approach for 1-clock DTA fails in the *multi*-clock setting as in case of a reset edge, it cannot be determined to which time point it will jump (unlike $x = 0$ in the 1-clock case). These time points, however, can be approximated by discretization as shown below. W.l.o.g. assume there are two clocks x, y. The maximal constants c_x, c_y to which x and y are compared in the DTA are discretized by $h = \frac{1}{N}$ for some a priori fixed $N \in \mathbb{N}_{>0}$. As a result, there are $c_x c_y / h^2 = N^2 c_x c_y$ areas (or grids). The behavior of all the points in one grid can be regarded as approximately the same. For grid (i, j), it can either delay to $(i+1, j+1)$, or jump to $(0, j)$ (resp. $(i, 0)$) by resetting clock x (resp. y) or to $(0, 0)$ by resetting both clocks. The following definition originates from [12].

Definition 7 (Product of CTMC and DTA). *Let* $\mathcal{C} = (S, \mathrm{AP}, L, s_0, \mathbf{P}, E)$ *be a CTMC and* $\mathcal{A} = (2^{\mathrm{AP}}, \mathcal{X}, Q, q_0, Q_F, \rightarrow)$ *be a DTA. Let* $\mathcal{C} \otimes \mathcal{A} = (Loc, \mathcal{X}, \ell_0, Loc_F, E, \rightsquigarrow)$, *where* $Loc = S \times Q$; $\ell_0 = (s_0, q_0)$; $E(s, q) = E(s)$; $Loc_F := S \times Q_F$, *and* $\rightsquigarrow$ *is the smallest relation defined by:*

$$\frac{\mathbf{P}(s, s') > 0 \;\wedge\; q \xrightarrow{L(s), g, X} q'}{s, q \xrightarrow{g, X} \pi} \; such\ that\ \pi(s', q') = \mathbf{P}(s, s').$$

Note that $\pi \in Distr(S \times Q)$ is a probability distribution. The symbolic edge $\ell \xrightarrow{g, X} \pi$ with distribution π induces transitions of the form $\ell \xmapsto[p]{g, X} \ell'$ with $p = \pi(\ell')$. $\mathcal{C} \otimes \mathcal{A}$ is a stochastic process that can be equipped with a probability measure on infinite paths, cf. [12]. We approximate the stochastic behavior by a discrete-time Markov chain (DTMC). This is done by discretizing clock values in equidistant step-sizes of size $h = 1/N$.

Definition 8 (Discretization). *Let* $\mathcal{C} \otimes \mathcal{A} = (Loc, \mathcal{X}, \ell_0, Loc_F, E, \rightsquigarrow)$, $k = |\mathcal{X}|$, C *be the largest constant in* $\mathcal{A}$, $\mathbf{C} = C^k$ *be a vector with all elements equal to* C, $h = 1/N$ *for some* $N \in \mathbb{N}_{>0}$ *and* $\mathbf{h} = h^k$. *The (unlabeled) DTMC* $\mathcal{D}_h = (S, s_0, \mathbf{P})$ *is given by* $S = Loc \times (\{i \cdot h \mid 0 \leqslant i \leqslant C \cdot N + 1\})^k$, $s_0 = (\ell_0, \mathbf{0})$ *and* $\mathbf{P}((\ell, \eta), (\ell', \eta'))$ *is given by:*

$- \; p \cdot \left(1 - \frac{1 - e^{-E(\ell) \cdot h}}{h \cdot E(\ell)} \right)$ iff $\ell \xmapsto[p]{g, X} \ell' \;\wedge\; \eta \models g \wedge \eta' = \eta[X := 0] \wedge \eta \neq \mathbf{C} + \mathbf{h}$;

$- \; p \cdot \left(\frac{1}{hE(\ell)} - (1 + \frac{1}{hE(\ell)}) e^{-E(\ell)h} \right)$ iff $\ell \xmapsto[p]{g, X} \ell' \quad \wedge \quad \eta \models g \;\wedge\; \eta' = (\eta + h)[X := 0] \;\wedge\; \eta \neq \mathbf{C} + \mathbf{h}$;

- $e^{-E(\ell) \cdot h}$ iff $\ell = \ell' \wedge \eta' = \eta + h \wedge \eta \neq C + \boldsymbol{h}$;
- p iff $\ell \xmapsto[p]{g,X} \ell' \wedge \eta \models g \wedge \eta' = \eta[X := 0] \wedge \eta = C + \boldsymbol{h}$;
- 0 *otherwise.*

The number of states in the derived DTMC is $|S| \cdot |Q| \cdot (C \cdot N + 1)^k$. The following result states that for h approaching zero, $Prob(\mathcal{C} \models \mathcal{A})$ equals the reachability probability in the DTMC $\mathcal{D}_h$ of a state of the form $(\ell, \cdot)$ with $\ell \in Loc_F$.

Theorem 4. *Let $P_h^{\mathcal{D}}(s_0, \Diamond F)$ be the reachability probability in the DTMC $\mathcal{D}_h$ to reach a state $(\ell, \cdot)$ with $\ell \in Loc_F$. Then $\lim_{h \to 0} P_h^{\mathcal{D}}(s_0, \Diamond F) = Prob(\mathcal{C} \models \mathcal{A})$.*

To illustrate the performance of the discretization technique, we add a clock y to the robot example, where y is never reset. The time constraint $y < T_y$ is added to all incoming edges of q_B. The results are shown in Table 5 with $h = 5 \cdot 10^{-2}$. Although the resulting DTMC size is quite large, the computation times are still acceptable, as computing reachability probabilities in a DTMC is rather fast. For the sake of comparison, we applied the discretization technique to the polling server system with a 1-clock DTA objective. We obtain the same probabilities as before; results are given in Table 6 with precision 0.001.

Table 5. Experimental results for the robot example with 2-clock DTA ($T_y = 3$)

N	CTMC size	DMTA size	DTMC size	time (s)
10	100	105	865305	79
20	400	475	3914475	412
30	900	1003	8265723	868
40	1600	1669	13754229	1605
50	2500	2356	19415796	2416
60	3600	3411	28110051	3559
70	4900	4850	39968850	22427

Table 6. Polling example with 1-clock DTA using discretization

N	CTMC size	DMTA size	DTMC size	time (s)
2	16	31	31031	15.89
3	48	89	89089	52.66
4	128	229	229229	152.2
5	320	557	557557	407.4
6	768	1309	1310309	1042
7	1792	3005	3008005	2577
8	4096	6781	6787781	6736
9	9216	15101	15116101	30865

6 Conclusion

We have presented a practical approach to verifying CTMCs against DTA objectives. First, we showed that a standard region construction on DTA suffices. For 1-clock DTA, we showed that the graph decomposition approach of [12] offers rather straightforward possibilities for optimizations, viz. bisimulation minimization and parallelization. Several experiments substantiate this claim. The main result of this paper is that it shows that 1-clock DTA objectives can be handled by completely standard means: region construction of timed automata, transient analysis of CTMCs, graph analysis, and solving linear equation systems. Our approach clearly outperforms alternative techniques for CSL^{TA} [3,14], and allows for the verification of objectives that cannot be treated with other CTMC model checkers. Our prototype is available at `http://moves.rwth-aachen.de/CoDeMoC`. Finally, we remark that although we only considered finite acceptance conditions in this paper, our approach can easily be extended to DTA with Rabin acceptance conditions.

Acknowledgement. We thank Verena Wolf (Saarland University) for providing us with the biology case study. This research is funded by the DFG research training group 1295 AlgoSyn, the SRO DSN project of CTIT, University of Twente, the EU FP7 project QUASIMODO and the DFG-NWO ROCKS project.

References

1. PRISM website, http://www.prismmodelchecker.org
2. Alur, R., Dill, D.L.: A theory of timed automata. TCS 126(2), 183–235 (1994)
3. Amparore, E.G., Donatelli, S.: Model checking CSL^{TA} with deterministic and stochastic Petri Nets. In: Dependable Systems and Networks (DSN), pp. 605–614 (2010)
4. Baier, C., Bertrand, N., Bouyer, P., Brihaye, T., Grösser, M.: Almost-sure model checking of infinite paths in one-clock timed automata. In: LICS, pp. 217–226 (2008)
5. Baier, C., Cloth, L., Haverkort, B.R., Kuntz, M., Siegle, M.: Model checking Markov chains with actions and state labels. IEEE TSE 33(4), 209–224 (2007)
6. Baier, C., Haverkort, B.R., Hermanns, H., Katoen, J.-P.: Model-checking algorithms for continuous-time Markov chains. IEEE TSE 29(6), 524–541 (2003)
7. Baier, C., Haverkort, B.R., Hermanns, H., Katoen, J.-P.: Performance evaluation and model checking join forces. Commun. of the ACM 53(9), 74–85 (2010)
8. Bérard, B., Petit, A., Diekert, V., Gastin, P.: Characterization of the expressive power of silent transitions in timed automata. Fund. Inf. 36(2-3), 145–182 (1998)
9. Bertrand, N., Bouyer, P., Brihaye, T., Markey, N.: Quantitative model-checking of one-clock timed automata under probabilistic semantics. In: QEST, pp. 55–64 (2008)
10. Blom, S., Haverkort, B.R., Kuntz, M., van de Pol, J.: Distributed Markovian bisimulation reduction aimed at CSL model checking. ENTCS 220(2), 35–50 (2008)
11. Brázdil, T., Krčál, J., Křetínský, J., Kučera, A., Řehák, V.: Stochastic real-time games with qualitative timed automata objectives. In: Gastin, P., Laroussinie, F. (eds.) CONCUR 2010. LNCS, vol. 6269, pp. 207–221. Springer, Heidelberg (2010)
12. Chen, T., Han, T., Katoen, J.-P., Mereacre, A.: Quantitative model checking of continuous-time Markov chains against timed automata specification. In: LICS, pp. 309–318 (2009)
13. Derisavi, S., Hermanns, H., Sanders, W.H.: Optimal state-space lumping in Markov chains. Inf. Process. Lett. 87(6), 309–315 (2003)
14. Donatelli, S., Haddad, S., Sproston, J.: Model checking timed and stochastic properties with CSL^{TA}. IEEE TSE 35(2), 224–240 (2009)
15. Goldstein, B., Faeder, J.R., Hlavacek, W.S.: Mathematical and computational models of immune-receptor signalling. Nat. Reviews Immunology 4, 445–456 (2004)
16. Haverkort, B.R.: Performance evaluation of polling-based communication systems using SPNs. In: Appl. of Petri Nets to Comm. Networks, pp. 176–209 (1999)
17. Katoen, J.-P., Hahn, E.M., Hermanns, H., Jansen, D.N., Zapreev, I.: The ins and outs of the probabilistic model checker MRMC. In: QEST, pp. 167–176 (2009)
18. Katoen, J.-P., Kemna, T., Zapreev, I., Jansen, D.: Bisimulation minimisation mostly speeds up probabilistic model checking. In: Grumberg, O., Huth, M. (eds.) TACAS 2007. LNCS, vol. 4424, pp. 87–101. Springer, Heidelberg (2007)
19. Valmari, A., Franceschinis, G.: Simple $O(m \log n)$ time markov chain lumping. In: Esparza, J., Majumdar, R. (eds.) TACAS 2010. LNCS, vol. 6015, pp. 38–52. Springer, Heidelberg (2010)
20. Vardi, M.Y.: Automatic verification of probabilistic concurrent finite-state programs. In: FOCS, pp. 327–338 (1985)

Efficient Interpolant Generation in Satisfiability Modulo Linear Integer Arithmetic

Alberto Griggio[1,*], Thi Thieu Hoa Le[2], and Roberto Sebastiani[2,**]

[1] FBK-Irst, Trento, Italy
[2] DISI, University of Trento, Italy

Abstract. The problem of computing Craig interpolants in SAT and SMT has recently received a lot of interest, mainly for its applications in formal verification. Efficient algorithms for interpolant generation have been presented for some theories of interest —including that of equality and uninterpreted functions ($\mathcal{EUF}$), linear arithmetic over the rationals ($\mathcal{LA}(\mathbb{Q})$), and their combination— and they are successfully used within model checking tools. For the theory of linear arithmetic over the integers ($\mathcal{LA}(\mathbb{Z})$), however, the problem of finding an interpolant is more challenging, and the task of developing efficient interpolant generators for the full theory $\mathcal{LA}(\mathbb{Z})$ is still the objective of ongoing research.

In this paper we try to close this gap. We build on previous work and present a novel interpolation algorithm for SMT($\mathcal{LA}(\mathbb{Z})$), which exploits the full power of current state-of-the-art SMT($\mathcal{LA}(\mathbb{Z})$) solvers. We demonstrate the potential of our approach with an extensive experimental evaluation of our implementation of the proposed algorithm in the MATHSAT SMT solver.

1 Motivations, Related Work and Goals

Given two formulas A and B such that $A \wedge B$ is inconsistent, a *Craig interpolant* (simply "interpolant" hereafter) for (A, B) is a formula I s.t. A entails I, $I \wedge B$ is inconsistent, and all uninterpreted symbols of I occur in both A and B.

Interpolation in both SAT and SMT has been recognized to be a substantial tool for formal verification. For instance, in the context of software model checking based on counter-example-guided-abstraction-refinement (CEGAR) interpolants of quantifier-free formulas in suitable theories are computed for automatically refining abstractions in order to rule out spurious counterexamples. Consequently, the problem of computing interpolants in SMT has received a lot of interest in the last years (e.g., [14,17,19,11,4,10,13,7,8,3,12]). In the recent years, efficient algorithms and tools for interpolant generation for quantifier-free formulas in SMT have been presented for some theories of interest, including that of equality and uninterpreted functions ($\mathcal{EUF}$) [14,7], linear arithmetic over the rationals ($\mathcal{LA}(\mathbb{Q})$) [14,17,4], and for their combination [19,17,4,8], and they are successfully used within model-checking tools.

For the theory of linear arithmetic over the *integers* ($\mathcal{LA}(\mathbb{Z})$), however, the problem of finding an interpolant is more challenging. In fact, it is not always possible to

---

* Supported by the European Community's FP7/2007-2013 under grant agreement Marie Curie FP7 - PCOFUND-GA-2008-226070 "progetto Trentino", project Adaptation.

** Supported by SRC under GRC Custom Research Project 2009-TJ-1880 WOLFLING.

P.A. Abdulla and K.R.M. Leino (Eds.): TACAS 2011, LNCS 6605, pp. 143–157, 2011.

obtain quantifier-free interpolants starting from quantifier-free input formulas in the standard signature of $\mathcal{LA}(\mathbb{Z})$ (consisting of Boolean connectives, integer constants and the symbols $+, \cdot, \leq, =$) [14]. For instance, there is no quantifier-free interpolant for the $\mathcal{LA}(\mathbb{Z})$-formulas $A \overset{\text{def}}{=} (2x - y + 1 = 0)$ and $B \overset{\text{def}}{=} (y - 2z = 0)$.

In order to overcome this problem, different research directions have been explored. One is to restrict to important *fragments* of $\mathcal{LA}(\mathbb{Z})$ where the problem does not occur. To this extent, efficient interpolation algorithms for the Difference Logic ($\mathcal{DL}$) and Unit-Two-Variables-Per-Inequality ($\mathcal{UTVPI}$) fragments of $\mathcal{LA}(\mathbb{Z})$ have been proposed in [4]. Another direction is to extend the signature of $\mathcal{LA}(\mathbb{Z})$ to contain *modular equalities* $=_c$ (or, equivalently, *divisibility predicates*), so that it is possible to compute quantifier-free $\mathcal{LA}(\mathbb{Z})$ interpolants by means of quantifier elimination —which is however prohibitively expensive in general, both in theory and in practice. For instance, $I \overset{\text{def}}{=} (-y + 1 =_2 0) \equiv \exists x.(2x - y + 1 = 0)$ is an interpolant for the formulas (A, B) above. Using modular equalities, Jain et al. [10] developed polynomial-time interpolation algorithms for linear equations and their negation and for linear modular equations. A similar algorithm was also proposed in [13]. The work in [3] was the first to present an interpolation algorithm for the full $\mathcal{LA}(\mathbb{Z})$ (augmented with divisibility predicates) which was not based on quantifier elimination. Finally, an alternative algorithm, exploiting efficient interpolation procedures for $\mathcal{LA}(\mathbb{Q})$ and for linear equations in $\mathcal{LA}(\mathbb{Z})$, has been recently presented in [12].

The obvious limitation of the first research direction is that it does not cover the full $\mathcal{LA}(\mathbb{Z})$. For the second direction, the approaches so far seem to suffer from some drawbacks. In particular, some of the interpolation rules of [3] might result in an exponential blow-up in the size of the interpolants wrt. the size of the proofs of unsatisfiability from which they are generated. The algorithm of [12] avoids this, but at the cost of significantly restricting the heuristics commonly used in state-of-the-art SMT solvers for $\mathcal{LA}(\mathbb{Z})$ (e.g. in the framework of [12] both the use of Gomory cuts [18] and of "cuts from proofs" [5] is not allowed). More in general, the important issue of how to efficiently integrate the presented techniques into a state-of-the-art SMT($\mathcal{LA}(\mathbb{Z})$) solver is not immediate to foresee from the papers.

In this paper we try to close this gap. After recalling the necessary background knowledge (§2), we present our contribution, which is twofold.

First (§3) we show how to extend the state-of-the art $\mathcal{LA}(\mathbb{Z})$-solver of MATHSAT [9] in order to implement interpolant generation on top of it without affecting its efficiency. To this extent, we combine different algorithms corresponding to the different submodules of the $\mathcal{LA}(\mathbb{Z})$-solver, so that each of the submodules requires only minor modifications, and implement them in MATHSAT (MATHSAT-MODEQ hereafter). An extensive empirical evaluation (§5) shows that MATHSAT-MODEQ outperforms in efficiency all existing interpolant generators for $\mathcal{LA}(\mathbb{Z})$.

Second (§4), we propose a novel and general interpolation algorithm for $\mathcal{LA}(\mathbb{Z})$, independent from the architecture of MATHSAT, which overcomes the drawbacks of the current approaches. The key idea is to extend both the signature and the domain of $\mathcal{LA}(\mathbb{Z})$: we extend the signature by adding the *ceiling function* $\lceil \cdot \rceil$ to it, and the domain by allowing non-variable terms to be non-integers. This greatly simplifies the interpolation procedure, and allows for producing interpolants which are much more compact than those generated by the algorithm of [3]. Also this novel technique was easily implemented on top of the $\mathcal{LA}(\mathbb{Z})$-solver of MATHSAT without affecting its efficiency. (We call this implementation MATHSAT-CEIL.) An extensive empirical evaluation (§5) shows that MATHSAT-CEIL drastically outperforms MATHSAT-MODEQ, and hence all other existing interpolant generators for $\mathcal{LA}(\mathbb{Z})$, for both efficiency and size of the final interpolant.

2   Background: SMT($\mathcal{LA}(\mathbb{Z})$)

2.1   Generalities

Satisfiability Modulo Theory – SMT. Our setting is standard first order logic. We use the standard notions of theory, satisfiability, validity, logical consequence. We call *Satisfiability Modulo (the) Theory* $\mathcal{T}$, SMT($\mathcal{T}$), the problem of deciding the satisfiability of quantifier-free formulas wrt. a background theory $\mathcal{T}$.[1] Given a theory $\mathcal{T}$, we write $\phi \models_{\mathcal{T}} \psi$ (or simply $\phi \models \psi$) to denote that the formula ψ is a logical consequence of ϕ in the theory $\mathcal{T}$. With $\phi \preceq \psi$ we denote that all uninterpreted (in $\mathcal{T}$) symbols of ϕ appear in ψ. With a little abuse of notation, we might sometimes denote conjunctions of literals $l_1 \wedge \ldots \wedge l_n$ as sets $\{l_1, \ldots, l_n\}$ and vice versa. If η is the set $\{l_1, \ldots, l_n\}$, we might write $\neg\eta$ to mean $\neg l_1 \vee \ldots \vee \neg l_n$.

We call $\mathcal{T}$-solver a procedure that decides the consistency of a conjunction of literals in $\mathcal{T}$. If S is a set of literals in $\mathcal{T}$, we call $\mathcal{T}$-*conflict set w.r.t.* S any subset η of S which is inconsistent in $\mathcal{T}$. We call $\neg\eta$ a $\mathcal{T}$-*lemma* (notice that $\neg\eta$ is a $\mathcal{T}$-valid clause).

A standard technique for solving the SMT($\mathcal{T}$) problem is to integrate a DPLL-based SAT solver and a $\mathcal{T}$-solver in a *lazy* manner (see, e.g., [2] for a detailed description). DPLL is used as an enumerator of truth assignments for the propositional abstraction of the input formula. At each step, the set of $\mathcal{T}$-literals in the current assignment is sent to the $\mathcal{T}$-solver to be checked for consistency in $\mathcal{T}$. If S is inconsistent, the $\mathcal{T}$-solver returns a conflict set η, and the corresponding $\mathcal{T}$-lemma $\neg\eta$ is added as a blocking clause in DPLL, and used to drive the backjumping and learning mechanism.

Interpolation in SMT. We consider the SMT($\mathcal{T}$) problem for some background theory $\mathcal{T}$. Given an ordered pair (A, B) of formulas such that $A \wedge B \models_{\mathcal{T}} \bot$, a *Craig interpolant* (simply "interpolant" hereafter) is a formula I s.t. (i) $A \models_{\mathcal{T}} I$, (ii) $I \wedge B$ is $\mathcal{T}$-inconsistent, and (iii) $I \preceq A$ and $I \preceq B$.

Following [14], an interpolant for (A, B) in SMT($\mathcal{T}$) can be generated by combining a propositional interpolation algorithm for the Boolean structure of the formula $A \wedge B$ with a $\mathcal{T}$-specific interpolation procedure that deals only with negations of $\mathcal{T}$-lemmas, that is, with $\mathcal{T}$-inconsistent conjunctions of $\mathcal{T}$-literals. Therefore, in the rest of the paper, we shall consider algorithms for conjunctions/sets of literals only, which can be extended to general formulas by simply "plugging" them into the algorithm of [14].

---

[1] The general definition of SMT deals also with quantified formulas. Nevertheless, in this paper we restrict our interest to quantifier-free formulas.

2.2 Efficient SMT($\mathcal{LA}(\mathbb{Z})$) Solving

In this section, we describe our algorithm for efficiently solving SMT($\mathcal{LA}(\mathbb{Z})$) problems, as implemented in the MATHSAT 5 SMT solver [9]. They key feature of our solver is an extensive use of *layering* and *heuristics* for combining different known techniques, in order to exploit the strengths and to overcome the limitations of each of them. Both the experimental results of [9] and the latest SMT solvers competition SMT-COMP'10[2] demonstrate that it represents the current state of the art in SMT($\mathcal{LA}(\mathbb{Z})$).

The architecture of the solver is outlined in Fig. 1. It is organized as a layered hierarchy of submodules, with cheaper (but less powerful) ones invoked earlier and more often. The general strategy used for checking the consistency of a set of $\mathcal{LA}(\mathbb{Z})$-constraints is as follows.

First, the rational relaxation of the problem is checked, using a Simplex-based $\mathcal{LA}(\mathbb{Q})$-solver similar to that described in [6]. If no conflict is detected, the model returned by the $\mathcal{LA}(\mathbb{Q})$-solver is examined to check whether all integer variables are assigned to an integer value. If this happens, the $\mathcal{LA}(\mathbb{Q})$-model is also a $\mathcal{LA}(\mathbb{Z})$-model, and the solver can return sat.

Otherwise, the specialized module for handling linear $\mathcal{LA}(\mathbb{Z})$ equations (Diophantine equations) is invoked. This module is similar to the first part of the Omega test described in [16]: it takes all the equations in the input problem, and tries to eliminate them by computing a parametric solution of the system and then substituting each variable in the inequalities with its parametric expression. If the system of equations itself is infeasible, this module is also able to detect the inconsistency.

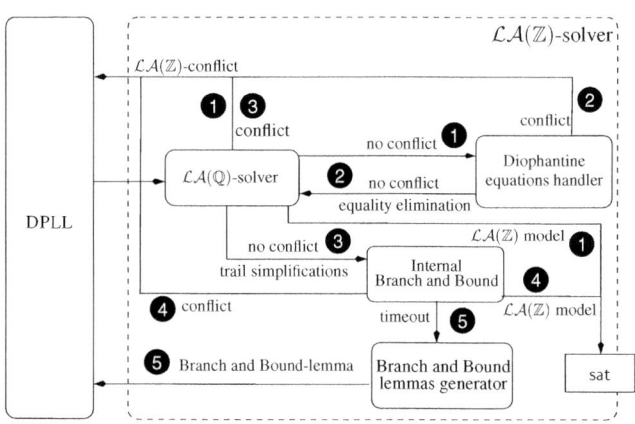

Fig. 1. Architecture of the $\mathcal{LA}(\mathbb{Z})$-solver of MATHSAT

Otherwise, the inequalities obtained by substituting the variables with their parametric expressions are normalized, tightened and then sent to the $\mathcal{LA}(\mathbb{Q})$-solver, in order to check the $\mathcal{LA}(\mathbb{Q})$-consistency of the new set of constraints.

If no conflict is detected, the branch and bound module is invoked, which tries to find a $\mathcal{LA}(\mathbb{Z})$-solution via branch and bound [18]. This module is itself divided into two submodules operating in sequence. First, the "internal" branch and bound module is activated, which performs case splits directly within the $\mathcal{LA}(\mathbb{Z})$-solver. The internal search is performed only for a bounded (and small) number of branches, after which the "external" branch and bound module is called. This works in cooperation with the DPLL engine, using the "splitting on-demand" approach of [1]: case splits are delegated to DPLL, by sending to it $\mathcal{LA}(\mathbb{Z})$-valid clauses of the form $(t-c \leq 0) \vee (-t+c+1 \leq 0)$ (called branch-and-bound lemmas) that encode the required splits. Such clauses are generated with the "cuts from proofs" algorithm of [5]: "normal" branch-and-bound steps – splitting cases on an individual variable – are interleaved with "extended" steps, in which branch-and-bound lemmas involve an arbitrary linear combination of variables, generated by computing proofs of unsatisfiability of particular systems of Diophantine equations.

3   From $\mathcal{LA}(\mathbb{Z})$-Solving to $\mathcal{LA}(\mathbb{Z})$-Interpolation

Our objective is that of devising an interpolation algorithm that could be implemented on top of the $\mathcal{LA}(\mathbb{Z})$-solver described in the previous section without affecting its efficiency. To this end, we combine different algorithms corresponding to the different submodules of the $\mathcal{LA}(\mathbb{Z})$-solver, so that each of the submodules requires only minor modifications.

3.1   Interpolation for Diophantine Equations

An interpolation procedure for systems of Diophantine equations was given by Jain et al. in [10]. The procedure starts from a proof of unsatisfiability expressed as a linear combination of the input equations whose result is an equation $(\sum c_i x_i + c = 0)$ in which the greatest common divisor (GCD) of the coefficients c_i of the variables is not a divisor of the constant term c.

Given a proof of unsatisfiability for a system of equations partitioned into A and B, let $(\sum_{x_i \in A \cap B} c_i x_i + \sum_{y_j \notin B} a_j y_j + c = 0)$ be the linear combination of the equations from A with the coefficients given by the proof of unsatisfiability. Then, $I \stackrel{def}{=} \sum_{x_i \in A \cap B} c_i x_i + c =_g 0$, where g is any integer that divides $GCD(a_j)$, is an interpolant for (A, B) [10]. Jain et al. show that a proof of unsatisfiability can be obtained by computing the Hermite Normal Form [18] of the system of equations. However, this is only one possible way of obtaining such proof. In particular, as shown in [9], the submodule of our $\mathcal{LA}(\mathbb{Z})$-solver that deals with Diophantine equations can already produce proofs of unsatisfiability directly. Therefore, we can apply the interpolation algorithm of [10] without any modification to the solver.

3.2   Interpolation for Inequalities

The second submodule of our $\mathcal{LA}(\mathbb{Z})$-solver checks the $\mathcal{LA}(\mathbb{Q})$-consistency of a set of inequalities, some of which obtained by *substitution* and *tightening* [9]. In this case, we produce interpolants starting from proofs of unsatisfiability in the *cutting-plane proof system*, a complete proof system for $\mathcal{LA}(\mathbb{Z})$ [18]. Similarly to previous work on $\mathcal{LA}(\mathbb{Q})$ and $\mathcal{LA}(\mathbb{Z})$ [14,3], we produce interpolants by annotating each step of the proof of unsatisfiability of $A \wedge B$, such that the annotation for the root of the proof (deriving an inconsistent inequality $(c \leq 0)$ with $c \in \mathbb{Z}^{>0}$) is an interpolant for (A, B).

Definition 1 (Valid annotated sequent). *An* annotated sequent *is a sequent in the form* $(A, B) \vdash (t \leq 0)[I]$ *where A and B are conjunctions of equalities and inequalities in* $\mathcal{LA}(\mathbb{Z})$, *and where I (called* annotation*) is a set of pairs $\langle (t_i \leq 0), E_i \rangle$ in which E_i is a (possibly empty) conjunction of equalities and modular equalities. It is said to be* valid *when:*

1. $A \models \bigvee_{\langle t_i \leq 0, E_i \rangle \in I}((t_i \leq 0) \wedge E_i)$;
2. *For all* $\langle t_i \leq 0, E_i \rangle \in I$, $B \wedge E_i \models (t - t_i \leq 0)$;
3. *For every element* $\langle (t_i \leq 0), E_i \rangle$ *of* I, $t_i \preceq A$, $(t - t_i) \preceq B$, $E_i \preceq A$ *and* $E_i \preceq B$.

Definition 2 (Interpolating Rules). *The* $\mathcal{LA}(\mathbb{Z})$-*interpolating inference rules that we use are the following:*

Hyp-A $\dfrac{}{(A, B) \vdash (t \leq 0)[\{\langle t \leq 0, \top \rangle\}]}$ $\quad$ *if* $(t \leq 0) \in A$ *or* $(t = 0) \in A$

Hyp-B $\dfrac{}{(A, B) \vdash (t \leq 0)[\{\langle 0 \leq 0, \top \rangle\}]}$ $\quad$ *if* $(t \leq 0) \in B$ *or* $(t = 0) \in B$

Comb $\dfrac{(A, B) \vdash t_1 \leq 0[I_1] \quad (A, B) \vdash t_2 \leq 0[I_2]}{(A, B) \vdash (c_1 t_1 + c_2 t_2 \leq 0)[I]}$ *where:*
- $c_1, c_2 > 0$
- $I \stackrel{\text{def}}{=} \{\langle c_1 t_1' + c_2 t_2' \leq 0, E_1 \wedge E_2 \rangle \mid \langle t_1' \leq 0, E_1 \rangle \in I_1 \text{ and } \langle t_2' \leq 0, E_2 \rangle \in I_2\}$

Strengthen $\dfrac{(A, B) \vdash \sum_i c_i x_i + c \leq 0[\{\langle t' \leq 0, \top \rangle\}]}{(A, B) \vdash \sum_i c_i x_i + c + k \leq 0[I]}$ *where:*
- $k \stackrel{\text{def}}{=} d\left\lceil \dfrac{c}{d} \right\rceil - c$, *and $d > 0$ is an integer that divides all the c_i's;*
- $I \stackrel{\text{def}}{=} \{\langle t' + j \leq 0, \exists(x \notin B).(t' + j = 0) \rangle \mid 0 \leq j < k\} \cup \{\langle t' + k \leq 0, \top \rangle\}$; *and*
- $\exists(x \notin B).(t' + j = 0)$ *denotes the result of the existential elimination from $(t' + j = 0)$ of all and only the variables $x_1, ..., x_n$ not occurring in B.*[3]

Theorem 1. *All the interpolating rules preserve the validity of the sequents.*

Corollary 1. *If we can derive a valid sequent $(A, B) \vdash c \leq 0[I]$ with $c \in \mathbb{Z}^{>0}$, then* $\varphi_I \stackrel{\text{def}}{=} \bigvee_{\langle t_i \leq 0, E_i \rangle \in I}((t_i \leq 0) \wedge E_i)$ *is an interpolant for (A, B).*

Notice that the first three rules correspond to the rules for $\mathcal{LA}(\mathbb{Q})$ given in [14], whereas Strengthen is a reformulation of the k-Strengthen rule given in [3]. Moreover, although the rules without annotations are refutationally complete for $\mathcal{LA}(\mathbb{Z})$, in the above formulation the annotation of Strengthen might prevent its applicability, thus losing completeness. In particular, it only allows to produce proofs with at most one strengthening per branch. Such restriction has been put only for simplifying the proofs of correctness, and it is not present in the original k-Strengthen of [3]. However, for our purposes this is not a problem, since we use the above rules only in the second submodule of our $\mathcal{LA}(\mathbb{Z})$-solver, which always produces proofs with at most one strengthening per branch.

---

[3] We recall that $\exists(x_1, \ldots, x_n).(\sum_i c_i x_i + \sum_j d_j y_j + c = 0) \equiv (\sum_j d_j y_j + c =_{GCD(c_i)} 0)$, and that $(t =_0 0) \equiv (t = 0)$.

Generating cutting-plane proofs in the $\mathcal{LA}(\mathbb{Z})$-solver. The equality elimination and tightening step generates new inequalities $(t' + c' + k \leq 0)$ starting from a set of input equalities $\{e_1 = 0, \ldots, e_n = 0\}$ and an input inequality $(t+c \leq 0)$. Thanks to its proof-production capabilities [9], we can extract from the Diophantine equations submodule the coefficients $\{c_1, \ldots, c_n\}$ such that $(\sum_i c_i e_i + t + c \leq 0) \equiv (t' + c' \leq 0)$. Thus, we can generate a proof of $(t' + c' \leq 0)$ by using the Comb and Hyp rules. We then use the Strengthen rule to obtain a proof of $(t' + c' + k \leq 0)$. The new inequalities generated are then added to the $\mathcal{LA}(\mathbb{Q})$-solver. If a $\mathcal{LA}(\mathbb{Q})$-conflict is found, then, the $\mathcal{LA}(\mathbb{Q})$-solver produces a $\mathcal{LA}(\mathbb{Q})$-proof of unsatisfiability (as described in [4]) in which some of the leaves are the new inequalities generated by equality elimination and tightening. We can then simply replace such leaves with the corresponding cutting-plane proofs to obtain the desired cutting-plane unsatisfiability proof.

3.3   Interpolation with Branch-and-Bound

Interpolation via splitting on-demand. In the splitting on-demand approach, the $\mathcal{LA}(\mathbb{Z})$ solver might not always detect the unsatisfiability of a set of constraints by itself; rather, it might cooperate with the DPLL solver by asking it to perform some case splits, by sending to DPLL some additional $\mathcal{LA}(\mathbb{Z})$-lemmas encoding the different case splits. In our interpolation procedure, we must take this possibility into account.

Let $(t - c \leq 0) \vee (-t + c + 1 \leq 0)$ be a branch-and-bound lemma added to the DPLL solver by the $\mathcal{LA}(\mathbb{Z})$-solver, using splitting on-demand. If $t \preceq A$ or $t \preceq B$, then we can exploit the Boolean interpolation algorithm also for computing interpolants in the presence of splitting-on-demand lemmas. The key observation is that the lemma $(t - c \leq 0) \vee (-t + c + 1 \leq 0)$ is a *valid clause* in $\mathcal{LA}(\mathbb{Z})$. Therefore, we can add it to any formula without affecting its satisfiability. Thus, if $t \preceq A$ we can treat the lemma as a clause from A, and if $t \preceq B$ we can treat it as a clause from B.[4]

Thanks to the observation above, in order to be able to produce interpolants with splitting on-demand the only thing we need is to make sure that we do not generate lemmas containing AB-mixed terms.[5] This is always the case for "normal" branch-and-bound lemmas (since they involve only one variable), but this is not true in general for "extended" branch-and-bound lemmas generated from proofs of unsatisfiability using the "cuts from proofs" algorithm of [5]. The following example shows one such case.

Example 1. Let A and B be defined as

$$A \stackrel{\text{def}}{=} (x - 2y \leq 0) \wedge (2y - x \leq 0), \quad B \stackrel{\text{def}}{=} (x - 2z - 1 \leq 0) \wedge (2z + 1 - x \leq 0)$$

When solving $A \wedge B$ using extended branch and bound, we might generate the following AB-mixed lemma: $(y - z \leq 0) \vee (-y + z + 1 \leq 0)$.

Since we want to be able to reuse the Boolean interpolation algorithm also for splitting on-demand, we want to avoid generating AB-mixed lemmas. However, we would still like to exploit the cuts from proofs algorithm of [5] as much as possible. We describe how we do this in the following.

---

[4] If both $t \preceq A$ and $t \preceq B$, we are free to choose between the two alternatives.

[5] That is, terms t such that $t \npreceq A$ and $t \npreceq B$.

The cuts from proofs algorithm in a nutshell. The core of the cuts from proofs algorithm is the identification of the *defining constraints* of the current solution of the rational relaxation of the input set of $\mathcal{LA}(\mathbb{Z})$ constraints. A defining constraint is an input constraint $\sum_i c_i x_i + c \bowtie 0$ (where $\bowtie \in \{\leq, =\}$) such that $\sum_i c_i x_i + c$ evaluates to zero under the current solution for the rational relaxation of the problem. After having identified the defining constraints D, the cuts from proofs algorithm checks the satisfiability of the system of Diophantine equations $D_E \stackrel{\text{def}}{=} \{\sum_i c_i x_i + c = 0 \mid (\sum_i c_i x_i + c \bowtie 0) \in D\}$. If D_E is unsatisfiable, then it is possible to generate a proof of unsatisfiability for it. The root of such proof is an equation $\sum_i c_i' x_i + c' = 0$ such that the GCD g of the c_i''s does not divide c'. From such equation, it is generated the extended branch and bound lemma:

$$(\sum_i \frac{c_i'}{g} x_i \leq \left\lceil \frac{-c_i'}{g} \right\rceil - 1) \vee (\left\lceil \frac{-c_i'}{g} \right\rceil \leq \sum_i \frac{c_i'}{g} x_i).$$

Avoiding AB-mixed lemmas. If $\sum_i \frac{c_i'}{g} x_i$ is not AB-mixed, we can generate the above lemma also when computing interpolants. If $\sum_i \frac{c_i'}{g} x_i$ is AB-mixed, instead, we generate a different lemma, still exploiting the unsatisfiability of (the equations corresponding to) the defining constraints. Since D_E is unsatisfiable, we know that the current rational solution μ is not compatible with the current set of defining constraints. If the defining constraints were all equations, the submodule for handling Diophantine equations would have detected the conflict. Therefore, there is at least one defining constraint $\sum_i \bar{c}_i x_i + \bar{c} \leq 0$. Our idea is that of *splitting* this constraint into $(\sum_i \bar{c}_i x_i + \bar{c} + 1 \leq 0)$ and $(\sum_i \bar{c}_i x_i + \bar{c} = 0)$, by generating the lemma

$$\neg(\sum_i \bar{c}_i x_i + \bar{c} \leq 0) \vee (\sum_i \bar{c}_i x_i + \bar{c} + 1 \leq 0) \vee (\sum_i \bar{c}_i x_i + \bar{c} = 0).$$

In this way, we are either "moving away" from the current bad rational solution μ (when $(\sum_i \bar{c}_i x_i + \bar{c} + 1 \leq 0)$ is set to true), or we are forcing one more element of the set of defining constraints to be an equation (when $(\sum_i \bar{c}_i x_i + \bar{c} = 0)$ is set to true): if we repeat the splitting, then, eventually all the defining constraints for the bad solution μ will be equations, thus allowing the Diophantine equations handler to detect the conflict without the need of generating more branch-and-bound lemmas. Since the set of defining constraints is a subset of the input constraints, lemmas generated in this way will never be AB-mixed.

It should be mentioned that this procedure is very similar to the algorithm used in the recent work [12] for avoiding the generation of AB-mixed cuts. However, the criterion used to select which inequality to split and how to split it is different (in [12] such inequality is selected among those that are violated by the closest integer solution to the current rational solution). Moreover, we don't do this systematically, but rather only if the cuts from proofs algorithm is not able to generate a non-AB-mixed lemma by itself. In a sense, the approach of [12] is "pessimistic" in that it systematically excludes certain kinds of cuts, whereas our approach is more "optimistic".

Interpolation for the internal branch-and-bound module. From the point of view of interpolation the subdivision of the branch-and-bound module in an "internal" and an "external" part poses no difficulty. The only difference between the two is that in the former the case splits are performed by the $\mathcal{LA}(\mathbb{Z})$-solver instead of DPLL. However, we can still treat such case splits as if they were performed by DPLL, build a Boolean resolution proof for the $\mathcal{LA}(\mathbb{Z})$-conflicts discovered by the internal branch-and-bound procedure, and then apply the propositional interpolation algorithm as in the case of splitting on-demand.

4   A Novel General Interpolation Technique for Inequalities

The use of the Strenghen rule allows us to produce interpolants with very little modifications to the $\mathcal{LA}(\mathbb{Z})$-solver (we only need to enable the generation of cutting-plane proofs), which in turn result in very little overhead *at search time*. However, the Strengthen rule might cause a very significant overhead *when generating the interpolant* from a proof of unsatisfiability. In fact, even a single Strengthen application results in a disjunction whose size is proportional to the *value of the constant k in the rule*. The following example, taken from [12], illustrates the problem.

Example 2. Consider the following (parametric) interpolation problem [12]:

$$A \stackrel{\text{def}}{=} (-y - 2nx - n + 1 \leq 0) \wedge (y + 2nx \leq 0)$$

$$B \stackrel{\text{def}}{=} (-y - 2nz + 1 \leq 0) \wedge (y + 2nz - n \leq 0)$$

where the parameter n is an integer constant greater than 1. Using the rules of §3.2, we can construct the following annotated cutting-plane proof of unsatisfiability:

$$\frac{\begin{array}{cc} y + 2nx \leq 0 & -y - 2nz + 1 \leq 0 \\ {[\{\langle y + 2nx \leq 0, \top \rangle\}]} & {[\{\langle 0 \leq 0, \top \rangle\}]} \end{array}}{\begin{array}{c} 2nx - 2nz + 1 \leq 0 \\ {[\{\langle y + 2nx \leq 0, \top \rangle\}]} \end{array}}$$

$$\frac{2nx - 2nz + 1 + (2n - 1) \leq 0}{[\{\langle y + 2nx + j \leq 0, \quad \exists x.(y + 2nx + j = 0)\rangle \mid \atop 0 \leq j < 2n - 1\} \cup \atop \{\langle y + 2nx + 2n - 1 \leq 0, \top \rangle\}]}$$

$$\frac{\begin{array}{cc} -y - 2nx - n + 1 \leq 0 & y + 2nz - n \leq 0 \\ {[\{\langle -y - 2nx - n + 1 \leq 0, \top \rangle\}]} & {[\{\langle 0 \leq 0, \top \rangle\}]} \end{array}}{\begin{array}{c} -2nx + 2nz - 2n + 1 \leq 0 \\ {[\{\langle -y - 2nx - n + 1 \leq 0, \top \rangle\}]} \end{array}}$$

$$\frac{}{1 \leq 0 \quad \begin{array}{l} {[\{\langle j - n + 1 \leq 0, \exists x.(y + 2nx + j = 0)\rangle \mid 0 \leq j < 2n - 1\} \cup} \\ {\{\langle (2n - 1) - n + 1 \leq 0, \top \rangle\}]} \end{array}}$$

By observing that $(j - n + 1 \leq 0) \models \bot$ when $j \geq n$, the generated interpolant is:

$$(y =_{2n} -n + 1) \vee (y =_{2n} -n + 2) \vee \ldots \vee (y =_{2n} 0),$$

whose size is linear in n, and thus exponential wrt. the size of the input problem. In fact, in [12], it is said that this is the only (up to equivalence) interpolant for (A, B) that can be obtained by using only interpreted symbols in the signature $\Sigma \stackrel{\text{def}}{=} \{=, \leq, +, \cdot\} \cup \mathbb{Z} \cup \{=_g \mid g \in \mathbb{Z}^{>0}\}$.

In order to overcome this drawback, we present a novel and very effective way of computing interpolants in $\mathcal{LA}(\mathbb{Z})$, which is inspired by a result by Pudlák [15]. The key idea is *to extend both the signature and the domain* of the theory by explicitly introducing the *ceiling function* $\lceil \cdot \rceil$ and by allowing non-variable terms to be non-integers.

As in the previous Section, we use the annotated rules Hyp-A, Hyp-B and Comb. However, in this case the annotations are *single* inequalities in the form $(t \le 0)$ rather than (possibly large) sets of inequalities and equalities. Moreover, we replace the Strenghten rule with the equivalent Division rule:

Division

$$\frac{(A, B) \vdash \sum_i a_i x_i + \sum_j c_j y_j + \sum_k b_k z_k + c \le 0 \; [\sum_i a_i x_i + \sum_j c'_j y_j + c' \le 0]}{(A, B) \vdash \sum_i \frac{a_i}{d} x_i + \sum_j \frac{c_j}{d} y_j + \sum_k \frac{b_k}{d} z_k + \left\lceil \frac{c}{d} \right\rceil \le 0 \; [\sum_i \frac{a_i}{d} x_i + \left\lceil \frac{\sum_j c'_j y_j + c'}{d} \right\rceil \le 0]}$$

where:
- $x_i \notin B$, $y_j \in A \cap B$, $z_k \notin A$
- $d > 0$ divides all the a_i's, c_j's and b_k's

As before, if we ignore the presence of annotations, the rules Hyp-A, Hyp-B, Comb and Division form a complete proof systems for $\mathcal{LA}(\mathbb{Z})$ [18]. Notice also that all the rules Hyp-A, Hyp-B, Comb and Division preserve the following invariant: the coefficients a_i of the A-local variables are always the same for the implied inequality and its annotation. This makes the Division rule always applicable. Therefore, the above rules can be used to annotate any cutting-plane proof. In particular, this means that our new technique can be applied also to proofs generated by other $\mathcal{LA}(\mathbb{Z})$ techniques used in modern SMT solvers, such as those based on Gomory cuts or on the Omega test [16].

Definition 3. *An annotated sequent* $(A, B) \vdash (t \le 0)[(t' \le 0)]$ *is* valid *when:*

1. $A \models (t' \le 0)$;
2. $B \models (t - t' \le 0)$;
3. $t' \preceq A$ and $(t - t') \preceq B$.

Theorem 2. *All the interpolating rules preserve the validity of the sequents.*

Corollary 2. *If we can derive a valid sequent* $(A, B) \vdash c \le 0[t \le 0]$ *with* $c > 0$, *then* $(t \le 0)$ *is an interpolant for* (A, B).

Example 3. Consider the following interpolation problem:

$$A \overset{\text{def}}{=} (y = 2x), \qquad B \overset{\text{def}}{=} (y = 2z + 1).$$

The following is an annotated cutting-plane proof of unsatisfiability for $A \wedge B$:

$$\frac{\dfrac{y = 2x}{y - 2x \le 0[y - 2x \le 0]} \quad \dfrac{y = 2z + 1}{2z + 1 - y \le 0[0 \le 0]}}{\dfrac{2z - 2x + 1 \le 0[y - 2x \le 0]}{z - x + 1 \le 0[-x + \lceil \frac{y}{2} \rceil \le 0]}} \quad \frac{\dfrac{y = 2x}{2x - y \le 0} \quad \dfrac{y = 2z + 1}{y - 2z - 1 \le 0}}{\dfrac{[2x - y \le 0] \quad [0 \le 0]}{2x - 2z - 1 \le 0[2x - y \le 0]}}$$

$$1 \le 0[-y + 2 \lceil \tfrac{y}{2} \rceil \le 0]$$

Then, $(-y + 2 \left\lceil \dfrac{y}{2} \right\rceil \le 0)$ is an interpolant for (A, B).

Using the ceiling function, we do not incur in any blowup of the size of the generated interpolant wrt. the size of the proof of unsatisfiability.[6] In particular, by using the ceiling function we might produce interpolants which are up to exponentially smaller than those generated using modular equations. The intuition is that the use of the ceiling function in the annotation of the Division rule allows for expressing *symbolically* the case distinction that the Strengthen rule of §3.2 was expressing *explicitly* as a disjunction of modular equations, which was the source of the exponential blowup.

Example 4. Consider again the parametric interpolation problem of Example 2:

$$A \stackrel{\text{def}}{=} (-y - 2nx - n + 1 \le 0) \wedge (y + 2nx \le 0)$$

$$B \stackrel{\text{def}}{=} (-y - 2nz + 1 \le 0) \wedge (y + 2nz - n \le 0)$$

Using the ceiling function, we can generate the following annotated proof:

$$
\cfrac{
  \cfrac{
    \cfrac{
      \cfrac{y + 2nx \le 0 \quad -y - 2nz + 1 \le 0}{2nx - 2nz + 1 \le 0} \quad [y + 2nx \le 0] \quad [0 \le 0]
    }{2n \cdot (x - z + 1 \le 0)} \quad [y + 2nx \le 0]
  }{[x + \lceil \frac{y}{2n} \rceil \le 0]}
  \quad
  \cfrac{
    \cfrac{-y - 2nx - n + 1 \le 0 \quad y + 2nz - n \le 0}{-2nx + 2nz - 2n + 1 \le 0} \quad [-y - 2nx - n + 1 \le 0] \quad [0 \le 0]
  }{[-y - 2nx - n + 1 \le 0]}
}{1 \le 0 \; [2n \lceil \frac{y}{2n} \rceil - y - n + 1 \le 0]}
$$

The interpolant corresponding to such proof is then $(2n \lceil \frac{y}{2n} \rceil - y - n + 1 \le 0)$, whose size is linear in the size of the input.

Solving and interpolating formulas with ceilings. Any SMT solver supporting $\mathcal{LA}(\mathbb{Z})$ can be easily extended to support formulas containing ceilings. In fact, we notice that we can eliminate ceiling functions from a formula φ with a simple preprocessing step as follows:

1. Replace every term $\lceil t_i \rceil$ occurring in φ with a fresh integer variable $x_{\lceil t_i \rceil}$;
2. Set φ to $\varphi \wedge \bigwedge_i \{ (x_{\lceil t_i \rceil} - 1 < t_i \le x_{\lceil t_i \rceil}) \}$.

Moreover, we remark that for using ceilings we must only be able to *represent* non-variable terms with rational coefficients, but we don't need to extend our $\mathcal{LA}(\mathbb{Z})$-solver to support Mixed Rational/Integer Linear Arithmetic. This is because, after the elimination of ceilings performed during preprocessing, we can multiply both sides of the introduced constraints $(x_{\lceil t_i \rceil} - 1 < t_i)$ and $(t_i \le x_{\lceil t_i \rceil})$ by the least common multiple of the rational coefficients in t_i, thus obtaining two $\mathcal{LA}(\mathbb{Z})$-inequalities.

For interpolation, it is enough to preprocess A and B separately, so that the elimination of ceilings will not introduce variables common to A and B.

5 Experimental Evaluation

The techniques presented in previous sections have been implemented within the MATH-SAT 5 SMT solver [9]. In this section, we experimentally evaluate our approach.

---

[6] However, we remark that, in general, cutting-plane proofs of unsatisfiability can be exponentially large wrt. the size of the input problem [18,15].

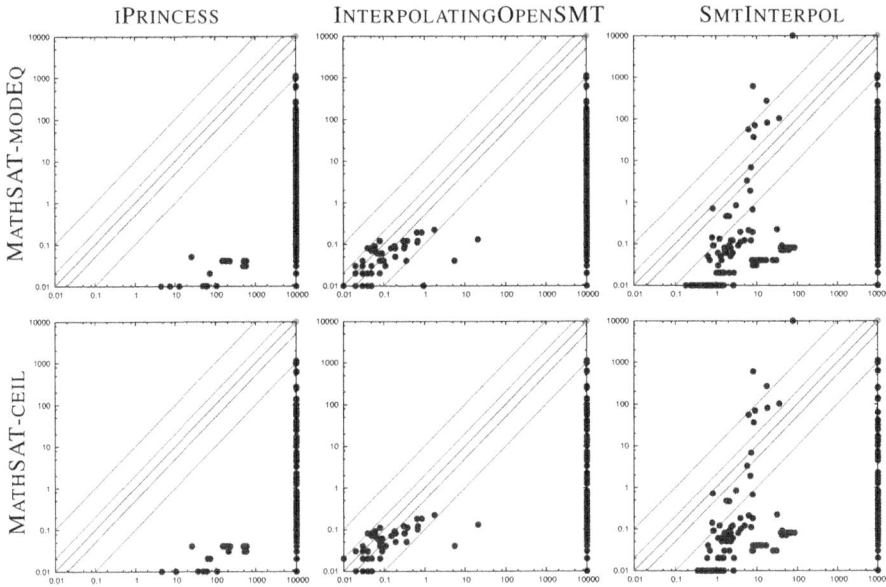

Fig. 2. Comparison between MATHSAT and the other $\mathcal{LA}(\mathbb{Z})$-interpolating tools, execution time

5.1 Description of the Benchmark Sets

We have performed our experiments on a subset of the benchmark instances in the QF_LIA ("quantifier-free $\mathcal{LA}(\mathbb{Z})$") category of the SMT-LIB.[7] More specifically, we have selected the subset of $\mathcal{LA}(\mathbb{Z})$-unsatisfiable instances whose rational relaxation is (easily) satisfiable, so that $\mathcal{LA}(\mathbb{Z})$-specific interpolation techniques are put under stress. In order to generate interpolation problems, we have split each of the collected instances in two parts A and B, by collecting about 40% of the toplevel conjuncts of the instance to form A, and making sure that A contains some symbols not occurring in B (so that A is never a "trivial" interpolant). In total, our benchmark set consists of 513 instances.

We have run the experiments on a machine with a 2.6 GHz Intel Xeon processor, 16 GB of RAM and 6 MB of cache, running Debian GNU/Linux 5.0. We have used a time limit of 1200 seconds and a memory limit of 3 GB.

All the benchmark instances and the executable of MATHSAT used to perform the experiments are available at `http://es.fbk.eu/people/griggio/papers/tacas11_experiments.tar.bz2`

5.2 Comparison with the State-of-the-Art Tools Available

We compare MATHSAT with all the other interpolant generators for $\mathcal{LA}(\mathbb{Z})$ which are available (to the best of our knowledge): IPRINCESS [3],[8] INTERPOLATINGOPENSMT

---

[7] `http://smtlib.org`
[8] `http://www.philipp.ruemmer.org/iprincess.shtml`

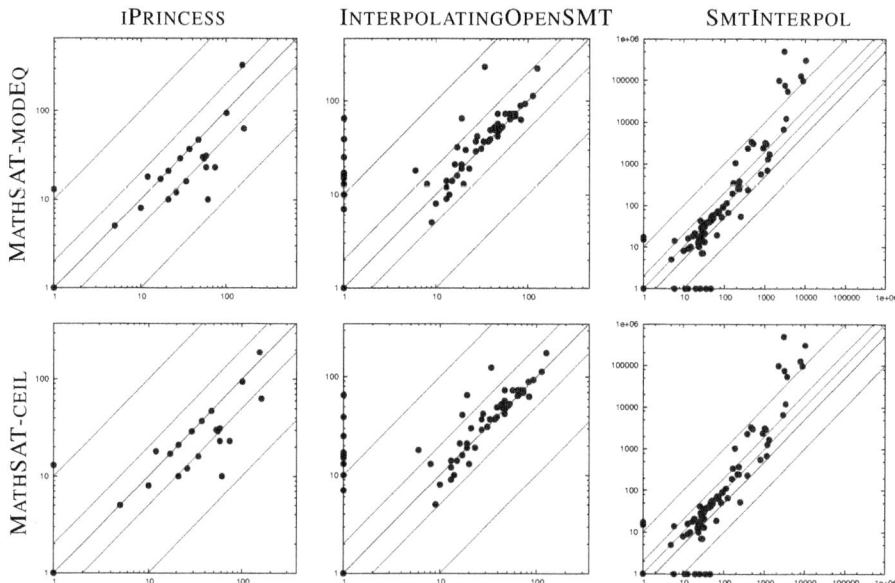

Fig. 3. Comparison between MATHSAT and the other $\mathcal{LA}(\mathbb{Z})$-interpolating tools, interpolants size (measured in number of nodes in the DAG of the interpolant). (See also footnote 11).

[12],[9] and SMTINTERPOL[10]. We compare not only the execution times for generating interpolants, but also the size of the generated formulas (measured in terms of number of nodes in their DAG representation).

For MATHSAT, we use two configurations: MATHSAT-MODEQ, which produces interpolants with modular equations using the Strengthen rule of §3, and MATHSAT-CEIL, which uses the ceiling function and the Division rule of §4.

Results on execution times for generating interpolants are reported in Fig. 2. Both MATHSAT-MODEQ and MATHSAT-CEIL could successfully generate an interpolant for 478 of the 513 interpolation problems (timing out on the others), whereas IPRINCESS, INTERPOLATINGOPENSMT and SMTINTERPOL were able to successfully produce an interpolant in 62, 192 and 217 cases respectively. Therefore, MATHSAT can solve more than twice as many instances as its closer competitor SMTINTERPOL, and in most cases with a significantly shorter execution time (Fig. 2).

For the subset of instances which could be solved by at least one other tool, therefore, the two configurations of MATHSAT seem to perform equally well. The situation is the same also when we compare the sizes of the produced interpolants, measured in number of nodes in a DAG representation of formulas. Comparisons on interpolant size are reported in Fig. 3, which shows that, on average, the interpolants produced by MATH-SAT are comparable to those produced by other tools. In fact, there are some cases in

---

[9] http://www.philipp.ruemmer.org/interpolating-opensmt.shtml
[10] http://swt.informatik.uni-freiburg.de/research/tools/
   smtinterpol. We are not aware of any publication describing the tool.

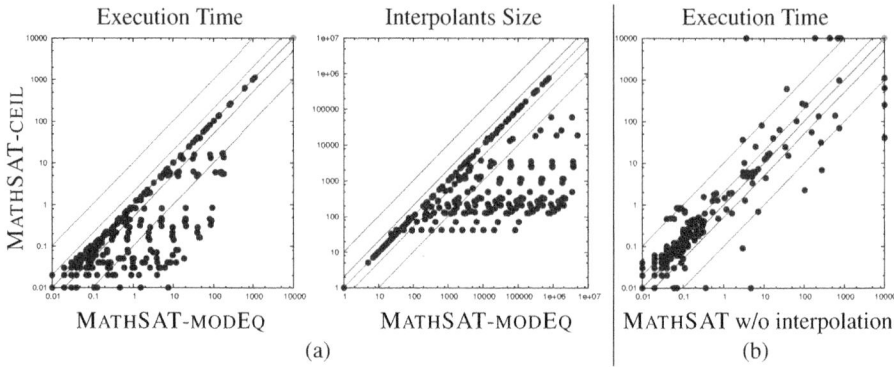

Fig. 4. (a) Comparison between MATHSAT-MODEQ and MATHSAT-CEIL configurations for interpolation. (b) Execution time overhead for interpolation with MATHSAT-CEIL.

which SMTINTERPOL produces significantly-smaller interpolants, but we remark that MATHSAT can solve 261 more instances than SMTINTERPOL.[11]

The differences between MATHSAT-MODEQ and MATHSAT-CEIL become evident when we compare the two configurations directly. The plots in Fig. 4(a) show that MATHSAT-CEIL is dramatically superior to MATHSAT-MODEQ, with gaps of up to two orders of magnitude in execution time, and up to four orders of magnitude in the size of interpolants. Such differences are solely due to the use of the ceiling function in the generated interpolants, which prevents the blow-up of the formula wrt. the size of the proof of unsatisfiability. Since most of the differences between the two configurations occur in benchmarks that none of the other tools could solve, the advantage of using ceilings was not visible in Figs. 2 and 3.

Finally, in Fig. 4(b) we compare the execution time of producing interpolants with MATHSAT-CEIL against the solving time of MATHSAT with interpolation turned off. The plot shows that the restriction on the kind of extended branch-and-bound lemmas generated when computing interpolants (see §3.3) can have a significant impact on individual benchmarks. However, on average MATHSAT-CEIL is not worse than the "regular" MATHSAT, and the two can solve the same number of instances, in approximately the same total execution time.

6   Conclusions

In this paper, we have presented a novel interpolation algorithm for $\mathcal{LA}(\mathbb{Z})$ that allows for producing interpolants from arbitrary cutting-plane proofs without the need of performing quantifier elimination. We have also shown how to exploit this algorithm, in combination with other existing techniques, in order to implement an efficient

---

[11] The plots of Fig. 3 show also some apparently-strange outliers in the comparison with INTERPOLATINGOPENSMT. A closer analysis revealed that those are instances for which INTERPOLATINGOPENSMT was able to detect that the inconsistency of $A \wedge B$ was due solely to A or to B, and thus could produce a trivial interpolant $\bot$ or $\top$, whereas the proof of unsatisfiability produced by MATHSAT involved both A and B. An analogous situation is visible also in the comparison between MATHSAT and SMTINTERPOL, this time in favor of MATHSAT.

interpolation procedure on top of a state-of-the-art SMT($\mathcal{LA}(\mathbb{Z})$)-solver, with almost no overhead in search, and with up to orders of magnitude improvements – both in execution time and in formula size – wrt. existing techniques for computing interpolants from arbitrary cutting-plane proofs.

References

1. Barrett, C., Nieuwenhuis, R., Oliveras, A., Tinelli, C.: Splitting on Demand in SAT Modulo Theories. In: Hermann, M., Voronkov, A. (eds.) LPAR 2006. LNCS (LNAI), vol. 4246, pp. 512–526. Springer, Heidelberg (2006)
2. Barrett, C.W., Sebastiani, R., Seshia, S.A., Tinelli, C.: Satisfiability Modulo Theories. In: Handbook of Satisfiability, ch. 25. IOS Press, Amsterdam (2009)
3. Brillout, A., Kroening, D., Rümmer, P., Wahl, T.: An Interpolating Sequent Calculus for Quantifier-Free Presburger Arithmetic. In: Giesl, J., Hähnle, R. (eds.) IJCAR 2010. LNCS, vol. 6173, pp. 384–399. Springer, Heidelberg (2010)
4. Cimatti, A., Griggio, A., Sebastiani, R.: Efficient Generation of Craig Interpolants in Satisfiability Modulo Theories. ACM Trans. Comput. Logic 12(1) (October 2010)
5. Dillig, I., Dillig, T., Aiken, A.: Cuts from Proofs: A Complete and Practical Technique for Solving Linear Inequalities over Integers. In: Bouajjani, A., Maler, O. (eds.) CAV 2009. LNCS, vol. 5643, pp. 233–247. Springer, Heidelberg (2009)
6. Dutertre, B., de Moura, L.: A Fast Linear-Arithmetic Solver for DPLL(T). In: Ball, T., Jones, R.B. (eds.) CAV 2006. LNCS, vol. 4144, pp. 81–94. Springer, Heidelberg (2006)
7. Fuchs, A., Goel, A., Grundy, J., Krstić, S., Tinelli, C.: Ground interpolation for the theory of equality. In: Kowalewski, S., Philippou, A. (eds.) TACAS 2009. LNCS, vol. 5505, pp. 413–427. Springer, Heidelberg (2009)
8. Goel, A., Krstić, S., Tinelli, C.: Ground Interpolation for Combined Theories. In: Schmidt, R.A. (ed.) CADE-22. LNCS, vol. 5663, pp. 183–198. Springer, Heidelberg (2009)
9. Griggio, A.: A Practical Approach to SMT($\mathcal{LA}(\mathbb{Z})$). In: SMT 2010 Workshop (July 2010)
10. Jain, H., Clarke, E.M., Grumberg, O.: Efficient Craig Interpolation for Linear Diophantine (Dis)Equations and Linear Modular Equations. In: Gupta, A., Malik, S. (eds.) CAV 2008. LNCS, vol. 5123, pp. 254–267. Springer, Heidelberg (2008)
11. Kapur, D., Majumdar, R., Zarba, C.G.: Interpolation for data structures. In: FSE 2005. ACM, New York (2006)
12. Kroening, D., Leroux, J., Rümmer, P.: Interpolating Quantifier-Free Presburger Arithmetic. In: Fermüller, C.G., Voronkov, A. (eds.) LPAR-17. LNCS, vol. 6397, pp. 489–503. Springer, Heidelberg (2010), http://www.philipp.ruemmer.org/publications.shtml
13. Lynch, C., Tang, Y.: Interpolants for Linear Arithmetic in SMT. In: Cha, S(S.), Choi, J.-Y., Kim, M., Lee, I., Viswanathan, M. (eds.) ATVA 2008. LNCS, vol. 5311, pp. 156–170. Springer, Heidelberg (2008)
14. McMillan, K.L.: An interpolating theorem prover. Theor. Comput. Sci. 345(1) (2005)
15. Pudlák, P.: Lower bounds for resolution and cutting planes proofs and monotone computations. J. of Symb. Logic 62(3) (1997)
16. Pugh, W.: The Omega test: a fast and practical integer programming algorithm for dependence analysis. In: SC (1991)
17. Rybalchenko, A., Sofronie-Stokkermans, V.: Constraint solving for interpolation. J. Symb. Comput. 45(11) (2010)
18. Schrijver, A.: Theory of Linear and Integer Programming. Wiley, Chichester (1986)
19. Yorsh, G., Musuvathi, M.: A combination method for generating interpolants. In: Nieuwenhuis, R. (ed.) CADE 2005. LNCS (LNAI), vol. 3632, pp. 353–368. Springer, Heidelberg (2005)

Generalized Craig Interpolation for Stochastic Boolean Satisfiability Problems[*]

Tino Teige and Martin Fränzle

Carl von Ossietzky Universität, Oldenburg, Germany
{teige,fraenzle}@informatik.uni-oldenburg.de

Abstract. The stochastic Boolean satisfiability (SSAT) problem has been introduced by Papadimitriou in 1985 when adding a probabilistic model of uncertainty to propositional satisfiability through randomized quantification. SSAT has many applications, among them bounded model checking (BMC) of symbolically represented Markov decision processes. This paper identifies a notion of *Craig interpolant* for the SSAT framework and develops an algorithm for computing such interpolants based on SSAT resolution. As a potential application, we address the use of interpolation in SSAT-based BMC, turning the falsification procedure into a verification approach for probabilistic safety properties.

1 Introduction

Papadimitriou [1] has proposed the idea of modeling uncertainty within propositional satisfiability (SAT) by adding *randomized* quantification to the problem description. The resultant *stochastic Boolean satisfiability* (SSAT) problems consist of a quantifier prefix followed by a propositional formula. The quantifier prefix is an alternating sequence of existentially quantified variables and variables bound by randomized quantifiers. The meaning of a randomized variable x is that x takes value `true` with a certain probability p and value `false` with the complementary probability $1 - p$. Due to the presence of such probabilistic assignments, the semantics of an SSAT formula Φ no longer is qualitative in the sense that Φ is satisfiable or unsatisfiable, but rather *quantitative* in the sense that we are interested in the *maximum probability of satisfaction* of Φ. Intuitively, a solution of Φ is a strategy for assigning the existential variables, i.e. a tree of assignments to the existential variables depending on the probabilistically determined values of preceding randomized variables, such that the assignments maximize the probability of satisfying the propositional formula.

In recent years, the SSAT framework has attracted interest within the Artificial Intelligence community, as many problems from that area involving uncertainty have concise descriptions as SSAT problems, in particular probabilistic planning problems [2,3,4]. Inspired by that work, other communities have started

---

[*] This work has been supported by the German Research Council (DFG) as part of the Transregional Collaborative Research Center "Automatic Verification and Analysis of Complex Systems" (SFB/TR 14 AVACS, www.avacs.org).

P.A. Abdulla and K.R.M. Leino (Eds.): TACAS 2011, LNCS 6605, pp. 158–172, 2011.
© Springer-Verlag Berlin Heidelberg 2011

to exploit SSAT and closely related formalisms within their domains. The Constraint Programming community is working on *stochastic constraint satisfaction* problems [5,6] to address, a.o., multi-objective decision making under uncertainty [7]. Recently, a technique for the symbolic analysis of probabilistic hybrid systems based on stochastic satisfiability has been suggested by the authors [8,9]. To this end, SSAT has been extended by embedded theory reasoning over arithmetic theories, as known from *satisfiability modulo theories* (SMT) [10], which yields the notion of *stochastic satisfiability modulo theories* (SSMT). By the expressive power of SSMT, bounded probabilistic reachability problems of uncertain hybrid systems can be phrased symbolically as SSMT formulae yielding the same probability of satisfaction. As this bounded model checking approach yields valid lower bounds *lb* for the probability of reaching undesirable system states along unbounded runs, it is able to *falsify* probabilistic safety requirements of shape "a system error occurs with probability at most 0.1‰".

Though the general SSAT problem is PSPACE-complete, the plethora of real-world applications calls for practically efficient algorithms. The first SSAT algorithm, suggested by Littman [11], extends the Davis-Putnam-Logemann-Loveland (DPLL) procedure [12,13] for SAT with appropriate quantifier handling and algorithmic optimizations like *thresholding*. Majercik improved the DPLL-style SSAT algorithm by *non-chronological backtracking* [14]. Unlike these explicit tree-traversal approaches and motivated by work on *resolution* for propositional and first-order formulae [15] and for QBF formulae [16], the authors have recently developed an alternative SSAT procedure based on resolution [17].

In this paper, we investigate the concept of Craig interpolation for SSAT. Given two formulae A and B for which $A \Rightarrow B$ is true, a *Craig interpolant* [18] $\mathcal{I}$ is a formula over variables common to A and B that "lies in between" A and B in the sense that $A \Rightarrow \mathcal{I}$ and $\mathcal{I} \Rightarrow B$. In the automatic hardware and software verification communities, Craig interpolation has found widespread use in model checking algorithms, both as a means of extracting reasons for non-concretizability of a counterexample obtained on an abstraction as well as for obtaining a symbolic description of reachable state sets. In McMillan's approach [19,20], interpolants are used to symbolically describe an overapproximation of the step-bounded reachable state set. If the sequence of interpolants thus obtained stabilizes eventually, i.e. no additional state is found to be reachable, then the corresponding state-set predicate R has all reachable system states as its models. The safety property that states satisfying B (where B is a predicate) are never reachable, is then verified by checking $R \wedge B$ for unsatisfiability.

Given McMillan's verification procedure for non-probabilistic systems, it is natural to ask whether a corresponding probabilistic counterpart can be developed, i.e. a *verification procedure for probabilistic systems based on Craig interpolation for stochastic SAT*. Such an approach would complement the aforementioned falsification procedure for probabilistic systems based on SSAT/SSMT. In this paper, we suggest a solution to the issue above. After a formal introduction of SSAT in Section 2, we recall (and adapt slightly) the resolution calculus for SSAT from [17] in Section 3. We suggest a generalization of the notion of Craig interpolants

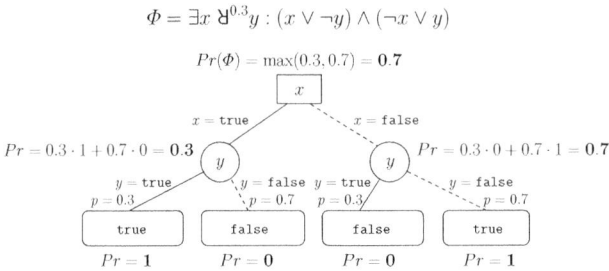

$$\Phi = \exists x \, \Uparrow^{0.3} y : (x \vee \neg y) \wedge (\neg x \vee y)$$

$Pr(\Phi) = \max(0.3, 0.7) = \mathbf{0.7}$

Fig. 1. Semantics of an SSAT formula depicted as a tree

suitable for SSAT as well as an algorithm based on SSAT resolution to compute such generalized interpolants (Section 4). Finally, we propose an interpolation-based approach to probabilistic model checking that is able to *verify* probabilistic safety requirements (Section 5) and illustrate the applicability of this verification procedure on a small example.

2   Stochastic Boolean Satisfiability

A *stochastic Boolean satisfiability* (SSAT) formula is of the form $\Phi = \mathcal{Q} : \varphi$ with a prefix $\mathcal{Q} = Q_1 x_1 \ldots Q_n x_n$ of quantified propositional variables x_i, where Q_i is either an existential quantifier $\exists$ or a randomized quantifier $\Uparrow^{p_i}$ with a rational constant $0 < p_i < 1$, and a propositional formula φ s.t. $Var(\varphi) \subseteq \{x_1, \ldots, x_n\}$, where $Var(\varphi)$ denotes the set of all (necessarily free) variables occurring in φ. W.l.o.g., we can assume that φ is in *conjunctive normal form* (CNF), i.e. a conjunction of disjunctions of propositional literals. A *literal* ℓ is a propositional variable, i.e. $\ell = x_i$, or its negation, i.e. $\ell = \neg x_i$. A *clause* is a disjunction of literals. Throughout the paper and w.l.o.g., we require that a clause does not contain the same literal more than once as $\ell \vee \ell \equiv \ell$. Consequently, we may also identify a clause with its set of literals. The semantics of Φ, as illustrated in Fig. 1, is defined by the *maximum probability of satisfaction* $Pr(\Phi)$ as follows.

$$Pr(\varepsilon : \varphi) \quad = \begin{cases} 0 & \text{if } \varphi \text{ is logically equivalent to } \mathtt{false} \\ 1 & \text{if } \varphi \text{ is logically equivalent to } \mathtt{true} \end{cases}$$
$$Pr(\exists x \; \mathcal{Q} : \varphi) = \max(Pr(\mathcal{Q} : \varphi[\mathtt{true}/x]), Pr(\mathcal{Q} : \varphi[\mathtt{false}/x]))$$
$$Pr(\Uparrow^p x \; \mathcal{Q} : \varphi) = p \cdot Pr(\mathcal{Q} : \varphi[\mathtt{true}/x]) + (1-p) \cdot Pr(\mathcal{Q} : \varphi[\mathtt{false}/x])$$

Note that the semantics is well-defined as Φ has no free variables s.t. all variables have been substituted by the constants $\mathtt{true}$ and $\mathtt{false}$ when reaching the quantifier-free base case.

3   Resolution for SSAT

As basis of the SSAT interpolation procedure introduced in Section 4, we recall the sound and complete resolution calculus for SSAT from [17], subsequently called *S-resolution*. In contrast to SSAT algorithms implementing a DPLL-based backtracking procedure, thereby explicitly traversing the tree given by the quantifier prefix and recursively computing the individual satisfaction probabilities for each subtree by the scheme illustrated in Fig. 1, S-resolution follows the idea of *resolution* for propositional and first-order formulae [15] and for QBF formulae [16] by deriving new clauses c^p annotated with probabilities $0 \leq p \leq 1$. S-resolution differs from non-stochastic resolution, as such derived clauses c^p need not be implications of the given formula, but are just entailed with some probability. Informally speaking, the derivation of a clause c^p means that under SSAT formula $\mathcal{Q} : \varphi$, the clause c is violated with a maximum probability at most p, i.e. the satisfaction probability of $\mathcal{Q} : (\varphi \wedge \neg c)$ is at most p. More intuitively, the minimum probability that clause c is implied by φ is at least $1 - p$.[1] Once an annotated empty clause $\emptyset^p$ is derived, it follows that the probability of the given SSAT formula is at most p, i.e. $Pr(\mathcal{Q} : (\varphi \wedge \neg \mathtt{false})) = Pr(\mathcal{Q} : \varphi) \leq p$.

The following presentation of S-resolution differs slightly from [17] in order to avoid overhead in interpolant generation incurred when employing the original definition, like the necessity of enforcing particular resolution sequences. For readers familiar with [17], the particular modifications are: 1) derived clauses c^p may also carry value $p = 1$, 2) former rules R.2 and R.5 are joined into the new rule R.2, and 3) former rules R.3 and R.4 are collapsed into rule R.3. These modifications do not affect soundness and completeness of S-resolution (cf. Corollary 1 and Theorem 1). The advantage of the modification is that derivable clauses c^p are forced to have a tight bound p in the sense that under each assignment which falsifies c, the satisfaction probability of the remaining subproblem *exactly* is p (cf. Lemma 1). This fact confirms the conjecture from [17, p. 14] about the existence of such clauses $(c \vee \ell)^p$ and allows for a generalized clause learning scheme to be integrated into DPLL-SSAT solvers: the idea is that under a partial assignment falsifying c, one may directly propagate literal ℓ as the satisfaction probability of the other branch, for which the negation of ℓ holds, is known to be p already.

In the sequel, let $\Phi = \mathcal{Q} : \varphi$ be an SSAT formula with φ in CNF. W.l.o.g., φ contains only non-tautological clauses[2], i.e. $\forall c \in \varphi : \not\models c$. Let $\mathcal{Q} = Q_1 x_1 \ldots Q_n x_n$ be the quantifier prefix and φ be some propositional formula with $Var(\varphi) \subseteq \{x_1, \ldots, x_n\}$. The quantifier prefix $\mathcal{Q}(\varphi)$ is defined to be shortest prefix of $\mathcal{Q}$ that contains all variables from φ, i.e. $\mathcal{Q}(\varphi) = Q_1 x_1 \ldots Q_i x_i$ where $x_i \in Var(\varphi)$ and for each $j > i : x_j \notin Var(\varphi)$. Let further be $Var(\varphi) \downarrow_k := \{x_1, \ldots, x_k\}$ for each integer $0 \leq k \leq n$. For a non-tautological clause c, i.e. if $\not\models c$, we define the unique assignment $f\!f_c$ that falsifies c as the mapping

$$f\!f_c : Var(c) \to \mathbb{B} \text{ such that } \forall x \in Var(c) : f\!f_c(x) = \begin{cases} \mathtt{true} & ; \; \neg x \in c, \\ \mathtt{false} & ; \; x \in c. \end{cases}$$

Consequently, c evaluates to $\mathtt{false}$ under assignment $f\!f_c$.

---

[1] We remark that $Pr(\mathcal{Q} : \psi) = 1 - Pr(\mathcal{Q}' : \neg\psi)$, where $\mathcal{Q}'$ arises from $\mathcal{Q}$ by replacing existential quantifiers by universal ones, where universal quantifiers call for *minimizing* the satisfaction probability.

[2] Tautological clauses c, i.e. $\models c$, are redundant, i.e. $Pr(\mathcal{Q} : (\varphi \wedge c)) = Pr(\mathcal{Q} : \varphi)$.

Starting with clauses in φ, *S-resolution* is given by the consecutive application of rules R.1 to R.3 to derive new clauses c^p with $0 \leq p \leq 1$. Rule R.1 derives a clause c^0 from an original clause c in φ. Referring to the definition of $Pr(\Phi)$ in Section 2, R.1 corresponds to the quantifier-free base case where φ is equivalent to `false` under any assignment that falsifies c.

(R.1)
$$\frac{c \in \varphi}{c^0}$$

Similarly, R.2 reflects the quantifier-free base case in which φ is equivalent to `true` under any assignment τ' that is conform to the partial assignment τ since $\models \varphi[\tau(x_1)/x_1]\ldots[\tau(x_i)/x_i]$. The constructed clause c^1 then encodes the opposite of this satisfying (partial) assignment τ. We remark that finding such a τ in the premise of R.2 is NP-hard (equivalent to finding a solution of a propositional formula in CNF). This strong condition on τ is not essential for soundness and completeness and could be removed[3] but, as mentioned above, facilitates a less technical presentation of generalized interpolation in Section 4. Another argument justifying the strong premise of R.2 is a potential integration of S-resolution into DPLL-based SSAT solvers since whenever a satisfying (partial) assignment τ of φ is found by an SSAT solver then τ meets the requirements of R.2.

(R.2)
$$\frac{\begin{array}{c}c \subseteq \{x, \neg x | x \in Var(\varphi)\}, \not\models c, \mathcal{Q}(c) = Q_1 x_1 \ldots Q_i x_i, \\ \text{for each } \tau : Var(\varphi) \downarrow_i \to \mathbb{B} \text{ with } \forall x \in Var(c) : \tau(x) = \mathit{ff}_c(x) : \\ \models \varphi[\tau(x_1)/x_1]\ldots[\tau(x_i)/x_i]\end{array}}{c^1}$$

Rule R.3 finally constitutes the actual resolution rule as known from the nonstochastic case. Depending on whether an existential or a randomized variable is resolved upon, the probability value of the resolvent clause is computed according to the semantics $Pr(\Phi)$ defined in Section 2.

(R.3)
$$\frac{\begin{array}{c}(c_1 \vee \neg x)^{p_1}, (c_2 \vee x)^{p_2}, Qx \in \mathcal{Q}, Qx \notin \mathcal{Q}(c_1 \vee c_2), \not\models (c_1 \vee c_2), \\ p = \begin{cases} \max(p_1, p_2) & ; Q = \exists \\ p_x \cdot p_1 + (1 - p_x) \cdot p_2 & ; Q = \exists^{p_x} \end{cases}\end{array}}{(c_1 \vee c_2)^p}$$

The derivation of a clause c^p by R.1 from c, by R.2, and by R.3 from $c_1^{p_1}, c_2^{p_2}$ is denoted by $c \vdash_{\mathsf{R.1}} c^p$, by $\vdash_{\mathsf{R.2}} c^p$, and by $(c_1^{p_1}, c_2^{p_2}) \vdash_{\mathsf{R.3}} c^p$, respectively. Given rules R.1 to R.3, S-resolution is sound and complete in the following sense.

Lemma 1. *Let clause c^p be derivable by S-resolution and let $\mathcal{Q}(c) = Q_1 x_1 \ldots Q_i x_i$. For each $\tau : Var(\varphi) \downarrow_i \to \mathbb{B}$ with $\forall x \in Var(c) : \tau(x) = \mathit{ff}_c(x)$ it holds that $Pr(Q_{i+1} x_{i+1} \ldots Q_n x_n : \varphi[\tau(x_1)/x_1]\ldots[\tau(x_i)/x_i]) = p$.*

Corollary 1 (Soundness of S-resolution). *If the empty clause $\emptyset^p$ is derivable by S-resolution from a given SSAT formula $\mathcal{Q} : \varphi$ then $Pr(\mathcal{Q} : \varphi) = p$.*

---

[3] Then, Lemma 1 must be weakened (as for original S-resolution [17]) to $Pr(Q_{i+1} x_{i+1} \ldots Q_n x_n : \varphi[\tau(x_1)/x_1]\ldots[\tau(x_i)/x_i]) \leq p$.

Theorem 1 (Completeness of S-resolution). *If $Pr(\mathcal{Q} : \varphi) = p$ for some SSAT formula $\mathcal{Q} : \varphi$ then the empty clause $\emptyset^p$ is derivable by S-resolution.*

The formal proofs of above and subsequent results can be found in [21].

Example. Consider the SSAT formula $\Phi = \mathcal{H}^{0.8} x_1 \, \exists x_2 \, \mathcal{H}^{0.3} x_3 : ((x_1 \vee x_2) \wedge (\neg x_2) \wedge (x_2 \vee x_3))$ with $Pr(\Phi) = 0.24$. Clauses $(x_1 \vee x_2)^0$, $(\neg x_2)^0$, $(x_2 \vee x_3)^0$ are then derivable by R.1. As $x_1 = \texttt{true}, x_2 = \texttt{false}, x_3 = \texttt{true}$ is a satisfying assignment, $\vdash_{\mathsf{R.2}} (\neg x_1 \vee x_2 \vee \neg x_3)^1$. Then, $((\neg x_1 \vee x_2 \vee \neg x_3)^1, (x_2 \vee x_3)^0) \vdash_{\mathsf{R.3}} (\neg x_1 \vee x_2)^{0.3}, ((\neg x_2)^0, (\neg x_1 \vee x_2)^{0.3}) \vdash_{\mathsf{R.3}} (\neg x_1)^{0.3}, ((\neg x_2)^0, (x_1 \vee x_2)^0) \vdash_{\mathsf{R.3}} (x_1)^0$, and finally $((\neg x_1)^{0.3}, (x_1)^0) \vdash_{\mathsf{R.3}} \emptyset^{0.24}$.

4   Interpolation for SSAT

Craig interpolation [18] is a well-studied notion in formal logics which has several applications in Computer Science, among them model checking [19,20]. Given two formulae φ and ψ such that $\varphi \Rightarrow \psi$ is valid, a *Craig interpolant* for (φ, ψ) is a formula $\mathcal{I}$ which refers only to common variables of φ and ψ, and $\mathcal{I}$ is "intermediate" in the sense that $\varphi \Rightarrow \mathcal{I}$ and $\mathcal{I} \Rightarrow \psi$. Such interpolants do trivially exist in all logics permitting quantifier elimination, e.g. in propositional logic. Using the observation that $\varphi \Rightarrow \psi$ holds iff $\varphi \wedge \neg \psi$ is unsatisfiable, this gives rise to an equivalent definition which we refer to in the rest of the paper:[4] given an unsatisfiable formula $\varphi \wedge \neg \psi$, formula $\mathcal{I}$ is an interpolant for (φ, ψ) iff both $\varphi \wedge \neg \mathcal{I}$ and $\mathcal{I} \wedge \neg \psi$ are unsatisfiable and $\mathcal{I}$ mentions only common variables.

In the remainder of this section, we investigate the issue of interpolation for stochastic SAT. We propose a generalization of Craig interpolants suitable for SSAT and show the general existence of such interpolants, alongside with an automatic method for computing them based on S-resolution.

4.1   Generalized Craig Interpolants

When approaching a reasonable definition of interpolants for SSAT, the semantics of the non-classical quantifier prefix poses problems: Let $\Phi = \mathcal{Q} : (A \wedge B)$ be an SSAT formula. Each variable in $A \wedge B$ is bound by $\mathcal{Q}$, which provides the probabilistic interpretation of the variables that is lacking without the quantifier prefix. This issue can be addressed by considering the quantifier prefix $\mathcal{Q}$ as the global setting that serves to interpret the quantifier-free part, and consequently interpreting the interpolant also within the scope of $\mathcal{Q}$, thus reasoning about $\mathcal{Q} : (A \wedge \neg \mathcal{I})$ and $\mathcal{Q} : (\mathcal{I} \wedge B)$. A more fundamental problem is that a classical Craig interpolant for Φ only exists if $Pr(\Phi) = 0$, since $A \wedge B$ has to be unsatisfiable by definition of a Craig interpolant which applies iff $Pr(\mathcal{Q} : (A \wedge B)) = 0$.

---

[4] This is of technical nature as SSAT formulae are interpreted by the maximum probability of satisfaction. As the *maximum* probability that an implication $\varphi \Rightarrow \psi$ holds is inappropriate for our purpose, we reason about the maximum satisfaction probability p of the negated implication, i.e. of $\varphi \wedge \neg \psi$, instead. The latter coincides with the *minimum* probability $1 - p$ that $\varphi \Rightarrow \psi$ holds, which is the desired notion.

The precondition that $Pr(\mathcal{Q} : (A \wedge B)) = 0$ would be far too restrictive for application of interpolation, as the notion of unsatisfiability of $A \wedge B$ is naturally generalized to satisfiability with insufficient probability, i.e. $Pr(\mathcal{Q} : (A \wedge B))$ being "sufficiently small", in the stochastic setting. Such relaxed requirements actually appear in practice, e.g. in probabilistic verification where safety properties like "a fatal system error is never reachable" are frequently replaced by probabilistic ones like "a fatal system error is reachable only with (sufficiently small) probability of at most 0.1‰". Motivated by above facts, interpolants for SSAT should also exist when $A \wedge B$ is satisfiable with reasonably low probability.

The resulting notion of interpolation, which is to be made precise in Definition 1, matches the following intuition. In classical Craig interpolation, when performed in logics permitting quantifier elimination, the Craig interpolants of $(A, \neg B)$ form a lattice with implication as its ordering, $A^{\exists} = \exists a_1, \ldots a_\alpha : A$ as its bottom element and $\overline{B}^{\forall} = \neg \exists b_1, \ldots b_\beta : B$ as its top element, where the a_i and b_i are the local variables of A and of B, respectively. In the generalized setting required for SSAT[5], $A \Rightarrow \neg B$ and thus $A^{\exists} \Rightarrow \overline{B}^{\forall}$ may no longer hold such that the above lattice can collapse to the empty set. To preserve the overall structure, it is however natural to use the lattice of propositional formulae "in between" $A^{\exists} \wedge \overline{B}^{\forall}$ as bottom element and $A^{\exists} \vee \overline{B}^{\forall}$ as top element instead. This lattice is non-empty and coincides with the classical one whenever $A \wedge B$ is unsatisfiable.

Definition 1 (Generalized Craig interpolant). *Let A, B be propositional formulae and $V_A := Var(A) \setminus Var(B) = \{a_1, \ldots, a_\alpha\}$, $V_B := Var(B) \setminus Var(A) = \{b_1, \ldots, b_\beta\}$, $V_{A,B} := Var(A) \cap Var(B)$, $A^{\exists} = \exists a_1, \ldots, a_\alpha : A$, and $\overline{B}^{\forall} = \neg \exists b_1, \ldots, b_\beta : B$. A propositional formula $\mathcal{I}$ is called* generalized Craig interpolant *for (A, B) iff $Var(\mathcal{I}) \subseteq V_{A,B}$, $\left(A^{\exists} \wedge \overline{B}^{\forall}\right) \Rightarrow \mathcal{I}$, and $\mathcal{I} \Rightarrow \left(A^{\exists} \vee \overline{B}^{\forall}\right)$.*

Given any two propositional formulae A and B, the four quantifier-free propositional formulae equivalent to $A^{\exists} \wedge \overline{B}^{\forall}$, to $A^{\exists}$, to $\overline{B}^{\forall}$, and to $A^{\exists} \vee \overline{B}^{\forall}$, are generalized Craig interpolants for (A, B). These generalized interpolants always exist since propositional logic has quantifier elimination.

While Definition 1 motivates the generalized notion of Craig interpolant from a model-theoretic perspective, we will state an equivalent definition of generalized Craig interpolants in Lemma 2 that substantiates the intuition of generalized interpolants and allows for an illustration of their geometric shape. Given two formulae A and B, the idea of generalized Craig interpolant is depicted in Fig. 2. The set of solutions of A is defined by the rectangle on the $V_A, V_{A,B}$-plane with a cylindrical extension in V_B-direction as A does not contain variables in V_B. Similarly, the solution set of B is given by the triangle on the $V_B, V_{A,B}$-plane and its cylinder in V_A-direction. The solution set of $A \wedge B$ is then determined by the intersection of both cylinders. Since $A \wedge B \wedge \neg(A \wedge B)$ is unsatisfiable, the sets $A \wedge \neg(A \wedge B)$ and $B \wedge \neg(A \wedge B)$ are disjoint. This gives us the possibility to talk about interpolants wrt. these sets. However, a formula $\mathcal{I}$ over only common variables in $V_{A,B}$ may not exist when demanding $A \wedge \neg(A \wedge B) \wedge \neg \mathcal{I}$ and

---

[5] Though the concept seems to be more general, this paper addresses SSAT only.

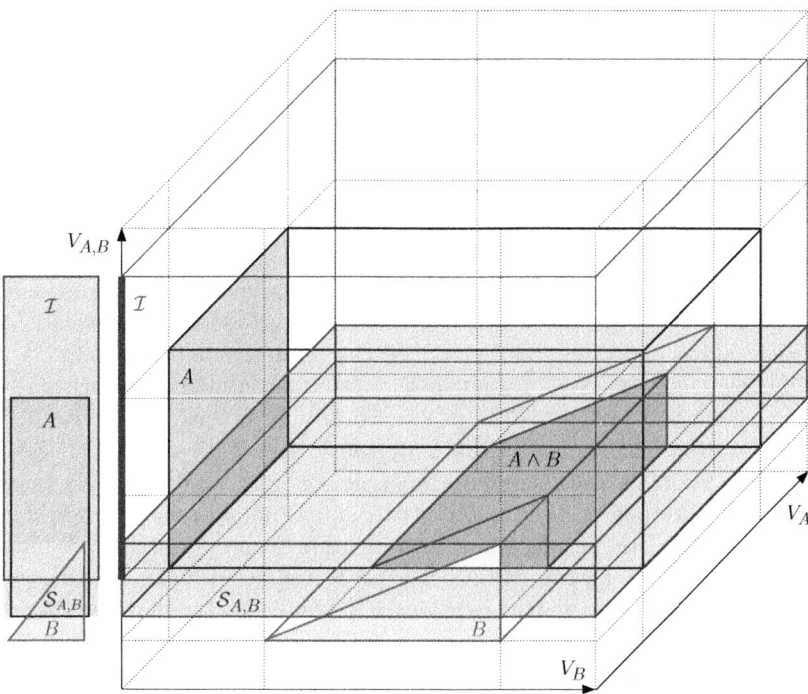

Fig. 2. Geometric interpretation of a generalized Craig interpolant $\mathcal{I}$. V_A-, V_B-, and $V_{A,B}$-axes denote assignments of variables occurring only in A, only in B, and in both A and B, respectively.

$\mathcal{I} \wedge B \wedge \neg(A \wedge B)$ to be unsatisfiable. This is indicated by Fig. 2 and proven by the simple example $A = (a)$, $B = (b)$. As $V_{A,B} = \emptyset$, $\mathcal{I}$ is either true or false. In first case, $\text{true} \wedge (b) \wedge \neg(a \wedge b)$ is satisfiable, while $(a) \wedge \neg(a \wedge b) \wedge \neg\text{false}$ is in second case. If we however project the solution set of $A \wedge B$ onto the $V_{A,B}$-axis and subtract the resulting hyperplane $\mathcal{S}_{A,B}$ from A and B then such a formula $\mathcal{I}$ over $V_{A,B}$-variables exists. The next lemma formalizes such generalized interpolants $\mathcal{I}$ and shows their equivalence to the ones from Definition 1.

Lemma 2 (Generalized Craig interpolant for SSAT). *Let $\Phi = \mathcal{Q} : (A \wedge B)$ be some SSAT formula, V_A, V_B, $V_{A,B}$ be defined as in Definition 1, and $\mathcal{S}_{A,B}$ be a propositional formula with $Var(\mathcal{S}_{A,B}) \subseteq V_{A,B}$ s.t. $\mathcal{S}_{A,B} \equiv \exists a_1, \ldots, a_\alpha, b_1, \ldots, b_\beta : (A \wedge B)$. Then, a propositional formula $\mathcal{I}$ is a generalized Craig interpolant for (A, B) iff the following properties are satisfied.*

1. *$Var(\mathcal{I}) \subseteq V_{A,B}$*
2. *$Pr(\mathcal{Q} : (A \wedge \neg\mathcal{S}_{A,B} \wedge \neg\mathcal{I})) = 0$*
3. *$Pr(\mathcal{Q} : (\mathcal{I} \wedge B \wedge \neg\mathcal{S}_{A,B})) = 0$*

We remark that the concept of generalized Craig interpolants is a generalization of Craig interpolants in the sense that whenever $A \wedge B$ is unsatisfiable, i.e. when

$Pr(Q : (A \land B)) = 0$, then each generalized Craig interpolant $\mathcal{I}$ for (A, B) actually is a Craig interpolant for A and B since $\mathcal{S}_{A,B} \equiv \texttt{false}$.

4.2 Computation of Generalized Craig Interpolants

In this subsection, we proceed to the efficient computation of generalized Craig interpolants. The remark following Definition 1 shows that generalized interpolants can in principle be computed by *explicit* quantifier elimination methods, like Shannon's expansion or BDDs. We aim at a more efficient method based on S-resolution, akin to resolution-based Craig interpolation for propositional SAT by Pudlák [22], as has been integrated into DPLL-based SAT solvers featuring conflict analysis and successfully applied to symbolic model checking [19,20].

Observe that on SSAT formulae $Q : (A \land B)$, Pudlák's algorithm, which has unsatisfiability of $A \land B$ as precondition, will not work in general. When instead considering the unsatisfiable formula $A \land B \land \neg \mathcal{S}_{A,B}$ with $\neg \mathcal{S}_{A,B}$ in CNF then Pudlák's method would be applicable and would actually produce a generalized interpolant. The main drawback of this approach, however, is the explicit construction of $\neg \mathcal{S}_{A,B}$, again calling for explicit quantifier elimination.

We now propose an algorithm based on S-resolution for computing generalized Craig interpolants which operates directly on $A \land B$ without adding $\neg \mathcal{S}_{A,B}$, and thus does not comprise any preprocessing involving quantifier elimination. For this purpose, rules of S-resolution are enhanced to deal with pairs (c^p, I) of annotated clauses c^p and propositional formulae I. Such formulae I are in a certain sense *intermediate* generalized interpolants, i.e. generalized interpolants for subformulae arising from instantiating some variables by partial assignments that falsify c (cf. Lemma 3). Once a pair $(\emptyset^p, I)$ comprising the empty clause is derived, I thus is a generalized Craig interpolant for the given SSAT formula. This augmented S-resolution, which we call *interpolating S-resolution*, is defined by rules R'.1, R'.2, and R'.3. The construction of intermediate interpolants I in R'.1 and R'.3 coincides with the classical rules by Pudlák [22], while R'.2 misses a corresponding counterpart. The rationale is that R'.2 (or rather R.2) refers to satisfying valuations τ of $A \land B$, which do not exist in classical interpolation. As $A \land B$ becomes a tautology after substituting the partial assignment τ from R.2 into it, its quantified variant $\mathcal{S}_{A,B} = \exists a_1, \ldots, b_1, \ldots : A \land B$ also becomes tautological under the same substitution $\mathcal{S}_{A,B}[\tau(x_1)/x_1, \ldots, \tau(x_i)/x_i]$. Consequently, $\neg \mathcal{S}_{A,B}[\tau(x_1)/x_1, \ldots, \tau(x_i)/x_i]$ is unsatisfiable, and so are $(A \land \neg \mathcal{S}_{A,B})[\tau(x_1)/x_1, \ldots, \tau(x_i)/x_i]$ and $(B \land \neg \mathcal{S}_{A,B})[\tau(x_1)/x_1, \ldots, \tau(x_i)/x_i]$. This implies that the actual intermediate interpolant in R'.2 can be chosen arbitrarily over variables in $V_{A,B}$. This freedom will allow us to control the geometric extent of generalized interpolants within the "don't care"-region provided by the models of $\mathcal{S}_{A,B}$ (cf. Corollary 3).

$$(R'.1) \qquad \frac{c \vdash_{R.1} c^p, I = \begin{cases} \texttt{false} \; ; \; c \in A \\ \texttt{true} \;\; ; \; c \in B \end{cases}}{(c^p, I)}$$

(R'.2)
$$\frac{\vdash_{\text{R.2}} c^p, I \text{ is any formula over } V_{A,B}}{(c^p, I)}$$

(R'.3)
$$\frac{\begin{array}{c} ((c_1 \vee \neg x)^{p_1}, I_1), ((c_2 \vee x)^{p_2}, I_2), \\ ((c_1 \vee \neg x)^{p_1}, (c_2 \vee x)^{p_2}) \vdash_{\text{R.3}} (c_1 \vee c_2)^p, \\ I = \left\{ \begin{array}{ll} I_1 \vee I_2 & ; \ x \in V_A \\ I_1 \wedge I_2 & ; \ x \in V_B \\ (\neg x \vee I_1) \wedge (x \vee I_2) & ; \ x \in V_{A,B} \end{array} \right. \end{array}}{((c_1 \vee c_2)^p, I)}$$

The following lemma establishes the theoretical foundation of computing generalized Craig interpolants by interpreting the derived pairs (c^p, I).

Lemma 3. *Let $\Phi = \mathcal{Q} : (A \wedge B)$ with $\mathcal{Q} = Q_1 x_1 \ldots Q_n x_n$ be some SSAT formula, and the pair (c^p, I) be derivable from Φ by interpolating S-resolution, where $\mathcal{Q}(c) = Q_1 x_1 \ldots Q_i x_i$. Then, for each $\tau : Var(A \wedge B) \downarrow_i \to \mathbb{B}$ with $\forall x \in Var(c) : \tau(x) = f\!\!f_c(x)$ it holds that*

1. *$Var(I) \subseteq V_{A,B}$,*
2. *$Pr(Q_{i+1} x_{i+1} \ldots Q_n x_n : (A \wedge \neg \mathcal{S}_{A,B} \wedge \neg I)[\tau(x_1)/x_1] \ldots [\tau(x_i)/x_i]) = 0$, and*
3. *$Pr(Q_{i+1} x_{i+1} \ldots Q_n x_n : (I \wedge B \wedge \neg \mathcal{S}_{A,B})[\tau(x_1)/x_1] \ldots [\tau(x_i)/x_i]) = 0$.*

Completeness of S-resolution, as stated in Theorem 1, together with the above Lemma, applied to the derived pair $(\emptyset^p, I)$, yields

Corollary 2 (Generalized Craig interpolants computation). *If $\Phi = \mathcal{Q} : (A \wedge B)$ is an SSAT formula then a generalized Craig interpolant for (A, B) can be computed by interpolating S-resolution.*

Note that computation of generalized interpolants does not depend on the actual truth state of $A \wedge B$. The next observation facilitates to effectively control the geometric extent of generalized Craig interpolants within the "don't care"-region $\mathcal{S}_{A,B}$. This result will be useful in probabilistic model checking in Section 5.

Corollary 3 (Controlling generalized Craig interpolants computation). *If $I = \texttt{true}$ is used within each application of rule R'.2 then $Pr(\mathcal{Q} : (A \wedge \neg \mathcal{I})) = 0$. Likewise, if $I = \texttt{false}$ is used in rule R'.2 then $Pr(\mathcal{Q} : (\mathcal{I} \wedge B)) = 0$.*

Observe that the special interpolants $\mathcal{I}$ from Corollary 3 relate to the classical strongest and weakest Craig interpolants $A^{\exists}$ and $\overline{B}^{\forall}$, resp., in the following sense: $Pr(\mathcal{Q} : (A \wedge \neg \mathcal{I})) = 0$ iff $\models A \Rightarrow \mathcal{I}$ iff $\models \forall a_1, \ldots, a_\alpha : (A \Rightarrow \mathcal{I})$ iff $\models (A^{\exists} \Rightarrow \mathcal{I})$, as $a_1, \ldots, a_\alpha$ do not occur in $\mathcal{I}$. Analogously, $Pr(\mathcal{Q} : (\mathcal{I} \wedge B)) = 0$ iff $\models \mathcal{I} \Rightarrow \neg B$ iff $\models \forall b_1, \ldots, b_\beta : (\mathcal{I} \Rightarrow \neg B)$ iff $\models \mathcal{I} \Rightarrow \overline{B}^{\forall}$.

5   Interpolation-Based Probabilistic Model Checking

In this section, we investigate the application of generalized Craig interpolation in probabilistic model checking. As a model we consider symbolically represented finite-state Markov decision processes (MDPs), which we check wrt. to probabilistic state reachability properties. That is, given a set of target states T in an MDP $\mathcal{M}$, the goal is to maximize the probability $P_{\mathcal{M}}^{\pi}(T)$ of reaching T over all policies π resolving the non-determinism in $\mathcal{M}$. When considering T as *bad* states of $\mathcal{M}$, e.g. as fatal system errors, this maximum probability $P_{\mathcal{M}}^{\max}(T) := \max_{\pi} P_{\mathcal{M}}^{\pi}(T)$ reveals the *worst-case* probability of bad system behavior. A *safety property* for $\mathcal{M}$ requires that this worst-case probability does not exceed some given threshold value $\theta \in [0,1)$, i.e. $P_{\mathcal{M}}^{\max}(T) \leq \theta$.

In [8], we proposed a symbolic *falsification* procedure for such safety properties. Though the approach in [8] is based on SSMT (arithmetic extension of SSAT) and considers the more general class of discrete-time probabilistic hybrid systems, which roughly are MDPs with arithmetic-logical transition guards and actions, the same procedure restricted to SSAT is applicable for finite-state MDPs. The key idea here is to adapt *bounded model checking* (BMC) [23] to the probabilistic case by encoding step-bounded reachability as an SSAT problem: like in classical BMC, the initial states, the transition relation, and the target states of an MDP are symbolically encoded by propositional formulae in CNF, namely by $Init(\boldsymbol{s})$, $Trans(\boldsymbol{s}, \boldsymbol{nt}, \boldsymbol{pt}, \boldsymbol{s}')$, and $Target(\boldsymbol{s})$, resp., where the propositional variable vector $\boldsymbol{s}$ represents the system state before and $\boldsymbol{s}'$ after a transition step. To keep track of the *non-deterministic* and *probabilistic* selections of transitions in $Trans(\boldsymbol{s}, \boldsymbol{nt}, \boldsymbol{pt}, \boldsymbol{s}')$, we further introduce propositional variables $\boldsymbol{nt}$ and $\boldsymbol{pt}$ to encode non-deterministic and probabilistic transition choices, respectively. Assignments to these variables determine which of possibly multiple available transitions departing from $\boldsymbol{s}$ is taken. In contrast to traditional BMC, all variables are quantified: all state variables $\boldsymbol{s}$ and $\boldsymbol{s}'$ are existentially quantified in the prefixes $\mathcal{Q}_{\boldsymbol{s}}$ and $\mathcal{Q}_{\boldsymbol{s}'}$. The transition-selection variables $\boldsymbol{nt}$ encoding non-deterministic choice are *existentially quantified* by $\mathcal{Q}_{\boldsymbol{nt}}$, while the probabilistic selector variables $\boldsymbol{pt}$ are bound by *randomized quantifiers* in $\mathcal{Q}_{\boldsymbol{pt}}$.[6] Let be $\boldsymbol{t} := \boldsymbol{nt} \cup \boldsymbol{pt}$ and $\mathcal{Q}_{\boldsymbol{t}} := \mathcal{Q}_{\boldsymbol{nt}}\mathcal{Q}_{\boldsymbol{pt}}$. Due to [8, Proposition 1], the maximum probability of reaching the target sates in a given MDP from the initial states within k transition steps is equivalent to the satisfaction probability

$$(1) \quad lb_k := Pr\Big(\mathcal{Q}(k) : \Big(\overbrace{Init(\boldsymbol{s}_0) \wedge \bigwedge_{i=1}^{k} Trans(\boldsymbol{s}_{i-1}, \boldsymbol{t}_i, \boldsymbol{s}_i)}^{\text{states reachable within } k \text{ steps}} \wedge \Big(\overbrace{\bigvee_{i=0}^{k} Target(\boldsymbol{s}_i)}^{\text{hit target states}} \Big) \Big) \Big)$$

with $\mathcal{Q}(k) := \mathcal{Q}_{\boldsymbol{s}_0}\mathcal{Q}_{\boldsymbol{t}_1}\mathcal{Q}_{\boldsymbol{s}_1} \ldots \mathcal{Q}_{\boldsymbol{s}_{k-1}}\mathcal{Q}_{\boldsymbol{t}_k}\mathcal{Q}_{\boldsymbol{s}_k}$. The probability lb_k can be determined by SSAT solvers. This symbolic approach, called *probabilistic bounded model checking* (PBMC), produces valid *lower* bounds lb_k for $P_{\mathcal{M}}^{\max}(T)$. Thus, PBMC is able to *falsify* a safety property once $lb_k > \theta$ for some k. However, the development of a corresponding counterpart based on SSAT that is able to compute safe *upper* bounds ub_k for $P_{\mathcal{M}}^{\max}(T)$ was left as an open question. Such an approach would permit to *verify* a safety property once $ub_k \leq \theta$ for some k.

We now propose such a verification procedure based on generalized Craig interpolation. The algorithm proceeds in two phases. Phase 1 computes a symbolic representation of an *overapproximation of the backward reachable state set*. This can be integrated into PBMC, as used to falsify the probabilistic safety property.

---

[6] Non-deterministic branching of n alternatives can be represented by a binary tree of depth $\lceil \log_2 n \rceil$ and probabilistic branching by a sequence of at most $n-1$ binary branches, yielding $\lceil \log_2 n \rceil$ existential and $n-1$ randomized quantifiers, respectively.

Whenever this falsification fails for a given step depth k, we apply generalized Craig interpolation to the (just failed) PBMC proof to compute a *symbolic over-approximation of the backward reachable state set* at depth k and then proceed to PBMC at some higher depth $k' > k$. As an alternative to the integration into PBMC, interpolants describing the backward reachable state sets can be successively extended by "stepping" them by prepending another transition, as explained below. In either case, phase 1 ends when the backward reachable state set becomes stable, in which case we have computed a symbolic overapproximation of the whole backward-reachable state set. In the second phase, we construct an SSAT formula with parameter k that forces the system to *stay within the backward reachable state set* for k steps. The maximum satisfaction probability of that SSAT formula then gives an upper bound on the maximum probability of reaching the target states. The rationale is that system executions leaving the backward reachable states will never reach the target states.

Phase 1. Given an SSAT encoding of an MDP $\mathcal{M}$ as above, the state-set predicate $\mathcal{B}^k(s)$ for $k \in \mathbb{N}_{\geq 0}$ over state variables s is inductively defined as $\mathcal{B}^0(s) := Target(s)$, and $\mathcal{B}^{k+1}(s) := \mathcal{B}^k(s) \vee \mathcal{I}^{k+1}(s)$ where $\mathcal{I}^{k+1}(s_{j-1})$ is a generalized Craig interpolant for $(Trans(s_{j-1}, t_j, s_j) \wedge \mathcal{B}^k(s_j), Init(s_0) \wedge \bigwedge_{i=1}^{j-1} Trans(s_{i-1}, t_i, s_i))$ with $j \geq 1$ wrt. to SSAT formula

$$(2) \quad \mathcal{Q}(j) : \left(Init(s_0) \wedge \overbrace{\bigwedge_{i=1}^{j-1} Trans(s_{i-1}, t_i, s_i)}^{j-1 \text{ steps "forward"} (=B)} \wedge \overbrace{Trans(s_{j-1}, t_j, s_j) \wedge \mathcal{B}^k(s_j)}^{\text{one step "backward"} (=A)} \right)$$

Observe that each $\mathcal{I}^{k+1}(s)$ can be computed by interpolating S-resolution if we rewrite $\mathcal{B}^k(s)$ into CNF (which is always possible in linear time by adding auxiliary V_A-variables). When doing so, we take $I = \texttt{true}$ in each application of R'.2 such that $\mathcal{B}^k(s)$ overapproximates all system states backward reachable from target states within k steps due to Corollary 3. Whenever $\mathcal{B}^k(s)$ stabilizes, i.e. $\mathcal{B}^{k+1}(s) \Rightarrow \mathcal{B}^k(s)$, we can be sure that $\widehat{BReach}(s) := \mathcal{B}^k(s)$ overapproximates all backward reachable states.

Note that parameter $j \geq 1$ can be chosen arbitrarily, i.e. the system may execute any number of transitions until state s_{j-1} is reached since this does not destroy the "backward-overapproximating" property of $\mathcal{B}^k(s)$. The rationale of having parameter j is the additional freedom in constructing $\mathcal{I}^k(s)$ as j may influence the shape of $\mathcal{I}^k(s)$ (cf. example below).

Phase 2. Having symbolically described all backward reachable states by $\widehat{BReach}(s)$, we are able to compute *upper* bounds ub_k of the maximum probability $P_{\mathcal{M}}^{\max}(T)$ of reaching target states T by SSAT solving applied to

$$(3) \quad ub_k := Pr\left(\mathcal{Q}(k) : \left(Init(s_0) \wedge \overbrace{\bigwedge_{i=1}^{k} Trans(s_{i-1}, t_i, s_i)}^{\text{states reachable within } k \text{ steps}} \wedge \overbrace{\bigwedge_{i=0}^{k} \widehat{BReach}(s_i)}^{\text{stay in back-reach set}} \right) \right)$$

First observe that the formula above excludes all system runs that leave the set of backward reachable states. This is sound since leaving $\widehat{BReach}(s)$ means to never reach the $Target(s)$ states. Second, the system behavior becomes more and more constrained for increasing k, i.e. the ub_k's are monotonically decreasing.

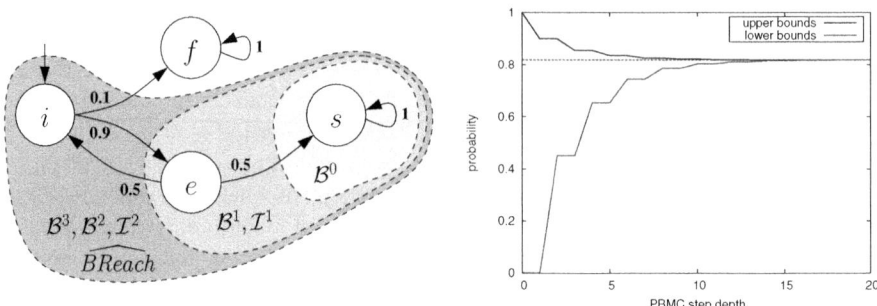

Fig. 3. A simple DTMC $\mathcal{M}$ (left) and lower and upper bounds on probability of reaching s over the PBMC step depth (right)

Regarding model checking, a safety property $P_{\mathcal{M}}^{\max}(T) \leq \theta$ is verified by the procedure above once $ub_k \leq \theta$ is computed for some k.

Example. Consider the simple *discrete-time Markov chain* (DTMC)[7] $\mathcal{M}$ on the left of Fig. 3, with s as the only target state. The (maximum) probability of reaching s from the initial state i clearly is $P = 0.9 \cdot (0.5 + 0.5 \cdot P) = \frac{9}{11}$. Applying the generalized interpolation scheme (2) for $k = 0$ with parameter $j = 1$ yields the interpolant $\mathcal{I}^1(s) = \neg i$. If we proceed then the set $\widehat{BReach}(s)$ covers all states though state f is not backward reachable. Thus, $j = 1$ is not a good choice as each ub_k in scheme (3) will then be 1. For $j = 2$, scheme (2) produces: $\mathcal{B}^0(s) = Target(s) = s$, $\mathcal{I}^1(s) = \neg i \wedge \neg f$, $\mathcal{B}^1(s) = \mathcal{B}^0(s) \vee \mathcal{I}^1(s) = (\neg i \vee s) \wedge (\neg f \vee s)$, $\mathcal{I}^2(s) = (\neg i \vee \neg e) \wedge \neg f$, $\mathcal{B}^2(s) = \mathcal{B}^1(s) \vee \mathcal{I}^2(s) = (\neg i \vee \neg e \vee s) \wedge (\neg f \vee s)$, $\mathcal{I}^3(s) = (\neg i \vee \neg e) \wedge \neg f$, $\mathcal{B}^3(s) = \mathcal{B}^2(s) \vee \mathcal{I}^3(s) = \mathcal{B}^2(s)$ (cf. left of Fig. 3). Thus, $\widehat{BReach}(s) = \mathcal{B}^2(s)$.

We are now able to compute upper bounds of the reachability probability $P = \frac{9}{11}$ by scheme (3). The results are shown on the right of Fig. 3 where the lower bounds are computed according to scheme (1). The figure indicates a fast convergence of the lower and upper bounds: for depth $k = 20$ the difference $ub_k - lb_k$ is below 10^{-3} and for $k = 100$ below 10^{-15}.

All 200 SSAT formulae were solved by the SSMT tool SiSAT [9] in 41.3 sec on a 1.83 GHz Intel Core 2 Duo machine with 1 GByte physical memory running Linux. For the moment, the SSAT encoding of $\mathcal{M}$ and all computations of generalized interpolants from PBMC proofs were performed manually. Details can be found in [21]. While the runtime of this first prototype does not compare favorably to value or policy iteration procedures, it should be noted that this is a first step towards a procedure embedding the same interpolation process into SSMT [8] and thereby directly addressing probabilistic hybrid systems, where the iteration procedures are only applicable after finite-state abstraction. It should also be noted that the symbolic procedures provided by SSAT and SSMT support compact representations of concurrent systems, thus alleviating the state explosion problem [9].

---

[7] A DTMC is an MDP without non-determinism.

6   Conclusion and Future Work

In this paper, we elaborated on the idea of Craig interpolation for stochastic SAT. In consideration of the difficulties that arise in this stochastic extension of SAT, we first proposed a suitable definition of a generalized Craig interpolant and second presented an algorithm to automatically compute such interpolants. For the latter purpose, we enhanced the SSAT resolution calculus by corresponding rules for the construction of generalized interpolants. We further demonstrated an application of generalized Craig interpolation in probabilistic model checking. The resulting procedure is able to verify probabilistic safety requirements of the form "the worst-case probability of reaching undesirable system states is smaller than a given threshold". This complements the existing SSAT-based bounded model checking approach, which mechanizes falsification of such properties.

An essential issue for future work is the practical evaluation of interpolation-based probabilistic model checking on realistic case studies. This involves, a.o., the integration of interpolating S-resolution into DPLL-based SSAT solvers and a thorough empirical study of the results obtained from the interpolation scheme (2). The latter includes the size and shape of generalized Craig interpolants as well as the computational effort for computing them in practice.

Another interesting future direction is the adaptation of generalized Craig interpolation to SSMT [8], i.e. the extension of SSAT with arithmetic theories. Computing Craig interpolants for SSMT would lift schemes (2) and (3) to SSMT problems, thus establishing a symbolic verification procedure for discrete-time probabilistic hybrid systems.

References

1. Papadimitriou, C.H.: Games against nature. J. Comput. Syst. Sci. 31(2), 288–301 (1985)
2. Littman, M.L., Majercik, S.M., Pitassi, T.: Stochastic Boolean Satisfiability. Journal of Automated Reasoning 27(3), 251–296 (2001)
3. Majercik, S.M., Littman, M.L.: MAXPLAN: A New Approach to Probabilistic Planning. In: Artificial Intelligence Planning Systems, pp. 86–93 (1998)
4. Majercik, S.M., Littman, M.L.: Contingent Planning Under Uncertainty via Stochastic Satisfiability. Artificial Intelligence Special Issue on Planning With Uncertainty and Incomplete Information 147(1-2), 119–162 (2003)
5. Walsh, T.: Stochastic constraint programming. In: Proc. of the 15th European Conference on Artificial Intelligence (ECAI 2002). IOS Press, Amsterdam (2002)
6. Balafoutis, T., Stergiou, K.: Algorithms for Stochastic CSPs. In: Benhamou, F. (ed.) CP 2006. LNCS, vol. 4204, pp. 44–58. Springer, Heidelberg (2006)
7. Bordeaux, L., Samulowitz, H.: On the stochastic constraint satisfaction framework. In: SAC, pp. 316–320. ACM, New York (2007)
8. Fränzle, M., Hermanns, H., Teige, T.: Stochastic satisfiability modulo theory: A novel technique for the analysis of probabilistic hybrid systems. In: Egerstedt, M., Mishra, B. (eds.) HSCC 2008. LNCS, vol. 4981, pp. 172–186. Springer, Heidelberg (2008)

9. Teige, T., Eggers, A., Fränzle, M.: Constraint-based analysis of concurrent probabilistic hybrid systems: An application to networked automation systems. Nonlinear Analysis: Hybrid Systems (to appear, 2011)
10. Barrett, C., Sebastiani, R., Seshia, S.A., Tinelli, C.: Satisfiability modulo theories. In: Biere, A., Heule, M.J.H., van Maaren, H., Walsh, T. (eds.) Handbook of Satisfiability. Frontiers in Artificial Intelligence and Applications, vol. 185, pp. 825–885. IOS Press, Amsterdam (2009)
11. Littman, M.L.: Initial Experiments in Stochastic Satisfiability. In: Proc. of the 16th National Conference on Artificial Intelligence, pp. 667–672 (1999)
12. Davis, M., Putnam, H.: A Computing Procedure for Quantification Theory. Journal of the ACM 7(3), 201–215 (1960)
13. Davis, M., Logemann, G., Loveland, D.: A Machine Program for Theorem Proving. Communications of the ACM 5, 394–397 (1962)
14. Majercik, S.M.: Nonchronological backtracking in stochastic Boolean satisfiability. In: ICTAI, pp. 498–507. IEEE Computer Society, Los Alamitos (2004)
15. Robinson, J.A.: A machine-oriented logic based on the resolution principle. J. ACM 12(1), 23–41 (1965)
16. Büning, H.K., Karpinski, M., Flögel, A.: Resolution for quantified Boolean formulas. Inf. Comput. 117(1), 12–18 (1995)
17. Teige, T., Fränzle, M.: Resolution for stochastic Boolean satisfiability. In: Fermüller, C.G., Voronkov, A. (eds.) LPAR-17. LNCS, vol. 6397, pp. 625–639. Springer, Heidelberg (2010)
18. Craig, W.: Linear Reasoning. A New Form of the Herbrand-Gentzen Theorem. J. Symb. Log. 22(3), 250–268 (1957)
19. McMillan, K.L.: Interpolation and SAT-based model checking. In: Hunt Jr., W.A., Somenzi, F. (eds.) CAV 2003. LNCS, vol. 2725, pp. 1–13. Springer, Heidelberg (2003)
20. McMillan, K.L.: Applications of Craig interpolants in model checking. In: Halbwachs, N., Zuck, L.D. (eds.) TACAS 2005. LNCS, vol. 3440, pp. 1–12. Springer, Heidelberg (2005)
21. Teige, T., Fränzle, M.: Generalized Craig interpolation for stochastic Boolean satisfiability problems. Reports of SFB/TR 14 AVACS 67, SFB/TR 14 AVACS (2011) ISSN: 1860-9821, http://www.avacs.org
22. Pudlák, P.: Lower Bounds for Resolution and Cutting Plane Proofs and Monotone Computations. Journal of Symbolic Logic 62(3), 981–998 (1997)
23. Biere, A., Cimatti, A., Clarke, E.M., Zhu, Y.: Symbolic model checking without BDDs. In: Cleaveland, W.R. (ed.) TACAS 1999. LNCS, vol. 1579, pp. 193–207. Springer, Heidelberg (1999)

Specification-Based Program Repair Using SAT

Divya Gopinath, Muhammad Zubair Malik, and Sarfraz Khurshid

The University of Texas at Austin

Abstract. Removing bugs in programs – even when location of faulty statements is known – is tedious and error-prone, particularly because of the increased likelihood of introducing new bugs as a result of fixing known bugs. We present an automated approach for generating likely bug fixes using behavioral specifications. Our key insight is to replace a faulty statement that has deterministic behavior with one that has nondeterministic behavior, and to use the specification constraints to prune the ensuing nondeterminism and repair the faulty statement. As an enabling technology, we use the SAT-based Alloy tool-set to describe specification constraints as well as for solving them. Initial experiments show the effectiveness of our approach in repairing programs that manipulate structurally complex data. We believe specification-based automated debugging using SAT holds much promise.

1 Introduction

The process of debugging, which requires (1) *fault localization*, i.e., finding the location of faults, and (2) *program repair*, i.e., fixing the faults by developing correct statements and removing faulty ones, is notoriously hard and time consuming, and automation can significantly reduce the cost of debugging. Traditional research on automated debugging has largely focused on fault localization [3, 10, 20], whereas program repair has largely been manual. The last few years have seen the development of a variety of exciting approaches to automate program repair, e.g., using game theory [9], model checking [18], data structure repair [11] and evolutionary algorithms [19]. However, previous approaches do not handle programs that manipulate structurally complex data that pervade modern software, e.g., perform destructive updates on program heap subject to rich structural invariants.

This paper presents a novel approach that uses rich behavioral specifications to automate program repair using off-the-shelf SAT technology. Our key insight is to transform a faulty program into a *nondeterministic* program and to use SAT to prune the nondeterminism in the ensuing program to transform it into a correct program with respect to the given specification. The key novelty of our work is the support for rich behavioral specifications, which precisely specify expected behavior, e.g., the `remove` method of a binary search tree only removes the given key from the input tree and does not introduce spurious key into the tree, as well as preserves acyclicity of the tree and the other class invariants, and the use of these specifications in pruning the state space for efficient generation of program statements.

We present a framework that embodies our approach and provides repair of Java programs using specifications written in the Alloy specification language [6]. Alloy

P.A. Abdulla and K.R.M. Leino (Eds.): TACAS 2011, LNCS 6605, pp. 173–188, 2011.

is a relational, first-order language that is supported by a SAT-based tool-set, which provides fully automatic analysis of Alloy formulas. The Alloy tool-set is the core enabling technology of a variety of systematic, bounded verification techniques, such as TestEra [12] for bounded exhaustive testing, and JAlloy [7] and JForge [4] for scope-bounded static checking. Our framework leverages JForge to find a smallest fault revealing input, which yields an output structure that violates the post-condition. Given such a counterexample and a candidate list of faulty statements, we parameterize each statement with variables that take a nondeterministic value from the domain of their respective types. A conjunction of the fixed pre-state, nondeterministic code, and post-condition is solved using SAT to prune the nondeterminism. The solution generated by SAT is abstracted to a list of program expressions, which are then iteratively filtered using bounded verification.

This paper makes the following contributions:

- **Alloy (SAT) for program repair.** We leverage the SAT-based Alloy tool-set to repair programs that operate on structurally complex data.
- **Rich behavioral specifications for program repair.** Our support for rich behavioral specifications for synthesizing program statements enables precise repair for complex programs.
- **Framework.** We present an automated specification-based framework for program repair using SAT, and algorithms that embody the framework.
- **Evaluation.** We use a textbook data structure and an application to evaluate the effectiveness and efficiency of our approach. Experimental results show our approach holds much promise for efficient and accurate program repair.

2   Basic Principle

The basic concept of bounded verification of Java programs using SAT is based on the relational model of the Java Heap. First proposed in the JAlloy technique [7], this involves encoding the data, the data-flow and control-flow of a program in relational logic. Every user-defined class and in-built type is represented as a set or a domain containing a bounded number of atoms(objects). A field of a class is encoded as a binary function that maps from the class to the type of the field. Local variables and arguments are encoded as singleton scalars. Data-flow is encoded as relational operations on the sets and relations. For instance, a field deference x.f, where x is an object of class A and f is a field of type B, is encoded as a relational join of the scalar 'x' and the binary relation f : A→B. Encoding control-flow is based on the computation graph obtained by unrolling the loops in the control flow graph a specified number of times. The control flow from one statement to the next is viewed as a relational implication, while at branch statements, the two branch edges are viewed as relational disjunctions. The entire code of a method can thus be represented as a formula in first-order relational logic such as Alloy [6], $P(s_t, s_t')$ relating the output state s_t' to its input s_t. The conjunction of the pre-condition, the formula representing the code and the negation of the post-condition specification of the method is fed into a relational engine such as Kodkod [16]. The formula is translated into boolean logic and off-the-shelf SAT solvers

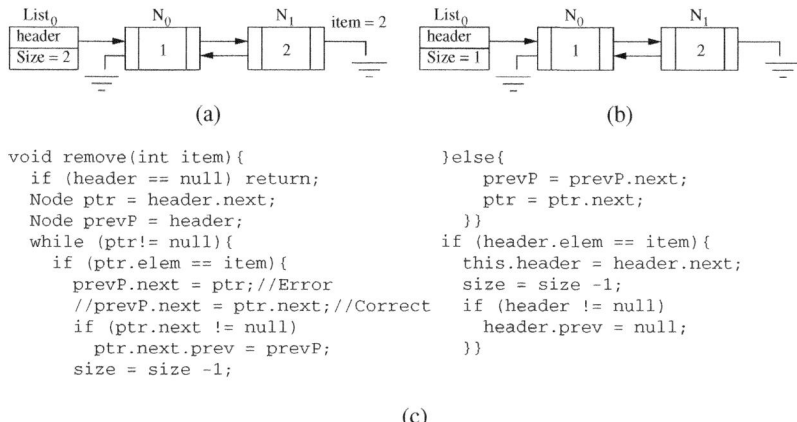

(a)   (b)

```
void remove(int item){                    }else{
  if (header == null) return;                 prevP = prevP.next;
  Node ptr = header.next;                     ptr = ptr.next;
  Node prevP = header;                      }}
  while (ptr!= null){                     if (header.elem == item){
    if (ptr.elem == item){                  this.header = header.next;
      prevP.next = ptr;//Error              size = size -1;
      //prevP.next = ptr.next;//Correct     if (header != null)
      if (ptr.next != null)                   header.prev = null;
        ptr.next.prev = prevP;              }}
      size = size -1;
```

(c)

Fig. 1. Faulty remove from Doubly Linked List(DLL) (a) Pre-state with input argument 2 (b) Post-state, size updated but node not removed. (c) Faulty DLL remove method.

are used to solve it. A solution represents a counter-example to the correctness specification, showing the presence of an input satisfying the pre-condition, tracing a valid path through the code and producing an output that does not satisfy the post-condition. Our technique builds on this idea to employ SAT to automatically perform alterations to program statements which would correct program errors leading to the specification violations. The technique can be best explained using a simple example.

The code snippet in Fig. 1c shows a faulty remove method of a doubly linked list data structure. The data structure is made up of nodes, each comprising of a next pointer, pointing to the subsequent node in the list, a prev pointer pointing to the previous node and an integer element. The header points to the first node in the list and size keeps a count of the number of nodes. The remove method removes the node which has its element value equal to input value. It uses two local pointers, ptr and prevP, to traverse the list and bypass the relevant node by setting the next of prevP to the next of ptr. A typical correctness specification (post-condition) for this method would check that the specified value has been removed from the list in addition to ensuring that the invariants such as acyclicity of the list are maintained. In the erroneous version, prevP.next is assigned to ptr instead of ptr.next. A scope bounded verification technique ([4]) can verify this method against the post-condition (specified in Alloy) to yield the smallest input structure which exposes the error. Figures 1a and 1b show the fault revealing input and the erroneous output. The pre-state is a doubly linked list with two nodes ($List_0$ is an instance of a list from the $List$ domain, N_0 and N_1 are instances from the $Node$ domain). In the corresponding erroneous post-state, node N_1 with element value 2 is still present in the list since the next pointer of N_0 points to N_1 rather than $null$. Let us assume the presence of a fault localization technique which identifies the statement prevP.next = ptr as being faulty. Our aim is to correct this assignment such that for the given input in the pre-state, the relational constraints of the path through the code yield an output that satisfies the post-condition. To accomplish this, we replace the operands of the assignment operator with new variables which can take any value from the Node domain or can be *null*. The statement would get altered to V_{lhs}.next $= V_{rhs}$, where V_{lhs} and V_{rhs} are the newly introduced variables. We then use SAT to find a solution for the conjunction of the new formula corresponding to the altered code and the post-condition specification. The values of the relations in the pre-state are fixed to correspond to the fault revealing input as follows, (header=($List_0$, N_0), next=(N_0, N_1), prev=(N_1, N_0), elem=(N_0,1), (N_1,2)). The solver searches for suitable valuations to the new variables such that an output is produced that satisfies the post-condition for the given fixed pre-state. In our example, these would be $V_{lhs} = N_0$ and $V_{rhs} = null$. These concrete state values are then abstracted to programming language expressions. For instance, in the example, the value of the local variables before the erroneous statement would be, ptr $= N_1$,prevP $= N_0$ and t $= List_0$. The expressions yielding the value for $V_{rhs} = null$ would be prevP.prev, prevP.next.next, ptr.next, ptr.prev.prev, t.header.prev, t.header.next.next. Similarly, a list of possible expressions can be arrived at for $V_{lhs} = N_0$. Each of these expressions yield an altered program statement yielding correct output for the specific fault revealing input. The altered program is then validated for all inputs using scope bounded verification to filter out wrong candidates. In our example, counter-examples would get thrown when any of the following expressions prevP.prev, ptr.prev.prev, t.header.prev, and t.header.next.next(for scope greater than 2) are used on the right hand side of the assignment. When no counter-example is obtained, it indicates that the program statement is correct for all inputs within the bounds specified. In this example, prevP.next = prevP.next.next and prevP.next = ptr.next would emerge as the correct statements satisfying all inputs.

3   Our Framework

Fig. 2 gives an overview of our framework for program repair. We assume the presence of a verification module indicating the presence of faults in the program. In our implementation, we have employed bounded verification (Forge framework [4]) which takes in a Java method, bounds on the input size(scope) and number of times to unroll loops and inline recursive calls(unrolls) and generates the smallest fault revealing input(s_t) within the given scope that satisfies the pre-condition of the method and yields an output(s_t') that violates the post-condition. A trace, comprising of the code statements in the path traversed for the given input, is also produced. We also assume the presence of a fault localization scheme, which yields a minimal list of possibly faulty statements. Please note that the technique works on the Control Flow Graph(CFG) representation of the program and the relational model view of the state.

Given a counter-example from the verification module and a list of suspicious statements $S_1, \ldots, S_m$, the first step performed by the repair module is to parameterize each of these statements. The operands in the statement are replaced with new variables. For instance, consider an assignment statement of the form, $x.\{f_1.f_2.\cdots.f_{n-1}\}.f_n = y$. The presence of a commission error locally in this statement indicates that either the source variable x, one or more of the subsequently de-referenced fields $f_1, f_2, \ldots, f_{n-1}$ or the target variable y have been specified wrongly. If the field being updated, f_n, has

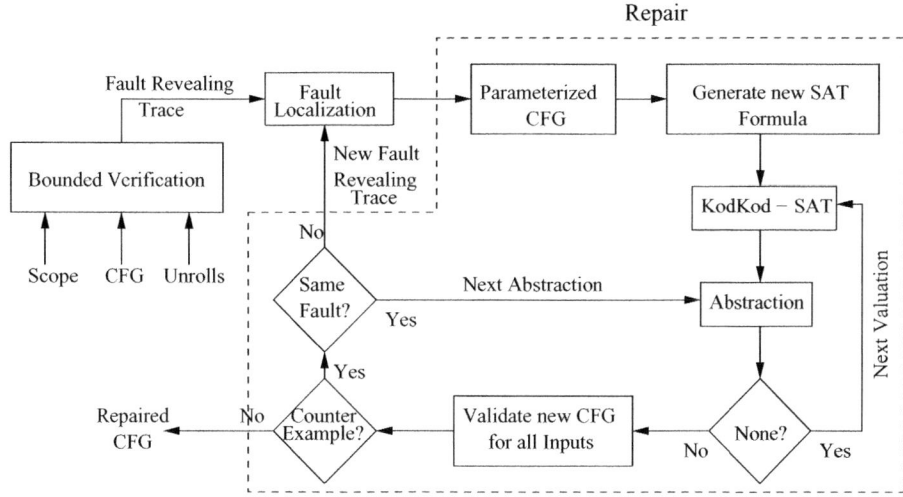

Fig. 2. Overview of our repair framework

been specified erroneously, it indicates that the statement to update the correct field (say f'_n) has been omitted. Handling of omission errors requires additional user input and is explained later in this section. The altered statement would be $V_{lhs}.f_n = V_{rhs}$, where V_{lhs} and V_{rhs} can be *null* or can take any value from the domains A and B respectively, if f_n is a relation mapping type A to B. Similarly a branch statement such as $x > y$, would be altered to $V_{lhs} > V_{rhs}$ and a variable update $y = x$ to $y = V_{rhs}$. Please note that we attempt to correct a statement by changing only the values read or written by it and not by altering the operators or constant values specified in the statement. We hypothesize that such syntactic and semantic errors can be caught by the programmer during his initial unit testing phase itself, whereas detection of errors relating to the state depends on the input being used. This may get missed during manual review or testing and a bounded verification technique which systematically checks for all possible input structures is adept in detecting such errors. However, if by altering the placement of the operands, the semantics of a wrongly specified operator can be made equivalent to the correct operator, then we can correct such errors. For instance, if $x < y$ has been wrongly specified as $x > y$, the corrected statement would be $y > x$. A new set of variables would thus get defined for each statement in suspect list, $\{V_{lhs}^{S_i}, V_{rhs}^{S_i}\}, \forall i \in \{1, \ldots, m\}$.

New constraints for the code with the newly introduced variables are generated in the next step. The input is fixed to the fault revealing structure by binding the pre-state relations exactly to the values in s_t when supplying the formula to the Kodkod engine. A formula made up of the conjunction of the fixed pre-state, the new code formula and the post-condition, $pre\text{-}state \wedge code\text{-}constraints \wedge post\text{-}condition$, is fed to the SAT solver. The solver looks for suitable valuations to the new variables such that a valid trace through the code is produced for the specified input(s_t) that yields an output(s''_t) that satisfies the post-condition. If $V_{lhs}^{S_i}$ is a variable of type Node, then a possible solution could bind it to any concrete state value in its domain, $V_{lhs}^{S_i} = n, \forall n \in \{N_0, \ldots, N_{scope-1}\}$. Hence every new variable in every statement in the suspect list would be bound to a concrete state value as follows, $V_{lhs}^{S_i} = C_{lhs}^{S_i}, V_{rhs}^{S_i} = C_{rhs}^{S_i}, \forall i \in \{1, \ldots, m\}$, where $C_{lhs}^{S_i}$ and $C_{rhs}^{S_i}$ are the state values are every statement.

The ensuing abstraction module attempts to map every concrete valuation to local variables or expressions that contain that value at that particular program point. The expressions are constructed by applying iterative field de-references to local variables. For example, if t is a local variable containing an instance of a Tree and the left pointer of its root node contains the required concrete value, then t.root.left would be the expression constructed. We ensure that we do not run into loops by hitting the same object repeatedly. Stating it formally, if for variable v, and fields $f_1, \ldots, f_k$, $v.f_1 \cdots f_k$ evaluates to v, then for any expression e that evaluates to v, the generated expression will not take the form $e.f_1 \cdots f_k.e'$, rather it will take the form $v.e'$. Please note that when a parameterized statement occurs inside a loop there would be several occurrences of it in the unrolled computation graph. Each occurrence could have different state valuations, however, their abstractions should yield the same expression. Hence we only consider the last occurrence to perform the abstraction. A new CFG is generated with the altered statements. In the event that no abstractions can be produced for a particular concrete valuation, we invoke SAT again to check if there are alternate correct valuations for the given input.

Please note that if SAT is unable to find suitable valuations that yield a satisfying solution for the given input or none of the valuations produced by SAT can be abstracted, it indicates that the fault cannot be corrected by altering the statements in the suspect list. The reason for this could either be that some statements had been omitted in the original code or that the initial list of suspicious statements is inaccurate. At this stage, manual inspection is required to determine the actual reason. Omission errors could be corrected by the tool using "templates" of the missing statements inserted by the user at appropriate locations. In case of the suspect list being incomplete, a feedback could be provided to the localization module to alter the set of possibly faulty statements.

In the validation phase, we use scope bounded checking to systematically ensure that all inputs covering the altered statements yield outputs satisfying the post-condition. If a counter-example is detected, the new fault revealing input and the corresponding trace, along with the altered CFG are fed back into the fault localization module. Based on the faults causing the new counter-example, an altered or the same set of statements would get produced and the process is repeated until no counter-example is detected in the validation phase. There could be faults in the code not detected during the initial verification or during the above mentioned validation process. These may be present in a portion of code not covered by the traces of the previously detected faults. Hence as a final step, we verify that the CFG is correct for all inputs covering the entire code. Thus, the repair process iteratively corrects all faults in the code. In case, all the faulty statements can be guaranteed to be present in the suspect list, the entire CFG can be checked whenever an abstraction is generated. When a counter-example is detected, the subsequent abstractions for the same set of statements could be systematically checked until one that satisfies all inputs is produced.

```
bool insert(Tree t, int k){                       void addChild(Tree t){
  Node y = null;                                     if ( t == null} return;
  Node x = t.root;                                   BaseTree childTree = (BaseTree)t;
  while (x != null)                                  if (childTree.isNil()) {
  //while(x.left != null) (4c)                          if (this.children != null &&
  { y = x;                                                   this.children == childTree.children)
    //y = t.root; (3b)(8)(9)                                throw new RuntimeException("...");
    if (k < x.key)                                       if (childTree.children != null) {
    //if(k < t.root.key) (9)(13)                           int n = childTree.children.size();
      x = x.left;                                          for (int i = 0; i < n; i++) {
    //x = x.right; (3a)                                    //for (int i = 0,j = 0; i < n; (i = j + 1)) { (1)
    //x = x.left.right (13)                                  Tree c = childTree.children.get(i);//list get
    else                                                     this.children.add(c);//list add
    { if (k > x.key)                                         c.setParent(this);
      //if (k < x.key)       (4a)                            //j = i + 1; (1)}}
      //if(k > t.root.key) (4b)                            else this.children = childTree.children;}
        x = x.right;                                     else {
      else                                                 if (children == null)
        return false;}}                                    //if (childTree.children == null) (2)
  x = new Node();                                            children = createChildrenList();
  x.key = k;                                               children.add(t); //list add
  if (y == null)                                           childTree.setParent(this);
  //if(x != null) (10)                                     //childTree.setParent(childTree); (2)}}
    t.root = x;
  else
  { if (k < y.key)
    //if(k > y.key) (2a)
    //if(k < x.key) (2b)(5)(11)(12)
      y.left = x;
    //y.left = y; (6)(10)(11)
    else
      y.right = x;}
    //y.right = y;}(10)(12)
  x.parent = y;
  //x.parent = x; (1)(7)(8)
  //y.parent = x; (5)(6)(10)
  /*x.parent = y;*/ //Omission Err(14)
  t.size += 1;
  return true;}
```

  (a)   (b)

Fig. 3. Code Snippets of (a) `BST.insert` (b) ANTLR `BaseTree.addChild` with error numbers. Repaired version produced in the CFG form, manually mapped back to source code.

4   Evaluation

This section presents two case studies on (i) Binary Search Tree `insert` method and (ii) `addChild` method of the ANTLR application [13]. The aim of the first study was to simulate an exhaustive list of error scenarios and measure the ability of the technique to repair faulty statements accurately and efficiently. The second evaluation focuses on studying the efficacy of the technique to scale to real world program sizes.

4.1   Candidates

Binary Search Tree (BST) is a commonly used text-book data structure. BST has complex structural invariants, making it non-trivial to perform correct repair actions to programs that manipulate its structure. The tree is made up of nodes, each comprising of an integer `key`, `left`, `right` and a `parent` pointer. Every tree has a `root` node which has no `parent` and a `size` field which stores the number of nodes reachable from the `root`. The data structure invariants include uniqueness of each element, acyclicity with respect to `left`,`right` and `parent` pointers, size and search constraints (`key` value of every node needs to be greater than the `key` values of all nodes in its `left` sub-tree and lesser than all keys in its `right` sub-tree). `Insert(int item)` is one of the most important methods of the class, which adds a node to the tree based on the input item value. The post-condition specification checks if the functionality of the method is met in addition to ensuring that the data structure invariants are preserved. It has considerable structural and semantic complexity, performing destructive updates to all fields of the tree class and comprising of 20 lines of code (4 branch statements and 1 loop).

ANother Tool for Language Recognition (ANTLR) [13], a part of the DaCapo benchmark [2], is a tool used for building recognizers, interpreters, compilers and translators from grammars. Rich tree data structures form the backbone of this application, which can be major sources of errors when manipulated incorrectly. Hence ANTLR is an excellent case study for repair on data structure programs. `BaseTree` class implements a generic tree structure to parse input grammars. Each tree instance maintains a list of children as its successors. Each child node is in turn a tree and has a pointer to its `parent` and a `childIndex` representing its position in the children list. Every node may also contain a `token` field which represents the payload of the node. Based on the documentation and the program logic, we derived invariants for the `BaseTree` data structure such as acyclicity, accurate parent-child relations and correct value of child indices. The `addChild(Tree node)` is the main method used to build all tree structures in ANTLR. The respective post-conditions check that the provided input has been added without any unwarranted perturbations to the tree structure.

4.2 Metrics

The efficiency of the technique was measured by the total time taken to repair a program starting from the time a counter-example and suspect list of statements were fed into the repair module. This included the time to correct all faults (commission) in the code satisfying all inputs within the supplied scope. Since the time taken by the SAT solver to systematically search for correct valuations and validate repair suggestions is a major factor adding to the repair time, we also measured the number of calls to the SAT solver. The repaired statements were manually verified for accuracy. They were considered to be correct if they were semantically similar to the statements in the correct implementation of the respective algorithms.

Our repair technique is implemented on top of the Forge framework [4]. The pre, post-conditions and CFG of the methods were encoded in the Forge Intermediate Representation(FIR). Scope bounded verification using Forge was used to automatically detect code faults. The suspect list of possibly faulty statements was generated manually. The input scope and number of loop unrolling were manually fixed to the maximum values required to produce repaired statements that would be correct for all inputs irrespective of the bounds. MINISAT was the SAT solver used. We ran the experiments on a system with 2.50GHz Core 2 Duo processor and 4.00GB RAM running Windows 7.

4.3   Results

Table 1 enlists the different types of errors seeded into the BST insert and ANTLR addChild methods. Figures 3a and 3b show the code snippets of the methods with the errors seeded. The errors have been categorized into 3 scenarios as described below.

Table 1. Case Study Results: P1 - BST.insert, P2 - ANTLR BaseTree.addChild. Errors categorized into 3 scenarios described in section 4.3. The number of actually faulty and correct statements in the suspect list of fault localization(FL) scheme are enumerated. Description highlights the type of the faulty statements. Efficiency measured by Repair Time and number of SAT Calls. Accuracy measured by (i) whether a fix was obtained, (ii) was the repaired statement exactly same as in correct implementation. Every result is an average of 5 runs(rounded to nearest whole number).

Name	Scr#	Error#	FL Scheme Output		Type of Stmts	Repair Time(secs)	# SAT Calls	Accuracy
			# Faulty	# Correct				
P1	1	1	1	0	Assign Stmt	3	2	✓, Same
		2a	1	0	Branch stmt	34	114	✓, Diff
		2b	1	0	Branch stmt	4	2	✓, Same
		3a	1	0	Assign stmt	5	2	✓, Diff
		3b	1	0	Assign stmt	5	4	✓, Same
		4a	1	0	Branch stmt	12	96	✓, Diff
		4b	1	0	Branch stmt	4	2	✓, Same
		4c	1	0	Loop condition	1	2	✓, Same
		5	2	0	Branch, Assign stmts	7	5	✓, Same
		6	2	0	Assign stmts	5	3	✓, Same
	2	7	1	2	Branch, Assign stmts	15	21	✓, Same
		8	2	1	Branch, Assign stmts	6	2	✓, Same
		9	1	1	Assign stmts	11	2	✓, Same
	3	10	4	0	Branch, Assign stmts	6	8	✓, Same
		11	2	0	Branch, Assign stmts	26	9	✓, Same
		12	2	1	Branch, Assign stmts	33	14	✓, Same
		13	2	1	Assign, Branch stmts	14	24	✓, Same
		14	0	2	Omission error	NA	NA	NA
P2	1	1	2	0	Assign Stmt	71	2	✓, Diff
	2	2	2	2	Branch, Assign stmts	1	5	✓, Same

Binary Search Tree insert

Scenario 1: All code faults identified in the initial verification, suspect list contains the exact list of faulty statements. In the first eight errors only 1 statement is faulty. In errors 1 and 3, the fault lies in the operands of an assignment statement. Error 1 involves a wrong update to the parent field causing a cycle in the output tree. The faults in the latter case, assign wrong values to local variables x and y respectively, which impact the program logic leading to errors in the output structure. For instance, in error 3a, the variable x is assigned to x.right instead of x.left if the input value k < x.key inside the loop. This impacts the point at which the new node gets added thus breaking the constraints on the key field. The repaired statement produced is x = y.left which is semantically similar to the expected expression, since the variable y is always assigned to x before this statement executes. In error 3, both the faults being inside a loop, require a higher scope to get detected as compared to the fault in error 1. This increases the search space for the solver to find correct valuations for the fault revealing input, resulting in higher repair time. Errors 2 and 4 involve faulty branch statements present outside and inside a loop respectively. In Errors 2a and 4a, the comparison operator is wrong and both the operands are parameterized as $V_{lhs} > V_{rhs}$. The search is on the integer domain and passes through many iterations before resulting in valuations that produce an abstraction satisfying all inputs. The final expression is semantically same as the correct implementation, with the operands interchanged ($y.key > k$ vs $k < y.key$). In errors 2b and 4b, the expression specified in the branch condition is faulty. Hence only the variable in the expression is parameterized ($k < V_{rhs}.key$). The search on the node domain results in correctly repaired statements with lesser number of SAT calls than a search on the bigger integer domain. In errors 5 and 6, combinations of branch and assignment statements are simultaneously faulty. Since the number of newly introduced variables are higher, the repair times are higher than the previous errors. Also, when there are more than one statements in fault, combinations of abstractions and valuations corresponding to each statement, need to be parsed to look for a solution satisfying all inputs.

Scenario 2: Fault localization scheme produces a possibly faulty list. In this scenario, the suspect list also includes statements which are actually correct. For instance, in error 7, the fault lies in an assignment statement which wrongly updates the `parent` of the inserted node similar to error 1, but the suspect list also includes 2 branch statements before this statement. In this case, all the operands of the 3 statements are parameterized leading to an increase in the state space and hence the repair time. It can be observed that as the percentage of actually faulty statements increases, the number of SAT calls decreases. In error 9, an assignment statement inside a loop wrongly updates a variable which is used by a subsequent branch statement. Owing to the data dependency between the statements, the number of possible combinations of correct valuations are less resulting in just 2 SAT calls, however, the large size of the fault revealing input increases the search time. The results were manually verified to ensure that the expressions assigned to the actually correct statements in the final repaired program were the same as before.

Scenario 3: There could be other faults than those revealed by the initial verification. Not all code faults may get detected in the initial verification stage. For instance, in error 10, an input structure as small as an empty tree can expose the fault in the branch statement. However, when the correction for this fault is validated using a higher scope, wrong updates to the `parent` and the `right` fields get detected. The new fault-revealing input and the CFG corrected for the first fault are fed back into fault localization scheme and the process repeats until a program which is correct for all inputs passing through the repaired statements is produced. However, when the entire program is checked, a wrong update to the `left` field of the inserted node gets detected. Since this assignment statement is control dependent on the branch which directs flow towards the update of the `right` field, it is not covered by the traces of the previous fault revealing inputs. The process repeats as explained before until the last fault is corrected. This explains the 8 calls to SAT to correct this error scenario. Owing to the small size of the fault revealing inputs, the suspect list containing only the erroneous statements and the fact that in most runs, the first abstraction for every faulty statement happened to be the correct ones, the total repair took only 5 seconds. However, correction of subsequent errors necessitate more number of iterations leading to higher repair times, which got exacerbated when the suspect list also included statements not in error. In the last case (error 14), the statement to update the `parent` field is omitted from the code, the localization scheme wrongly outputs the statements which update the `left` and `right` fields as being possibly faulty. The repair module is unable to find a valuation which produces a valid output for the fault revealing input. It thus displays a message stating that it could possibly be an omission error or the statements output by the localization scheme could be wrong.

ANTLR addChild: The `addChild` method is much more complex than BST insert consisting of calls to 4 other methods, nested branch structures and total source code LOC of 45 (including called methods). The first fault consists of a faulty increment to the integer index of a loop. An extra local variable `j` is assigned the value of integer index (`i`) plus 1 inside the loop. After every iteration, `i` is assigned `j + 1`, erroneously incrementing the index by 2 instead of 1. This fault requires an input of size 4 nodes and 2 unrolls of the loop to get detected. The error can be corrected by assigning `i` instead of `i + 1` to `j`, or `j` to `i` after every iteration, or `i + 1` to `i`(original implementation). We assumed that the suspect list includes both the assignment statements. In the repaired version, `i` was assigned `i + 1` after every iteration and the assignment to `j` was not altered. The correction was performed in just 2 SAT calls but consumed 71 seconds due to the large state space. The second scenario simulates a case wherein both a branch statement checking whether the current tree has any children or not and a statement updating the `parent` field of the added child tree are faulty. These faults are detected with a scope of 3 and unrolls 1. Two actually correct update statements are also included in the suspect list. Our technique is able to repair all the statements in one second.

5   Discussion

As can be observed from the results of the evaluation, our repair technique was success-ful in correcting faults in different types of statements and programming constructs. The technique was able to correct up to 4 code faults for `BST insert` within a worst case time of 33 seconds. The technique consumed a little more than a minute (worst case) to correct faults in `ANTLR addChild`, highlighting its ability to scale to code sizes in real world applications. Overall, we can infer that the repair time is more impacted by the size of the fault revealing input rather than the number of lines of source code. It has been empirically validated that faults in most data structure programs can be detected using small scopes [1]. This is an indication that our technique has the potential to be applicable for many real world programs using data structures.

One of the shortcomings of the technique is that the accuracy of the repaired state-ments is very closely tied to the correctness and completeness of the post-condition constraints and the maximum value of scope used in the validation phase. For instance, in the second error scenario of `ANTLR addChild`(3b), when the post-condition just checked the addition of a child tree, an arbitrary expression yielding `true` was substi-tuted in place of the erroneous branch condition. However, only when the constraints were strengthened to ensure that the child had been inserted in the correct position in the children list of the current tree, an accurate branch condition same as that in the original correct implementation was obtained. This limitation is inherited from specification-based bounded verification techniques which consider a program to be correct as long as the output structure satisfies the user-specified post-condition constraints. Hence, the technique can be relied upon to produce near accurate repair suggestions, which need to be manually verified by the user. However, the user can either refine the constraints or the scope supplied to the tool to refine the accuracy of the corrections. Following points are avenues for improvements in the repair algorithm; Erroneous constant values in a statement could be corrected by parameterizing them to find correct replacements but avoiding the abstraction step. Erroneous operators in a statement could be handled with the help of user-provided templates of possibly correct operators, similar to the handling of omission errors as described in Section 3. The number of iterations required to validate the abstractions for a particular valuation can be decreased by always starting with those involving direct use of local variables declared in the methods. Methods that manipulate input object graphs often use local variables as pointers into the input graphs for traversing them and accessing their desired components. Hence the probability of them being the correct abstractions would increase.

6   Correctness Argument

In this section, we present a brief argument that our repair algorithm is accurate, terminates and handles a variety of errors. We make the assumptions that the faults are not present in the operator or constant values specified in a statement. The basic intuition behind parameterizing possibly faulty statements is that any destructive update to a linked data structure should be through wrong updates to one or more fields of the objects of the structure. Hence we try to correct the state by altering the values read and written by statements directly updating fields. We also cover cases where wrong local variable values may indirectly lead to wrong field updates. We also consider erroneous branch condition expressions which may direct the flow towards a wrong path. The variables introduced during parameterizations are constrained exactly at the location of the faulty statements. Further, the state values assigned to them need to be reachable from the local variables at that program point. The fact that the input is fixed to a valid structure and that the statements before and after the faulty statements are correct, ensures that finding an abstraction at the respective program points should lead to a valid update to the state such that the output apart from being structurally correct, is in alignment with the intended program logic. An abstraction yielding a correct output for a particular input structure need not be correct for all input structures covering that statement. There could be bigger structures with higher depth or traversing a different path before reaching the repaired statement, which may require a different abstraction. There could also be more than one correct valuation which yield correct output for a particular input. For instance, when both operands of a branch statement are parameterized, the variables could take any value that results in the correct boolean output to direct the control flow in the right direction. These cases are detected in the abstraction and validation stages and other valuations and/or abstractions for a particular valuation are systematically checked. Provided that there are no new code faults and all the faulty statements are guaranteed to be in the suspect list, a valuation and/or abstraction satisfying all inputs should be in the list obtained from the initial fault revealing input. Hence, the repair process would terminate.

The guarantee on the completeness and correctness of the repaired statements is the same as that provided by other scope bounded verification techniques. The repaired statements are correct with respect to the post-condition specifications for all inputs within the highest scope used to validate the repaired program.

7 Related Work

We first discuss two techniques most closely related to our work: program sketching using SAT [15] (Section 7.1) and program repair based on data structure repair [11] (Section 7.2). We next discuss other related work (Section 7.3).

7.1 Program Sketching Using SAT

Program synthesis using sketching [15] employs SAT solvers to generate parts of programs. The user gives partial programs that define the basic skeleton of the code. A SAT solver completes the implementation details by generating expressions to fill the "holes" of the partial program. A basic difference between our work and program sketching stems from the intention. We are looking to perform repair or modifications to statements identified as being faulty while sketching aims to complete a given partial but fixed sketch of a program. A key technical difference is our support for specifications or constraints on the concrete state of the program. Therefore, the search space for the SAT solver is the bounded state space of the program, instead of the domain of all possible program expressions as in sketching. To illustrate the difference between repair and sketching, consider the doubly linked list example. The sketching process starts with a random input structure. Assume that this is the same input structure with two nodes as in Fig. 1. Assume also that the user specified the details for the entire `remove` method except for the statement that updates `prevP.next`). The user would then need to specify a generator stating all possible program expressions that can occur at the "hole" (RHS of the assignment statement). Following the sketch language described in the synthesis paper, this would be specified using generators as shown:

$$\#define \ \mathrm{LOC} = \{(\mathrm{head}|\mathrm{ptr}|\mathrm{prevP}).(\mathrm{next}|\mathrm{prev})?.(\mathrm{next}|\mathrm{prev})?|\mathrm{null}\}$$
```
prevP.ptr = LOC;
```

This bounds the possible expressions to be all values obtained by starting from `header`, `ptr`, `prevP` and subsequent de-referencing of `next` or `prev` pointers (`header`, `header.next`, `ptr.next.prev`, `prevP.prev.prev` so on). Even for this small example with just 1 unknown in 1 statement, the number of possible program expressions to be considered is more than the list obtained by mapping back from the correct concrete state. As the program size and complexity increases, the set of all possible program expressions that could be substituted for all the holes can become huge. Further, to obtain a correct program, the user needs to estimate and accurately specify all the possible expressions. In such cases, it would be faster to look for a correct structure in the concrete state space. The number of expressions mapping to the correct state value at a program point would be much lesser in number. Our technique automatically infers the correct expressions at a point and does not require the user to specify them. The synthesis technique employs an iterative process of validation and refinement (CEGIS) to filter out sketches that do not satisfy all inputs, similar to our validation phase. However since synthesis starts with a random input which may have poor coverage, it would require higher number of iterations to converge to a correct sketch. Since we start with the input that covers the faulty statement, the iterations should converge to a correct abstraction faster. Further, if the statements produced by fault localization can be guaranteed to be the only ones faulty, the correct expression must be present in the list of abstractions for the first input. Hence scanning this list would be faster than having to feedback every new counter-example into the system.

7.2   Program Repair Using Data Structure Repair

On-the-fly repair of erroneous data structures is starting to gain more attention in the research community. Juzi [5] is a tool which performs symbolic execution of the class invariant to determine values for fields that would yield a structurally correct program state. A recent paper [11] presents a technique on how the repair actions on the state could be abstracted to program statements and thus aid in repair of the program. The main drawback of this technique is that it focuses on class invariants and does not handle the specification of a particular method. Hence the repaired program would yield an output that satisfies the invariants but may be fairly different from the intended behavior of the specific method. Further, in cases where the reachability of some nodes in the structure gets broken, Juzi may be unable to parse to the remaining nodes of the tree and hence fail to correct the structure. Error 4c (3a) in the `BST insert` method, highlights such a scenario wherein the loop condition used to parse into the tree structure being wrong, the new node gets wrongly added as the root of the tree. Our technique looks for a structure that satisfies the specific post-condition of method, which requires that the nodes in the pre-state also be present in the output structure. Hence a correct output tree structure gets produced.

7.3   Other Recent Work on Program Correction

Our technique bears similarities with that of Jobstmann et al [9]. Given a list of suspect statements, their technique replaces the expressions of the statements with unknowns. It then uses a model checker to find a strategy that can yield a solution to the product of a model representing the broken program and an automaton of its linear time logic (LTL) specifications. Though the basic method of formulating the problem is similar, their problem space is very different from ours since their technique is targeted and evaluated on concurrent, finite state transition systems against properties written in temporal logic, whereas our technique is developed for sequential programs that operate on complex data structures. Extending their approach to the domain of complex structures is nontrivial due to the dissimilarity in the nature of programs under repair as well as the specifications. Moreover, our approach of bounded modular debugging of code using declarative specifications likely enables more efficient modeling and reasoning about linked structures. Additionally, translation of the specifications and the entire code into a SAT formula aids in exploiting the power of SAT technology in handling the size and complexity of the underlying state space.

In contrast to specification-based repair, some recently developed program repair techniques do not require users to provide specifications and instead multiple passing and failing runs with similar coverage. Machine learning [8] and genetic programming based techniques [19] fall in this category. However, these techniques have a high sensitivity to the quality of the given set of tests and do not address repairing programs that operate on structurally complex data. The recently proposed AutoFix-E tool [17] attempts to bridge the gap between specification-based and test-based repair. Boolean queries of a class are used to build an abstract notion of the state which forms the basis to represent contracts of the class, fault profile of failing tests and a behavioral model based on passing tests. A comparison between failing and passing profiles is performed for fault localization and a subsequent program synthesis effort generates the repaired statements. This technique however only corrects violations of simple assertions, which can be formulated using boolean methods already present in the class, but not rich post-conditions.

8   Conclusion

This paper presents a novel technique to repair data structure programs using the SAT-based Alloy tool-set. Given a fault revealing input and a list of possibly faulty statements, the technique introduces nondeterminism in the statements and employs SAT to prune the nondeterminism to generate an output structure that satisfies the post-condition. The SAT solution is abstracted to program expressions, which yield a list of possible modifications to the statement, which are then iteratively refined using bounded verification. We performed an experimental validation using a prototype implementation of our approach to cover a variety of different types and number of faults on the subject programs. The experimental results demonstrate the efficiency, accuracy of our approach and the promise it holds to scale to real world applications.

Our tool can act as a semi-automated method to produce feedback to fault localization schemes such as [14] to refine the accuracy of localization. The concept of searching a bounded state space to generate program statements can also be used to develop more efficient program synthesis techniques. The idea of introducing nondeterminism in code can aid in improving the precision of data structure repair as well. We also envisage a possible integration of our technique with other contract and test suite coverage based techniques like AutoFix-E [17]. We believe our approach of modeling and reasoning about programs using SAT can aid in improving the repair efficiency and accuracy of existing tools.

Acknowledgements

This material is based upon work partially supported by the NSF under Grant Nos. IIS-0438967 and CCF-0845628, and AFOSR grant FA9550-09-1-0351.

References

1. Andoni, A., Daniliuc, D., Khurshid, S., Marinov, D.: Evaluating the "Small Scope Hypothesis". Technical report, MIT CSAIL (2003)
2. Blackburn, S.M., et al.: The DaCapo benchmarks: Java benchmarking development and analysis. In: OOPSLA (2006)

3. Collofello, J.S., Cousins, L.: Towards automatic software fault location through decision-to-decision path analysis
4. Dennis, G., Chang, F.S.-H., Jackson, D.: Modular verification of code with SAT. In: ISSTA (2006)
5. Elkarablieh, B., Khurshid, S.: Juzi: A tool for repairing complex data structures. In: ICSE (2008)
6. Jackson, D.: Software Abstractions: Logic, Language and Analysis. MIT-P, Cambridge (2006)
7. Jackson, D., Vaziri, M.: Finding bugs with a constraint solver. In: ISSTA (2000)
8. Jeffrey, D., Feng, M., Gupta, N., Gupta, R.: BugFix: A learning-based tool to assist developers in fixing bugs. In: ICPC (2009)
9. Jobstmann, B., Griesmayer, A., Bloem, R.: Program repair as a game. In: Etessami, K., Rajamani, S.K. (eds.) CAV 2005. LNCS, vol. 3576, pp. 226–238. Springer, Heidelberg (2005)
10. Jones, J.A.: Semi-Automatic Fault Localization. PhD thesis, Georgia Institute of Technology (2008)
11. Malik, M.Z., Ghori, K., Elkarablieh, B., Khurshid, S.: A case for automated debugging using data structure repair. In: ASE (2009)
12. Marinov, D., Khurshid, S.: TestEra: A novel framework for automated testing of Java programs (2001)
13. Parr, T., et al.: Another tool for language recognition, http://www.antlr.org/
14. Renieris, M., Reiss, S.P.: Fault localization with nearest neighbor queries. In: ASE (2003)
15. Solar-Lezama, A.: The sketching approach to program synthesis. In: Hu, Z. (ed.) APLAS 2009. LNCS, vol. 5904, pp. 4–13. Springer, Heidelberg (2009)
16. Torlak, E., Jackson, D.: Kodkod: A relational model finder. In: Grumberg, O., Huth, M. (eds.) TACAS 2007. LNCS, vol. 4424, pp. 632–647. Springer, Heidelberg (2007)
17. Wei, Y., Pei, Y., Furia, C.A., Silva, L.S., Buchholz, S., Meyer, B., Zeller, A.: Automated fixing of programs with contracts. In: ISSTA (2010)
18. Weimer, W.: Patches as better bug reports. In: GPCE (2006)
19. Weimer, W., Nguyen, T., Goues, C.L., Forrest, S.: Automatically finding patches using genetic programming. In: ICSE (2009)
20. Weiser, M.: Programmers use slices when debugging. Commun. ACM (1982)

Optimal Base Encodings for
Pseudo-Boolean Constraints[*]

Michael Codish[1], Yoav Fekete[1], Carsten Fuhs[2], and Peter Schneider-Kamp[3]

[1] Department of Computer Science, Ben Gurion University of the Negev, Israel
[2] LuFG Informatik 2, RWTH Aachen University, Germany
[3] IMADA, University of Southern Denmark, Denmark

Abstract. This paper formalizes the *optimal base problem*, presents an algorithm to solve it, and describes its application to the encoding of Pseudo-Boolean constraints to SAT. We demonstrate the impact of integrating our algorithm within the Pseudo-Boolean constraint solver MINISAT$^+$. Experimentation indicates that our algorithm scales to bases involving numbers up to 1,000,000, improving on the restriction in MINISAT$^+$ to prime numbers up to 17. We show that, while for many examples primes up to 17 do suffice, encoding with respect to optimal bases reduces the CNF sizes and improves the subsequent SAT solving time for many examples.

1 Introduction

The optimal base problem is all about finding an efficient representation for a given collection of positive integers. One measure for the efficiency of such a representation is the sum of the digits of the numbers. Consider for example the decimal numbers $S = \{16, 30, 54, 60\}$. The sum of their digits is 25. Taking binary representation we have $S_{(2)} = \{10000, 11110, 110110, 111100\}$ and the sum of digits is 13, which is smaller. Taking ternary representation gives $S_{(3)} = \{121, 1010, 2000, 2020\}$ with an even smaller sum of digits, 12. Considering the *mixed radix* base $B = \langle 3, 5, 2, 2 \rangle$, the numbers are represented as $S_{(B)} = \{101, 1000, 1130, 10000\}$ and the sum of the digits is 9. The optimal base problem is to find a (possibly mixed radix) base for a given sequence of numbers to minimize the size of the representation of the numbers. When measuring size as "sum of digits", the base B is indeed optimal for the numbers of S. In this paper we present the optimal base problem and illustrate why it is relevant to the encoding of Pseudo-Boolean constraints to SAT. We also present an algorithm and show that our implementation is superior to current implementations.

Pseudo-Boolean constraints take the form $a_1 x_1 + a_2 x_2 + \cdots + a_n x_n \geq k$, where $a_1, \ldots, a_n$ are integer coefficients, $x_1, \ldots, x_n$ are Boolean literals (i.e., Boolean variables or their negation), and k is an integer. We assume that constraints are in Pseudo-Boolean normal form [3], that is, the coefficients a_i and k are always positive and Boolean variables occur at most once in $a_1 x_1 + a_2 x_2 + \cdots + a_n x_n$.

---

[*] Supported by GIF grant 966-116.6 and the Danish Natural Science Research Council.

P.A. Abdulla and K.R.M. Leino (Eds.): TACAS 2011, LNCS 6605, pp. 189–204, 2011.
© Springer-Verlag Berlin Heidelberg 2011

Pseudo-Boolean constraints are well studied and arise in many different contexts, for example in verification [6] and in operations research [5]. Typically we are interested in the satisfiability of a conjunction of Pseudo-Boolean constraints. Since 2005 there is a series of Pseudo-Boolean Evaluations [11] which aim to assess the state of the art in the field of Pseudo-Boolean solvers. We adopt these competition problems as a benchmark for the techniques proposed in this paper.

Pseudo-Boolean constraint satisfaction problems are often reduced to SAT. Many works describe techniques to encode these constraints to propositional formulas [1,2,9]. The Pseudo-Boolean solver MiniSat$^+$ ([9], cf. `http://minisat.se`) chooses between three techniques to generate SAT encodings for Pseudo-Boolean constraints. These convert the constraint to: (a) a BDD structure, (b) a network of sorters, and (c) a network of (binary) adders. The network of adders is the most concise encoding, but it has the weakest propagation properties and often leads to higher SAT solving times than the BDD based encoding, which, on the other hand, generates the largest encoding. The encoding based on sorting networks is often the one applied and is the one we consider in this paper.

To demonstrate how sorters can be used to translate Pseudo-Boolean constraints, consider the constraint $\psi = x_1 + x_2 + x_3 + 2x_4 + 3x_5 \geq 4$ where the sum of the coefficients is 8. On the right, we illustrate an 8×8 sorter where x_1, x_2, x_3 are each fed into a single input, x_4 into two of the inputs, and x_5 into three of the inputs. The outputs are

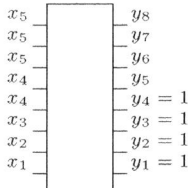

$y_1, \ldots, y_8$. First, we represent the sorting network as a Boolean formula, φ, which in general, for n inputs, will be of size $O(n \log^2 n)$ [4]. Then, to assert ψ we take the conjunction of φ with the formula $y_1 \wedge y_2 \wedge y_3 \wedge y_4$.

But what happens if the coefficients in a constraint are larger than in this example? How should we encode $16x_1 + 30x_2 + 54x_3 + 30x_4 + 60x_5 \geq 87$? How should we handle very large coefficients (larger than 1,000,000)? To this end, the authors in [9] generalize the above idea and propose to decompose the constraint into a number of interconnected sorting networks. Each sorter represents a digit in a mixed radix base. This construction is governed by the choice of a suitable mixed radix base and the objective is to find a base which minimizes the size of the sorting networks. Here the optimal base problem comes in, as the size of the networks is directly related to the size of the representation of the coefficients. We consider the sum of the digits (of coefficients) and other measures for the size of the representations and study their influence on the quality of the encoding.

In MiniSat$^+$ the search for an optimal base is performed using a brute force algorithm and the resulting base is constructed from prime numbers up to 17. The starting point for this paper is the following remark from [9] (Footnote 8):

> *This is an ad-hoc solution that should be improved in the future. Finding the optimal base is a challenging optimization problem in its own right.*

In this paper we take the challenge and present an algorithm which scales to find an optimal base consisting of elements with values up to 1,000,000. We illustrate that in many cases finding a better base leads also to better SAT solving time.

Section 2 provides preliminary definitions and formalizes the optimal base problem. Section 3 describes how MINISAT$^+$decomposes a Pseudo-Boolean constraint with respect to a given mixed radix base to generate a corresponding propositional encoding, so that the constraint has a solution precisely when the encoding has a model. Section 4 is about (three) alternative measures with respect to which an optimal base can be found. Sections 5–7 introduce our algorithm based on classic AI search methods (such as cost underapproximation) in three steps: Heuristic pruning, best-first branch and bound, and base abstraction. Sections 8 and 9 present an experimental evaluation and some related work. Section 10 concludes. Proofs are given in [8].

2   Optimal Base Problems

In the classic base r radix system, positive integers are represented as finite sequences of digits $\mathbf{d} = \langle d_0, \ldots, d_k \rangle$ where for each digit $0 \le d_i < r$, and for the most significant digit, $d_k > 0$. The integer value associated with $\mathbf{d}$ is $v = d_0 + d_1 r + d_2 r^2 + \cdots + d_k r^k$. A mixed radix system is a generalization where a base is an infinite radix sequence $B = \langle r_0, r_1, r_2, \ldots \rangle$ of integers where for each radix, $r_i > 1$ and for each digit, $0 \le d_i < r_i$. The integer value associated with $\mathbf{d}$ is $v = d_0 w_0 + d_1 w_1 + d_2 w_2 + \cdots + d_k w_k$ where $w_0 = 1$ and for $i \ge 0$, $w_{i+1} = w_i r_i$. The sequence $weights(B) = \langle w_0, w_1, w_2, \ldots \rangle$ specifies the weighted contribution of each digit position and is called the *weight sequence* of B. A finite mixed radix base is a finite sequence $B = \langle r_0, r_1, \ldots, r_{k-1} \rangle$ with the same restrictions as for the infinite case except that numbers always have $k + 1$ digits (possibly padded with zeroes) and there is no bound on the value of the most significant digit, d_k.

In this paper we focus on the representation of finite multisets of natural numbers in finite mixed radix bases. Let *Base* denote the set of finite mixed radix bases and $ms(\mathbb{N})$ the set of finite non-empty multisets of natural numbers. We often view multisets as ordered (and hence refer to their first element, second element, etc.). For a finite sequence or multiset S of natural numbers, we denote its length by $|S|$, its maximal element by $max(S)$, its ith element by $S(i)$, and the multiplication of its elements by $\prod S$ (if S is the empty sequence then $\prod S = 1$). If a base consists of prime numbers only, then we say that it is a prime base. The set of prime bases is denoted $Base_p$.

Let $B \in Base$ with $|B| = k$. We denote by $v_{(B)} = \langle d_0, d_1, \ldots, d_k \rangle$ the representation of a natural number v in base B. The most significant digit of $v_{(B)}$, denoted $msd(v_{(B)})$, is d_k. If $msd(v_{(B)}) = 0$ then we say that B is redundant for v. Let $S \in ms(\mathbb{N})$ with $|S| = n$. We denote the $n \times (k + 1)$ matrix of digits of elements from S in base B as $S_{(B)}$. Namely, the ith row in $S_{(B)}$ is the vector $S(i)_{(B)}$. The most significant digit column of $S_{(B)}$ is the $k + 1$ column of the matrix and denoted $msd(S_{(B)})$. If $msd(S_{(B)}) = \langle 0, \ldots, 0 \rangle^T$, then we say that B is redundant for S. This is equivalently characterized by $\prod B > max(S)$.

Definition 1 (non-redundant bases). *Let $S \in ms(\mathbb{N})$. We denote the set of non-redundant bases for S, $Base(S) = \left\{\, B \in Base \,\middle|\, \prod B \le max(S) \,\right\}$. The* set of non-redundant prime bases for S is denoted $Base_p(S)$. The set of non-redundant (prime) bases for S, containing elements no larger than ℓ, is denoted $Base^\ell(S)$ $(Base_p^\ell(S))$. The set of bases in $Base(S)/Base^\ell(S)/Base_p^\ell(S)$, is often viewed as a tree with root $\langle\,\rangle$ (the empty base) and an edge from B to B' if and only if B' is obtained from B by extending it with a single integer value.

Definition 2 (*sum_digits*). *Let $S \in ms(\mathbb{N})$ and $B \in Base$. The sum of the digits of the numbers from S in base B is denoted $sum\_digits(S_{(B)})$.*

Example 3. The usual binary "base 2" and ternary "base 3" are represented as the infinite sequences $B_1 = \langle 2, 2, 2, \ldots \rangle$ and $B_2 = \langle 3, 3, 3, \ldots \rangle$. The finite sequence $B_3 = \langle 3, 5, 2, 2 \rangle$ and the empty sequence $B_4 = \langle\,\rangle$ are also bases. The empty base is often called the "unary base" (every number in this base has a single digit). Let $S = \{16, 30, 54, 60\}$. Then, $sum\_digits(S_{(B_1)}) = 13$, $sum\_digits(S_{(B_2)}) = 12$, $sum\_digits(S_{(B_3)}) = 9$, and $sum\_digits(S_{(B_4)}) = 160$.

Let $S \in ms(\mathbb{N})$. A cost function for S is a function $cost_S : Base \to \mathbb{R}$ which associates bases with real numbers. An example is $cost_S(B) = sum\_digits(S_{(B)})$. In this paper we are concerned with the following *optimal base problem*.

Definition 4 (optimal base problem). *Let $S \in ms(\mathbb{N})$ and $cost_S$ a cost function. We say that a base B is an* optimal base *for S with respect to $cost_S$, if for all bases B', $cost_S(B) \leq cost_S(B')$. The corresponding* optimal base problem *is to find an optimal base B for S.*

The following two lemmata confirm that for the *sum_digits* cost function, we may restrict attention to non-redundant bases involving prime numbers only.

Lemma 5. *Let $S \in ms(\mathbb{N})$ and consider the sum_digits cost function. Then, S has an optimal base in $Base(S)$.*

Lemma 6. *Let $S \in ms(\mathbb{N})$ and consider the sum_digits cost function. Then, S has an optimal base in $Base_p(S)$.*

How hard is it to solve an instance of the optimal base problem (namely, for $S \in ms(\mathbb{N})$)? The following lemma provides a polynomial (in $max(S)$) upper bound on the size of the search space. This in turn suggests a pseudo-polynomial time brute force algorithm (to traverse the search space).

Lemma 7. *Let $S \in ms(\mathbb{N})$ with $m = max(S)$. Then, $\big|Base(S)\big| \leq m^{1+\rho}$ where $\rho = \zeta^{-1}(2) \approx 1.73$ and where ζ is the Riemann zeta function.*

Proof. Chor *et al.* prove in [7] that the number of ordered factorizations of a natural number n is less than n^ρ. The number of bases for all of the numbers in S is hence bounded by $\sum_{n \leq m} n^\rho$, which is bounded by $m^{1+\rho}$.

3   Encoding Pseudo-Boolean Constraints

This section presents the construction underlying the sorter based encoding of Pseudo-Boolean constraints applied in MINISAT$^+$[9]. It is governed by the choice of a mixed radix base B, the optimal selection of which is the topic of this paper. The construction sets up a series of sorting networks to encode the digits, in base

B, of the sum of the terms on the left side of a constraint $\psi = a_1x_1 + a_2x_2 + \cdots + a_nx_n \geq k$. The encoding then compares these digits with those from $k_{(B)}$ from the right side. We present the construction, step by step, through an example where $B = \langle 2, 3, 3 \rangle$ and $\psi = 2x_1 + 2x_2 + 2x_3 + 2x_4 + 5x_5 + 18x_6 \geq 23$.

Step one - representation in base

The coefficients of ψ form a multiset $S = \{2, 2, 2, 2, 5, 18\}$ and their representation in base B, a 6×4 matrix, $S_{(B)}$, depicted on the right. The rows of the matrix correspond to the representation of the coefficients in base B.

$$S_{(B)} = \begin{pmatrix} 0\ 1\ 0\ 0 \\ 0\ 1\ 0\ 0 \\ 0\ 1\ 0\ 0 \\ 0\ 1\ 0\ 0 \\ 1\ 2\ 0\ 0 \\ 0\ 0\ 0\ 1 \end{pmatrix}$$

Step two - counting: Representing the coefficients as four digit numbers in base $B = \langle 2, 3, 3 \rangle$ and considering the values $weights(B) = \langle 1, 2, 6, 18 \rangle$ of the digit positions, we obtain a decomposition for the left side of ψ:

$$2x_1 + 2x_2 + 2x_3 + 2x_4 + 5x_5 + 18x_6 =$$
$$\mathbf{1} \cdot (x_5) + \mathbf{2} \cdot (x_1 + x_2 + x_3 + x_4 + 2x_5) + \mathbf{6} \cdot (0) + \mathbf{18} \cdot (x_6)$$

To encode the sums at each digit position $(1, 2, 6, 18)$, we set up a series of four sorting networks as depicted below. Given values for the variables, the sorted outputs from these networks represented unary numbers d_1, d_2, d_3, d_4 such that the left side of ψ takes the value $1 \cdot d_1 + 2 \cdot d_2 + 6 \cdot d_3 + 18 \cdot d_4$.

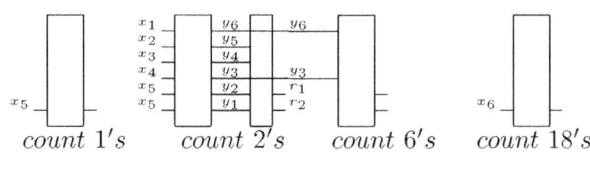

count 1's   *count 2's*   *count 6's*   *count 18's*

Step three - converting to base: For the outputs d_1, d_2, d_3, d_4 to represent the digits of a number in base $B = \langle 2, 3, 3 \rangle$, we need to encode also the "carry" operation from each digit position to the next. The first 3 outputs must represent valid digits for B, i.e., unary numbers less than $\langle 2, 3, 3 \rangle$ respectively. In our example the single potential violation to this restriction is d_2, which is represented in 6 bits. To this end we add two components to the encoding: (1) each third output of the second network (y_3 and y_6 in the diagram) is fed into the third network as an additional (carry) input; and (2) clauses are added to encode that the output of the second network is to be considered modulo 3. We call these additional clauses a *normalizer*. The normalizer defines two outputs $R = \langle r_1, r_2 \rangle$ and introduces clauses specifying that the (unary) value of R equals the (unary) value of $d_2 \bmod 3$.

count 1's   *count 2's*   *count 6's*   *count 18's*

Step four - comparison: The outputs from these four units now specify a number in base B, each digit represented in unary notation. This number is now compared (via an encoding of the lexicographic order) to $23_{(B)}$ (the value from the right-hand side of ψ).

4   Measures of Optimality

We now return to the objective of this paper: For a given Pseudo-Boolean constraint, how can we choose a mixed radix base with respect to which the encoding of the constraint via sorting networks will be optimal? We consider here three alternative cost functions with respect to which an optimal base can be found. These cost functions capture with increasing degree of precision the actual size of the encodings.

The first cost function, *sum_digits* as introduced in Definition 2, provides a coarse measure on the size of the encoding. It approximates (from below) the total number of input bits in the network of sorting networks underlying the encoding. An advantage in using this cost function is that there always exists an optimal base which is prime. The disadvantage is that it ignores the carry bits in the construction, and as such is not always a precise measure for optimality. In [9], the authors propose to apply a cost function which considers also the carry bits. This is the second cost function we consider and we call it *sum_carry*.

Definition 8 (cost function: *sum_carry*). *Let* $S \in ms(\mathbb{N})$, $B \in Base$ *with* $|B| = k$ *and* $S_{(B)} = (a_{ij})$ *the corresponding* $n \times (k+1)$ *matrix of digits. Denote the sequences* $\bar{s} = \langle s_0, s_1, \ldots, s_k \rangle$ *(sums) and* $\bar{c} = \langle c_0, c_1, \ldots, c_k \rangle$ *(carries) defined by:* $s_j = \sum_{i=1}^{n} a_{ij}$ *for* $0 \le j \le k$, $c_0 = 0$, *and* $c_{j+1} = (s_j + c_j)$ *div* $B(j)$ *for* $0 \le j \le k$. *The "sum of digits with carry" cost function is defined by the equation on the right.*

$$sum\_carry(S_{(B)}) = \sum_{j=0}^{k} (s_j + c_j)$$

The following example illustrates the *sum_carry* cost function and that it provides a better measure of base optimality for the (size of the) encoding of Pseudo-Boolean constraints.

Example 9. Consider the encoding of a Pseudo-Boolean constraint with coefficients $S = \{1, 3, 4, 8, 18, 18\}$ with respect to bases: $B_1 = \langle 2, 3, 3 \rangle$, $B_2 = \langle 3, 2, 3 \rangle$, and $B_3 = \langle 2, 2, 2, 2 \rangle$. Figure 1 depicts the sizes of the sorting networks for each of these bases. The upper tables illustrate the representation of the coefficients in the corresponding bases. In the lower tables, the rows labeled "sum" indicate the number of bits per network and (to their right) their total sum which is the *sum_digits* cost. With respect to the *sum_digits* cost function, all three bases are optimal for S, with a total of 9 inputs. The algorithm might as well return B_3.

The rows labeled "carry" indicate the number of carry bits in each of the constructions and (to their right) their totals. With respect to the *sum_carry* cost function, bases B_1 and B_2 are optimal for S, with a total of $9 + 2 = 11$ bits while B_3 involves $9 + 5 = 14$ bits. The algorithm might as well return B_1.

The following example shows that when searching for an optimal base with respect to the *sum_carry* cost function, one must consider also non-prime bases.

Example 10. Consider again the Pseudo Boolean constraint $\psi = 2x_1 + 2x_2 + 2x_3 + 2x_4 + 5x_5 + 18x_6 \ge 23$ from Section 3. The encoding with respect to

$B_1 = \langle 2,3,3 \rangle$

S	1's	2's	6's	18's	
1	1	0	0	0	
3	1	1	0	0	
4	0	2	0	0	
8	0	1	1	0	
18	0	0	0	1	
18	0	0	0	1	
sum	2	4	1	2	9
$carry$	0	1	1	0	2
$comp$	1	9	1	1	12

$B_2 = \langle 3,2,3 \rangle$

S	1's	3's	6's	18's	
1	1	0	0	0	
3	0	1	0	0	
4	1	1	0	0	
8	2	0	1	0	
18	0	0	0	1	
18	0	0	0	1	
sum	4	2	1	2	9
$carry$	0	1	1	0	2
$comp$	5	3	1	1	10

$B_3 = \langle 2,2,2,2 \rangle$

S	1's	2's	4's	8's	16's	
1	1	0	0	0	0	
3	1	1	0	0	0	
4	0	0	1	0	0	
8	0	0	0	1	0	
18	0	1	0	0	1	
18	0	1	0	0	1	
sum	2	3	1	1	2	9
$carry$	0	1	2	1	1	5
$comp$	1	5	3	1	3	13

Fig. 1. Number of inputs/carries/comparators when encoding $S = \{1, 3, 4, 8, 18, 18\}$ and three bases $B_1 = \langle 2, 3, 3 \rangle$, $B_2 = \langle 3, 2, 3 \rangle$, and $B_3 = \langle 2, 2, 2, 2 \rangle$

$B_1 = \langle 2, 3, 3 \rangle$ results in 4 sorting networks with 10 inputs from the coefficients and 2 carries. So a total of 12 bits. The encoding with respect to $B_2 = \langle 2, 9 \rangle$ is smaller. It has the same 10 inputs from the coefficients but no carry bits. Base B_2 is optimal and non-prime.

We consider a third cost function which we call the *num_comp* cost function. Sorting networks are constructed from "comparators" [10] and in the encoding each comparator is modeled using six CNF clauses. This function counts the number of comparators in the construction. Let $f(n)$ denote the number of comparators in an $n \times n$ sorting network. For small values of $0 \le n \le 8$, $f(n)$ takes the values $0, 0, 1, 3, 5, 9, 12, 16$ and 19 respectively which correspond to the sizes of the optimal networks of these sizes [10]. For larger values, the construction uses Batcher's odd-even sorting networks [4] for which $f(n) = n \cdot \lceil \log_2 n \rceil \cdot (\lceil \log_2 n \rceil - 1)/4 + n - 1$.

Definition 11 (cost function: *num_comp*). *Consider the same setting as in Definition 8. Then,*

$$num\_comp(S_{(B)}) = \sum_{j=0}^{k} f(s_j + c_j)$$

Example 12. Consider again the setting of Example 9. In Figure 1 the rows labeled "comp" indicate the number of comparators in each of the sorting networks and their totals. The construction with the minimal number of comparators is that obtained with respect to the base $B_2 = \langle 3, 2, 3 \rangle$ with 10 comparators.

It is interesting to remark the following relationship between the three cost functions: The *sum_digits* function is the most "abstract". It is only based on the representation of numbers in a mixed radix base. The *sum_carry* function considers also properties of addition in mixed-radix bases (resulting in the carry bits). Finally, the *num_comp* function considers also implementation details of the odd-even sorting networks applied in the underlying MINISAT$^+$ construction. In Section 8 we evaluate how the alternative choices for a cost function influence the size and quality of the encodings obtained with respect to corresponding optimal bases.

5   Optimal Base Search I: Heuristic Pruning

This section introduces a simple, heuristic-based, depth-first, tree search algorithm to solve the optimal base problem. The search space is the domain of non-redundant bases as presented in Definition 1. The starting point is the brute force algorithm applied in MINISAT$^+$. For a sequence of integers S, MINISAT$^+$ applies a depth-first traversal of $Base_p^{17}(S)$ to find the base with the optimal value for the cost function $cost_S(B) = sum\_carry(S_{(B)})$.

Our first contribution is to introduce a heuristic function and to identify branches in the search space which can be pruned early on in the search. Each tree node B encountered during the traversal is inspected to check if given the best node encountered so far, $bestB$, it is possible to determine that all descendants of B are guaranteed to be less optimal than $bestB$. In this case, the subtree rooted at B may be pruned. The resulting algorithm improves on the one of MINISAT$^+$ and provides the basis for the further improvements introduced in Sections 6 and 7. We need first a definition.

Definition 13 (base extension, partial cost, and admissible heuristic). *Let $S \in ms(\mathbb{N})$, $B, B' \in Base(S)$, and $cost_S$ a cost function. We say that: (1) B' extends B, denoted $B' \succ B$, if B is a prefix of B', (2) $\partial cost_S$ is a partial cost function for $cost_S$ if $\forall B' \succ B.\ cost_S(B') \geq \partial cost_S(B)$, and (3) h_S is an admissible heuristic function for $cost_S$ and $\partial cost_S$ if $\forall B' \succ B.\ cost_S(B') \geq \partial cost_S(B') + h_S(B') \geq \partial cost_S(B) + h_S(B)$.*

The intuition is that $\partial cost_S(B)$ signifies a part of the cost of B which will be a part of the cost of any extension of B, and that $h_S(B)$ is an under-approximation on the additional cost of extending B (in any way) given the partial cost of B. We denote $cost_S^\alpha(B) = \partial cost_S(B) + h_S(B)$. If $\partial cost_S$ is a partial cost function and h_S is an admissible heuristic function, then $cost_S^\alpha(B)$ is an under-approximation of $cost_S(B)$. The next lemma provides the basis for heuristic pruning using the three cost functions introduced above.

Lemma 14. *The following are admissible heuristics for the cases when:*

1. $cost_S(B) = sum\_digits(S_{(B)})$: $\quad \partial cost_S(B) = cost_S(B) - \sum msd(S_{(B)})$.
2. $cost_S(B) = sum\_carry(S_{(B)})$: $\quad \partial cost_S(B) = cost_S(B) - \sum msd(S_{(B)})$.
3. $cost_S(B) = num\_comp(S_{(B)})$: $\quad \partial cost_S(B) = cost_S(B) - f(s_{|B|} + c_{|B|})$.

In the first two settings we take $h_S(B) = \left|\left\{ x \in S \mid x \geq \prod B \right\}\right|$.
In the case of num_comp we take the trivial heuristic estimate $h_S(B) = 0$

The algorithm, which we call dfsHP for depth-first search with heuristic pruning, is now stated as Figure 2 where the input to the algorithm is a multiset of integers S and the output is an optimal base. The algorithm applies a depth-first traversal of $Base(S)$ in search of an optimal base. We assume given: a cost function $cost_S$, a partial cost function $\partial cost_S$ and an admissible heuristic h_S. We denote $cost_S^\alpha(B) = \partial cost_S(B) + h_S(B)$. The abstract data type base has two operations: extend(int) and extenders(multiset). For a base B and an

```
/*input*/  multiset S
/*init*/   base bestB = ⟨2, 2, ..., 2⟩
/*dfs*/    depth-first traverse Base(S)
               at each node B, for the next value p ∈ B.extenders(S) do
                   base newB = B.extend(p)
                   if (cost_S^α(newB) > cost_S(bestB)) prune
                   else if (cost_S(newB) < cost_S(bestB)) bestB = newB
/*output*/ return bestB;
```

Fig. 2. dfsHP: depth-first search for an optimal base with heuristic pruning

integer p, B.extend(p) is the base obtained by extending B by p. For a multiset S, B.extenders(S) is the set of integer values p by which B can be extended to a non-redundant base for S, i.e., such that $\prod$ B.extend(p) $\leq max(S)$. The definition of this operation may have additional arguments to indicate if we seek a prime base or one containing elements no larger than ℓ.

Initialization (/*init*/ in the figure) assigns to the variable bestB a finite binary base of size $\lfloor \log_2(max(S)) \rfloor$. This variable will always denote the best base encountered so far (or the initial finite binary base). Throughout the traversal, when visiting a node newB we first check if the subtree rooted at newB should be pruned. If this is not the case, then we check if a better "best base so far" has been found. Once the entire (with pruning) search space has been considered, the optimal base is in bestB.

To establish a bound on the complexity of the algorithm, denote the number of different integers in S by s and $m = max(S)$. The algorithm has space complexity $O(\log(m))$, for the depth first search on a tree with height bound by $\log(m)$ (an element of $Base(S)$ will have at most $\log_2(m)$ elements). For each base considered during the traversal, we have to calculate $cost_S$ which incurs a cost of $O(s)$. To see why, consider that when extending a base B by a new element giving base B', the first columns of $S_{(B')}$ are the same as those in $S_{(B)}$ (and thus also the costs incurred by them). Only the cost incurred by the most significant digit column of $S_{(B)}$ needs to be recomputed for $S_{(B')}$ due to base extension of B to B'. Performing the computation for this column, we compute a new digit for the s different values in S. Finally, by Lemma 7, there are $O(m^{2.73})$ bases and therefore, the total runtime is $O(s * m^{2.73})$. Given that $s \leq m$, we can conclude that runtime is bounded by $O(m^{3.73})$.

6   Optimal Base Search II: Branch and Bound

In this section we further improve the search algorithm for an optimal base. The search algorithm is, as before, a traversal of the search space using the same partial cost and heuristic functions as before to prune the tree. The difference is that instead of a depth first search, we maintain a priority queue of nodes for expansion and apply a best-first, branch and bound search strategy.

Figure 3 illustrates our enhanced search algorithm. We call it B&B. The abstract data type priority_queue maintains bases prioritized by the value of

```
base findBase(multiset S)
/*1*/    base bestB = ⟨2, 2, ..., 2⟩;  priority_queue  Q = {⟨ ⟩};
/*2*/    while (Q ≠ {} && costₛᵅ(Q.peek()) < costₛ(bestB))
/*3*/       base B = Q.popMin();
/*4*/       foreach (p ∈ B.extenders(S))
/*5*/          base newB = B.extend(p);
/*6*/          if (costₛᵅ(newB) ≤ costₛ(bestB))
/*7*/             { Q.push(newB); if (costₛ(newB) < costₛ(bestB)) bestB = newB; }
/*8*/    return bestB;
```

Fig. 3. Algorithm B&B: best-first, branch and bound

$cost_S^\alpha$. Operations popMin(), push(base) and peek() (peeks at the minimal entry) are the usual. The reason to box the text "priority_queue" in the figure will become apparent in the next section.

On line /*1*/ in the figure, we initialize the variable bestB to a finite binary base of size $\lfloor \log_2(max(S)) \rfloor$ (same as in Figure 2) and initialize the queue to contain the root of the search space (the empty base). As long as there are still nodes to be expanded in the queue that are potentially interesting (line /*2*/), we select (at line /*3*/) the best candidate base B from the frontier of the tree in construction for further expansion. Now the search tree is expanded for each of the relevant integers (calculated at line /*4*/). For each child newB of B (line /*5*/), we check if pruning at newB should occur (line /*6*/) and if not we check if a better bound has been found (line /*7*/) Finally, when the loop terminates, we have found the optimal base and return it (line /*8*/).

7   Optimal Base Search III: Search Modulo Product

This section introduces an abstraction on the search space, classifying bases according to their product. Instead of maintaining (during the search) a priority queue of all bases (nodes) that still need to be explored, we maintain a special priority queue in which there will only ever be at most one base with the same product. So, the queue will never contain two different bases B_1 and B_2 such that $\prod B_1 = \prod B_2$. In case a second base, with the same product as one already in, is inserted to the queue, then only the base with the minimal value of $cost_S^\alpha$ is maintained on the queue. We call this type of priority queue a *hashed priority queue* because it can conveniently be implemented as a hash table.

The intuition comes from a study of the *sum_digits* cost function for which we can prove the following **Property 1** on bases: Consider two bases B_1 and B_2 such that $\prod B_1 = \prod B_2$ and such that $cost_S^\alpha(B_1) \leq cost_S^\alpha(B_2)$. Then for any extension of B_1 and of B_2 by the same sequence C, $cost_S(B_1C) \leq cost_S(B_2C)$. In particular, if one of B_1 or B_2 can be extended to an optimal base, then B_1 can. A direct implication is that when maintaining the frontier of the search space as a priority queue, we only need one representative of the class of bases which have the same product (the one with the minimal value of $cost_S^\alpha$).

A second **Property 2** is more subtle and true for any cost function that has the first property: Assume that in the algorithm described as Figure 3 we at some stage remove a base B_1 from the priority queue. This implies that if in the future we encounter any base B_2 such that $\prod B_1 = \prod B_2$, then we can be sure that $cost_S(B_1) \leq cost_S(B_2)$ and immediately prune the search tree from B_2.

Our third and final algorithm, which we call hashB&B (best-first, branch and bound, with hash priority queue) is identical to the algorithm presented in Figure 3, except that the the boxed priority queue introduced at line /*1*/ is replaced by a $\boxed{\texttt{hash\_priority\_queue}}$.

The abstract data type hash_priority_queue maintains bases prioritized by the value of $cost_S^\alpha$. Operations popMin() and peek() are as usual. Operation push(B_1) works as follows: (a) if there is no base B_2 in the queue such that $\prod B_1 = \prod B_2$, then add B_1. Otherwise, (b) if $cost_S^\alpha(B_2) \leq cost_S^\alpha(B_1)$ then do not add B_1. Otherwise, (c) remove B_2 from the queue and add B_1.

Theorem 15
*(1) The sum_digits cost function satisfies **Property 1**; and (2) the hashB&B algorithm finds an optimal base for any cost function which satisfies **Property 1**.*

We conjecture that the other cost functions do not satisfy **Property 1**, and hence cannot guarantee that the hashB&B algorithm always finds an optimal base. However, in our extensive experimentation, all bases found (when searching for an optimal prime base) are indeed optimal.

A direct implication of the above improvements is that we can now provide a tighter bound on the complexity of the search algorithm. Let us denote the number of different integers in S by s and $m = max(S)$. First note that in the worst case the hashed priority queue will contain m elements (one for each possible value of a base product, which is never more than m). Assuming that we use a Fibonacci Heap, we have a $O(\log(m))$ cost (amortized) per popMin() operation and in total a $O(m * \log(m))$ cost for popping elements off the queue during the search for an optimal base.

Now focus on the cost of operations performed when extracting a base B from the queue. Denoting $\prod B = q$, B has at most m/q children (integers which extend it). For each child we have to calculate $cost_S$ which incurs a cost of $O(s)$ and possibly to insert it to the queue. Pushing an element onto a hashed priority queue (in all three cases) is a constant time operation (amortized), and hence the total cost for dealing with a child is $O(s)$.

Finally, consider the total number of children created during the search which corresponds to the following sum:

$$O(\sum_{q=1}^{m} m/q) = O(m \sum_{q=1}^{m} 1/q) = O(m * \log(m))$$

So, in total we get $O(m * \log(m)) + O(m * \log(m) * s) \leq O(m^2 * \log(m))$. When we restrict the extenders to be prime numbers then we can further improve this bound to $O(m^2 * \log(\log(m)))$ by reasoning about the density of the primes. A proof can be found in [8].

8   Experiments

Experiments are performed using an extension to MiniSat$^+$ [9] where the only change to the tool is to plug in our optimal base algorithm. The reader is invited to experiment with the implementation via its web interface.[1] All experiments are performed on a Quad-Opteron 848 at 2.2 GHz, 16 GB RAM, running Linux.

Our benchmark suite originates from 1945 Pseudo-Boolean Evaluation [11] instances from the years 2006–2009 containing a total of 74,442,661 individual Pseudo-Boolean constraints. After normalizing and removing constraints with $\{0,1\}$ coefficients we are left with 115,891 different optimal base problems where the maximal coefficient is $2^{31} - 1$. We then focus on 734 PB instances where at least one optimal base problem from the instance yields a base with an element that is non-prime or greater than 17. When solving PB instances, in all experiments, a 30 minute timeout is imposed as in the Pseudo-Boolean competitions. When solving an optimal base problem, a 10 minute timeout is applied.

Experiment 1 (Impact of optimal bases): The first experiment illustrates the advantage in searching for an optimal base for Pseudo-Boolean solving. We compare sizes and solving times when encoding w.r.t. the binary base vs. w.r.t. an optimal base (using the hashB&B algorithm with the *num_comp* cost function). Encoding w.r.t. the binary base, we solve 435 PB instances (within the time limit) with an average time of 146 seconds and average CNF size of 1.2 million clauses. Using an optimal base, we solve 445 instances with an average time of 108 seconds, and average CNF size of 0.8 million clauses.

Experiment 2 (Base search time): Here we focus on the search time for an optimal base in six configurations using the *sum_carry* cost function. Configurations M17, dfsHP17, and B&B17, are respectively, the MiniSat$^+$ implementation, our dfsHP and our B&B algorithms, all three searching for an optimal base from $Base_p^{17}$, i.e., with prime elements up to 17. Configurations hashB&B1,000,000, hashB&B10,000, and hashB&B17 are our hashB&B algorithm searching for a base from $Base_p^{\ell}$ with bounds of $\ell = 1,000,000$, $\ell = 10,000$, and $\ell = 17$, respectively.

Results are summarized in Fig. 4 which is obtained as follows. We cluster the optimal base problems according to the values $\lceil \log_{1.9745} M \rceil$ where M is the maximal coefficient in a problem. Then, for each cluster we take the average runtime for the problems in the cluster. The value 1.9745 is chosen to minimize the standard deviation from the averages (over all clusters). These are the points on the graphs. Configuration M17 times out on 28 problems. For dfsHP17, the maximal search time is 200 seconds. Configuration B&B17 times out for 1 problem. The hashB&B configurations have maximal runtimes of 350 seconds, 14 seconds and 0.16 seconds, respectively for the bounds 1,000,000, 10,000 and 17.

Fig. 4 shows that: (left) even with primes up to 1,000,000, hashB&B is faster than the algorithm from MiniSat$^+$ with the limit of 17; and (right) even with primes up to 10,000, the search time using hashB&B is essentially negligible.

---

[1] http://aprove.informatik.rwth-aachen.de/forms/unified_form_PBB.asp

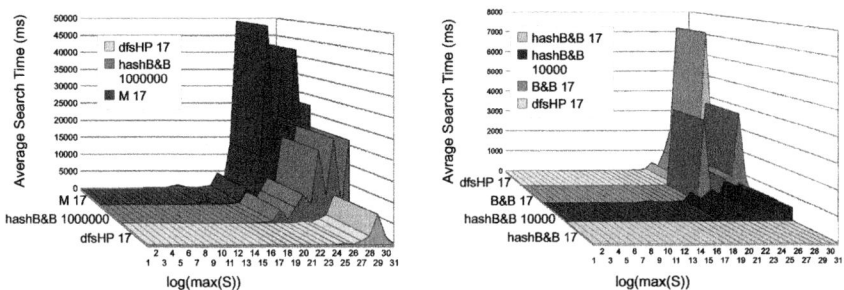

Fig. 4. Experiment 2: the 3 slowest configurations (left) (from back to front) M17(blue), hashB&B1,000,000(orange) and dfsHP17(yellow). The 4 fastest configurations (right) (from back to front) dfsHP17(yellow), B&B17(green), hashB&B10,000(brown) and hashB&B17(azure). Note the change of scale for the y-axis with 50k ms on the left and 8k ms on the right. Configuration dfsHP17 (yellow) is lowest on left and highest on right, setting the reference point to compare the two graphs.

Experiment 3 (Impact on PB solving): Fig. 5 illustrates the influence of improved base search on SAT solving for PB Constraints. Both graphs depict the number of instances solved (the x-axis) within a time limit (the y-axis). On the left, total solving time (with base search), and on the right, SAT solving time only.

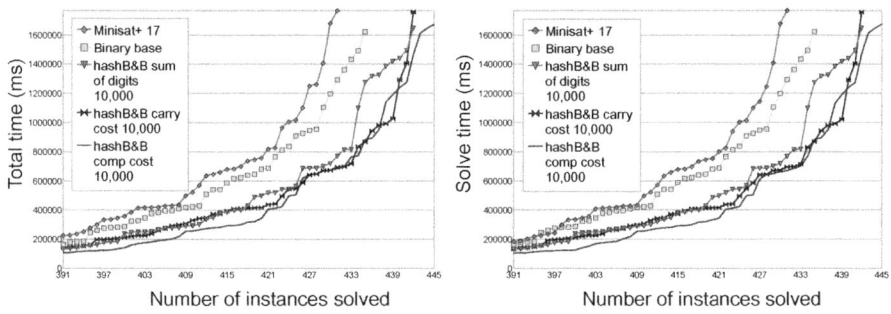

Fig. 5. Experiment 3: total times (left), solving times (right)

Both graphs consider the 734 instances of interest and compare SAT solving times with bases found using five configurations. The first is MINISAT$^+$ with configuration M17, the second is with respect to the binary base, the third to fifth are hashB&B searching for bases from $Base_p^{10,000}(S)$ with cost functions: *sum_digits*, *sum_carry*, and *num_comp*, respectively. The average total/solve run-times (in sec) are 150/140, 146/146, 122/121, 116/115 and 108/107 (left to right). The total number of instances solved are 431, 435, 442, 442 and 445 (left to right). The average CNF sizes (in millions of clauses) for the entire test set/the set where all algorithms solved/the set where no algorithm solved are 7.7/1.0/18, 9.5/1.2/23, 8.4/1.1/20, 7.2/0.8/17 and 7.2/0.8/17 (left to right).

The graphs of Fig. 5 and average solving times clearly show: **(1)** SAT solving time dominates base finding time, **(2)** MINISAT$^+$ is outperformed by the trivial binary base, **(3)** total solving times with our algorithms are faster than with the binary base, and **(4)** the most specific cost function (comparator cost) outperforms the other cost functions both in solving time and size. Finally, note that sum of digits with its nice mathematical properties, simplicity, and application independence solves as many instances as cost carry.

Experiment 4 (Impact of high prime factors): This experiment is about the effects of restricting the maximal prime value in a base (i.e. the value $\ell = 17$ of MINISAT$^+$). An analysis of the our benchmark suite indicates that coefficients with small prime factors are overrepresented. To introduce instances where coefficients have larger prime factors we select 43 instances from the suite and multiply their coefficients to introduce the prime factor 31 raised to the power $i \in \{0, \ldots, 5\}$. We also introduce a slack variable to avoid gcd-based simplification. This gives us a collection

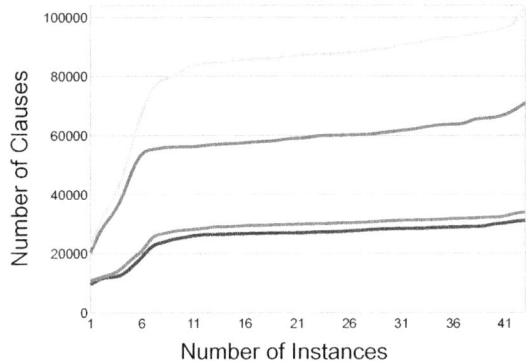

Fig. 6. Experiment 4: Number (x-axis) of instances encoded within number of clauses (y-axis) on 4 configurations. From top line to bottom: (yellow) $\ell = 17$, $i = 5$, (red) $\ell = 17$, $i = 2$, (green) $\ell = 31$, $i = 5$, and (blue) $\ell \in \{17, 31\}$, $i = 0$.

of 258 new instances. We used the B&B algorithm with the *sum_carry* cost function applying the limit $\ell = 17$ (as in MINISAT$^+$) and $\ell = 31$. Results indicate that for $\ell = 31$, both CNF size and SAT-solving time are independent of the factor 31^i introduced for $i > 0$. However, for $\ell = 17$, both measures increase as the power i increases. Results on CNF sizes are reported in Fig. 6 which plots for 4 different settings the number of instances encoded (x-axis) within a CNF with that many clauses (y-axis).

9 Related Work

Recent work [2] encodes Pseudo-Boolean constraints via "totalizers" similar to sorting networks, determined by the representation of the coefficients in an underlying base. Here the authors choose the standard base 2 representation of numbers. It is straightforward to generalize their approach for an arbitrary mixed base, and our algorithm is directly applicable. In [12] the author considers the *sum_digits* cost function and analyzes the size of representing the natural numbers up to n with (a particular class of) mixed radix bases. Our Lemma 6 may lead to a contribution in that context.

10   Conclusion

It has been recognized now for some years that decomposing the coefficients in a Pseudo-Boolean constraint with respect to a mixed radix base can lead to smaller SAT encodings. However, it remained an open problem to determine if it is feasible to find such an optimal base for constraints with large coefficients. In lack of a better solution, the implementation in the MiniSat$^+$ tool applies a brute force search considering prime base elements less than 17.

To close this open problem, we first formalize the optimal base problem and then significantly improve the search algorithm currently applied in MiniSat$^+$. Our algorithm scales and easily finds optimal bases with elements up to 1,000,000. We also illustrate that, for the measure of optimality applied in MiniSat$^+$, one must consider also non-prime base elements. However, choosing the more simple *sum_digits* measure, it is sufficient to restrict the search to prime bases.

With the implementation of our search algorithm it is possible, for the first time, to study the influence of basing SAT encodings on optimal bases. We show that for a wide range of benchmarks, MiniSat$^+$ does actually find an optimal base consisting of elements less than 17. We also show that many Pseudo-Boolean instances have optimal bases with larger elements and that this does influence the subsequent CNF sizes and SAT solving times, especially when coefficients contain larger prime factors.

Acknowledgement. We thank Daniel Berend and Carmel Domshlak for useful discussions.

References

1. Bailleux, O., Boufkhad, Y., Roussel, O.: A translation of pseudo boolean constraints to SAT. Journal on Satisfiability, Boolean Modeling and Computation (JSAT) 2(1-4), 191–200 (2006)
2. Bailleux, O., Boufkhad, Y., Roussel, O.: New encodings of pseudo-boolean constraints into CNF. In: Kullmann, O. (ed.) SAT 2009. LNCS, vol. 5584, pp. 181–194. Springer, Heidelberg (2009)
3. Barth, P.: Logic-based 0-1 constraint programming. Kluwer Academic Publishers, Norwell (1996)
4. Batcher, K.E.: Sorting networks and their applications. In: AFIPS Spring Joint Computing Conference. AFIPS Conference Proceedings, vol. 32, pp. 307–314. Thomson Book Company, Washington, D.C (1968)
5. Bixby, R.E., Boyd, E.A., Indovina, R.R.: MIPLIB: A test set of mixed integer programming problems. SIAM News 25, 16 (1992)
6. Bryant, R.E., Lahiri, S.K., Seshia, S.A.: Deciding CLU logic formulas via boolean and pseudo-boolean encodings. In: Proc. Intl. Workshop on Constraints in Formal Verification, CFV 2002 (2002)
7. Chor, B., Lemke, P., Mador, Z.: On the number of ordered factorizations of natural numbers. Discrete Mathematics 214(1-3), 123–133 (2000)
8. Codish, M., Fekete, Y., Fuhs, C., Schneider-Kamp, P.: Optimal Base Encodings for Pseudo-Boolean Constraints. Technical Report, arXiv:1007.4935 [cs.DM], http://arxiv.org/abs/1007.4935

9. Eén, N., Sörensson, N.: Translating pseudo-boolean constraints into SAT. Journal on Satisfiability, Boolean Modeling and Computation (JSAT) 2(1-4), 1–26 (2006)
10. Knuth, D.: The Art of Computer Programming, Volume III: Sorting and Searching. Addison-Wesley, Reading (1973)
11. Manquinho, V.M., Roussel, O.: The first evaluation of Pseudo-Boolean solvers (PB 2005). Journal on Satisfiability, Boolean Modeling and Computation (JSAT) 2(1-4), 103–143 (2006)
12. Sidorov, N.: Sum-of-digits function for certain nonstationary bases. Journal of Mathematical Sciences 96(5), 3609–3615 (1999)

Predicate Generation for Learning-Based Quantifier-Free Loop Invariant Inference[*]

Yungbum Jung[1], Wonchan Lee[1], Bow-Yaw Wang[2], and Kwangkuen Yi[1]

[1] Seoul National University
[2] INRIA and Academia Sinica

Abstract. We address the predicate generation problem in the context of loop invariant inference. Motivated by the interpolation-based abstraction refinement technique, we apply the interpolation theorem to synthesize predicates implicitly implied by program texts. Our technique is able to improve the effectiveness and efficiency of the learning-based loop invariant inference algorithm in [14]. Experiments excerpted from Linux, SPEC2000, and Tar source codes are reported.

1 Introduction

One way to prove that an annotated loop satisfies its pre- and post-conditions is by giving loop invariants. In an annotated loop, pre- and post-conditions specify intended effects of the loop. The actual behavior of the annotated loop however does not necessarily conform to its specification. Through loop invariants, verification tools can check whether the annotated loop fulfills its specification automatically [10,5].

Finding loop invariants is tedious and sometimes requires intelligence. Recently, an automated technique based on algorithmic learning and predicate abstraction is proposed [14]. Given a fixed set of atomic predicates and an annotated loop, the learning-based technique can infer a quantifier-free loop invariant generated by the given atomic predicates. By employing a learning algorithm and a mechanical teacher, the new technique is able to generate loop invariants without constructing abstract models nor computing fixed points. It gives a new invariant generation framework that can be less sensitive to the number of atomic predicates than traditional techniques.

As in other techniques based on predicate abstraction, the selection of atomic predicates is crucial to the effectiveness of the learning-based technique. Oftentimes, users extract atomic predicates from program texts heuristically. If this simple strategy does not yield necessary atomic predicates to express loop invariants, the loop invariant inference algorithm will not be able to infer a loop invariant. Even when the heuristic does give necessary atomic predicates, it may select too many redundant predicates and impede the efficiency of loop invariant inference algorithm.

---

[*] This work was supported by the Engineering Research Center of Excellence Program of Korea Ministry of Education, Science and Technology(MEST) / National Research Foundation of Korea(NRF) (Grant 2010-0001717), National Science Council of Taiwan Grant Numbers 97-2221-E-001-003-MY3 and 97-2221-E-001-006-MY3, the FORMES Project within LIAMA Consortium, and the French ANR project SIVES ANR-08-BLAN-0326-01.

P.A. Abdulla and K.R.M. Leino (Eds.): TACAS 2011, LNCS 6605, pp. 205–219, 2011.

One way to circumvent this problem is to generate atomic predicates by need. Several techniques have been developed to synthesize atomic predicates by interpolation [11, 12, 17, 18, 9]. Let A and B be logic formulae. An interpolant I of A and B is a formula such that $A \Rightarrow I$ and $I \wedge B$ is inconsistent. Moreover, the non-logical symbols in I must occur in both A and B. By Craig's interpolation theorem, an interpolant I always exists for any first-order formulae A and B when $A \wedge B$ is inconsistent [6]. The interpolant I can be seen as a concise summary of A with respect to B. Indeed, interpolants have been used to synthesize atomic predicates for predicate abstraction refinement in software model checking [11, 12, 17, 18, 9].

Inspired by the refinement technique in software model checking, we develop an interpolation-based technique to synthesize atomic predicates in the context of loop invariant inference. Our algorithm does not add new atomic predicates by interpolating invalid execution paths in control flow graphs. We instead interpolate the loop body with purported loop invariants from the learning algorithm. Our technique can improve the effectiveness and efficiency of the learning-based loop invariant inference algorithm in [14]. Constructing sets of atomic predicates can be fully automatic and on-demand.

Example. Consider the following annotated loop:

$$\{\, n \geq 0 \wedge x = n \wedge y = n \,\} \texttt{ while } x > 0 \texttt{ do } x = x - 1; \; y = y - 1 \texttt{ done } \{\, x + y = 0 \,\}$$

Assume that variables x and y both have the value $n \geq 0$ before entering the loop. In the loop body, each variable is decremented by one until the variable x is zero. We want to show that $x + y$ is zero after executing the loop. Note that the predicate $x = y$ is implicitly implied by the loop. The program text however does not reveal this equality explicitly. Moreover, atomic predicates from the program text can not express loop invariants that establish the specification. Using atomic predicates in the program text does not give necessary atomic predicates.

Any loop invariant must be weaker than the pre-condition and stronger than the disjunction of the loop guard and the post-condition. We use the atomic predicates in an interpolant of $n \geq 0 \wedge x = n \wedge y = n$ and $\neg(x + y = 0 \vee x > 0)$ to obtain the initial atomic predicates $\{x = y, 2y \geq 0\}$. Observe that the interpolation theorem is able to synthesize the implicit predicate $x = y$. In fact, $x = y \wedge x \geq 0$ is a loop invariant that establishes the specification of the loop.

Related Work. Loop invariant inference using algorithmic learning is introduced in [14]. In [15], the learning-based technique is extended to quantified loop invariants. Both algorithms require users to provide atomic predicates. The present work addresses this problem for the case of quantifier-free loop invariants.

Many interpolation algorithms and their implementations are available [17, 3, 7]. Interpolation-based techniques for predicate refinement in software model checking are proposed in [11, 12, 18, 9, 13]. Abstract models used in these techniques however may require excessive invocations to theorem provers. Another interpolation-based technique for first-order invariants is developed in [19]. The paramodulation-based technique does not construct abstract models. It however only generates invariants in first-order logic with equality. A template-based predicate generation technique for quantified invariants is proposed [20]. The technique reduces the invariant inference problem to constraint programming and generates predicates in user-provided templates.

This paper is organized as follows. After Introduction, preliminaries are given in Section 2. We review the learning-based loop invariant inference framework in Section 3. Our technical results are presented in Section 4. Section 5 gives the loop invariant inference algorithm with automatic predicate generation. We report our experimental results in Section 6. Section 7 concludes this work.

2 Preliminaries

Let QF denote the quantifier-free logic with equality, linear inequality, and uninterpreted functions [17, 18]. Define the *domain* $\mathbb{D} = \mathbb{Q} \cup \mathbb{B}$ where $\mathbb{Q}$ is the set of rational numbers and $\mathbb{B} = \{F, T\}$ is the Boolean domain. Fix a set X of variables. A *valuation* over X is a function from X to $\mathbb{D}$. The class of valuations over X is denoted by Val_X. For any formula $\theta \in QF$ and valuation ν over free variables in θ, θ is *satisfied* by ν (written $\nu \models \theta$) if θ evaluates to T under ν; θ is *inconsistent* if θ is not satisfied by any valuation. Given a formula $\theta \in QF$, a *satisfiability modulo theories (SMT) solver* returns a satisfying valuation ν of θ if θ is not inconsistent [8, 16].

For $\theta \in QF$, we denote the set of non-logical symbols occurred in θ by $\sigma(\theta)$. Let $\Theta = [\theta_1, \ldots, \theta_m]$ be a sequence with $\theta_i \in QF$ for $1 \le i \le m$. The sequence Θ is *inconsistent* if $\theta_1 \wedge \theta_2 \wedge \cdots \wedge \theta_m$ is inconsistent. The sequence $\Lambda = [\lambda_0, \lambda_1, \ldots, \lambda_m]$ of quantifier-free formulae is an *inductive interpolant* of Θ if

- $\lambda_0 = T$ and $\lambda_m = F$;
- for all $1 \le i \le m$, $\lambda_{i-1} \wedge \theta_i \Rightarrow \lambda_i$; and
- for all $1 \le i < m$, $\sigma(\lambda_i) \subseteq \sigma(\theta_i) \cap \sigma(\theta_{i+1})$.

The interpolation theorem states that an inductive interpolant exists for any inconsistent sequence [6, 17, 18]. We consider the following imperative language in this paper:

$$\mathsf{Stmt} \triangleq \mathsf{nop} \mid \mathsf{Stmt}; \mathsf{Stmt} \mid x := \mathsf{Exp} \mid x := \mathsf{nondet} \mid \mathtt{if}\ \mathsf{BExp}\ \mathtt{then}\ \mathsf{Stmt}\ \mathtt{else}\ \mathsf{Stmt}$$
$$\mathsf{Exp} \triangleq n \mid x \mid \mathsf{Exp} + \mathsf{Exp} \mid \mathsf{Exp} - \mathsf{Exp}$$
$$\mathsf{BExp} \triangleq F \mid x \mid \neg\mathsf{BExp} \mid \mathsf{BExp} \wedge \mathsf{BExp} \mid \mathsf{Exp} < \mathsf{Exp} \mid \mathsf{Exp} = \mathsf{Exp}$$

Two basic types are available: natural numbers and Booleans. A term in Exp is a natural number; a term in BExp is of Boolean type. The keyword $\mathtt{nondet}$ denotes an arbitrary value in the type of the assigned variable. An *annotated loop* is of the form:

$$\{\delta\}\ \mathtt{while}\ \kappa\ \mathtt{do}\ S_1; S_2; \cdots; S_m\ \mathtt{done}\ \{\epsilon\}$$

The BExp formula κ is the *loop guard*. The BExp formulae δ and ϵ are the *precondition* and *postcondition* of the annotated loop respectively.

Define $X^{\langle k \rangle} = \{x^{\langle k \rangle} : x \in X\}$. For any term e over X, define $e^{\langle k \rangle} = e[X \mapsto X^{\langle k \rangle}]$. A *transition formula* $[\![S]\!]$ for a statement S is a first-order formula over variables $X^{\langle 0 \rangle} \cup X^{\langle 1 \rangle}$ defined as follows.

$$[\![\text{nop}]\!] \triangleq \bigwedge_{x \in X} x^{\langle 1 \rangle} = x^{\langle 0 \rangle} \qquad [\![x := \text{nondet}]\!] \triangleq \bigwedge_{y \in X \setminus \{x\}} y^{\langle 1 \rangle} = y^{\langle 0 \rangle}$$

$$[\![x := e]\!] \triangleq x^{\langle 1 \rangle} = e^{\langle 0 \rangle} \wedge \bigwedge_{y \in X \setminus \{x\}} y^{\langle 1 \rangle} = y^{\langle 0 \rangle}$$

$$[\![S_0; S_1]\!] \triangleq \exists X. [\![S_0]\!][X^{\langle 1 \rangle} \mapsto X] \wedge [\![S_1]\!][X^{\langle 0 \rangle} \mapsto X]$$

$$[\![\text{if } p \text{ then } S_0 \text{ else } S_1]\!] \triangleq (p^{\langle 0 \rangle} \wedge [\![S_0]\!]) \vee (\neg p^{\langle 0 \rangle} \wedge [\![S_1]\!])$$

Let ν and ν' be valuations, and S a statement. We write $\nu \xrightarrow{S} \nu'$ if $[\![S]\!]$ evaluates to true by assigning $\nu(x)$ and $\nu'(x)$ to $x^{\langle 0 \rangle}$ and $x^{\langle 1 \rangle}$ for each $x \in X$ respectively. Given a sequence of statements $S_1; S_2; \cdots ; S_m$, a *program execution* $\nu_0 \xrightarrow{S_1} \nu_1 \xrightarrow{S_2} \cdots \xrightarrow{S_m} \nu_m$ is a sequence $[\nu_0, \nu_1, \ldots, \nu_m]$ of valuations such that $\nu_i \xrightarrow{S_i} \nu_{i+1}$ for $0 \le i < m$.

A *precondition* $Pre(\theta : S_1; S_2; \cdots ; S_m)$ for $\theta \in QF$ with respect to the statement $S_1; S_2; \cdots ; S_m$ is a first-order formula that entails θ after executing the statement $S_1; S_2; \cdots ; S_m$. Given an annotated loop $\{\delta\}$ while κ do $S_1; S_2; \cdots ; S_m$ done $\{\epsilon\}$, the *loop invariant inference problem* is to compute a formula $\iota \in QF$ satisfying (1) $\delta \Rightarrow \iota$; (2) $\iota \wedge \neg \kappa \Rightarrow \epsilon$; and (3) $\iota \wedge \kappa \Rightarrow Pre(\iota : S_1; S_2; \cdots ; S_m)$. Observe that the condition (2) is equivalent to $\iota \Rightarrow \epsilon \vee \kappa$. The first two conditions specify necessary and sufficient conditions of any loop invariants respectively. The formulae δ and $\epsilon \vee \kappa$ are called the *strongest* and *weakest approximations* to loop invariants respectively.

3   Inferring Loop Invariants with Algorithmic Learning

Given an annotated loop $\{\delta\}$ while κ do $S_1; S_2; \cdots ; S_m$ done $\{\epsilon\}$, we would like to infer a loop invariant to establish the pre- and post-conditions. Given a set P of atomic predicates, the work in [14] shows how to apply a learning algorithm for Boolean formulae to infer quantifier-free loop invariants freely generated by P. The authors first adopt predicate abstraction to relate quantifier-free and Boolean formulae. They then design a mechanical teacher to guide the learning algorithm to a Boolean formula whose concretization is a loop invariant.

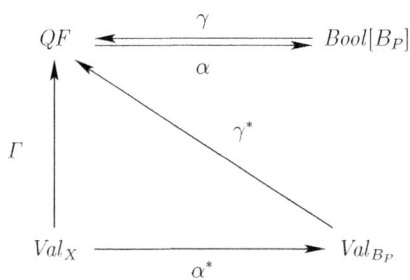

Fig. 1. Relating QF and $Bool[B_P]$

Let $QF[P]$ denote the set of quantifier-free formulae generated from the set of atomic predicates P. Consider the set of Boolean formulae $Bool[B_P]$ generated by the set of Boolean variables $B_P \triangleq \{b_p : p \in P\}$. An *abstract valuation* is a function from B_P to $\mathbb{B}$. We write Val_{B_P} for the set of abstract valuations. A Boolean formula in $Bool[B_P]$ is a *canonical monomial* if it is a conjunction of literals, where each Boolean variable in B_P occurs exactly once. Formulae in $QF[P]$ and $Bool[B_P]$ are related by the following functions [14] (Figure 1):

$$\gamma(\beta) \overset{\triangle}{=} \beta[B_P \mapsto P]$$
$$\alpha(\theta) \overset{\triangle}{=} \bigvee\{\beta \in Bool[B_P] : \beta \text{ is a canonical monomial and } \theta \wedge \gamma(\beta) \text{ is satisfiable}\}$$
$$\gamma^*(\mu) \overset{\triangle}{=} \bigwedge_{\mu(b_p)=T} \{p\} \wedge \bigwedge_{\mu(b_p)=F} \{\neg p\}$$
$$\alpha^*(\nu) \overset{\triangle}{=} \mu \text{ where } \mu(b_p) = \begin{cases} T & \text{if } \nu \models p \\ F & \text{if } \nu \not\models p \end{cases}$$

Consider, for instance, $P = \{n \geq 0, x = n, y = n\}$ and $B_P = \{b_{n \geq 0}, b_{x=n}, b_{y=n}\}$. We have $\gamma(b_{n \geq 0} \wedge \neg b_{x=n}) = n \geq 0 \wedge \neg(x = n)$ and

$$\alpha(\neg(x = y)) = \begin{array}{l} (b_{n\geq0} \wedge b_{x=n} \wedge \neg b_{y=n}) \vee (b_{n\geq0} \wedge \neg b_{x=n} \wedge b_{y=n}) \vee \\ (b_{n\geq0} \wedge \neg b_{x=n} \wedge \neg b_{y=n}) \vee (\neg b_{n\geq0} \wedge b_{x=n} \wedge \neg b_{y=n}) \vee \\ (\neg b_{n\geq0} \wedge \neg b_{x=n} \wedge b_{y=n}) \vee (\neg b_{n\geq0} \wedge \neg b_{x=n} \wedge \neg b_{y=n}). \end{array}$$

Moreover, $\alpha^*(\nu)(b_{n\geq0}) = \alpha^*(\nu)(b_{x=n}) = \alpha^*(\nu)(b_{y=n}) = T$ when $\nu(n) = \nu(x) = \nu(y) = 1$. And $\gamma^*(\mu) = n \geq 0 \wedge x = n \wedge \neg(y = n)$ when $\mu(b_{n\geq0}) = \mu(b_{x=n}) = T$ but $\mu(b_{y=n}) = F$. Observe that the pair (α, γ) forms the Galois correspondence in Cartesian predicate abstraction [2].

After formulae in QF and valuations in Val_X are abstracted to $Bool[B_P]$ and Val_{B_P} respectively, a learning algorithm is used to infer abstractions of loop invariants. Let ξ be an unknown *target* Boolean formula in $Bool[B_P]$. A *learning algorithm* computes a representation of the target ξ by interacting with a teacher. The *teacher* should answer the following queries [1,4]:

- *Membership queries.* Let $\mu \in Val_{B_P}$ be an abstract valuation. The membership query $MEM(\mu)$ asks if the unknown target ξ is satisfied by μ. If so, the teacher answers *YES*; otherwise, *NO*.
- *Equivalence queries.* Let $\beta \in Bool[B_P]$ be an *abstract conjecture*. The equivalence query $EQ(\beta)$ asks if β is equivalent to the unknown target ξ. If so, the teacher answers *YES*. Otherwise, the teacher gives an abstract valuation μ such that the exclusive disjunction of β and ξ is satisfied by μ. The abstract valuation μ is called an *abstract counterexample*.

With predicate abstraction and a learning algorithm for Boolean formulae at hand, it remains to design a mechanical teacher to guide the learning algorithm to the abstraction of a loop invariant. The key idea in [14] is to exploit approximations to loop invariants. An *under-approximation* to loop invariants is a quantifier-free formula $\underline{\iota}$ which is stronger than some loop invariants of the given annotated loop; an *over-approximation* is a quantifier-free formula $\overline{\iota}$ which is weaker than some loop invariants.

To see how approximations to loop invariants can be used in the design of the mechanical teacher, let us consider an equivalence query $EQ(\beta)$. On the abstract conjecture $\beta \in Bool[B_P]$, the mechanical teacher computes the corresponding quantifier-free formula $\theta = \gamma(\beta)$. It then checks if θ is a loop invariant. If so, we are done. Otherwise, the algorithm compares θ with approximations to loop invariants. If θ is stronger than the under-approximation or weaker than the over-approximation, a valuation ν satisfying $\neg(\underline{\iota} \Rightarrow \theta)$ or $\neg(\theta \Rightarrow \overline{\iota})$ can be obtained from an SMT solver. The abstract

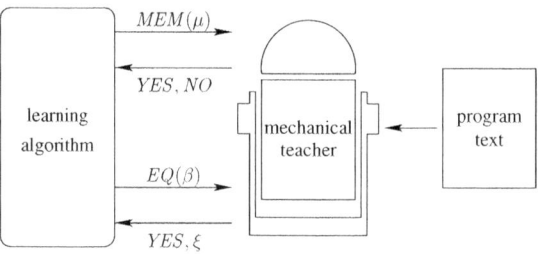

Fig. 2. Learning-based Framework

valuation $\alpha^*(\nu)$ gives an abstract counterexample. Approximations to loop invariants can also be used to answer membership queries. For a membership query $MEM(\mu)$ with $\mu \in Val_{B_P}$, the mechanical teacher computes its concretization $\theta = \gamma^*(\mu)$. It returns YES if $\theta \Rightarrow \iota$; it returns NO if $\theta \not\Rightarrow \bar{\iota}$. Otherwise, a random answer is returned.

Figure 2 shows the learning-based loop invariant inference framework. In the framework, a learning algorithm is used to drive the search of loop invariants. It "learns" an unknown loop invariant by inquiring a mechanical teacher. The mechanical teacher of course does not know any loop invariant. It nevertheless can try to answer these queries by the information derived from program texts. In this case, approximations to loop invariants are used. Observe the simplicity of the learning-based framework. By employing a learning algorithm, it suffices to design a mechanical teacher to find loop invariants. Moreover, the new framework does not construct abstract models nor compute fixed points. It can be more scalable than traditional techniques.

4   Predicate Generation by Interpolation

One drawback in the learning-based approach to loop invariant inference is to require a set of atomic predicates. It is essential that at least one quantifier-free loop invariant is representable by the given set P of atomic predicates. Otherwise, concretization of formulae in $Bool[B_P]$ cannot be loop invariants. The mechanical teacher never answers YES to equivalence queries. To address this problem, we will synthesize new atomic predicates for the learning-based loop invariant inference framework progressively.

The interpolation theorem is essential to our predicate generation technique [6, 19, 18,12]. Let $\Theta = [\theta_1, \theta_2, \ldots, \theta_m]$ be an inconsistent sequence of quantifier-free formula and $\Lambda = [\lambda_0, \lambda_1, \lambda_2, \ldots, \lambda_m]$ its inductive interpolant. By definition, $\theta_1 \Rightarrow \lambda_1$. Assume $\theta_1 \wedge \theta_2 \wedge \cdots \wedge \theta_i \Rightarrow \lambda_i$. We have $\theta_1 \wedge \theta_2 \wedge \cdots \wedge \theta_{i+1} \Rightarrow \lambda_{i+1}$ since $\lambda_i \wedge \theta_{i+1} \Rightarrow \lambda_{i+1}$. Thus, λ_i is an over-approximation to $\theta_1 \wedge \theta_2 \wedge \cdots \wedge \theta_i$ for $0 \leq i \leq m$. Moreover, $\sigma(\lambda_i) \subseteq \sigma(\theta_i) \cap \sigma(\theta_{i+1})$. Hence λ_i can be seen as a concise summary of $\theta_1 \wedge \theta_2 \wedge \cdots \wedge \theta_i$ with restricted symbols. Since each λ_i is written in a less expressive vocabulary, new atomic predicates among variables can be synthesized. We therefore apply the interpolation theorem to synthesize new atomic predicates and refine the abstraction.

Our predicate generation technique consists of three components. Before the learning algorithm is invoked, an initial set of atomic predicates is computed (Section 4.1). When the learning algorithm is failing to infer loop invariants, new atomic predicates are generated to refine the abstraction (Section 4.2). Lastly, conflicting answers to queries may incur from predicate abstraction. We further refine the abstraction with these conflicting answers (Section 4.3). Throughout this section, we consider the annotated loop $\{\delta\}$ while κ do $S_1; S_2; \cdots ; S_m$ done $\{\epsilon\}$ with the under-approximation $\underline{\iota}$ and over-approximation $\overline{\iota}$.

4.1 Initial Atomic Predicates

The under- and over-approximations to loop invariants must satisfy $\underline{\iota} \Rightarrow \overline{\iota}$. Otherwise, there cannot be any loop invariant ι such that $\underline{\iota} \Rightarrow \iota$ and $\iota \Rightarrow \overline{\iota}$. Thus, the sequence $[\underline{\iota}, \neg\overline{\iota}]$ is inconsistent. For any interpolant $[T, \lambda, F]$ of $[\underline{\iota}, \neg\overline{\iota}]$, we have $\underline{\iota} \Rightarrow \lambda$ and $\lambda \Rightarrow \overline{\iota}$. The quantifier-free formula λ can be a loop invariant if it satisfies $\lambda \wedge \kappa \Rightarrow Pre(\lambda : S_1; S_2; \cdots ; S_m)$. It is however unlikely that λ happens to be a loop invariant. Yet our loop invariant inference algorithm can generalize λ by taking the atomic predicates in λ as the initial atomic predicates. The learning algorithm will try to infer a loop invariant freely generated by these atomic predicates.

4.2 Atomic Predicates from Incorrect Conjectures

Consider an equivalence query $EQ(\beta)$ where $\beta \in Bool[B_P]$ is an abstract conjecture. If the concretization $\theta = \gamma(\beta)$ is not a loop invariant, we interpolate the loop body with the incorrect conjecture θ. For any quantifier-free formula θ over variables $X^{\langle 0 \rangle} \cup X^{\langle 1 \rangle}$, define $\theta^{\langle k \rangle} = \theta[X^{\langle 0 \rangle} \mapsto X^{\langle k \rangle}, X^{\langle 1 \rangle} \mapsto X^{\langle k+1 \rangle}]$. The *desuperscripted* form of a quantifier-free formula λ over variables $X^{\langle k \rangle}$ is $\lambda[X^{\langle k \rangle} \mapsto X]$. Moreover, if ν is a valuation over $X^{\langle 0 \rangle} \cup \cdots \cup X^{\langle m \rangle}$, $\nu\downarrow_{X^{\langle k \rangle}}$ represents a valuation over X such that $\nu\downarrow_{X^{\langle k \rangle}} (x) = \nu(x^{\langle k \rangle})$ for $x \in X$. Let ϕ and ψ be quantifier-free formulae over X. Define the following sequence:

$$\Xi(\phi, S_1, \ldots, S_m, \psi) \triangleq [\phi^{\langle 0 \rangle}, [\![S_1]\!]^{\langle 0 \rangle}, [\![S_2]\!]^{\langle 1 \rangle}, \ldots, [\![S_m]\!]^{\langle m-1 \rangle}, \neg\psi^{\langle m \rangle}].$$

Observe that

- $\phi^{\langle 0 \rangle}$ and $[\![S_1]\!]^{\langle 0 \rangle}$ share the variables $X^{\langle 0 \rangle}$;
- $[\![S_m]\!]^{\langle m-1 \rangle}$ and $\neg\psi^{\langle m \rangle}$ share the variables $X^{\langle m \rangle}$; and
- $[\![S_i]\!]^{\langle i-1 \rangle}$ and $[\![S_{i+1}]\!]^{\langle i \rangle}$ share the variables $X^{\langle i \rangle}$ for $1 \le i < m$.

Starting from the program states satisfying $\phi^{\langle 0 \rangle}$, the formula

$$\phi^{\langle 0 \rangle} \wedge [\![S_1]\!]^{\langle 0 \rangle} \wedge [\![S_2]\!]^{\langle 1 \rangle} \wedge \cdots \wedge [\![S_i]\!]^{\langle i-1 \rangle}$$

characterizes the images of $\phi^{\langle 0 \rangle}$ during the execution of $S_1; S_2; \cdots ; S_i$.

Lemma 1. *Let X denote the set of variables in the statement $S_1; S_2; \cdots ; S_i$, and ϕ a quantifier-free formula over X. For any valuation ν over $X^{\langle 0 \rangle} \cup X^{\langle 1 \rangle} \cup \cdots \cup X^{\langle i \rangle}$, the formula $\phi^{\langle 0 \rangle} \wedge [\![S_1]\!]^{\langle 0 \rangle} \wedge [\![S_2]\!]^{\langle 1 \rangle} \wedge \cdots \wedge [\![S_i]\!]^{\langle i-1 \rangle}$ is satisfied by ν if and only if $\nu\downarrow_{X^{\langle 0 \rangle}} \xrightarrow{S_1} \nu\downarrow_{X^{\langle 1 \rangle}} \xrightarrow{S_2} \cdots \xrightarrow{S_i} \nu\downarrow_{X^{\langle i-1 \rangle}}$ is a program execution and $\nu\downarrow_{X^{\langle 0 \rangle}} \models \phi$.*

By definition, $\phi \Rightarrow Pre(\psi : S_1; S_2; \cdots ; S_m)$ implies that the image of ϕ must satisfy ψ after the execution of $S_1; S_2; \cdots ; S_m$. The sequence $\Xi(\phi, S_1, \ldots, S_m, \psi)$ is inconsistent if $\phi \Rightarrow Pre(\psi : S_1; S_2; \cdots ; S_m)$. The following proposition will be handy.

Proposition 1. *Let $S_1; S_2; \cdots ; S_m$ be a sequence of statements. For any ϕ with $\phi \Rightarrow Pre(\psi : S_1; S_2; \cdots ; S_m)$, $\Xi(\phi, S_1, \ldots, S_m, \psi)$ has an inductive interpolant*[1].

Let $\Lambda = [T, \lambda_1, \lambda_2, \ldots, \lambda_{m+1}, F]$ be an inductive interpolant of $\Xi(\phi, S_1, \ldots, S_m, \psi)$. Recall that λ_i is a quantifier-free formula over $X^{\langle i-1 \rangle}$ for $1 \le i \le m + 1$. It is also an over-approximation to the image of ϕ after executing $S_1; S_2; \cdots ; S_{i-1}$. Proposition 1 can be used to generate new atomic predicates. One simply finds a pair of quantifier-free formulae ϕ and ψ with $\phi \Rightarrow Pre(\psi : S_1; S_2; \cdots ; S_m)$, applies the interpolation theorem, and collects desuperscripted atomic predicates in an inductive interpolant of $\Xi(\phi, S_1, \ldots, S_m, \psi)$. In the following, we show how to obtain such pairs with under- and over-approximations to loop invariants.

Interpolating Over-Approximation. It is not hard to see that an over-approximation to loop invariants characterizes loop invariants after the execution of the loop body. Recall that $\iota \Rightarrow \bar{\iota}$ for some loop invariant ι. Moreover, $\iota \wedge \kappa \Rightarrow Pre(\iota : S_1; S_2; \cdots ; S_m)$. By the monotonicity of $Pre(\bullet : S_1; S_2; \cdots ; S_m)$, we have $\iota \wedge \kappa \Rightarrow Pre(\bar{\iota} : S_1; S_2; \cdots ; S_m)$.

Proposition 2. *Let $\bar{\iota}$ be an over-approximation to loop invariants of the annotated loop $\{\delta\}$ while κ do $S_1; S_2; \cdots ; S_m$ done $\{\epsilon\}$. For any loop invariant ι with $\iota \Rightarrow \bar{\iota}$, $\iota \wedge \kappa \Rightarrow Pre(\bar{\iota} : S_1; S_2; \cdots ; S_m)$.*

Proposition 2 gives a necessary condition to loop invariants of interest. Recall that $\theta = \gamma(\beta)$ is an incorrect conjecture of loop invariants. If $\nu \models \neg(\theta \wedge \kappa \Rightarrow Pre(\bar{\iota} : S_1; S_2; \cdots ; S_m))$, the mechanical teacher returns the abstract counterexample $\alpha^*(\nu)$. Otherwise, Proposition 1 is applicable with the pair $\theta \wedge \kappa$ and $\bar{\iota}$.

Corollary 1. *Let $\bar{\iota}$ be an over-approximation to loop invariants of the annotated loop $\{\delta\}$ while κ do $S_1; S_2; \cdots ; S_m$ done $\{\epsilon\}$. For any θ with $\theta \wedge \kappa \Rightarrow Pre(\bar{\iota} : S_1; S_2; \cdots ; S_m)$, the sequence $\Xi(\theta \wedge \kappa, S_1, S_2, \ldots, S_m, \bar{\iota})$ has an inductive interpolant.*

Interpolating Under-Approximation. For under-approximations, there is no necessary condition. Nevertheless, Proposition 1 is applicable with the pair $\underline{\iota} \wedge \kappa$ and θ.

Corollary 2. *Let $\underline{\iota}$ be an under-approximation to loop invariants of the annotated loop $\{\delta\}$ while κ do $S_1; S_2; \cdots ; S_m$ done $\{\epsilon\}$. For any θ with $\underline{\iota} \wedge \kappa \Rightarrow Pre(\theta : S_1; S_2; \cdots ; S_m)$, the sequence $\Xi(\underline{\iota} \wedge \kappa, S_1, S_2, \ldots, S_m, \theta)$ has an inductive interpolant.*

Generating atomic predicates from an incorrect conjecture θ should now be clear (Algorithm 1). Assuming that the incorrect conjecture satisfies the necessary condition in Proposition 2, we simply collect all desuperscripted atomic predicates in an inductive interpolant of $\Xi(\theta \wedge \kappa, S_1, S_2, \ldots, S_m, \bar{\iota})$ (Corollary 1). More atomic predicates can be obtained from an inductive interpolant of $\Xi(\underline{\iota} \wedge \kappa, S_1, S_2, \ldots, S_m, \theta)$ if additionally $\underline{\iota} \wedge \kappa \Rightarrow Pre(\theta : S_1; S_2; \cdots ; S_m)$ (Corollary 2).

---

[1] The existential quantifiers in $[\![S; S']\!]$ are eliminated by introducing fresh variables.

```
/* {δ} while κ do S₁;···;Sₘ done {ε} : an annotated loop        */
/* ι,ῑ : under- and over-approximations to loop invariants */
```
Input: a formula $\theta \in QF[P]$ such that $\theta \wedge \kappa \Rightarrow Pre(\bar{\iota} : S_1; S_2; \cdots ; S_m)$
Output: a set of atomic predicates
$I :=$ an inductive interpolant of $\Xi(\theta \wedge \kappa, S_1, S_2, \ldots, S_m, \bar{\iota})$;
$Q :=$ desuperscripted atomic predicates in I;
if $\underline{\iota} \wedge \kappa \Rightarrow Pre(\theta : S_1; S_2; \cdots ; S_m)$ **then**
    $J :=$ an inductive interpolants of $\Xi(\underline{\iota} \wedge \kappa, S_1, S_2, \ldots, S_m, \theta)$;
    $R :=$ desuperscripted atomic predicates in J;
    $Q := Q \cup R$;
end
return Q

Algorithm 1. PredicatesFromConjecture (θ)

4.3 Atomic Predicates from Conflicting Abstract Counterexamples

Because of the abstraction, conflicting abstract counterexamples may be given to the learning algorithm. Consider the example in Section 1. Recall that $n \geq 0 \wedge x = n \wedge y = n$ and $x + y = 0 \vee x > 0$ are the under- and over-approximations respectively. Suppose there is only one atomic predicate $y = 0$. The learning algorithm tries to infer a Boolean formula $\lambda \in Bool[b_{y=0}]$. Let us resolve the equivalence queries $EQ(T)$ and $EQ(F)$. On the equivalence query $EQ(F)$, we check if F is weaker than the under-approximation by an SMT solver. It is not, and the SMT solver gives the valuation $\nu_0(n) = \nu_0(x) = \nu_0(y) = 1$ as a witness. Applying the abstraction function α^* to ν_0, the mechanical teacher returns the abstract counterexample $b_{y=0} \mapsto F$. The abstract counterexample is intended to notify that the target formula λ and F have different truth values when $b_{y=0}$ is F. That is, λ is satisfied by the valuation $b_{y=0} \mapsto F$.

On the equivalence query $EQ(T)$, the mechanical teacher checks if T is stronger than the over-approximation. It is not, and the SMT solver now returns the valuation $\nu_1(x) = 0, \nu_1(y) = 1$ as a witness. The mechanical teacher in turn computes $b_{y=0} \mapsto F$ as the corresponding abstract counterexample. The abstract counterexample notifies that the target formula λ and T have different truth values when $b_{y=0}$ is F. That is, λ is not satisfied by the valuation $b_{y=0} \mapsto F$. Yet the target formula λ cannot be satisfied and unsatisfied by the valuation $b_{y=0} \mapsto F$. We have conflicting abstract counterexamples.

Such conflicting abstract counterexamples arise because the abstraction is too coarse. This gives us another chance to refine the abstraction. Define

$$\Gamma(\nu) \stackrel{\triangle}{=} \bigwedge_{x \in X} x = \nu(x).$$

The function $\Gamma(\nu)$ specifies the valuation ν in QF (Figure 1). For distinct valuations ν and ν', $\Gamma(\nu) \wedge \Gamma(\nu')$ is inconsistent. For instance, $\Gamma(\nu_0) = (n = 1) \wedge (x = 1) \wedge (y = 1)$, $\Gamma(\nu_1) = (x = 0) \wedge (y = 1)$, and $\Gamma(\nu_1) \wedge \Gamma(\nu_0)$ is inconsistent.

Algorithm 2 generates atomic predicates from conflicting abstract counterexamples. Let ν and ν' be distinct valuations in Val_X. We compute formulae $\chi = \Gamma(\nu)$ and $\chi' = \Gamma(\nu')$. Since ν and ν' are conflicting, they correspond to the same abstract valuation $\alpha^*(\nu) = \alpha^*(\nu')$. Let $\rho = \gamma^*(\alpha^*(\nu))$. We have $\chi \Rightarrow \rho$ and $\chi' \Rightarrow \rho$ [14]. Recall that

Input: distinct valuations ν and ν' such that $\alpha^*(\nu) = \alpha^*(\nu')$
Output: a set of atomic predicates
$\chi := \Gamma(\nu)$;
$\chi' := \Gamma(\nu')$;
```
/* χ ∧ χ' is inconsistent                                  */
```
$\rho := \gamma^*(\alpha^*(\nu))$;
$Q :=$ atomic predicates in an inductive interpolant of $[\chi, \chi' \vee \neg\rho]$;
return Q;

Algorithm 2. PredicatesFromConflict (ν, ν')

$\chi \wedge \chi'$ is inconsistent. $[\chi, \chi' \vee \neg\rho]$ is also inconsistent for $\chi \Rightarrow \rho$. Algorithm 2 returns atomic predicates in an inductive interpolant of $[\chi, \chi' \vee \neg\rho]$.

5   Algorithm

Our loop invariant inference algorithm is given in Algorithm 3. For an annotated loop $\{\delta\}$ while κ do $S_1; S_2; \cdots; S_m$ done $\{\epsilon\}$, we heuristically choose $\delta \vee \epsilon$ and $\epsilon \vee \kappa$ as the under- and over-approximations respectively. Note that the under-approximation is different from the strongest approximation δ. It is reported that the approximations $\delta \vee \epsilon$ and $\epsilon \vee \kappa$ are more effective in resolving queries [14].

```
/* {δ} while κ do S₁; S₂; ⋯ ; Sₘ done {ε} : an annotated loop    */
```
Output: a loop invariant for the annotated loop
$\underline{\iota} := \delta \vee \epsilon$;
$\overline{\iota} := \epsilon \vee \kappa$;
$P :=$ atomic predicates in an inductive interpolant of $[\underline{\iota}, \neg\overline{\iota}]$;
repeat
    try
        call a learning algorithm for Boolean formulae where membership and
        equivalence queries are resolved by Algorithms 4 and 5 respectively;
    catch conflict abstract counterexamples $\rightarrow$
        find distinct valuations ν and ν' such that $\alpha^*(\nu) = \alpha^*(\nu')$;
        $P := P \cup$ PredicatesFromConflict(ν, ν');
until *a loop invariant is found* ;

Algorithm 3. Loop Invariant Inference

We compute the initial atomic predicates by interpolating $\underline{\iota}$ and $\neg\overline{\iota}$ (Section 4.1). The main loop invokes a learning algorithm. It resolves membership and equivalence queries from the learning algorithm by under- and over-approximations (detailed later). If there is a conflict, the loop invariant inference algorithm adds more atomic predicates by Algorithm 2. Then the main loop reiterates with the new set of atomic predicates.

For membership queries, we compare the concretization of the abstract valuation with approximations to loop invariants (Algorithm 4). The mechanical teacher returns NO when the concretization is inconsistent. If the concretization is stronger than the under-approximation, the mechanical teacher returns YES; if the concretization is

```
/* ι,ῑ : under- and over-approximations to loop invariants */
```
Input: a membership query $MEM(\mu)$ with $\mu \in Val_{B_P}$
Output: YES or NO
$\theta := \gamma^*(\mu)$;
if θ *is inconsistent* **then return** NO;
if $\theta \Rightarrow \iota$ **then return** YES;
if $\nu \models \neg(\theta \Rightarrow \bar{\iota})$ **then return** NO;
return YES *or* NO *randomly*;

Algorithm 4. Membership Query Resolution

```
/* τ : a threshold to generate new atomic predicates     */
/* {δ} while κ do S₁;S₂;···;Sₘ done {ε} : an annotated loop   */
/* ι,ῑ : under- and over-approximations to loop invariants */
```
Input: an equivalence query $EQ(\beta)$ with $\beta \in Bool[B_P]$
Output: YES or an abstract counterexample
$\theta := \gamma(\beta)$;
if $\delta \Rightarrow \theta$ *and* $\theta \Rightarrow \epsilon \lor \kappa$ *and* $\theta \land \kappa \Rightarrow Pre(\theta : S_1; S_2; \cdots; S_m)$ **then return** YES;
if $\nu \models \neg(\iota \Rightarrow \theta)$ *or* $\nu \models \neg(\theta \Rightarrow \bar{\iota})$ *or* $\nu \models \neg(\theta \land \kappa \Rightarrow Pre(\bar{\iota} : S_1; S_2; \cdots; S_m))$ **then**
    record ν; **return** $\alpha^*(\nu)$;
if *the number of random abstract counterexamples* $\leq \tau$ **then**
    return *a random abstract counterexample*;
else
    $P := P \cup \text{PredicatesFromConjecture}(\theta)$; $\tau := \lceil 1.3^{|P|} \rceil$; reiterate the main loop;
end

Algorithm 5. Equivalence Query Resolution

weaker than the over-approximation, it returns NO. Otherwise, a random answer is returned [14].

The equivalence query resolution algorithm is given in Algorithm 5. For any equivalence query, the mechanical teacher checks if the concretization of the abstract conjecture is a loop invariant. If so, it returns YES and concludes the loop invariant inference algorithm. Otherwise, the mechanical teacher compares the concretization of the abstract conjecture with approximations to loop invariants. If the concretization is stronger than the under-approximation, weaker than the over-approximation, or it does not satisfy the necessary condition given in Proposition 2, an abstract counterexample is returned after recording the witness valuation [14, 15]. The witnessing valuations are needed to synthesize atomic predicates when conflicts occur.

If the concretization is not a loop invariant and falls between both approximations to loop invariants, there are two possibilities. The current set of atomic predicates is sufficient to express a loop invariant; the learning algorithm just needs a few more iterations to infer a solution. Or, the current atomic predicates are insufficient to express any loop invariant; the learning algorithm cannot derive a solution with these predicates. Since we cannot tell which scenario arises, a threshold is deployed heuristically. If the number of random abstract counterexamples is less than the threshold, we give the learning algorithm more time to find a loop invariant. Only when the number of random abstract counterexamples exceeds the threshold, can we synthesize more atomic predicates for abstraction refinement. Intuitively, the current atomic predicates are likely to be insufficient if lots of random abstract counterexamples have been generated. In this case, we invoke Algorithm 2 to synthesize more atomic predicates from the incorrect conjecture, update the threshold to $\lceil 1.3^{|P|} \rceil$, and then restart the main loop.

6   Experimental Results

We have implemented the proposed technique in OCaml[2]. In our implementation, the SMT solver YICES and the interpolating theorem prover CSISAT [3] are used for query resolution and interpolation respectively. In addition to the examples in [14], we add two more examples: `riva` is the largest loop expressible in our simple language from Linux[3], and `tar` is extracted from Tar[4]. All examples are translated into annotated loops manually. Data are the average of 100 runs and collected on a 2.4GHz Intel Core2 Quad CPU with 8GB memory running Linux 2.6.31 (Table 1).

Table 1. Experimental Results.
P : # of atomic predicates, MEM : # of membership queries, EQ : # of equivalence queries, RE : # of the learning algorithm restarts, T : total elapsed time (s).

case	$SIZE$	PREVIOUS [14]					CURRENT					BLAST [18]	
		P	MEM	EQ	RE	T	P	MEM	EQ	RE	T	P	T
ide-ide-tape	16	6	13	7	1	0.05	4	6	5	1	0.05	21	1.31(1.07)
ide-wait-ireason	9	5	790	445	33	1.51	5	122	91	7	1.09	9	0.19(0.14)
parser	37	17	4,223	616	13	13.45	9	86	32	1	0.46	8	0.74(0.49)
riva	82	20	59	11	2	0.51	7	14	5	1	0.37	12	1.50(1.17)
tar	7	6	∞	∞	∞	∞	2	2	5	1	0.02	10	0.20(0.17)
usb-message	18	10	21	7	1	0.10	3	7	6	1	0.04	4	0.18(0.14)
vpr	8	5	16	9	2	0.05	1	1	3	1	0.01	4	0.13(0.10)

In the table, the column PREVIOUS represents the work in [14] where atomic predicates are chosen heuristically. Specifically, all atomic predicates in pre- and post-conditions, loop guards, and conditions of `if` statements are selected. The column CURRENT gives the results for our automatic predicate generation technique. Interestingly, heuristically chosen atomic predicates suffice to infer loop invariants for all examples except `tar`. For the `tar` example, the learning-based loop invariant inference algorithm fails to find a loop invariant due to ill-chosen atomic predicates. In contrast, our new algorithm is able to infer a loop invariant for the `tar` example in 0.02s. The number of atomic predicates can be significantly reduced as well. Thanks to a smaller number of atomic predicates, loop invariant inference becomes more economical in these examples. Without predicate generation, four of the six examples take more than one second. Only one of these examples takes more than one second using the new technique. Particularly, the `parser` example is improved in orders of magnitude.

---

[2] Available at http://ropas.snu.ac.kr/tacas11/ap-gen.tar.gz
[3] In Linux 2.6.30 drivers/video/riva/riva_hw.c:nv10CalcArbitration()
[4] In Tar 1.13 src/mangle.c:extract_mangle()

The column BLAST gives the results of lazy abstraction technique with interpolants implemented in BLAST [18]. In addition to the total elapsed time, we also show the preprocessing time in parentheses. Since the learning-based framework does not construct abstract models, our new technique outperforms BLAST in all cases but one (ide-wait-ireason). If we disregard the time for preprocessing in BLAST, the learning-based technique still wins three cases (ide-ide-tape, tar, vpr) and ties one (usb-message). Also note that the number of atomic predicates generated by the new technique is always smaller except parser. Given the simplicity of the learning-based framework, our preliminary experimental results suggest a promising outlook for further optimizations.

6.1 tar from Tar

This simple fragment is excerpted from the code for copying two buffers. M items in the source buffer are copied to the target buffer that already has N items. The variable $size$ keeps the number of remaining items in the source buffer and $copy$ denotes the number of items in the target buffer after the last copy. In each iteration, an arbitrary number of items are copied and the values of $size$ and $copy$ are updated accordingly.

Observe that the atomic predicates in the program text cannot express any loop invariant that proves the specification. However, our new algorithm successfully finds the following loop invariant in this example:

```
{ size = M ∧ copy = N }
1 while size > 0 do
2     available := nondet;
3     if available > size then
4         copy := copy + available;
5         size := size − available;
6 done
{ size = 0 ⟹ copy = M + N }
```

Fig. 3. A Sample Loop in Tar

$$M+N \le copy+size \land copy+size \le M+N$$

The loop invariant asserts that the number of items in both buffers is equal to $M + N$. It requires atomic predicates unavailable from the program text. Predicate generation is essential to find loop invariants for such tricky loops.

6.2 parser from SPEC2000 Benchmarks

For the parser example (Figure 4), 9 atomic predicates are generated. These atomic predicates are a subset of the 17 atomic predicates from the program text. Every loop invariant found by the loop invariant inference algorithm contains all 9 atomic predicates. This suggests that there are no redundant predicates. Few atomic predicates make loop invariants easier to comprehend. For instance, the following loop invariant summarizes the condition when $success$ or $give\_up$ is true:

$$(success \lor give\_up) \Rightarrow$$
$$(valid \ne 0 \lor cutoff = maxcost \lor words < count) \land$$
$$(\neg search \lor valid \ne 0 \lor words < count) \land$$
$$(linkages = canonical \land linkages \ge valid \land linkages \le 5000)$$

$\{\ phase = \mathrm{F} \wedge success = \mathrm{F} \wedge give\_up = \mathrm{F} \wedge cutoff = 0 \wedge count = 0 \ \}$

```
1  while ¬(success ∨ give_up) do
2     entered_phase := F;
3     if ¬phase then
4        if cutoff = 0 then cutoff := 1;
5        else if cutoff = 1 ∧ maxcost > 1 then cutoff := maxcost;
6           else phase := T; entered_phase := T; cutoff := 1000;
7        if cutoff = maxcost ∧ ¬search then give_up := T;
8     else
9        count := count + 1;
10       if count > words then give_up := T;
11    if entered_phase then count := 1;
12    linkages := nondet;
13    if linkages > 5000 then linkages := 5000;
14    canonical := 0; valid := 0;
15    if linkages ≠ 0 then
16       valid := nondet;
17       assume 0 ≤ valid ∧ valid ≤ linkages;
18       canonical := linkages;
19    if valid > 0 then success := T;
20 done
```

$\{\ (valid > 0 \vee count > words \vee (cutoff = maxcost \wedge \neg search)) \wedge$
$valid \leq linkages \wedge canonical = linkages \wedge linkages \leq 5000 \ \}$

Fig. 4. A Sample Loop in SPEC2000 Benchmark PARSER

Fewer atomic predicates also lead to a smaller standard deviation of the execution time. The execution time now ranges from 0.36s to 0.58s with the standard deviation equal to 0.06. In contrast, the execution time for [14] ranges from 1.20s to 80.20s with the standard deviation equal to 14.09. By Chebyshev's inequality, the new algorithm infers a loop invariant in one second with probability greater than 0.988. With a compact set of atomic predicates, loop invariant inference algorithm performs rather predictably.

7   Conclusions

A predicate generation technique for learning-based loop invariant inference was presented. The technique applies the interpolation theorem to synthesize atomic predicates implicitly implied by program texts. To compare the efficiency of the new technique, examples excerpted from Linux, SPEC2000, and Tar source codes were reported. The learning-based loop invariant inference algorithm is more effective and performs much better in these realistic examples.

More experiments are always needed. Especially, we would like to have more realistic examples which require implicit predicates unavailable in program texts. Additionally, loops manipulating arrays often require quantified loop invariants with linear inequalities. Extension to quantified loop invariants is also important.

Acknowledgment. The authors would like to thank Wontae Choi, Soonho Kong, and anonymous referees for their comments in improving this work.

References

1. Angluin, D.: Learning regular sets from queries and counterexamples. Information and Computation 75(2), 87–106 (1987)
2. Ball, T., Podelski, A., Rajamani, S.K.: Boolean and cartesian abstraction for model checking c programs. In: Margaria, T., Yi, W. (eds.) TACAS 2001. LNCS, vol. 2031, pp. 268–283. Springer, Heidelberg (2001)
3. Beyer, D., Zufferey, D., Majumdar, R.: CSISAT: Interpolation for LA+EUF. In: Gupta, A., Malik, S. (eds.) CAV 2008. LNCS, vol. 5123, pp. 304–308. Springer, Heidelberg (2008)
4. Bshouty, N.H.: Exact learning boolean functions via the monotone theory. Information and Computation 123, 146–153 (1995)
5. Canet, G., Cuoq, P., Monate, B.: A value analysis for c programs. In: Source Code Analysis and Manipulation, pp. 123–124. IEEE, Los Alamitos (2009)
6. Craig, W.: Linear reasoning. a new form of the herbrand-gentzen theorem. J. Symb. Log. 22(3), 250–268 (1957)
7. D'Silva, V., Kroening, D., Purandare, M., Weissenbacher, G.: Interpolant strength. In: Barthe, G., Hermenegildo, M. (eds.) VMCAI 2010. LNCS, vol. 5944, pp. 129–145. Springer, Heidelberg (2010)
8. Dutertre, B., Moura, L.D.: The Yices SMT solver. Technical report, SRI International (2006)
9. Esparza, J., Kiefer, S., Schwoon, S.: Abstraction refinement with craig interpolation and symbolic pushdown systems. In: Hermanns, H. (ed.) TACAS 2006. LNCS, vol. 3920, pp. 489–503. Springer, Heidelberg (2006)
10. Filliâtre, J.C., Marché, C.: Multi-prover verification of C programs. In: Davies, J., Schulte, W., Barnett, M. (eds.) ICFEM 2004. LNCS, vol. 3308, pp. 15–29. Springer, Heidelberg (2004)
11. Henzinger, T.A., Jhala, R., Majumdar, R., McMillan, K.L.: Abstractions from proofs. In: POPL 2004, pp. 232–244. ACM, New York (2004)
12. Jhala, R., Mcmillan, K.L.: A practical and complete approach to predicate refinement. In: Hermanns, H. (ed.) TACAS 2006. LNCS, vol. 3920, pp. 459–473. Springer, Heidelberg (2006)
13. Jhala, R., McMillan, K.L.: Array abstractions from proofs. In: Damm, W., Hermanns, H. (eds.) CAV 2007. LNCS, vol. 4590, pp. 193–206. Springer, Heidelberg (2007)
14. Jung, Y., Kong, S., Wang, B.Y., Yi, K.: Deriving invariants by algorithmic learning, decision procedures, and predicate abstraction. In: Barthe, G., Hermenegildo, M. (eds.) VMCAI 2010. LNCS, vol. 5944, pp. 180–196. Springer, Heidelberg (2010)
15. Kong, S., Jung, Y., David, C., Wang, B.Y., Yi, K.: Automatically inferring quantified loop invariants by algorithmic learning from simple templates. In: Ueda, K. (ed.) APLAS 2010. LNCS, vol. 6461, pp. 328–343. Springer, Heidelberg (2010)
16. Kroening, D., Strichman, O.: Decision Procedures an algorithmic point of view. EATCS. Springer, Heidelberg (2008)
17. McMillan, K.L.: An interpolating theorem prover. Theoretical Computer Science 345(1), 101–121 (2005)
18. McMillan, K.L.: Lazy abstraction with interpolants. In: Ball, T., Jones, R.B. (eds.) CAV 2006. LNCS, vol. 4144, pp. 123–136. Springer, Heidelberg (2006)
19. McMillan, K.L.: Quantified invariant generation using an interpolating saturation prover. In: Ramakrishnan, C.R., Rehof, J. (eds.) TACAS 2008. LNCS, vol. 4963, pp. 413–427. Springer, Heidelberg (2008)
20. Srivastava, S., Gulwani, S.: Program verification using templates over predicate abstraction. In: PLDI, pp. 223–234. ACM, New York (2009)

Next Generation LearnLib[*]

Maik Merten[1], Bernhard Steffen[1], Falk Howar[1], and Tiziana Margaria[2]

[1] Technical University Dortmund, Chair for Programming Systems, Dortmund,
D-44227, Germany
{maik.merten,steffen,falk.howar}@cs.tu-dortmund.de
[2] University Potsdam, Chair for Service and Software Engineering, Potsdam,
D-14482, Germany
tiziana.margaria@cs.uni-potsdam.de

Abstract. The Next Generation LearnLib (NGLL) is a framework for model-based construction of dedicated learning solutions on the basis of extensible component libraries, which comprise various methods and tools to deal with realistic systems including test harnesses, reset mechanisms and abstraction/refinement techniques. Its construction style allows application experts to control, adapt, and evaluate complex learning processes with minimal programming expertise.

1   Introduction

Creating behavioral models of un(der)specified systems, e.g., for documentation- or verification-purposes, using (semi-)automated learning algorithms, has become a viable method for quality assurance. Its practical impact increases with the advances in computer resources and, in particular, with the ability to exploit application-specific frame conditions for optimization. Still, creating fitting learning setups is laborious, in part because available learning methods in practice are not engineered to be versatile, often being hard-coded for specific use cases and thus showing limited potential for adaptability towards new fields of application. The Next Generation LearnLib (NGLL) is designed to ease this task by offering an extensive and extensible component library comprising various methods and tools to deal with realistic systems including test harnesses, reset mechanisms and abstraction/refinement techniques. A modeling layer based on the NGLL allows for model-based construction of easily refinable learning solutions. Being internet-enabled, NGLL supports the integration of remote components. Thus learning solutions can be composed of mixtures of components running locally or anywhere in the world, a fact that can in particular be exploited to learn remote systems or to flexibly distribute the learning effort on distributed resources.

In the remainder of this paper, we will describe the technology underlying the NGLL in Section 2, present model-driven creation of learning setups in Section 3, and outline the usefulness in a competitive environment in Section 4, before we conclude in Section 5.

---

[*] This work is supported by the European FP 7 project CONNECT (IST 231167).

P.A. Abdulla and K.R.M. Leino (Eds.): TACAS 2011, LNCS 6605, pp. 220–223, 2011.
© Springer-Verlag Berlin Heidelberg 2011

2   Base Technology

The NGLL is the result of an extensive reengineering effort on the original LearnLib [7], which has originally been designed to systematically build finite state machine models of unknown real world systems (Telecommunications Systems, Web Services, etc.). The experience with the LearnLib soon led to the construction of a platform for experimentation with different learning algorithms and to statistically analyze their characteristics in terms of learning effort, run time and memory consumption. The underlying learning technology is active learning following the pattern of Angluin's L^* algorithm [2], which introduced active system interrogation to automata learning. One of the main obstacles in practical learning is the implementation of the idealized form of interrogation in terms of membership and equivalence queries proposed by Angluin. This requires an application-specific interplay of testing and abstraction technology, driving the reengineering effort that created the NGLL.

The foundation of NGLL is a new extensive Java framework of data structures and utilities, based on a set of interface agreements extensively covering concerns of active learning from constructing alphabets to tethering target systems. This supports the development of new learning components with little boilerplate code and the integration of third-party learning technology, such as libalf [3].

All learning solutions we know of, like libalf, focus on providing fixed sets of learning algorithms. In contrast, the component model of the NGLL extends into the core of the learning algorithms, enabling application-fit tailoring of learning algorithms, at design- as well as at runtime. In particular, it is unique in

- comprising features for addressing real-world or legacy systems, like instrumentation, abstraction, and resetting,
- resolving abstraction-based non-determinism by alphabet abstraction refinement, which would otherwise lead to the failure of learning attempts [4],
- supporting execution and systematic experimentation and evaluation, even including remote learning and evaluation components, and, most notably, in
- its high-level modeling approach described in the next section.

3   Modeling Learning Solutions

LearnLib Studio, which is based on jABC [9], our service-oriented framework for the modeling, development, and execution of complex applications and processes, is NGLL's graphical interface for designing and executing learning and experimentation setups.

A complete learning solution is usually composed of several components, some of which are optional: learning algorithms for various model types, system adapters, query filters and caches, model exporters, statistical probes, abstraction providers, handlers for counterexamples etc.. Many of these components are reusable in nature. NGLL makes them available as easy-to-use building blocks for the graphical composition of application-fit learning experiments.

Figure 1 illustrates the graphical modeling style typical for LearnLib Studio along a very basic learning scenario. One easily identifies a common three phase

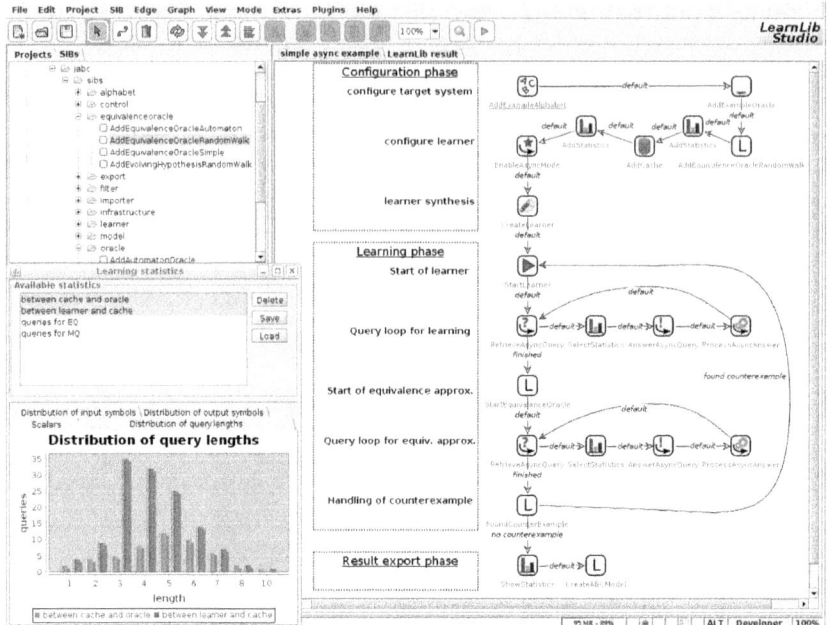

Fig. 1. Executable model of a simple learning experiment in LearnLib Studio

pattern recurring in most learning solutions: The learning process starts with a configuration phase, where in particular the considered alphabet and the system connector are selected, before the learner itself is created and started. The subsequent central learning phase is characterized by the L^*-typical iterations, which organize the test-based interrogation of the system to be learned. These iterations are structured in phases of exploration, which end with the construction of a hypothesis automaton, and the (approximate) realization of the so-called equivalence query, which in practice searches for counterexamples separating the hypothesis automaton from the system to be learned. If this search is successful, a new phase of exploration is started in order to take care of all the consequences implied by the counterexample. Otherwise the learning process terminates after some postprocessing in the third phase, e.g., to produce statistical data.

Most learning experiments follow this pattern, usually enriched by application-specific refinements. Our graphical modeling environment is designed for developing such kinds of refinements by supporting, e.g., component reuse, versioning, optimization and evaluation.

4 Fast-Cycle Experimentation: The ZULU Experience

The ability to quickly design and extend learning setups, coupled with statistical probes and visualizations, was invaluable during the ZULU competition [1]. Various learning setups, involving different learning algorithms and strategies for finding counterexamples, were evaluated in a time-saving manner in the

graphical environment. The NGLL allows one to configure whole experimentation series for automatic evaluation in batch mode, resulting in aggregated statistical charts highlighting the various profiles. This way we were able to identify
the winning setup for the ZULU competition, by playing with variants of finding and evaluating counterexamples [8,6] and combining them to a continuous
evolution process for the construction of learning hypotheses [5].

5   Conclusion

The NGLL provides a machine learning framework, designed for flexibility, ex-
tensibility and reusability. It comprises LearnLib Studio, which enables quick experimentation with learning methods in research and practice, and thus helps to
design fitting learning setups for application-specific contexts. Being executable
jABC graphs, learning setups in LearnLib Studio can use every facility of the
jABC-framework. This includes step-by-step execution, transformation of learning setups into standalone applications using code generation, parallel and hierarchical structuring of the models, model-checking, and automatic deployment
on various platforms. Many concepts only briefly mentioned, but not discussed
here in detail due to limited space, will be demonstrated during the tool demo.
In experiments the NGLL demonstrated the ability to learn models with approximately 40,000 systems states and 50 alphabet symbols. The NGLL is available
for download at the `http://www.learnlib.de` website.

References

1. Zulu - Active learning from queries competition (2010),
   `http://labh-curien.univ-st-etienne.fr/zulu/`
2. Angluin, D.: Learning Regular Sets from Queries and Counterexamples. Information
   and Computation 2(75), 87–106 (1987)
3. Bollig, B., Katoen, J.-P., Kern, C., Leucker, M., Neider, D., Piegdon, D.R.: `libalf`:
   The automata learning framework. In: Touili, T., Cook, B., Jackson, P. (eds.) CAV
   2010. LNCS, vol. 6174, pp. 360–364. Springer, Heidelberg (2010)
4. Howar, F., Steffen, B., Merten, M.: Automata Learning with Automated Alphabet
   Abstraction Refinement. In: Jhala, R., Schmidt, D. (eds.) VMCAI 2011. LNCS,
   vol. 6538, pp. 263–277. Springer, Heidelberg (2011)
5. Howar, F., Steffen, B., Merten, M.: From ZULU to RERS - Lessons learned in the
   ZULU challenge. In: Margaria, T., Steffen, B. (eds.) ISoLA 2010. LNCS, vol. 6415,
   Springer, Heidelberg (2010)
6. Kearns, M.J., Vazirani, U.V.: An Introduction to Computational Learning Theory.
   MIT Press, Cambridge (1994)
7. Raffelt, H., Steffen, B., Berg, T., Margaria, T.: LearnLib: a framework for extrapo-
   lating behavioral models. Int. J. Softw. Tools Technol. Transf. 11(5), 393–407 (2009)
8. Rivest, R.L., Schapire, R.E.: Inference of finite automata using homing sequences.
   Inf. Comput. 103(2), 299–347 (1993)
9. Steffen, B., Margaria, T., Nagel, R., Jörges, S., Kubczak, C.: Model-Driven Development with the jABC. In: Bin, E., Ziv, A., Ur, S. (eds.) HVC 2006. LNCS, vol. 4383,
   pp. 92–108. Springer, Heidelberg (2007)

Applying CEGAR to the Petri Net State Equation

Harro Wimmel and Karsten Wolf

Universität Rostock, Institut für Informatik

Abstract. We propose a reachability verification technique that combines the *Petri net state equation* (a linear algebraic overapproximation of the set of reachable states) with the concept of *counterexample guided abstraction refinement*. In essence, we replace the search through the set of reachable states by a search through the space of solutions of the state equation. We demonstrate the excellent performance of the technique on several real-world examples. The technique is particularly useful in those cases where the reachability query yields a negative result: While state space based techniques need to fully expand the state space in this case, our technique often terminates promptly. In addition, we can derive some diagnostic information in case of unreachability while state space methods can only provide witness paths in the case of reachability.

Keywords: Petri Net, Reachability Problem, Integer Programming, CEGAR, Structure Analysis, Partial Order Reduction.

1 Introduction

Reachability is *the* fundamental verification problem. For place/transition Petri nets (which may have infinitely many states), it is one of the hardest decision problems known among the naturally emerging yet decidable problems in computer science. General solutions have been found by Mayr [12] and Kosaraju [7] with later simplifications made by Lambert [9], but there are complexity issues. All these approaches use coverability graphs which can have a non-primitive-recursive size with respect to the corresponding Petri net. A new approach by Leroux [10] not using such graphs gives some hope, but a concrete upper bound for the worst case complexity so far eludes us. In a sense even worse, Lipton [11] has shown that the problem is EXPSPACE-hard, so any try at programming a tool efficiently solving this problem to the full extent must surely fail.

Nevertheless, efficient tools exist that are applicable to a considerable number of problem instances. Model checkers, symbolic [2] or with partial order reduction [17], have been used successfully to solve quite large reachability problems. On a positive answer, a model checker can typically generate a trace, i.e. a firing sequence leading to the final marking. In contrast, negative answers are usually not accompanied by any diagnostic information. Such information, i.e. a counterexample or reasoning why the problem has a negative solution would require a deep analysis of the structure of the Petri net. So far, no tools are known that analyze the structure of a net and allow for such reasoning.

This paper presents an approach to the reachability problem that combines two existing methods. First, we employ the *state equation* for Petri nets. This is a linear-algebraic

P.A. Abdulla and K.R.M. Leino (Eds.): TACAS 2011, LNCS 6605, pp. 224–238, 2011.

overapproximation on the set of reachable states. Second, we use the concept of *counterexample guided abstraction refinement* (CEGAR) [3] for enhancing the expressiveness of the state equation. In essence, we iteratively analyse spurious solutions of the state equation and add constraints that exclude a solution found to be spurious but do not exclude any real solution. The approach has several advantages compared to (explicit or symbolic) purely state space based verification techniques:

- The search is quite focussed from the beginning as we traverse the solution space of the state equation rather than the set of reachable states;
- The search is close to breadth-first traversal, so small witness traces are generated;
- The method may perform well on unreachable problem instances (where state space techniques compute maximum size state spaces);
- In several unreachable problem instances, some kind of diagnostic information can be provided;
- A considerable workload can be shifted to very mature tools for solving linear programming problems.

In Sect. 2 we give the basic definitions. Section 3 shows how to use integer programming tools to find candidates for a solution. Section 4 deals with the analysis of the Petri net structure that is needed to push the integer programming onto the right path. In Sect. 5 we use methods of partial order reduction to mold the results of the integer programming into firing sequences solving the reachability problem. Finally, section 6 compares the results of an implementation with another model checker, showing that structure analysis can compete with other approaches.

2   The Reachability Problem

Definition 1 (Petri net, marking, firing sequence). *A Petri net N is a tuple (S, T, F) with a set S of places, a set T of transitions, where $S \neq \emptyset \neq T$ and $S \cap T = \emptyset$, and a mapping $F \colon (S \times T) \cup (T \times S) \to \mathbb{N}$ defining arcs between places and transitions.*

A marking or state of a Petri net is a map $m \colon S \to \mathbb{N}$. A place s is said to contain k tokens under m if $m(s) = k$. A transition $t \in T$ is enabled under m, $m[t\rangle$, if $m(s) \geq F(s, t)$ for every $s \in S$. A transition t fires under m and leads to m', $m[t\rangle m'$, if additionally $m'(s) = m(s) - F(s, t) + F(t, s)$ for every $s \in S$.

A word $\sigma \in T^*$ is a firing sequence under m and leads to m', $m[\sigma\rangle m'$, if either $m = m'$ and $\sigma = \varepsilon$, the empty word, or $\sigma = wt$, $w \in T^*$, $t \in T$ and $\exists m'' \colon m[w\rangle m''[t\rangle m'$. A firing sequence σ under m is enabled under m, i.e. $m[\sigma\rangle$. The Parikh image of a word $\sigma \in T^*$ is the vector $\wp(\sigma) \colon T \to \mathbb{N}$ with $\wp(\sigma)(t) = \#_t(\sigma)$, where $\#_t(\sigma)$ is the number of occurrences of t in σ. For any firing sequence σ, we call $\wp(\sigma)$ realizable.

As usual, places are drawn as circles (with tokens as black dots inside them), transitions as rectangles, and arcs as arrows with $F(x, y) > 0$ yielding an arrow pointing from x to y. If an arc has a weight of more than one, i.e. $F(x, y) > 1$, the number $F(x, y)$ is written next to the arc. In case $F(x, y) = F(y, x) > 0$, we may sometimes draw a line with arrowheads at both ends.

Note, that the Parikh image is not an injective function. Therefore, $\wp(\sigma)$ can be realizable even if σ is not a firing sequence (provided there is another firing sequence σ' with $\wp(\sigma) = \wp(\sigma')$).

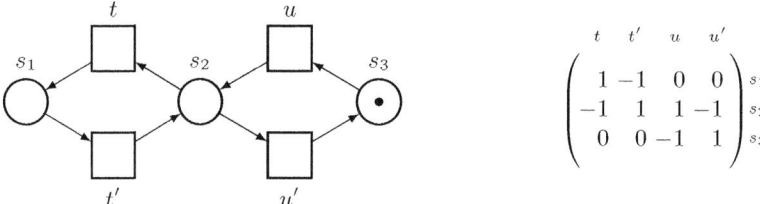

Fig. 1. The word tt' cannot fire, but we can borrow a token from the circle uu', so $utt'u'$ can fire and leads to the same marking as tt'. The incidence matrix of the net is shown on the right.

Definition 2 (Reachability problem). *A marking m' is* reachable *from a marking m in a net $N = (S, T, F)$ if there is a firing sequence $\sigma \in T^*$ with $m[\sigma\rangle m'$. A tuple (N, m, m') of a net and two markings is called a* reachability *problem and has the answer "yes" if and only if m' is reachable from m in N. The set $\mathsf{RP} = \{(N, m, m') \mid N$ is a Petri net, m' reachable from m in $N\}$ is generally called the* reachability problem, *for which membership is to be decided.*

It is well-known that a necessary condition for a positive answer to the reachability problem is the feasibility of the *state equation*.

Definition 3 (State equation). *For a Petri net $N = (S, T, F)$ let $C \in \mathbb{N}^{S \times T}$, defined by $C_{s,t} = F(t, s) - F(s, t)$, be the* incidence matrix *of N. For two markings m and m', the system of linear equations $m + Cx = m'$ is the* state equation *of N for m and m'. A vector $x \in \mathbb{N}^T$ fulfilling the equation is called a* solution.

Remark 1. For any firing sequence σ of a net $N = (S, T, F)$ leading from m to m', i.e. $m[\sigma\rangle m'$, holds $m + C\wp(\sigma) = m'$, i.e. the Parikh vector of σ is a solution of the state equation for N, m, and m'. This is just a reformulation of the firing condition for σ.

It is possible to have a sequence σ such that its Parikh image fulfills the state equation but it is not a firing sequence. The easiest example for this occurs in a net $N = (\{s\}, \{t\}, F)$ with $F(s, t) = 1 = F(t, s)$. Let m and m' be the *empty marking*, i.e. one with zero tokens overall, then $m[t\rangle m'$ is obviously wrong but $m + C\wp(\sigma) = m'$ holds since $C = (0)$. The effect can occur whenever the Petri net contains a cycle of transitions. Interestingly, certain cycles of transitions can also help to overcome this problem, see Fig. 1. Here, we would like to fire a word tt' from the marking m with $m(s_1) = m(s_2) = 0$ and $m(s_3) = 1$, but obviously, this is impossible. If we borrow a token from s_3, we can fire tt', or more precisely $utt'u'$. As we return the borrowed token to s_3 in the end we reach the same marking tt' would have reached (if enabledness were neglected).

Definition 4 (T-invariant). *Let $N = (S, T, F)$ be a Petri net and C its incidence matrix. A vector $x \in \mathbb{N}^T$ is called a* T-invariant *if $Cx = 0$.*

A realizable T-invariant corresponds to a cycle in the state space. Its occurrence does not change the marking. However, its interleaving with another sequence σ may turn σ from unrealizable to realizable. The reason is that the partial occurrence of the

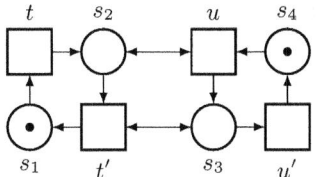

Fig. 2. Neither the T-invariant $\wp(tt')$ nor $\wp(uu')$ is realizable, but $\wp(tt'uu')$ is, by the sequence $tut'u'$

T-invariant may "lend" tokens to the otherwise blocked σ and can be completed after σ has produced another token on the same place later on.

Solving the state equation is a non-negative integer programming problem. From linear algebra we know that the solution space is semi-linear.

Corollary 1 (Solution space). *For a given state equation $m + Cx = m'$ over a net $N = (S, T, F)$, there are numbers $j, k \in \mathbb{N}$ and finite sets of vectors $B = \{b_i \in \mathbb{N}^T \mid 1 \leq i \leq j\}$ (base vectors) and $P = \{p_i \in \mathbb{N}^T \mid 1 \leq i \leq k\}$ (period vectors) such that:*

- *all $b_i \in B$ are pairwise incomparable (by standard componentwise comparison for vectors) and thus minimal solutions,*
- *P forms a basis for the non-negative solution space $P^* = \{\sum_{i=1}^{k} n_i p_i \mid n_i \in \mathbb{N}, p_i \in P\}$ of $Cx = 0$,*
- *for all solutions x there are $n_i \in \mathbb{N}$ for $1 \leq i \leq k$ and $n \in \{1, \dots, j\}$ such that $x = b_n + \sum_{i=1}^{k} n_i p_i$,*
- *for every solution x, all vectors of the set $x + P^*$ are solutions as well.*

Note that only linear combinations with nonnegative coefficients are considered in this representation.

So we know that all solutions can be obtained by taking a minimal solution b of the state equation and adding a linear combination of T-invariants from some basis P. Usually, not all the elements from B and P we use for a solution are realizable, though. While the sum of two realizable T-invariants remains realizable (just concatenate the according firing sequences as they have identical initial and final marking), the sum of two non-realizable T-invariants may well become realizable. This can be seen in Fig. 2, where neither $\wp(tt')$ nor $\wp(uu')$ is realizable under the marking m with $m(s_1) = m(s_4) = 1$ and $m(s_2) = m(s_3) = 0$, but the sequence $tut'u'$ realizes $\wp(tt'uu')$. The matter is even more complicated when a minimal solution from B is introduced, because positive minimal solutions are never T-invariants (unless $m = m'$), i.e. they change the marking of the net, so their realizations cannot just be concatenated.

3   Traversing the Solution Space

For solving the state equation an IP solver can be used. Fast IP solvers like *lp_solve* [1] allow to define an objective function – in our case to minimize the solution size and obtain firing sequences that are as short as possible – and yield a single solution, at least if a solution exists. Fortunately, we can force an IP solver to produce more than just one solution — this is the CEGAR part of our approach. If a solution found is not realizable, we may add an inequation to our state equation to forbid that solution. Starting the IP solver again will then lead to a different solution. The trick is, of course, to add inequations in such a way that no realizable solution is lost.

Definition 5 (Constraints). *Let $N = (S, T, F)$ be a Petri net. We define two forms of constraints, both being linear inequations over transitions:*

- *a* jump constraint *takes the form $t < n$ with $n \in \mathbb{N}$ and $t \in T$.*
- *an* increment constraint *takes the form $\sum_{i=1}^{k} n_i t_i \geq n$ with $n_i \in \mathbb{Z}, n \in \mathbb{N}$, and $t_i \in T$.*

Jump constraints can be used to switch (jump) to another base solution, exploiting the incomparability of different minimal base solutions, while increment constraints are used to force non-minimal solutions. To understand the idea for differentiating between these two forms of constraints, it is necessary to introduce the concept of a partial solution first. A partial solution is obtained from a solution of the state equation under given constraints by firing as many transitions as possible.

Definition 6 (Partial solution). *Let $N = (S, T, F)$ be a Petri net and Ω a total order over $\mathbb{N}^T$ that includes the partial order given by $x < y$ if $\sum_{t \in T} x(t) < \sum_{t \in T} y(t)$. A partial solution of a reachability problem (N, m, m') is a tuple $(\mathcal{C}, x, \sigma, r)$ of*

- *a family of (jump and increment) constraints $\mathcal{C} = (c_1, \ldots, c_n)$,*
- *the Ω-smallest solution x fulfilling the state equation of (N, m, m') and the constraints of $\mathcal{C}$,*
- *a firing sequence $\sigma \in T^*$ with $m[\sigma\rangle$ and $\wp(\sigma) \leq x$,*
- *a remainder r with $r = x - \wp(\sigma)$ and $\forall t \in T \colon (r(t) > 0 \implies \neg m[\sigma t\rangle)$.*

The vectors x and r are included for convenience only, they can be computed from $\mathcal{C}, \sigma, \Omega$, and the problem instance.

A full solution *is a partial solution $(\mathcal{C}, x, \sigma, r)$ with $r = 0$. In this case, σ is a firing sequence solving our reachability problem (with answer 'yes').*

We choose Ω such that an IP solver can be assumed to always produce the Ω-smallest solution that does not contradict its linear system of equations.

Corollary 2 (Realizable solutions are full solutions). *For any realizable solution x of the state equation we find a full solution $(\mathcal{C}, x, \sigma, \emptyset)$ where $\mathcal{C}$ consists of constraints $t \geq x(t)$ for every t with $x(t) > 0$, and $\wp(\sigma) = x$.*

Note, that x is the smallest solution fulfilling c and therefore also the Ω-smallest solution.

By adding a constraint to a partial solution we may obtain new partial solutions (or not, if the linear system becomes infeasible). Any full solution can eventually be reached by consecutively extending an Ω-minimal partial solution with constraints.

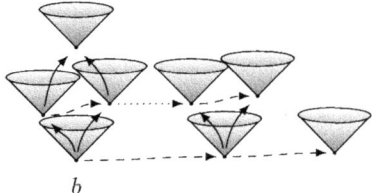

b

Fig. 3. Paths from the Ω-minimal solution b to any solution. Black dots represent solutions, cones stand for linear solution spaces over such solutions, which may or may not intersect or include each other. Normal arrows increment a solution by adding a T-invariant, dashed arrows are jumps to an incomparable Ω-greater solution. Such jumps can also occur on higher levels of linear solution spaces, shown by the dotted arrow.

Lemma 1 (A path to a full solution). *Let b be the Ω-minimal solution of the state equation of a reachability problem (N, m, m') and $ps' = ((c_j)_{1 \le j \le \ell}, b' + \sum_{i=1}^{k} n_i p_i, \sigma', 0)$ a full solution of the problem. For $0 \le n \le \ell$, there are partial solutions $ps_n = ((c_j)_{1 \le j \le n}, x_n, \sigma_n, r_n)$ with $ps_0 = (\emptyset, b, \sigma_0, r_0)$, $ps_\ell = ps'$, and $x_{n_1} \le_\Omega x_{n_2}$ for $n_1 \le n_2$.*

Proof. Let $\mathcal{C}_n = (c_j)_{1 \le j \le n}$. If ps_{n_1}, ps_{n_2} are two partial solutions (with $1 \le n_1 < n_2 \le \ell$) then x_{n_2} is a solution of the state equation plus $\mathcal{C}_{n_1}$, since it even fulfills the state equation plus $\mathcal{C}_{n_2}$ with $\mathcal{C}_{n_1} \subseteq \mathcal{C}_{n_2}$. As x_{n_1} is the Ω-smallest solution of the state equation plus $\mathcal{C}_{n_1}$, $x_{n_1} \le_\Omega x_{n_2}$ holds. Therefore, $b \le_\Omega x_1 \le_\Omega \ldots \le_\Omega x_\ell$. Since $x_\ell = b' + \sum_{i=1}^{k} n_i p_i$ is an existing solution of the strictest system, i.e. state equation plus $\mathcal{C}_\ell$, each system of state equation plus one family of constraints $\mathcal{C}_n$ is solvable. As a σ_n can be determined by just firing transitions as long as possible, all the partial solutions ps_n exist.

Now, let us assume a partial solution $ps = (\mathcal{C}, x, \sigma, r)$ that is not a full solution, i.e. $r \ne 0$. Obviously, some transitions cannot fire often enough. There are three possible remedies for this situation:

1. If x is realizable, we can find a full solution $ps' = (\mathcal{C}, x, \sigma', 0)$ with $\wp(\sigma') = x$.
2. We can add a jump constraint to obtain an Ω-greater solution vector for a different partial solution.
3. If $r(t) > 0$ for some transition t, we can add an increment constraint to increase the maximal number of tokens available on a place in the preset of t. Since the final marking remains the same, this means to borrow tokens for such a place. This can be done by adding a T-invariant containing the place to the solution.

A visualization of these ideas can be seen in Fig. 3 where b denotes the Ω-smallest solution. The cone over b represents all solutions $b + P^*$ with P being the set of period vectors, i.e. T-invariants. Jump constraints lead along the dashed or dotted lines to the next Ω-minimal solution while normal arrows representing increment constraints lead upwards to show the addition of a T-invariant. How to build constraints doing just what we want them to do is the content of the next section.

4 Building Constraints

Let us first argue that for a state equation, any of the minimal solution vectors in B can be obtained by using jump constraints.

Lemma 2 (Jumps to minimal solutions). *Let $b, b' \in B$ be base vectors of the solution space of the state equation $m + Cx = m'$ plus some set of constraints C. Assume b to be the Ω-minimal solution of the system. Then, we can obtain b' as output of our IP solver by consecutively adding jump constraints of the form $t_i < n_i$ with $n_i \in \mathbb{N}$ to C.*

Proof. We know $b \leq_\Omega b'$ holds, but since b' is a minimal solution, $b \leq b'$ cannot hold. Therefore, a transition t with $b'(t) < b(t)$ must exist. After adding the constraint $t < b(t)$ to C the IP solver can no longer generate b as a solution. Assume b'' is the newly generated solution. If $b' = b''$ we are done. Otherwise, since b' fulfills $t < b(t)$, it is still a solution of our system, and also a minimal one as the solution space is restricted by the added constraint. Thus, $b'' \leq_\Omega b'$ holds and we may recursively use the same argument as above for $b := b''$. Since there are only finitely many solutions Ω-smaller than b', the argument must terminate reaching b'.

Non-minimal solutions may not be reachable this way, since the argument "$b'(t) < b(t)$ for some t" does not necessarily hold. We will need increment constraints for this, but unluckily, increment constraints and jump constraints may contradict each other. Assume our state equation has a solution of the form $b' + p$ with a period vector $p \in P$ and to obtain $b' \in B$ from the Ω-minimal solution $b \in B$ we need to add (at least) a jump constraint $t_i < n_i$ to the state equation. If p contains t_i often enough, we will find that $(b' + p)(t_i) \geq n_i$ holds. Therefore, $b' + p$ is not a solution of the state equation plus the constraint $t_i < n_i$, i.e. adding an increment constraint demanding enough occurrences of t_i for $b' + p$ will render the linear equation system infeasible. The only way to avoid this problem is to remove the jump constraints before adding increment constraints.

Lemma 3 (Transforming jumps). *Let z be the Ω-minimal solution of the state equation $m + Cx = m'$ plus some constraints C. Let C' consist of all increment constraints of C plus a constraint $t \geq z(t)$ for each transition t. Then, for all $y \geq z$, y is a solution of $m + Cx = m'$ plus $C \cap C'$ if and only if y is a solution of $m + Cx = m'$ plus C'. Furthermore, no Ω-smaller solution of $m + Cx = m'$ plus C than z solves $m + Cx = m'$ plus C'.*

Proof. Let $y \geq z$ be a solution of $m + Cx = m'$ plus $C \cap C'$. The additional constraints in C' only demand $y(t) \geq z(t)$, which is obviously the case. The other direction is trivial. For the second part, let $z' \leq_\Omega z$ with $z \neq z'$ be some solution of $m + Cx = m'$ plus C. Since $\sum_t z'(t) \leq \sum_t z(t)$ (following from Ω) but $z \neq z'$, for at least one transition t holds $z'(t) < z(t)$. Due to the constraint $t \geq z(t)$ in C', z' cannot be a solution of $m + Cx = m'$ plus C'.

As a consequence, if we are only interested in solutions of the cone $z + P^*$ over z, we can add increment constraints guaranteeing solutions greater or equal than z and remove all jump constraints without any further restriction. Our IP solver will yield z

as the Ω-minimal solution for both families of constraints, $\mathcal{C}$ and $\mathcal{C}'$, and we can add further constraints leading us to any solution in the cone $z + P^*$ now.

Let $ps = (\mathcal{C}, x, \sigma, r)$ now be a partial solution with $r > 0$. We would like to determine sets of places that need additional tokens (and the number of these tokens) that would enable us to fire the remainder r of transitions. Obviously, this problem is harder than the original problem of finding out if a transition vector is realizable, i.e. just testing if zero additional tokens are sufficient. A recursive approach would probably be very inefficient as for every solution x there may be many different remainders r. Even though the remainders are smaller than the solution vector x, the number of recursion steps might easily grow exponentially with the size of x, i.e. $\sum_t x(t)$. We therefore adopt a different strategy, namely finding good heuristics to estimate the number of tokens needed. If a set of places actually needs n additional tokens with $n > 0$, our estimate may be any number from one to n. If we guess too low, we will obtain a new partial solution allowing us to make a guess once again, (more or less) slowly approaching the correct number. We propose a two-part algorithm, the first part dealing with sets of places and transitions that are of interest.

input: Reachability prob. (N, m, m'); partial solution $ps = (\mathcal{C}, x, \sigma, r)$
output: A set of tuples (S_i, T_i, X_i) with $S_i \subseteq S$, $T_i \cup X_i \subseteq T$
Determine $\hat{m}$ with $m[\sigma\rangle\hat{m}$;
Build a bipartite graph $G = (S_0 \cup T_0, E)$ with
$T_0 := \{t \in T \mid r(t) > 0\}$; $S_0 := \{s \in S \mid \exists t \in T_0 \colon F(s, t) > \hat{m}(s)\}$;
$E := \{(s, t) \in S_0 \times T_0 \mid F(s, t) > \hat{m}(s)\} \cup \{(t, s) \in T_0 \times S_0 \mid F(t, s) > F(s, t)\}$;
Calculate the strongly connected components $(SCCs)$ of G;
$i := 1$;
for each source SCC (i.e. one without incoming edges):
  $S_i := SCC \cap S_0$;
  $T_i := SCC \cap T_0$;
  $X_i := \{t \in T_0 \backslash SCC \mid \exists s \in S_i \colon (s, t) \in E\}$;
  $i := i + 1$;
end for

The edges of the graph G constructed in the algorithm have a different meaning depending on their direction. Edges from transitions to places signal that the transition would increase the number of tokens on the place upon firing, while edges in the other direction show the reason for the non-enabledness of the transition. A source SCC, i.e. a strongly connected component without incoming edges from other components, can therefore not obtain tokens by the firing of transitions from other SCCs. This means, tokens must come from somewhere else, that is, from firing transitions not appearing in the remainder r. For each set of places S_i such identified as non-markable by the remainder itself, there are two sets of transitions. If one transition from the set T_i would become firable, it is possible that all other transitions could fire as well, since the former transition effectively produces tokens on some place in the component. If the set T_i is empty (the SCC consisting of a single place), the token needs of all the transitions in X_i together must be fulfilled, since they cannot activate each other. We can thus calculate how many tokens we need at least:

input: A tuple (S_i, T_i, X_i); (N, m, m') and $\hat{m}$ from above
output: A number of tokens n (additionally needed for S_i)
if $T_i \neq \emptyset$
then $n := \min_{t \in T_i} (\sum_{s \in S_i} (F(s,t) - \hat{m}(s)))$
else sort X_i in groups $G_j := \{t \in X_i \mid F(t,s) = j\}$ (with $S_i = \{s\}$);
    $n := 0$; $c := 0$;
    for j with $G_j \neq \emptyset$ **downwards loop**
        $c := c - j * (|G_j| - 1) + \sum_{t \in G_j} F(s,t)$;
        if $c > 0$ **then** $n := n + c$ **end if**;
        $c := -j$
    end for
end if

Note that the transitions in X_i all effectively consume tokens from $s \in S_i$, but they may leave tokens on this place due to a loop. By firing those transitions with the lowest $F(t,s)$-values last, we minimize the leftover. Transitions with the same $F(t,s)$-value j can be processed together, each consuming effectively $F(s,t) - j$ tokens except for the "first" transition which will need j more tokens. If some group G_j of transitions leaves tokens on s, the next group can consume them, which is memorized in the variable c (for carryover or consumption). Observe, that the algorithm cannot return zero: There must be at least one transition in $T_i \cup X_i$, otherwise there would be no transition that cannot fire due to a place in S_i and the places in S_i would not have been introduced at all. If T_i is not empty, line 4 in the algorithm will minimize over positive values; if T_i is empty, line 8 will set c to a positive value at its first execution, yielding a positive value for n. Overall, our argumentation shows:

Corollary 3. *For each set of places S_i that need additional tokens according to the first part of the algorithm, the second part estimates that number of tokens to be in a range from one to the actual number of tokens necessary.*

We can thus try to construct a constraint from a set of places S_i generated by the first part of the algorithm and the token number calculated in the second part. Since our state equation has transitions as variables, we must transform our condition on places into one on transitions first.

Corollary 4. *Let $N = (S, T, F)$ be a Petri net, (N, m, m') the reachability problem to be solved, $ps = (\mathcal{C}, x, \sigma, r)$ a partial solution with $r > 0$, and $\hat{m}$ the marking reached by $m[\sigma\rangle\hat{m}$. Let S_i be a set of places and n a number of tokens to be generated on S_i. Further, let $T_i := \{t \in T \mid r(t) = 0 \wedge \sum_{s \in S_i} (F(t,s) - F(s,t)) > 0\}$. We define a constraint c by*

$$\sum_{t \in T_i} \sum_{s \in S_i} (F(t,s) - F(s,t))t \geq n + \sum_{t \in T_i} \sum_{s \in S_i} (F(t,s) - F(s,t))\wp(\sigma)(t).$$

Then, for the system $m + Cx = m'$ plus $\mathcal{C}$ plus c, if our IP solver can generate a solution $x + y$ (y being a T-invariant) we can obtain a partial solution $ps' = (\mathcal{C} \cup \{c\}, x + y, \sigma\tau, r+z)$ with $\wp(\tau) + z = y$. Furthermore, $\sum_{t \in T} \sum_{s \in S_i} (F(t,s) - F(s,t))y(t) \geq n$.

First, note that T_i contains the transitions that produce more on S_i than they consume, but we have explicitly excluded all transitions of the remainder r, since we do not want the IP solver to increase the token production on S_i by adding transitions that could not fire anyway. I.e., we would like to have a chance to fire the additional transitions in y at some point, though there are no guarantees. The left hand side of c contains one instance of a transition t for each token that t effectively adds to S_i. If we apply some transition vector x to the left hand side of c, we therefore get the number of tokens added to S_i by the transitions from T_i in x. Of course, other transitions in x might reduce this number again. For the right hand side of c, we calculate how many tokens are actually added to S_i by the transitions from T_i in the firing sequence σ (and therefore also in the solution x) and increase that number by the n extra tokens we would like to have. Since the extra tokens cannot come from x in a solution $x + y$, they must be produced by y, i.e. $\sum_{t \in T} \sum_{s \in S_i} (F(t,s) - F(s,t))y(t) \geq n$. We might be able to fire some portion of y after σ, resulting in the obvious $\wp(\tau) + z = y$. When we apply our constraint we might get less or more than the n extra tokens, depending on the T-invariants in the net. Further constraints may or may not help. At this point we can state:

Theorem 1 (Reachability of solutions). *Every realizable solution of the solution space of a state equation can be reached by consecutively adding constraints to the system of equations, always transforming jump constraints before adding increment constraints.*

5   Finding Partial Solutions

Producing partial solutions $ps = (\mathcal{C}, x, \sigma, r)$ from a solution x of the state equation (plus $\mathcal{C}$) is actually quite easily done by brute force. We can build a tree with marking-annotated nodes and the firing of transitions as edges, allowing at most $x(t)$ instances of a transition t on any path from the root of the tree to a leaf. Any leaf is a new partial solution from which we may generate new solutions by adding constraints to the state equation and forwarding the evolving linear system to our IP solver. If we just make a depth-first-search through our tree and backtrack at any leaf, we build up all possible firing sequences realizable from x. This is obviously possible without explicitly building the whole tree at once, thus saving memory. Of course, the tree might grow exponentially in the size of the solution vector x and so some optimizations are in order to reduce the run-time. We would like to suggest a few ones here, especially partial order reductions.

1. The stubborn set method [8] determines a set of transitions that can be fired before all others by investigating conflicts and dependencies between transitions at the active marking. The stubborn set is often much smaller than the set of enabled transitions under the same marking, leading to a tree with a lower degree. In our case, in particular the version of [13] is useful as, using this method, the reduced state space contains, for each trace to the target marking, at least one permution of the same trace. Hence, the reduction is consistent with the given solution of the state equation.
2. Especially if transitions should fire multiple times ($x(t) > 1$) we observe that the stubborn set method alone is not efficient. The situation in Fig. 4 may occur quite

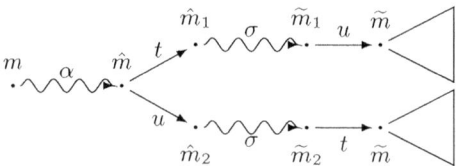

Fig. 4. If both sequences $\alpha t \sigma u$ and $\alpha u \sigma t$ can be fired, the subtrees after the nodes with marking $\tilde{m}$ are identical. Only one of the subtrees needs to be evaluated, the other one may be omitted. Snaked lines denote firing sequences.

often. Assume we reach some marking $\hat{m}$ by a firing sequence α, so that transitions t and u are enabled. After proceeding through the subtree behind t we backtrack to the same point and now fire u followed by some sequence σ after which t is enabled, leading to $m[\alpha\rangle\hat{m}[u\sigma t\rangle\tilde{m}$. If $\hat{m}[t\sigma u\rangle$ holds, we know that it reaches the same marking $\tilde{m}$ and the same remainder r of transitions still has to fire. Therefore, in both cases the future is identical. Since we have already investigated what happens after firing $\alpha t \sigma u$, we may backtrack now omitting the subtree after $\alpha u \sigma t$. Note that a test if $\hat{m}[t\sigma u\rangle$ holds is quite cheap, as only those places s with $C_{s,t} < C_{s,u}$ can prevent the sequence $t\sigma$. Enabledness of u after $t\sigma$ can be tested by reverse calculating $\tilde{m}_1 = \tilde{m} - Cu$ and checking whether $\tilde{m}_1$ is a marking and $\tilde{m}_1[u\rangle\tilde{m}$ holds.

3. There are situations where a leaf belongs to a partial solution ps' that cannot lead to a (new) full solution. In this case the partial solution does not need to be processed. If we already tried to realize x yielding a partial solution $ps = (\mathcal{C}, x, \sigma, r)$ and $ps' = (\mathcal{C} \cup \{c\}, x + y, \sigma, r + y)$ is our new partial solution with an increment constraint c and a T-invariant y, any realizable solution $x + y + z$ obtainable from ps' can also be reached from ps by first adding a constraint c' for the T-invariant z (and later c, y). If no transition of z can be fired after σ, $y + z$ is also not realizable after firing σ. We may be able to mingle the realization of z with the firing of σ, but that will be reflected by alternate partial solutions (compared to both, ps and ps'). Therefore, not processing ps' will not lose any full solutions.

4. A similar situation occurs for $ps' = (\mathcal{C} \cup \{c\}, x + y, \sigma\tau, r)$ with $\wp(\tau) = y$. There is one problem, though. Since we estimated a token need when choosing c and that estimate may be too low, it is possible that while firing τ we get closer to enabling some transition t in r without actually reaching that limit where t becomes firable. We thus have to check for such a situation (by counting the minimal number of missing tokens for firing t in the intermediate markings occurring when firing σ and τ). If τ does not help in approaching enabledness of some t in r, we do not need to process ps' any further.

5. Partial solutions should be memorized if possible to avoid using them as input for CEGAR again if they show up more than once.

6   Experimental Results

The algorithm presented here has been implemented in a tool named Sara [16]. We compare Sara to LoLA [17], a low level analyzer searching the (reduced) state space of a Petri net. According to independent reports, e.g. [15], LoLA performs very well on reachability queries and possibly is the fastest tool for standard low level Petri nets. The following tests, real-world examples as well as academic constructions, were run on a 2.6GHz PC with 4GB RAM under Windows XP and Cygwin. While the CPU had four cores, only one was used for the tools. Tests on a similar Linux system lead to comparable but slightly faster results.

- 590 business processes with about 20 up to 300 actions each were tested for "relaxed soundness". The processes were transformed into Petri nets and for each action a test was performed to decide if it was possible to execute the action and reach the final state of the process afterwards. Successful tests for all actions/transitions yield relaxed soundness. Sara was able to decide relaxed soundness for all of the 590 nets together (510 were relaxed sound) in 198 seconds, which makes about a third of a second per net. One business process was especially hard and took 12278 calls to lp_solve and 24 seconds before a decision could be made. LoLA was unable to solve 17 of the problems (including the one mentioned above) and took 24 minutes for the remaining 573.
- Four Petri nets derived in the context of verifying parameterized boolean programs (and published on a web page [6]) were presented to us to decide coverability. Sara needed less than one time slice of the CPU per net and solved all instances correctly. LoLA was not able to find the negative solution to one of the problems due to insufficient memory (here, tests were made with up to 32GB RAM), the remaining three problems were immediately solved.
- In 2003, H. Garavel [5] proposed a challenge on the internet to check a Petri net derived from a LOTOS specification for dead (i.e. never firable) transitions. The net consisted of 776 transitions and 485 places, so 776 tests needed to be made. Of the few tools that succeeded, LoLA was the fastest with about 10 minutes, but it was necessary to handle two of the transitions separately with a differently configured version of LoLA. In our setting, seven years later, LoLA needed 41 seconds to obtain the same result. Sara came to the same conclusions in 26 seconds. In most cases the first solution of lp_solve was sufficient, but for some transitions it could take up to 15 calls to lp_solve. Since none of the 776 transitions is dead, Sara also delivered 776 firing sequences to enable the transitions, with an average length of 15 and a longest sequence of 28 transitions. In 2003 the best upper bound for the sequences lengths was assumed to be 35, while LoLA found sequences of widely varying length, though most were shorter than 50 transitions.
- Using specifically constructed nets with increasing arc weights (and token numbers) it was possible to outsmart Sara – the execution times rose exponentially with linearly increasing arc weights, the first five times being 0.1, 3.3, 32, 180, and 699 seconds. LoLA, on the other hand, decided reachability in less than 3 seconds (seemingly constant time) in these cases.

We also checked our heuristics from Sec. 5 with some of the above nets by switching the former off and comparing the results (see Table 1). Our implementation needs both forms of constraints, jump and increment, to guarantee that all solutions of the state equation can be visited. Going through these solutions in a different order, e.g. the total order Ω, is difficult and a comparison was not possible so far.

Table 1. Results for shutting down one heuristic. Inst. is the number of problem instances to be solved for the net, Sol? the average solution length or "-" if no solution exists. Columns Full, ¬1, ¬2, ¬3/4, and ¬5 contain the result with all optimizations, without stubborn sets, without subtree cutting, without partial solution cutting, and without saving intermediate results (numbers are according to Sec. 5). Each entry shows the elapsed time and the number of necessary CEGAR steps (average), or NR if no result could be obtained in less than a day.

Net	Inst.	Sol?	Full	¬1	¬2	¬3/4	¬5
garavel	776	15	26s (0.11)	25s (0.11)	26s (0.11)	26s (0.11)	26s (0.11)
bad-bp	142	-	24s (85)	24s (85)	24s (85)	24s (85)	NR
good-bp	144	53	1.7s (0)	1.7s (0)	1.7s (0)	1.7s (0)	1.7s (0)
test7	10	175	29s (13)	990s (22)	NR	49s (14)	29s (13)
test8-1	1	40	0.1s (13)	0.35s (22)	49s (13)	0.2s (14)	0.11s (13)
test8-2	1	76	3.3s (21)	24s (51)	NR	11s (34)	3.8s (21)
test8-3	1	112	32s (27)	390s (80)	NR	175s (71)	33s (27)
test9	1	-	0.4s (53)	22s (464)	NR	NR	0.9s (65)

The nets tested fall in two categories. Garavel's net and the business processes are extensive nets with a low token count and without much concurrency that could be tackled by partial order reduction. The heuristics have no effect here, short runtimes result from finding a good solution to the state equation early on. Only for the hardest of the business processes (bad-bp) memorizing intermediate results to avoid checking the same partial solution over and over made sense – without it we did not get a result at all.

The other category are compact nets. In our test examples a high number of tokens is produced and then must be correctly distributed, before the tokens can be removed again to produce the final marking. With a high level of concurrency in the nets, partial order reduction is extremely useful, the cutting off of already seen subtrees(2) even more than the stubborn set method(1). In the last net (test9), the sought intermediate token distribution is unreachable but the state equation has infinitely many solutions. Only by cutting off infinite parts of the solution tree with the help of optimization 3 and 4 it becomes possible to solve the problem at all. Without them, the number of outstanding CEGAR steps reaches 1000 within less than a minute and continues to increase monotonically. The algorithm slows down more and more then as the solutions to the state equation and thus the potential firing sequences become larger.

Beyond what other tools can do, namely solving the problem and – in the positive case – present a witness path, i.e. firing sequence, Sara can also provide diagnostic information in the negative case as long as the state equation has a solution. This feature was tested e.g. with the hardest of the 590 business processes from above, which provides such a negative case for some of its 142 transitions. Since we cannot present such a large net here, a condensed version with the same important features is shown in Fig. 5.

Sara provides a partitioning of the net showing where the relaxed soundness test (for any of the transitions k_1, k_2, or x_2) fails, e.g. it is impossible to fire x_2 and afterwards reach the final marking with exactly one token on place o (other places being empty). The solution $d + k_1 + k_2 + x_2$ of the state equation can neither be realized nor extended to a "better" solution. The ascending pattern shows a region of the net (given by Sara)

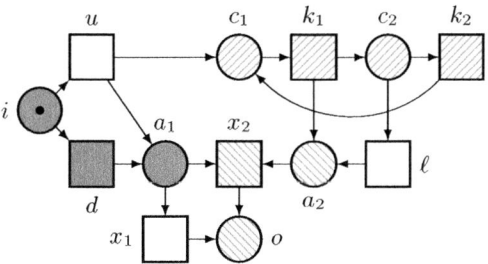

Fig. 5. A condensed, flawed business process. One token should flow from the initial place i to the output place o with all other places empty finally. Non-white transitions appear in Sara's solution to the state equation, but only the dark gray one is firable. Ascending stripes show the area with non-firable transitions where additional tokens could not be generated.

where tokens are needed but cannot be generated without violating the state equation. The descending pattern marks areas that are affected by the former ones, i.e. areas with also non-firable transitions. The gray transition d is the only firable transition occurring in the solution. When analyzing the net we can see that the cycle $c_1 - k_1 - c_2 - k_2$ indeed constitutes a flaw for a business process: if the cycle gets marked and then emptied later, at least two tokens must flow through a_2, one of which can never be removed. Using u instead of d is therefore impossible, i.e. dx_1 is the only firing sequence reaching the final marking.

7   Conclusion

We proposed a promising technique for reachability verification. For reachable problem instances, it tends to yield short witness paths. For unreachable instances, it may terminate early, without an exhaustive search. Furthermore, it may provide some diagnostic information in that case. Our approach applies the concept of counterexample guided abstraction refinement in a novel context: the abstraction is not given as a transition system but as a linear-algebraic overapproximation of the reachable states. In essence, we replace the search in the set of states by the more focussed search through the solutions of the state equation.

The state equation as such has been used earlier for verification purposes, see for instance [4]. In [14], it is used as an initial way of narrowing the state space exploration but not refined according to the CEGAR.

References

1. Berkelaar, M., Eikland, K., Notebaert, P.: lp_solve Reference Guide (2010),
   http://lpsolve.sourceforge.net/
2. Ciardo, G., Marmorstein, R., Siminiceanu, R.: The saturation algorithm for symbolic state space exploration. Software Tools for Technology Transfer 8(1), 4–25 (2006)

3. Clarke, E., Grumberg, O., Jha, S., Lu, Y., Veith, H.: Counterexample-guided abstraction refinement. In: Emerson, E.A., Sistla, A.P. (eds.) CAV 2000. LNCS, vol. 1855. Springer, Heidelberg (2000)
4. Esparza, J., Melzer, S., Sifakis, J.: Verification of safety properties using integer programming: Beyond the state equation. Formal Methods in System Design 16 (2), 159–189 (2000)
5. Garavel, H.: Efficient Petri Net tool for computing quasi-liveness (2003), http://www.informatik.uni-hamburg.de/cgi-bin/TGI/pnml/getpost?id=2003/07/2709
6. Geeraerts, G., Raskin, J.F., Van Begin, L.: Expand, enlarge and check (2010), http://www.ulb.ac.be/di/ssd/ggeeraer/eec/
7. Kosaraju, S.R.: Decidability of reachability in vector addition systems. In: Proceedings of the 14th Annual ACM STOC, pp. 267–281 (1982)
8. Kristensen, L.M., Schmidt, K., Valmari, A.: Question-guided Stubborn Set Methods for State Properties. Formal Methods in System Design 29(3), 215–251 (2006)
9. Lambert, J.L.: A structure to decide reachability in Petri nets. Theoretical Computer Science 99, 79–104 (1992)
10. Leroux, J.: The General Vector Addition System Reachability Problem by Presburger Inductive Invariants. In: Proceedings of the 24th Annual IEEE Symposium on Logic in Computer Science, pp. 4–13. IEEE Computer Society, Los Alamitos (2009)
11. Lipton, R.J.: The Reachability Problem Requires Exponential Space. Research Report 62 (1976)
12. Mayr, E.: An algorithm for the general Petri net reachability problem. SIAM Journal of Computing 13(3), 441–460 (1984)
13. Schmidt, K.: Stubborn sets for standard properties. In: Donatelli, S., Kleijn, J. (eds.) ICATPN 1999. LNCS, vol. 1639, pp. 46–65. Springer, Heidelberg (1999)
14. Schmidt, K.: Narrowing petri net state spaces using the state equation. Fundamenta Informaticae 47(3-4), 325–335 (2001)
15. Talcott, C., Dill, D.: The pathway logic assistent. In: Third International Workshop on Computational Methods in Systems Biology (2005)
16. Wimmel, H.: Sara – Structures for Automated Reachability Analysis (2010), http://service-technology.org/tools/download
17. Wolf, K.: Generating Petri net state spaces. In: Kleijn, J., Yakovlev, A. (eds.) ICATPN 2007. LNCS, vol. 4546, pp. 29–42. Springer, Heidelberg (2007)

Biased Model Checking Using Flows

Muralidhar Talupur[1] and Hyojung Han[2]

[1] Strategic CAD Labs, Intel
[2] University of Colorado at Boulder

Abstract. We describe two new state exploration algorithms, called biased-dfs and biased-bfs, that bias the search towards regions more likely to have error states using high level hints supplied by the user. These hints are in the form of priorities or markings describing which transitions are important and which aren't. We will then describe a natural way to mark the transitions using flows or partial orders on system events. Apart from being easy to understand, flows express succinctly the basic organization of a system. An advantage of this approach is that assigning priorities does not involve low level details of the system. Using flow-derived priorities we study the performance of the biased algorithms in the context of cache coherence protocols by comparing them against standard bfs, dfs and directed model checking. Preliminary results are encouraging with biased-bfs finding bugs about 3 times faster on average than standard bfs while returning shortest counter examples almost always. Biased-dfs on the other hand is couple of orders of magnitude faster than bfs and slightly faster than even standard dfs while being more robust than it.

1 Introduction

We present two new state exploration procedures, called *biased-bfs* and *biased-dfs*, that steer the search towards regions more likely to have error states by using high level user supplied hints. These hints take the form of marking transitions as important or not. In practice, model checking problems often have high level structure that is completely ignored by the standard model checkers. The basic premise behind our work is that this high level information, if used properly, can help model checkers scale to the complex real world systems.

More concretely, consider the class of distributed message passing protocols, an important and industrially relevant class of systems. These protocols are often built around a set of transactions or *flows* [12]. In other words, the protocol follows an implicit set of partial order on actions/events, such as sending and receipt of messages. Even very large industrial protocols have a fairly concise and easy to understand set of flows[1].

Empirically, not all the flows are equally critical to the functioning of a protocol. By identifying the important flows and eagerly exploring those regions of the state space where these flows are active we can increase the chances of hitting deep bugs. The information about important flows is transferred to the model checkers by marking (or assigning higher priorities) to actions involved in those flows.

---

[1] This is what makes designing a distributed protocol 5000 lines long possible in the first place.

P.A. Abdulla and K.R.M. Leino (Eds.): TACAS 2011, LNCS 6605, pp. 239–253, 2011.
© Springer-Verlag Berlin Heidelberg 2011

Given a set of marked transitions biased model checking algorithms work as follows. Biased-bfs, a variant of breadth first search(bfs), begins by exploring the model in standard breadth first manner. But when computing the next frontier, that is, states reachable from the current frontier, it accumulates all the states that have marked transitions enabled. For each such state s, biased-bfs explores the part of state space reachable from s using only the marked transitions. All the states encountered during this specialized sub-search are added to the next frontier and the search continues from the next frontier. As an effect of firing marked transitions eagerly, the states reachable using those transitions are explored much quicker than other states.

While biased-bfs does not assume anything about the underlying system, biased-dfs on the other hand is targeted specifically towards systems composed of multiple agents executing asynchronously. For these systems it helps to not just explore the marked transitions early, but also explore the different inter-leavings of marked transitions early. For instance, suppose we have a system with two agents A and B. A trace $t_1^A, t_2^B, t_3^A, t_4^B, ... t_n^A$ with inter-leaved firing of important transitions of A, B is much more likely to have a bug than a trace $t_1^A, t_2^A, ... t_m^A, t_{m+1}^B, ... t_n^B$ that has a series of marked transitions of A followed by a series of marked transitions of B. To reach such traces quickly, biased-dfs starts like standard dfs but it avoids switching to a different agent unless the currently executing agent has reached a state where it has some marked transitions enabled. That is, biased-dfs forces each agent to a state where it has a marked transition enabled. Once a state s that has a desired percentage of agents with marked actions enabled is reached, a specialized sub search which does a more thorough exploration is invoked on s. Thus, biased-dfs tries to co-ordinate execution of multiple agents. Like biased-bfs, biased-dfs will explore a system exhaustively if let to run to completion.

An advantage of biased model checking is that, unlike heuristic functions in dmc, priority functions in our case do not involve low level details of the system. They simply amount to saying which transitions, more specifically which syntactic *rules* occurring in the protocol description, are marked. As we will demonstrate in Section 6, figuring out which transitions to mark is not hard.

1.1 Applications to Distributed Message Passing Protocols

In this paper we apply the above general algorithms to an important class of systems namely distributed message passing protocols, in particular, cache coherence protocols. For these protocols, we show how to derive markings on transitions using *flows* or partial orders on system events.

As shown in [12,7], these systems have implicit partial orders on system events, like sending and receiving messages, built into them. In fact, the protocol architects explicitly reason in terms of flows and *they are readily available in the design documents.*

Given the flows of a protocol, we can identify which flows are important for finding violations to properties we are checking and within the flows which specific rules or transitions are important. This involves some human insight – just as is in dmc – but since flows are high level artifacts involving only names of *rules* they tend to be far easier to reason about than the protocol itself.

1.2   Evaluation

We introduced a series of bugs into German's and Flash cache coherence protocols and evaluated the biased model checkers by comparing them against bfs, dfs and also dmc.

Our preliminary results indicate that biased-bfs runs around 3 times faster than standard breadth first search on deep bugs and returns shortest counter examples in most cases.

Biased-dfs, is much faster – in many cases up to couple of orders of magnitude faster – than standard bfs and faster than even dfs in most cases. Moreover, unlike dfs the performance of biased-dfs is stable and does not fluctuate too much with the order of transitions in the system.

The rest of the paper is organized as follows. The next section covers the related work and the section after that introduces the preliminary concepts and the system model. Section 4 presents the biased-bfs and Section 5 presents biased-dfs. The section following that presents our experimental results and Section 8 concludes the paper.

2   Related Work

The idea of providing guidance to state exploration is a natural one and several forms of guided model checkers have been proposed over the years, including works by Hajek [4], by Yang and Dill [13], work by Ravi and Somenzi [9] and by Edelkamp et al. on symbolic dmc [10] and on explicit dmc [3,2].

The key idea in dmc, the most well known of the above approaches, is to adapt the A* algorithm to state exploration. The A* algorithm works by reordering the work queue (containing the states to be explored) according a cost function, which is a measure of the "goodness" of the state. For dmc, the user has to provide a cost function that gives for each state an approximate measure of how close it is to a buggy state. Tuning the cost function provides a way for controlling the search. In theory we can use cost functions on states to mark transitions as well (by making the cost function correspond to the guard of a transition). But this will be very tedious in practice as the guards can be complicated.

Another work that prioritizes finding bugs over full coverage is *iterative context bounded execution* [6], which explores a concurrent system starting from traces with the lowest number of context switches and working upwards towards those with more context switches. Biased-dfs generalizes the algorithm in [6]: the main exploration loop of biased-dfs avoids switching contexts (i.e. the executing agent) till it reaches a state s where multiple marked transitions are enabled. Once at s, a different exploration mechanism is invoked that does a thorough exploration of states reachable from s. Thus, the main outer search follows the context switching order but the sub-search does not. If we did not specify any special transitions then biased-dfs would be same as context bounded execution.

Recently randomized exploration routines, such as Cuzz [5], have been successfully applied to concurrent software. The idea is to sample the different inter-leavings (for the same input to the system) efficiently. These methods have not yet been extended to message passing systems as far as we are aware.

BT Murphi [1] also uses the notion of transactions, which are similar to flows, to speed up bug hunting. But the notion of transaction is central to their algorithm which bounds the search by limiting the number and types of transactions active. Biased-bfs and biased-dfs on the other hand do not limit the number of flows active but only guide the search to those regions where more flows are active.

3   Preliminaries

Given a set of state variables V, each system P we consider is given as a collection $\mathcal{R}$ of *guard-action* pairs (called *rules* in Murphi). Each rule is of the form

$$rl : \rho \to a$$

where rl is the name of the rule, ρ is a condition on the current state and action a is a piece of sequential code specifying next state values of the variables in V. A subset of these rules are *marked* to indicate that they can be used to bias the state exploration.

For biased-dfs, we require that P be composed of a collection of asynchronously executing agents, not necessarily all identical, with indices from an ordered set $\mathcal{I}$. The examples considered in this paper are directory based cache coherence protocols, which consist of a collection of asynchronously executing (and identical) caching agents. Thus, the set of state variables of V is $\bigcup_i V_i$ where V_i is the state variables of agent i.

Semantically viewed, each system P is a quadruple of form (S, I, T, T^*) where S is a collection of states, $I \subseteq S$ is the set of initial states and $T \subseteq S \times S$ is the transition relation. In addition to these standard components, a subset $T^* \subset T$ consisting of *marked* transitions is also specified. Each transition $t \in T$ corresponds to unique a $rl \in \mathcal{R}$.

For the biased-dfs procedure the underlying system P is a collection of agents. Thus, $S = \Pi_{i \in \mathcal{I}} S_i$ where each S_i is the set of states of agent i. We can then identify, for agent i, the set $T_i \subseteq T$ of transitions that change the state of agent i. Thus, $T = \bigcup_i T_i$. To keep the presentation of the biased-dfs algorithm simple, we assume that the $T_i's$ are disjoint, that is, each transition in T can affect the state of only one agent. Extension of our work to the general case is not difficult.

We denote by *next* the successor function that takes a collection of states $S' \subseteq S$ and a collection of transitions $T' \subseteq T$ and returns all states reachable from S' in one step using transitions from T'. If the set of transitions is not specified it is assumed to be T and we write $next(S')$ instead of $next(S', T)$. In case of singleton sets, we just write $next(s)$ instead of $next(\{s\}, T)$ and $next(s, t)$ instead of $next(\{s\}, \{t\})$.

4   Biased BFS Algorithm

Biased-bfs works by trying to reduce the number of times bfs has to compute the next frontier. The size of the new frontier in bfs increases very rapidly with the depth of the search and if biased-bfs is able to reduce this depth by even a small number, say 3 or 4, the savings in terms of states explored and time would be significant. This reduction in depth is achieved by eagerly exploring marked transitions before exploring others.

Given a system $P = (S, I, T, T^*)$ biased bfs works as follows: the main outer exploration loop, shown in Figure 1, does standard breadth first search starting with the initial set of states and computing the next frontier $next\_front$ from the current frontier $curr\_front$. While $next\_front$ is being computed biased-bfs collects states that have marked transitions enabled into another queue $front_{red}$. Before going onto the next frontier, the outer loop invokes a reduced breadth first search bfs_{red} on $front_{red}$ which does bfs search using only marked transitions. The set of states reached by this reduced search are added to the next frontier and the outer loop then proceeds as before.

Note that the search frontier during biased-bfs does not grow by the shortest path order but parts of state space reachable using the marked transitions are explored faster than the rest of the state space. Another way to view is to consider the marked transitions as having 0 weight while computing the path length. With this definition of shortest path the states are indeed explored according to the shortest path order.

The reduction in the search depth is obtained at the expense of making $next\_front$ larger (by adding states returned by bfs_{red}). Having more marked transitions means more transitions in a counter example trace are likely to be marked, which means it will be found in fewer iterations of the main **while** loop in Figure 1. But, as the number of markings goes up, the number of states bfs_{red} returns will also go up. This would make $next\_front$ large and negate the benefits of reduction in search depth. Thus, we have to mark as few rules as possible while marking as many of the important ones, that is, the ones that are likely to be present on counter example traces, as possible.

Theorem 1. *Biased bfs algorithm given above terminates for finite systems and fully explores the state space.*

Note that biased-bfs makes no assumption about the underlying system except for assuming that some of the transitions are marked. If no transitions are marked then it reduces to normal bfs and if all transitions are marked then its equivalent to doing bfs twice.

4.1 Optimizations

There are two different ways of realizing the basic outline presented above: we can either accumulate all the states in current frontier that have marked transitions enabled and call reduced bfs on them in one step (as shown in Figure 1) or call reduced bfs on the individual states as and when they are discovered. We found that the version presented is slightly faster than the greedy version.

Adding $front_{add}$ to $next\_front$ can make it grow quickly, especially if there are many states with marked rules enabled in the current frontier. This growth can be controlled by limiting how many states are added to $front_{red}$ and indirectly limiting the size of $front_{add}$. Remarkably, we found during our experiments that limiting the size of $front_{red}$ to just 5 states gives all the benefits of biasing!

5   Biased DFS Algorithm

A straight-forward analogue of biased-bfs would be to modify standard dfs so that it fires marked rules first before firing un-marked rules. This would be equivalent to sorting the list of transitions so that marked rules appear before unmarked rules. But most

Function $bfs_{red}(front_{red})$ {
  $curr\_front_{red}= front_{red}$
  $next\_front_{red}= \phi$
  $front_{return}= front_{red}$
  while $(curr\_front_{red}\neq \phi)$ **do**
    while $(curr\_front_{red}\neq \phi)$ **do**
      pick s from $curr\_front_{red}$
      if (s $\notin visited_{red}$) **then**
        add s to $visited_{red}$ and
          to $front_{return}$
        successors_set = next(s,$trans_{red}$)
        add successors_set
          to $next\_front_{red}$
      end if
    end while
    $curr\_front_{red}:= next\_front_{red}$
  end while
  return $front_{return}$
}

Procedure biased-bfs ( ) {
  $visited= \phi$
  {Set of visited states}
  $curr\_front=$ I
  {Queue holding states in the current
    frontier. Holds initial states at start}

$next\_front= \phi$
{Queue holding states in the
  next frontier}
$visited_{red}= \phi$
$curr\_front_{red}= \phi$
$next\_front_{red}= \phi$
while $(curr\_front\neq \phi)$ **do**
  while $(curr\_front\neq \phi)$ **do**
    pick s from $curr\_front$
    if (s $\notin visited$) **then**
      successors_set = next(s)
      add successors_set
        to $next\_front$
      if (s has marked transitions
        enabled) **then**
        add s to $front_{red}$
      end if
    end if
  end while
  $front_{add}= bfs_{red}(front_{red})$
  add $front_{add}$to the front
    of $next\_front$
  $curr\_front:= next\_front$
  $next\_front:= \phi$
end while
}

Fig. 1. Biased-bfs algorithm

bugs in concurrent protocols happen when multiple processes & multiple flows are active in a trace and this procedure would only ensure that marked rules for an agent get fired before other rules but would not ensure co-ordination between agents.

Consider for instance a system S with three processes A, B and C. Consider a state (a^*, b, c), where a, b, c are local states of A, B, C and $*$ denotes that agent A has a marked transition enabled. Naive biasing will result in marked transitions of A being greedily fired first. Consequently, the traces we will see first will have a stretch of marked transitions from A perhaps followed by a stretch of marked transitions from B and so on. The traces with tight interleaving of marked transitions from A, B, C won't be seen until the former traces have been explored whereas the correct thing to do would be to explore the interleaved traces first.

Biased-dfs presented in Figure 2 rectifies this by actively seeking out and exploring regions where multiple processes with marked rules enabled are present. It accomplishes this by having two sub-routines, namely, *bounded-execution*, which seeks out states with multiple processes having marked rules enabled, and *dfs-explore*, which does a thorough exploration once such a state is found.

To see how *bounded-execution* works consider a state (a_0, b_0, c_0) in which none of the processes has a marked transition enabled. *Bounded-execution* starts executing transitions of A until it reaches a state (a_m^*, b_0, c_0) where A has a marked rule enabled. Now it switches to executing transitions from the next process[2], B in this case, until the system reaches a state (a_m^*, b_m^*, c_0) where A, B both have marked transitions enabled. Depending on how many processes in marked states we consider sufficient, *bounded-execution* can either switch to executing transitions from process C or call *dfs-explore* to thoroughly explore inter-leavings of A, B starting from (a_m^*, b_m^*, c_0). Once *dfs-explore* is invoked on a state s it explores the descendants of s using all the transitions stopping only at those states in which none of the processes has any marked transitions enabled. All these border states at which *dfs-explore* stops are transferred over to *bounded-execution* which again tries to find interesting states starting from them.

Note that the set of transitions *bounded-execution* picks to explore at a state depends on the process it is currently executing. So in addition to storing states in the visited state set (*visited* in Figure 2) we also need to store for which ids a state can be considered visited. Thus, at a state $s = (a_m, b_n, c_k)$ if *bounded-execution* is currently executing transitions from A then (s, A) is added to *visited*. The exploration of transitions of B, C starting from s is deferred to later by adding $(s, B), (s, C)$ to $next_{stage}$. Intuitively, (s, B) and (s, C) would correspond to switching contexts at s and these are deferred to later.

Unlike *bounded-execution, dfs-explore* considers all the transitions enabled at a state. If any of them is a marked transition then all the transitions are taken. If it reaches a states $s = (a_m, b_n, c_k)$ where none of the processes has a marked transition enabled then *dfs-explore* adds $(s, A), (s, B)$ and (s, C) to $curr_{stage}$ for *bounded-execution* to explore further. Since *dfs-explore* does not track which process it is executing, once it is done all the states it visited will be added to *visited* paired with all possible ids.

Finally, the main procedure *biased-dfs* in Figure 2 initializes $curr_{stage}$ with all initial states paired with all process ids and invokes *bounded-execution* on them. Note that if no transition is marked then *dfs-explore* never gets invoked and the resulting search is just context bounded execution [6][3].

Theorem 2. *Biased dfs terminates and explores all the states of a given model.*

6   Flows and Markings

For an important class of systems, namely distributed message passing protocols, we show a simple way to derive markings on transitions by considering *flows* [12]. Recall that the systems we consider are described as a collection of guarded action *rules*. A flow [12] is simply a partial order on interesting system events or rule firings.

Figure 3 shows a couple of flows from German's cache coherence protocol. The *ReqShar* flow on left says that first event is agent i sending a request for shared access (by firing rule *SndReqS*). The second event is directory receiving that request (by firing *RecvReqS*), followed by directory granting the access to i (by firing *SndGntS*). The flow terminates with i firing *RecvGntS* to receive the grant message. The *ReqExcl* flow on right i similar.

---

[2] Bounded execution also switches to a different process if the current process reaches a deadend state.

[3] If all the transitions are marked then biased-dfs reduces to plain dfs.

Function bounded-execution (st, pid) {
  if ((st,pid) ∈ *visited*) **then**
    do nothing
  else
    if (number of processes with marked tran-
    sitions enabled in st crosses threshold)
      then dfs-explore(st)
    else
      let $trans_{en}$= all transitions enabled
        for process pid in state st
      if ($trans_{en}= \phi$) **then**
        if (pid is not the last id in $\mathcal{I}$ and
          next id is pid')
          for all(id ∈ *pids*   and
            id *notin* {pid,pid'}) **do**
            add (st,id) to $next_{stage}$
          end for
          add (st,pid) to *visited*
          bounded-execution(st,pid')
        else
          return
        end if
      else
        for all(id ∈ *pids*) **do**
          add (st,id) to $next_{stage}$
        end for
        add (st,pid) to *visited*
        for all(t ∈ $trans_{en}$) **do**
          nxt_st = next(st,t)
          bounded-execution(nxt_st,pid)
        end for
      end if
    end if
  end if
}

Function dfs-explore (st) {
  if (∀ id. (st,id) ∉ *visited*) **then**
    if (no process has a marked transition
      enabled in state st) **then**

      for all(id ∈ *pids*) **do**
        add (st,id) to $curr_{stage}$
      end for
    else
      for all(id ∈ *pids*) **do**
        add (st, id) to *visited*
      end for
      let $trans_{en}$= enabled transitions
        in state st
      for all(t ∈ $trans_{en}$) **do**
        nxt_st = next(st,t)
        dfs-explore(nxt_st)
      end for
    end if
  end if
}

Procedure biased-dfs ( ) {
  $visited= \phi$
  {Set of (state,id) pairs visited so far}
  $curr_{stage}= \phi$
  {Queue of (state,id) pairs to explore
  in the current stage}
  $next_{stage}= \phi$
  {Queue of pairs deferred to the next
    stage}
  for all(id ∈ *pids*) **do**
    for all(st_{init}∈ I) **do**
      add (st_{init}, id) to $curr_{stage}$
    end for
  end for
  while ($curr_{stage}\neq \phi$) **do**
    while ($curr_{stage}\neq \phi$) **do**
      pick (st,id) from $curr_{stage}$
      bounded-execution(st,id)
    end while
    $curr_{stage}= next_{stage}$
    $next_{stage}= \phi$
  end while
}

Fig. 2. Biased-dfs algorithm

It is easy to see that flows capture succinctly and intuitively the basic organization of
the protocol in terms of events such as sending and receiving of messages. In fact the
protocol designers explicitly use the flows during design and for the distributed protocols we are interested *the full set of flows are already available in design documents*.
So there is no effort on our part in coming up with the flows.

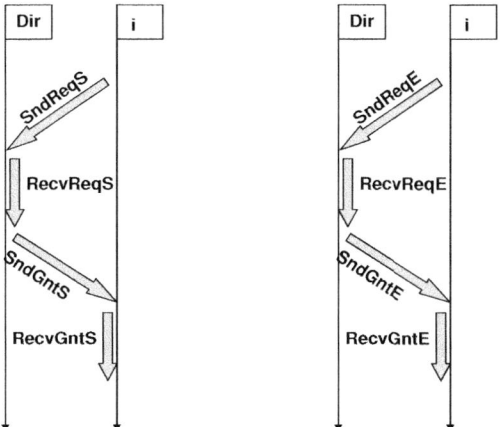

Fig. 3. Linear sequence events illustrating *ReqShar* (left) and *ReqExcl* (right) flows in German's protocol

Finally, even a large industrial protocol has only about 40 different flows and the full set of flows is orders of magnitude smaller than the protocol itself. Reasoning in terms of flows consequently tends to be a lot simpler than reasoning in terms of the protocol variables. A full treatment of flows is beyond the scope of this paper and the reader is referred to [12,7] for more information. For the rest of this section, we will assume we have the full set of flows available.

Using these flows, we identify which rules to mark as important and this in turn gives us markings for transitions with the interpretation that if a rule rl is marked then all the transitions in T corresponding to rl are marked.

Note that more the number of marked rules greater the overhead on biased model checkers. So we should to mark only a small number to reduce the overhead while ensuring that as many of the "important" ones as possible are marked. To mark rules we proceed backwards from the property which are verifying . The property we considered was the standard MESI property

$$\forall i, j.i \neq j \Rightarrow (Excl[i] \Rightarrow \neg(Shar[j] \vee Excl[j]))$$

where $Shar[i]$ mean i is in *shared* state and similarly for $Excl[i]$.

Markings for biased-bfs. Given flows and the property to be verified, we identify the flows that are likely to involved in the potential violations of the property. In our case, a counter example to safety property is most likely to involve the flow for getting exclusive access as at least one agent must be in exclusive state. For German's protocol this is $ReqExcl$ flow and the rules present in it, namely, $RecvReqE$, $SendGntE$ and $RecvGntE$ are marked. We did not mark the first rule, $SndReqS$, for sending request, because this rule can fire without any constraint, and the core of the flow begins with directory receiving the request.

For Flash protocol, $ni\_local\_getx\_putx$ (corresponding to the directory receiving a request for exclusive and sending a grant message) and ni_remote_putx (corresponding to an agent receiving a grant exclusive message) were marked. The first rule for sending the request is not marked again. Thus, for German's protocol 3 out of 12 rules were marked and for Flash 2 out 33 were marked. The Murphi models of the protocols along with the bugs we introduced are available at [11].

Markings for biased-dfs. Markings are derived slightly differently in this case. Here the focus is on multi-agent interactions that are likely to expose safety violations. So we identify flows whose interaction is likely to cause property violations. In our case, the violation to mutual exclusion will most likely involve flows for gaining exclusive and shared accesses, we mark rules appearing in these flows. For German's this means, in addition to the above marking, we should also mark $RecvReqS$, $SndGntS$, $RecvGntS$ and for Flash $ni\_local\_get\_put$ and ni_remote_put in addition to the ones above. To keep the experimental section brief we will use the markings used for biased-bfs for biased-dfs as well. Even with these sub-optimal markings biased-dfs performs extremely well.

Finally, note that to derive markings we are reasoning based on the property and not based on any specific bug. Though deriving markings requires user intuition, having flows makes the job quite a lot simpler.

7   Experimental Evaluation

We wrote prototype implementations of the biased procedures and also of the bfs and dfs routines in OCaml. While OCaml is relatively slow and not amenable to low level optimizations required to make model checkers run fast, it is adequate to study the relative performance of the biased procedures versus the standard procedures. There were no optimizations except for symmetry reduction which was common to all procedures. The input protocols models were also in OCaml – they were translated from Murphi to OCaml automatically using a translator we wrote.

To study the performance of the biased procedures, we introduced bugs into German's and Flash protocols. These are standard examples in protocol verification and are non-trivial protocols – in terms of lines of Murphi code German's protocol is about 150 lines and Flash about 1000 lines. The bugs were introduced consecutively and they all led to violations of the MESI property.

7.1   Results

Tables 1 and 2 summarize the results of our experiments for German's and Flash protocols respectively. Each entry is either of the form trace-length/time-in-seconds or gives the number of states in thousands (k) rounded to the nearest integer. Apart from considering different bugs, we also considered systems with 4, 5, and 6 agents to study the scaling behavior of the biased procedures. The columns *bbfs*, *bdfs* and *dmc* stand for biased-bfs, biased-dfs and directed model checking respectively. Fastest running times are shown in bold and the second fastest times are italicized. All the experiments were run on a 2.6 GHz and 8GB Intel Xeon machine with a time out of 10 mins.

Performance of bfs. Standard bfs finishes on all the examples and in one case (Flash bug 3) it is close to being fastest. Apart from returning the shortest counter example always, performance of bfs is predictable and does not vary much with the order in which the transitions are presented. The markings on transitions have no effect on the performance of bfs.

Performance of dfs. Standard dfs, when it finishes within the time out, finds bugs extremely quickly but the counter examples it returns are extremely long. The other major disadvantage of dfs is that its performance fluctuates wildly with the order in which the transitions are listed. For instance, when we reverse the order of the transitions for Flash protocol, bugs 1 and 2 were caught by dfs in under a second with relatively short counter examples (< 25 states long). But performance on bug 3 deteriorated dramatically, with dfs timing out on 6 agent system and on 5 agent system it returned a counter example that was more than 50K states long after 130 seconds.

Performance of biased-bfs. Biased-bfs also finishes on all the examples within time out and it returns the shortest counter examples in all the cases. On two bugs (Flash bugs 1 and 2) it is the second fastest.

Significantly, biased-bfs outperforms standard bfs on the deeper bugs (bugs 1 and 4 in German and bugs 1 and 2 in Flash) by a factor of around three and this factor gets better as the number of processes increases. On shallower bugs standard bfs is faster than biased bfs.

This result validates the basic premise behind biased-bfs: reducing the number of frontier expansions required by standard bfs search leads to bugs quicker but at the cost of increased frontier sizes. Thus, for deeper bugs we see big gains for biased-bfs but not so much for shallower bugs, where the increase in frontier sizes seems to outweigh the decrease in the exploration depth.

While depth-first routines outperform biased-bfs, performance of biased-bfs is more robust and predictable with all the bugs being found within the time out. Further, counter examples returned by biased-bfs are considerably shorter. In fact, on all the examples, biased bfs returns the shortest counter examples (though this need not be the case always).

Performance of biased-dfs. Biased-dfs is the fastest in most cases except for bug 2 in German (where it is the second fastest) and bug 3 in Flash. It finishes within time out on all cases except Flash bug 3 on system with 6 agents.

Further, biased-dfs is less fickle than standard dfs. To see if the order of transitions has any effect on biased-dfs, we reversed the transitions lists in both German and Flash protocol. Biased-dfs continued taking around the same time as before, indicating that it is relatively immune to changes in the transition order.

The only disadvantage of biased-dfs is that it does not return short counter examples.

Performance of dmc. The original paper on dmc [3] presented several different variations of dmc with varying performance different examples. We implemented the straight forward adaptation of A* algorithm. The priority of each state is given by 100 minus

Table 1. Results for German's protocol

Bug	procs	Trace length and run times					States explored		
		bfs	bbfs	dfs	bdfs	dmc	bfs	bbfs	dmc
1	4	12/7.8	12/2.8	70/*0.06*	>75/**0.05**	12/24	19k	6k	25k
	5	12/38.2	12/10.6	86/*0.09*	>91/**0.07**	12/111	60k	15k	97k
	6	12/249.3	12/55.6	102/*0.16*	>107/**0.1**	t.o	172k	37k	-
2	4	9/1.0	9/1.6	34/**0.05**	>23/*0.05*	13/6	2k	3k	9.2k
	5	9/3.14	9/5.4	35/**0.05**	>24/*0.06*	10/23.5	5k	8k	34k
	6	9/14.6	9/24.5	36/**0.11**	>25/*0.15*	13/135	10k	17k	118k
3	4	9/1.1	9/1.6	58/*0.06*	>28/**0.05**	13/4.4	2k	3k	7k
	5	9/3.5	9/5.3	71/*0.09*	>28/**0.06**	11/16	5k	8k	25k
	6	9/16.7	9/24.5	84/*0.15*	>28/**0.1**	14/82	12k	17k	92k
4	4	12/9.3	12/3.3	240/*0.17*	>31/ **0.1**	14/29	22k	7k	34k
	5	12/45.6	12/12.5	256/*0.22*	>31/ **0.13**	13/147	69k	17k	133k
	6	12/283.5	12/62.1	272/*0.36*	>31/ **0.2**	t.o	193k	40k	-

the number marked transitions enabled at the state. Thus, larger the number of marked transitions enabled the lower the priority and the earlier the state will be picked for consideration. As Tables 1 and 2 show dmc comes last in almost all the examples. The first four routines, bfs, bbfs, dfs, bdfs all use the same data structure to store the states and perform similar state related operations. So the time taken is also a good indication of number of states explored. But the underlying data structure used by dmc[4] is different and to get another view we compare the number of states explored by bbfs and dmc. We don't show numbers for dmc as in almost all the cases the entry for bdfs would be 0k after rounding to nearest integer. As the last two columns of the tables show dmc explores a lot more states than bbfs before finding the bug[5].

7.2 Biased Procedures in Murphi

Encouraged by the performance of OCaml implementations, we recently modified the Murphi model checker to perform biased-bfs, biased-dfs. This speeds up the performance considerably and also allows us to access a wider set of models.

We tried out modified Murphi on a large micro-architectural model of the core-ring interface (CRI) used in one of the multi-processors developed at Intel. The model, written in Murphi, captures the various in and out buffers, arbiters and other control structures managing the traffic between a core and the communication ring. The model also includes L1 and L2 caches of the core and is parameterized by the number of addresses

---

[4] It requires a structure for storing sets of states that supports membership queries and minimum priority extraction.

[5] The amount of computation for each visited state also differ with dmc taking the most resources.

Table 2. Results for Flash protocol

Bug	procs	Trace length and run times					States Explored		
		bfs	bbfs	dfs	bdfs	dmc	bfs	bbfs	dmc
1	4	9/2.0	9/*0.9*	t.o	>13/**0.04**	>8/2.6	2k	0k	3k
	5	9/7.3	9/*2.5*	t.o	>13/**0.05**	>8/7.4	3k	1k	8k
	6	9/23.7	9/*7.6*	t.o	>13/**0.1**	>8/30	7k	2k	22k
2	4	9/5.1	9/*1.8*	t.o	>23/**0.04**	>5/4.8	3k	1k	5.5k
	5	9/21.4	9/*5.8*	t.o	>23/**0.07**	>5/15.4	9k	2k	15k
	6	9/77.6	9/*20.6*	t.o	>23/**0.15**	>5/65	21k	5k	45k
3	4	7/*0.54*	7/1.4	119/**0.12**	>11K/6.7	7/1.1	0k	1k	1k
	5	7/*1.5*	7/4.6	476/**0.43**	>34K/ 34.9	7/3.3	1k	1k	3k
	6	7/*4.6*	7/13.5	1996/**2.8**	t.o	7/13	3k	3k	9k

in the system. The property we considered was the inclusion property, namely, if a data item is present in L1 cache then it is present in the L2 cache as well.

We asked the designer who wrote the model to specify 4 rules that he thought were most relevant to the inclusion property and also to introduce a bug into the model. Coming up with 4 rules took the designer, who did not know the details of our approach, just couple of minutes.

Table 3 shows the result of this experiment. We ran bfs, dfs and biased-bfs on models with 4, 6 and 8 addresses with a time out of 20 hrs. Biased-dfs is not applicable to this model which deals with a single agent system. The table shows the number of states explored by bfs, dfs and biased-bfs. With the exception of the model with 4 addresses, bfs runs out of memory on all cases (after exploring considerably more states than biased-bfs) even when Murphi is given 12GB of memory. Interestingly, dfs hits the bug fairly quickly after exploring between 2-4M states but it fails to generate a counter example trace even after 20 hrs. This is because Murphi has to undo the symmetry reduction applied during the search phase to generate a counter example and for large models with deep counter examples this can be a very expensive operation[6]. In contrast, even for the largest model, biased-bfs generates a counter example trace in under 3.5 hrs after exploring less than 20M states. This result confirms that biased model checking does scale to industrial strength systems.

7.3   Discussion

Experimental results presented above amply demonstrate that the biased procedures can be very effective in catching bugs. When dealing with systems with asynchronously executing agents, biased-dfs is probably the best choice for checking existence of bugs. Random simulation is often used during early debugging but, at least for concurrent

---

[6] This is a relatively unknown facet of dfs that only people who have dealt with large models seem to be aware of.

Table 3. Results for CRI model

Addr	bfs	dfs	bbfs
4	16M	t.o	12M
6	>25M	t.o	17M
8	>25M	t.o	18M

protocols, they are good only at finding shallow bugs and quickly get overwhelmed by the non-determinism.

For debugging, which requires short counter example traces, biased-bfs seems to be the best bet, especially for deep bugs. Apart from getting significant speed up compared to standard bfs and also standard dfs on some examples, biased bfs returns more or less the shortest counter example.

8   Conclusion

We have described two new biased model checking algorithms that use high level information implicit in protocols to reach error states faster. The mechanism of marking transitions as important or not is used to pass the high level information in flows to the biased model checkers.

Our early results indicates that the direction holds promise in tackling large model checking problems. A natural extension of this work is to combine priorities on transitions with priorities on states. This will provide us finer control on the exploration process potentially leading to an even more powerful method.

For instance, consider communication fabrics on a modern multi-core processor. One worrying scenario is the *high water-mark* scenario where all the buffers are full, potentially resulting in a deadlock. To bias the search towards such scenarios we can mark all those actions that *add* or *delete* some entry from a buffer. This will bias the search towards the corner cases where the buffer is full or empty. To gain finer control we can make the marking dependent on the state as well. Thus, an add action will be marked only in states where the buffer is already close to being full and a delete action will be marked only for those states where the buffer is close to empty. This will ensure that we hit the corner case scenarios and potential bugs even more quickly.

References

1. Chen, X., Gopalakrishnan, G.: Bt: a bounded transaction model checking for cache coherence protocols. Technical report UUCS-06-003, School of Computing, University of Utah (2006)
2. Edelkamp, S., Leue, S., Lluch-Lafuente, A.: Directed Explicit state model checking in validation of Communication protocols. In: Proceedings of STTT (2004)

3. Edelkamp, S., Lluch-Lafuente, A., Leue, S.: Protocol Verification with Heuristic Search. In: AAAI Symposium on Model-based validation of Intelligence (2001)
4. Hajek, J.: Automatically verified data transfer protocols. In: Proceedings of the International Computer Communications Conference (1978)
5. Musuvathi, M., Burckhardt, S., Kothari, P., Nagarakatte, S.: A randomized scheduler with probabilistic guarantees of finding bugs. In: ASPLOS (2010)
6. Musuvathi, M., Qadeer, S.: Iterative context bounding for systematic testing of multithreaded programs. In: PLDI (2007)
7. O'Leary, J., Talupur, M., Tuttle, M.R.: Protocol Verification using Flows: An Industrial Experience. In: Proc. FMCAD (2009)
8. Qadeer, S., Wu, D.: Kiss: keep it simple and sequential. In: PLDI (2004)
9. Ravi, K., Somenzi, F.: Hints to accelerate Symbolic Traversal. In: Proceedings of the 10th IFIP WG 10.5 Advanced Research Working Conference on Correct Hardware Design and Verification Methods (1999)
10. Reffel, F., Edelkamp, S.: Error detection with directed symbolic model checking. In: Woodcock, J.C.P., Davies, J. (eds.) FM 1999. LNCS, vol. 1708, pp. 195–211. Springer, Heidelberg (1999)
11. Talupur, M.: Murphi models of German's and Flash protocols along with the bugs, www.cs.cmu.edu/~tmurali/fmcad10
12. Talupur, M., Tuttle, M.R.: Going with the Flow: Parameterized Verification using Message Flows. In: Proc. FMCAD (2008)
13. Yang, C.H., Dill, D.L.: Validation with guided search of the state space. In: Proceedings of the 35th Annual Design Automation Conference (1998)

S-TaLiRo: A Tool for Temporal Logic Falsification for Hybrid Systems*

Yashwanth Annapureddy[1], Che Liu[1],
Georgios Fainekos[1], and Sriram Sankaranarayanan[2]

[1] Arizona State University, Tempe, AZ
{Yashwanthsingh.Annapureddy,Che.Liu,fainekos}@asu.edu
[2] University of Colorado, Boulder, CO
srirams@colorado.edu

Abstract. S-TaLiRo is a Matlab (TM) toolbox that searches for trajectories of minimal robustness in Simulink/Stateflow diagrams. It can analyze arbitrary Simulink models or user defined functions that model the system. At the heart of the tool, we use randomized testing based on stochastic optimization techniques including Monte-Carlo methods and Ant-Colony Optimization. Among the advantages of the toolbox is the seamless integration inside the Matlab environment, which is widely used in the industry for model-based development of control software. We present the architecture of S-TaLiRo and its working on an application example.

1 Introduction

Temporal verification involves the ability to prove as well as falsify temporal logic properties of systems. In this paper, we present our tool S-TaLiRo for temporal logic falsification. S-TaLiRo searches for counterexamples to *Metric Temporal Logic* (MTL) properties for non-linear hybrid systems through global minimization of a *robustness metric* [4]. The global optimization is carried out using stochastic optimization techniques that perform a random walk over the initial states, controls and disturbances of the system. In particular, the application of *Monte-Carlo techniques* that use sampling biased by robustness is described in our HSCC 2010 paper [6]. In [1], we report on our experience with other optimization techniques including *Ant-Colony Optimization*.

At its core, S-TaLiRo integrates robustness computation for traces of hybrid systems (TaLiRo) [4,6] with stochastic simulation [9]. The search returns the simulation trace with the smallest robustness value that was found. In practice, traces with negative robustness are falsifications of temporal logic properties. Alternatively, traces with positive - but low - robustness values are closer in distance to falsifying traces using a mathematically well defined notion of distance

---

* This work was partially supported by a grant from the NSF Industry/University Cooperative Research Center (I/UCRC) on Embedded Systems at Arizona State University and NSF awards CNS-1017074 and CNS-1016994.

P.A. Abdulla and K.R.M. Leino (Eds.): TACAS 2011, LNCS 6605, pp. 254–257, 2011.
© Springer-Verlag Berlin Heidelberg 2011

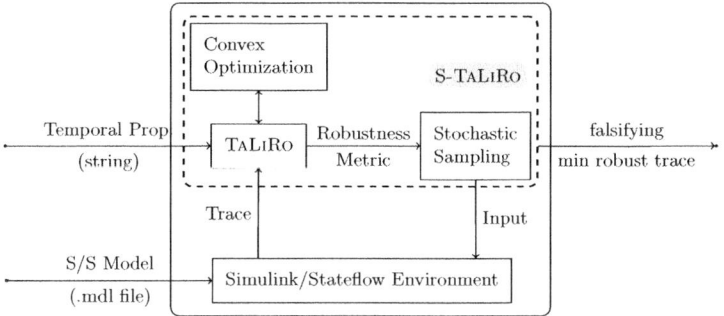

Fig. 1. The architecture of the S-TaLiRo tool

between trajectories and temporal logic properties. Such traces may provide valuable insight to the developer on why a given property holds or how to refocus a failed search for a counter-example.

S-TaLiRo supports systems implemented as Simulink/Stateflow (TM) models as well as general *m-functions* in Matlab. Other frameworks can be readily supported, provided that a Matlab (TM) interface is made available to their simulators. S-TaLiRo has been designed to be used by developers with some basic awareness of temporal logic specifications. Simulink/Stateflow (TM) models are the *de-facto* standard amongst developers of control software in many domains such as automotive control and avionics. S-TaLiRo also supports the easy input of MTL formulae through an in-built parser. It has been designed and packaged as a Matlab toolbox with a simple command line interface. S-TaLiRo also contains an optimized implementation of the computation of the robustness metric (TaLiRo) over the previous version [3] along with the ability to plug-in other stochastic optimization algorithms.

2   The S-TaLiRo Tool

Figure 1 shows the overall architecture of our toolbox. The toolbox consists of a temporal logic robustness analysis engine (TaLiRo) that is coupled with a stochastic sampler. The sampler suggests input signals/parameters to the Simulink/Stateflow (TM) simulator which returns an execution trace after the simulation. The trace is then analyzed by the robustness analyzer which returns a robustness value. The robustness is computed based on the results of convex optimization problems used to compute signed distances. In turn, the robustness score computed is used by the stochastic sampler to decide on a next input to analyze. If a falsifying trace is found in this process, it is reported to the user. The trace itself can be examined inside the Simulink/Stateflow modeling environment. If, on the other hand, the process times out, then the least robust trace found by the tool is output for user examination.

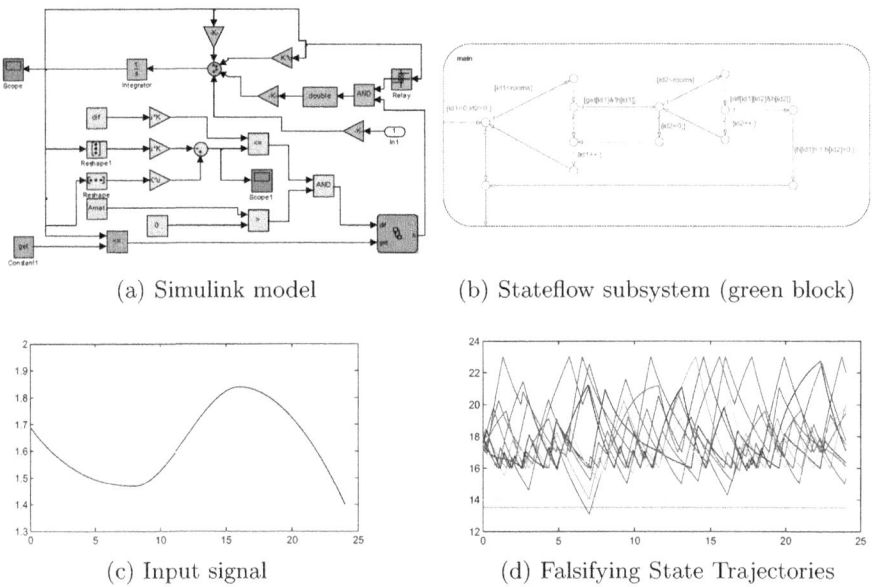

(a) Simulink model   (b) Stateflow subsystem (green block)

(c) Input signal   (d) Falsifying State Trajectories

Fig. 2. Room heating benchmark HEAT30 and results obtained from S-TaLiRo run

3   Usage

S-TaLiRo has been designed to be seamlessly integrated in the model based design process of Matlab/Simulink (TM). The user designs the model in the Simulink/Stateflow (TM) environment as before. At present, the only requirement is that input signals must be provided to the Simulink model through input ports. Then S-TaLiRo is executed with the name of the Simulink model as a parameter along with the set of initial conditions, the constraints on the input signals (if any) and the MTL specification. Currently, the user may select one of the two available stochastic optimization algorithms: Monte Carlo or Ant Colony Optimization. However, the architecture of S-TaLiRo is modular and, thus, any other stochastic optimization method can be readily implemented.

As a demonstration, we applied S-TaLiRo to the room heating benchmark from [5] (see Fig. 2). We chose the benchmark instance HEAT30. This is a hybrid system with 10 continuous variables (10 rooms) and 3360 discrete locations $(\binom{10}{4})2^4$ where 4 is the number of the heaters). The set of initial conditions is $[17, 18]^{10}$ and input signal u can range in $[1, 2]$. The goal is to verify that no room temperature drops below $[14.50\ 14.50\ 13.50\ 14.00\ 13.00\ 14.00\ 14.00\ 13.00\ 13.50\ 14.00]^T$. The input signal was parameterized using a piecewise cubic Hermite interpolating polynomial with 4 control points evenly distributed in the simulation time. S-TaLiRo found a falsifying trace with robustness value of -0.429. Figure 2 shows the trace and the input signal discovered by S-TaLiRo. In detail, the initial conditions were $x_0 = [17.4705\ 17.2197\ 17.0643\ 17.8663\ 17.4316\ 17.5354\ 17.9900\ 17.6599\ 17.8402\ 17.2036]^T$.

4   Related Work

The problem of testing hybrid systems has been investigated by many researchers (see the related research section in [6]). Most of the research focuses on *parameter estimation* [8,2]. Recently, however, the problem of temporal logic falsification for hybrid systems has received some attention [7,6]. Unfortunately, the publicly available tool support has been fairly low in this space. The only other publicly available toolbox that supports computation of robustness for temporal logic formulas with respect to real-valued signals is BREACH [2]. However, BREACH currently does not support temporal logic falsification for arbitrary Simulink/Stateflow models. Along the lines of commercial products, Mathworks provides a number of tools such as SystemTest[1] (TM) and Simulink Design Verifier[2] (TM). S-TaLiRo does not attempt to be a comprehensive test tool suite as the above, but rather to solve a very targeted problem, i.e., the problem of temporal logic falsification for hybrid systems. In the future, we hope to extend S-TaLiRo and the theory of robustness to estimate properties such as worst-case timings and integrate it into the statistical model checking framework.

References

1. Annapureddy, Y.S.R., Fainekos, G.: Ant colonies for temporal logic falsification of hybrid systems. In: Proceedings of the 36th Annual Conference of IEEE Industrial Electronics (2010)
2. Donzé, A.: Breach, A toolbox for verification and parameter synthesis of hybrid systems. In: Touili, T., Cook, B., Jackson, P. (eds.) CAV 2010. LNCS, vol. 6174, pp. 167–170. Springer, Heidelberg (2010)
3. Fainekos, G.E., Pappas, G.J.: A user guide for TaLiRo. Technical report, Dept. of CIS, Univ. of Pennsylvania (2008)
4. Fainekos, G.E., Pappas, G.J.: Robustness of temporal logic specifications for continuous-time signals. Theoretical Computer Science 410(42), 4262–4291 (2009)
5. Fehnker, A., Ivančić, F.: Benchmarks for hybrid systems verification. In: Alur, R., Pappas, G.J. (eds.) HSCC 2004. LNCS, vol. 2993, pp. 326–341. Springer, Heidelberg (2004)
6. Nghiem, T., Sankaranarayanan, S., Fainekos, G., Ivančić, F., Gupta, A., Pappas, G.: Monte-Carlo techniques for the falsification of temporal properties of non-linear systems. In: Hybrid Systems: Computation and Control, pp. 211–220. ACM Press, New York (2010)
7. Plaku, E., Kavraki, L.E., Vardi, M.Y.: Falsification of LTL Safety Properties in Hybrid Systems. In: Kowalewski, S., Philippou, A. (eds.) TACAS 2009. LNCS, vol. 5505, pp. 368–382. Springer, Heidelberg (2009)
8. Rizk, A., Batt, G., Fages, F., Soliman, S.: On a continuous degree of satisfaction of temporal logic formulae with applications to systems biology. In: Heiner, M., Uhrmacher, A.M. (eds.) CMSB 2008. LNCS (LNBI), vol. 5307, pp. 251–268. Springer, Heidelberg (2008)
9. Rubinstein, R.Y., Kroese, D.P.: Simulation and the Monte Carlo Method. Wiley Series in Probability and Mathematical Statistics (2008)

---

[1] http://www.mathworks.com/products/systemtest/
[2] http://www.mathworks.com/products/sldesignverifier/

GAVS+: An Open Platform for the Research of Algorithmic Game Solving

Chih-Hong Cheng[1], Alois Knoll[1], Michael Luttenberger[1], and Christian Buckl[2]

[1] Department of Informatics, Technische Universität München
Boltzmann Str. 3, Garching D-85748, Germany
[2] Fortiss GmbH, Guerickestr. 25, D-80805 München, Germany
`http://www6.in.tum.de/~chengch/gavs`

Abstract. This paper presents a major revision of the tool GAVS. First, the number of supported games has been greatly extended and now encompasses in addition many classes important for the design and analysis of programs, e.g., it now allows to explore concurrent / probabilistic / distributed games, and games played on pushdown graphs. Second, among newly introduced utility functions, GAVS+ includes features such that the user can now process and synthesize planning (game) problems described in the established STRIPS/PDDL language by introducing a slight extension which allows to specify a second player. This allows researchers in verification to profit from the rich collection of examples coming from the AI community.

1 Introduction

We present a major revision of the open-source tool GAVS[1], called GAVS+, which now targets to serve as an open platform for the research community in algorithmic game solving. In addition, GAVS+ is meant to serve as a playground for researchers thinking of mapping interesting problems to game solving.

GAVS+ has three main goals: (i) support of game types currently under active research; many times an implementation of a solver is only hard to come by, or only partial implementations exist; (ii) support of different input and output formats in order to allow for both interoperability with other tools and for easy access of existing collections of models, examples, and test cases in concrete application domains; (iii) ease of use by a unified graphical user interface (GUI) which allows to graphically specify the game and explore the computed solution. The last requirement is partially fulfilled by the previous version of GAVS+: the GUI allows to visualize two-player, turn-based games on finite graph, solve the game, and store intermediate results in order to visualize the algorithms step-by-step. This also makes it a very useful tool for teaching these algorithms. In this release, we have concentrated on goals (i) and (ii).

Regarding goal (i), we have added support for several games of practical relevance: we have introduced support for stochastic [10], concurrent [6], distributed [9] games, as well as games played on pushdown graphs [2]. We opted for these games as they arise

---

[1] Short for "Game Arena Visualization and Synthesis".

P.A. Abdulla and K.R.M. Leino (Eds.): TACAS 2011, LNCS 6605, pp. 258–261, 2011.

Game type (visualization)	Implemented algorithms
Fundamental game	Symbolic: (Co-)reachability, Büchi, Weak-parity, Staiger-Wagner
	Explicit state: Parity (discrete strategy improvement)
	Reduction: Muller, Streett
Concurrent game	**Sure reachability, Almost-sure reachability, Limit-sure reachability**
Pushdown game[‡]	**Reachability (positional min-rank strategy, PDS strategy),**
	Büchi (positional min-rank strategy), Parity (reduction)
Distributed game	**Reachability (bounded distributed positional strategy for player-0)**
Markov decision process	**Policy iteration, Value iteration, Linear programming (LP)**
Simple stochastic game	**Shapley (value iteration), Hoffman-Karp (policy iteration)**

Fig. 1. Game types and implemented algorithms in GAVS+ (games marked bold are extensions to the previous version [3]), where "[‡]" indicates that visualization is currently not available

either naturally in the context of parallel resp. recursive programs, or a widely used for systems with uncertainties.

As to goal (ii), one feature amongst newly introduced utilities in GAVS+ is the ability to be applied to existing planning problems formalized in PDDL (level 1; the STRIPS fragment) [8]. We consider it as an interesting feature, as it allows researchers in the verification community to access the huge collection of planning problems coming from the artificial intelligence (AI) and the robotics community.

Although a large collection of PDDL specific solvers is available, only one tool [7] is based on a similar approach combining planning and games. Unfortunately, this tool is not publicly available. Furthermore, the goal of [7] is to "study the application and extension of planning technology for general game playing" whereas our goal is rather to allow researchers in the verification community to profit from the rich collection of models coming from the AI community.

In the rest of the paper, first we give a very brief overview on the newly supported games, and then illustrate our integration and extension of STRIPS/PDDL into GAVS+. Due to space limit we only outline two tiny examples using GAVS+; for details and complete functionalities we refer interested readers to our website.

2   Supported Games in GAVS+

For a complete overview on all supported games we refer the reader to Figure 1, here, we only give a brief description of the newly added games.

Concurrent games [6] are used to capture the condition when the next location is based on the combined decision simultaneously made by control and environment. When considering randomized strategies in reachability games (CRG), efficient algorithms to compute sure, almost-sure, and limit-sure winning regions are available [6].

Stochastic games model systems with uncertainty. The classical model of *Markov decision process (MDP, $1\frac{1}{2}$-player game)* [12] is widely used in economics and machine learning and considers a single player who has to work against an environment exhibiting random behavior. Adding an opponent to MDPs, one obtains *stochastic ($2\frac{1}{2}$-player) games* [10]. Currently we focus on the subclass of simple stochastic games (SSG) [5]; many complicated games can be reduced to SSGs or solved by algorithms similar to algorithms solving SSG. For implemented algorithms for MDP and SSG, we refer readers to two survey papers [12,5] for details.

Games on pushdown graphs (APDS) arise naturally when recursive programs are considered. Symbolic algorithms exist for reachability and Büchi winning conditions [2], and for parity conditions, a reduction[2] to two-player, finite-state parity games based on summarization is possible [2]. An example for the interactive reachability simulation of an APDS using GAVS+ is illustrated with Figure 2: the user can act as the role of player-1 (environment) by selecting the rewriting rule, while GAVS+ updates the cost for the positional min-rank strategy and output the next move for player-0 (control).

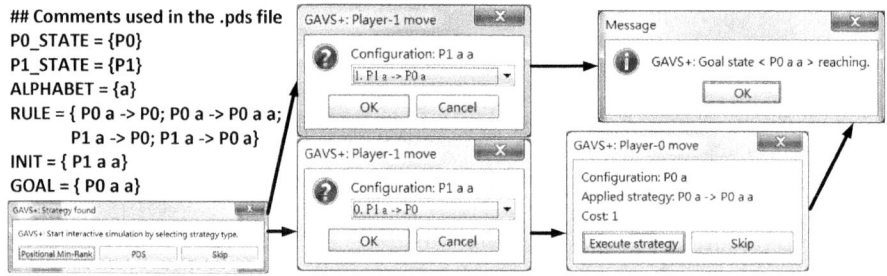

Fig. 2. An APDS in textual form and screenshots of the interactive simulation using GAVS

Distributed games [9] are games formulating multiple processes with no interactions among themselves but only with the environment. Generating strategies for such a game is very useful for distributed systems, as a strategy facilitates orchestration of interacting components. Although the problem is undecidable in general [9], finding a distributed positional strategy for player-0 of such a system ($PositionalDG_0$), if it exists, is a practical problem. As $PositionalDG_0$ is NP-complete for reachability games [4], we modify the SAT-based witness algorithms in [1] and implement a distributed version [4] for bounded reachability games.

3   Extension of STRIPS/PDDL for Game Solving: An Example

To connect the use of games with concrete application domains, GAVS+ offers the translation scheme from PDDL to symbolic representation of games. Here we consider a simplified scenario of a robot with two humanoid arms[3]. For details (game domain and problem as PDDL files, synthesized strategies), we refer readers to the GAVS+ software package in the website.

- **(Fault-free planning)** In the case of fault-free planning, GAVS+ reads the normal PDDL file, performs the forward symbolic search and generates a totally ordered plan (action sequence) similar to other tools.

---

[2] Currently, as the algorithm introduces an immediate exponential blowup in the graph, it is difficult to solve the game using the built-in algorithm specified in [11].

[3] Motivated by the scenario in our chair: http://www6.in.tum.de/Main/ResearchEccerobot

- **(Fault-tolerant strategy generation)** We outline steps for game creation and solving when faults are introduced in GAVS+ using this example.

  1. *(Fault modeling)* When an arm is out-of-service, the tension over the artificial muscle is freed, and the object under grasp falls down to the ground (fault-effect). It is also assumed that for the robot, at most one arm can be out-of-service during its operation cycle (fault-frequency). The above behavior can be modeled as a fault action in the PDDL domain.

  2. *(State Partitioning)* Currently in GAVS+, we simply introduce a binary predicate `POTRAN` on each action in the domain to partition player-0 and player-1 states and transitions.

  3. *(Synthesis)* By modeling the domain and problem using PDDL, GAVS+ synthesizes strategies to perform tasks while resisting the potential loss of one arm: the strategy is a FSM outputted to a separate file using Java-like formats.

References

1. Alur, R., Madhusudan, P., Nam, W.: Symbolic computational techniques for solving games. International Journal on Software Tools for Technology Transfer (STTT) 7(2), 118–128 (2005)
2. Cachat, T.: Games on Pushdown Graphs and Extensions. PhD thesis, RWTH Aachen (2003)
3. Cheng, C.-H., Buckl, C., Luttenberger, M., Knoll, A.: GAVS: Game arena visualization and synthesis. In: Bouajjani, A., Chin, W.-N. (eds.) ATVA 2010. LNCS, vol. 6252, pp. 347–352. Springer, Heidelberg (2010)
4. Cheng, C.-H., Rueß, H., Knoll, A., Buckl, C.: Synthesis of fault-tolerant embedded systems using games: From theory to practice. In: Jhala, R., Schmidt, D. (eds.) VMCAI 2011. LNCS, vol. 6538, pp. 118–133. Springer, Heidelberg (2011)
5. Condon, A.: On algorithms for simple stochastic games. Advances in Computational Complexity theory 13, 51–73 (1993)
6. De Alfaro, L., Henzinger, T., Kupferman, O.: Concurrent reachability games. Theoretical Computer Science 386(3), 188–217 (2007)
7. Edelkamp, S., Kissmann, P.: Symbolic exploration for general game playing in pddl. In: ICAPS-Workshop on Planning in Games (2007)
8. Fox, M., Long, D.: Pddl2.1: An extension to pddl for expressing temporal planning domains. Journal of Artificial Intelligence Research 20(1), 61–124 (2003)
9. Mohalik, S., Walukiewicz, I.: Distributed games. In: Pandya, P.K., Radhakrishnan, J. (eds.) FSTTCS 2003. LNCS, vol. 2914, pp. 338–351. Springer, Heidelberg (2003)
10. Shapley, L.: Stochastic games. Proceedings of the National Academy of Sciences of the United States of America 39, 1095 (1953)
11. Vöge, J., Jurdziński, M.: A discrete strategy improvement algorithm for solving parity games. In: Emerson, E.A., Sistla, A.P. (eds.) CAV 2000. LNCS, vol. 1855, pp. 202–215. Springer, Heidelberg (2000)
12. White, C., et al.: Markov decision processes. European Journal of Operational Research 39(1), 1–16 (1989)

Büchi Store: An Open Repository of Büchi Automata*

Yih-Kuen Tsay, Ming-Hsien Tsai, Jinn-Shu Chang, and Yi-Wen Chang

Department of Information Management, National Taiwan University, Taiwan

Abstract. We introduce the Büchi Store, an open repository of Büchi automata for model-checking practice, research, and education. The repository contains Büchi automata and their complements for common specification patterns and numerous temporal formulae. These automata are made as small as possible by various construction techniques, in view that smaller automata are easier to understand and often help in speeding up the model-checking process. The repository is open, allowing the user to add new automata or smaller ones that are equivalent to some existing automaton. Such a collection of Büchi automata is also useful as a benchmark for evaluating complementation or translation algorithms and as examples for studying Büchi automata and temporal logic.

1 Introduction

Büchi automata [1] are finite automata operating on infinite words. They play a fundamental role in the automata-theoretic approach to linear-time model checking [20]. In the approach, model checking boils down to testing the emptiness of an intersection automaton $A \cap B_{\neg\varphi}$, where A is a Büchi automaton modeling the system and $B_{\neg\varphi}$ is another Büchi automaton representing all behaviors not permitted by a temporal specification formula φ. In general, for a given system, the smaller $B_{\neg\varphi}$ is, the faster the model-checking process may be completed.

To apply the automata-theoretic approach, an algorithm for translating a temporal formula into an equivalent Büchi automaton is essential. There has been a long line of research on such translation algorithms, aiming to produce smaller automata. According to our experiments, none of the proposed algorithms outperforms the others for every temporal formula tested. The table below shows a comparison of some of the algorithms for three selected cases.

Formula	LTL2AUT[4]		Couvreur[3]		LTL2BA[5]		LTL2Buchi[6]		Spin[7]	
	state	tran.	state	tran.	state	tran.	state	tran.	state	tran.
$\neg p\,\mathcal{W}\,q$	4	16	3	12	3	12	3	12	4	16
$\square(p \to \Diamond q)$	4	30	3	20	6	41	3	20	4	28
$\Diamond\square p \vee \Diamond\square q$	8	38	5	28	5	28	5	28	3	12

---

* This work was supported in part by National Science Council, Taiwan, under the grant NSC97-2221-E-002-074-MY3.

P.A. Abdulla and K.R.M. Leino (Eds.): TACAS 2011, LNCS 6605, pp. 262–266, 2011.

Given that smaller automata usually expedite the model-checking process, it is certainly desirable that one is always guaranteed to get the smallest possible automaton for (the negation of) a specification formula. One way to provide the guarantee is to try all algorithms or even manual construction and take the best result. This simple-minded technique turns out to be feasible, as most specifications use formulae of the same patterns and the tedious work of trying all alternatives needs only be done once for a particular pattern instance.

To give the specification as a temporal formula sometimes may not be practical, if not impossible (using quantification over propositions). When the specification is given directly as an automaton, taking the complement of the specification automaton becomes necessary. Consequently, in parallel with the research on translation algorithms, there has also been substantial research on algorithms for Büchi complementation. The aim again is to produce smaller automata.

Several Büchi complementation algorithms have been proposed that achieve the lower bound of $2^{\Omega(n \log n)}$ [11]. However, the performances of these "optimal" algorithms differ from case to case, sometimes quite dramatically. The table below shows a comparison of some of the complementation algorithms for four selected Büchi automata (identified by equivalent temporal formulae). In the literature, evaluations of these algorithms usually stop at a theoretical-analysis level, partly due to the lack of or inaccessibility to actual implementations. This may be remedied if a suitable set of benchmark cases becomes available and subsequent evaluations are conducted using the same benchmark.

Formula	Safra[13] state tran.	Piterman[12] state	tran.	Rank-Based[9,14] state	tran.	Slice-Based[8] state	tran.
$\Box(p \to \Diamond(q \land \Diamond r))$	76 662	90	777	96	917	219	2836
$\Diamond\Box(\Diamond p \to q)$	35 188	13	62	13	72	24	119
$\Box(p \to p\,\mathcal{U}\,(q\,\mathcal{U}\,r))$	17 192	8	76	7	54	7	49
$p\,\mathcal{U}\,q \lor p\,\mathcal{U}\,r$	5 34	5	34	8	23	3	12

The Büchi Store was thus motivated and implemented as a website, accessible at http://buchi.im.ntu.edu.tw. One advantage for the Store to be on the Web is that the user always gets the most recent collection of automata. Another advantage is that it is easily made open for the user to contribute better (smaller) automata. The initial collection contains over six hundred Büchi automata. In the following sections we describe its implementation and main features, suggest three use cases, and then conclude by highlighting directions for improvement.

2   Implementation and Main Features

The basic client-server interactions in accessing the Büchi Store are realized by customizing the CodeIgniter [2], which is an open-source Web application framework. To perform automata and temporal formulae-related operations, such as equivalence checking and formula to automaton translation, the Store relies on the GOAL tool [19] and its recent extensions. One particularly important (and highly nontrivial) task is the classification of temporal formulae that identify the Büchi automata in the Store into the Temporal Hierarchy of Manna and Pnueli [10]. To carry out the task automatically, we implemented the classification algorithm described in the same paper, which is based on characterization of a Streett automaton equivalent to the temporal formula being classified.

The main features of the current Büchi Store include:

- **Search:** Every automaton in the Store is identified by a temporal formula (in a variant of QPTL [15,16], which is expressively equivalent to Büchi automata). The user may find the automata that accept a particular language by posing a query with an equivalent temporal formula. Propositions are automatically renamed to increase matches (semantic matching between whole formulae is not attempted due to its high cost). This is like asking for a translation from the temporal formula into an equivalent Büchi automaton. A big difference is that the answer automata, if any, are the best among the results obtained from a large number of translation algorithms, enhanced with various optimization techniques such as simplification by simulation [17] or even manually optimized (and machine-checked for correctness).
- **Browse:** The user may browse the entire collection of Büchi automata by having the collection sorted according to temporal formula length, number of states, class in the Temporal Hierarchy, or class in the Spec Patterns [18]. While classification in the Temporal Hierarchy has been automated, the classification for the last sorting option has not. Rather, the Store relies on the user to provide suggestions, based on which a final classification could be made. This may be useful for educational purposes.
- **Upload:** The user may upload a Büchi automaton for a particular temporal formula. The automaton is checked for correctness, i.e., if it is indeed equivalent to the accompanying temporal formula. If it is correct and smaller than the automata for the formula in the Store, the repository is updated accordingly, keeping only the three smallest automata.

3   Use Cases

We describe three cases that we expect to represent typical usages of the Store.

- **Linear-time model checking:** The user may shop in the Store for the automata that are equivalent (with probable propositions renaming) to the negations of the temporal formulae which he wants to verify. The automata may be downloaded in the PROMELA format, for model checking using SPIN.
- **Benchmark cases for evaluating complementation algorithms:** Every Büchi automaton in the initial collection has a complement, which is reasonably well optimized. A subset of the collection could serve as a set of benchmark cases for evaluating Büchi complementation algorithms. This use case can certainly be adapted for evaluating translation algorithms.
- **Classification of temporal formulae:** The look of a temporal formula may not tell immediately to which class it belongs in the Temporal Hierarchy. It should be educational to practice on the cases that do not involve the

complication of going through Streett automata. For example, $\Box(p \to \Diamond q)$ is a recurrence formula because it is equivalent to $\Box\Diamond(\neg p \, \mathcal{B} \, q)$ (where $\mathcal{B}$ means "back-to", the past version of wait-for or weak until).

Concluding Remarks. To further improve the Store, first of all, we as the developers will continue to expand the collection, besides hoping for the user to do the same. Explanatory descriptions other than temporal formulae should be helpful additions for searching and understanding. Automatic classification of temporal formulae into the various specification patterns should also be useful.

References

1. Büchi, J.R.: On a decision method in restricted second-order arithmetic. In: Int'l Congress on Logic, Methodology and Philosophy of Science, pp. 1–11 (1962)
2. CodeIgniter, http://codeigniter.com/
3. Couvreur, J.M.: On-the-fly verification of linear temporal logic. In: Woodcock, J.C.P., Davies, J. (eds.) FM 1999. LNCS, vol. 1708, pp. 253–271. Springer, Heidelberg (1999)
4. Daniele, M., Giunchiglia, F., Vardi, M.Y.: Improved automata generation for linear temporal logic. In: Halbwachs, N., Peled, D.A. (eds.) CAV 1999. LNCS, vol. 1633, pp. 249–260. Springer, Heidelberg (1999)
5. Gastin, P., Oddoux, D.: Fast LTL to Büchi automata translation. In: Berry, G., Comon, H., Finkel, A. (eds.) CAV 2001. LNCS, vol. 2102, pp. 53–65. Springer, Heidelberg (2001)
6. Giannakopoulou, D., Lerda, F.: From states to transitions: Improving translation of LTL formulae to Büchi automata. In: Peled, D.A., Vardi, M.Y. (eds.) FORTE 2002. LNCS, vol. 2529, pp. 308–326. Springer, Heidelberg (2002)
7. Holzmann, G.J.: The SPIN Model Checker: Primer and Reference Manual. Addison-Wesley, Reading (2003)
8. Kähler, D., Wilke, T.: Complementation, disambiguation, and determinization of Büchi automata unified. In: Aceto, L., Damgård, I., Goldberg, L.A., Halldórsson, M.M., Ingólfsdóttir, A., Walukiewicz, I. (eds.) ICALP 2008, Part I. LNCS, vol. 5125, pp. 724–735. Springer, Heidelberg (2008)
9. Kupferman, O., Vardi, M.Y.: Weak alternating automata are not that weak. ACM Transactions on Computational Logic 2(3), 408–429 (2001)
10. Manna, Z., Pnueli, A.: A hierarchy of temporal properties. In: PODC, pp. 377–408. ACM, New York (1990)
11. Michel, M.: Complementation is more difficult with automata on infinite words. In: CNET, Paris (1988)
12. Piterman, N.: From nondeterministic Büchi and Streett automata to deterministic parity automata. In: LICS, pp. 255–264. IEEE, Los Alamitos (2006)
13. Safra, S.: On the complexity of ω-automta. In: FOCS, pp. 319–327. IEEE, Los Alamitos (1988)
14. Schewe, S.: Büchi complementation made tight. In: STACS, pp. 661–672 (2009)
15. Sistla, A.P.: Theoretical Issues in the Design and Verification of Distributed Systems. PhD thesis, Harvard (1983)

16. Sistla, A.P., Vardi, M.Y., Wolper, P.: The complementation problem for Büchi automata with applications to temporal logic. TCS 49, 217–237 (1987)
17. Somenzi, F., Bloem, R.: Efficient Büchi automata from LTL formulae. In: Emerson, E.A., Sistla, A.P. (eds.) CAV 2000. LNCS, vol. 1855, pp. 248–263. Springer, Heidelberg (2000)
18. The Spec Patterns repository, http://patterns.projects.cis.ksu.edu/
19. Tsay, Y.-K., Chen, Y.-F., Tsai, M.-H., Chan, W.-C., Luo, C.-J.: GOAL extended: Towards a research tool for omega automata and temporal logic. In: Ramakrishnan, C.R., Rehof, J. (eds.) TACAS 2008. LNCS, vol. 4963, pp. 346–350. Springer, Heidelberg (2008)
20. Vardi, M.Y., Wolper, P.: An automata-theoretic approach to automatic program verification. In: LICS, pp. 332–344. IEEE, Los Alamitos (1986)

QUASY: Quantitative Synthesis Tool

Krishnendu Chatterjee[1], Thomas A. Henzinger[1,2],
Barbara Jobstmann[3], and Rohit Singh[4]

[1] Institute of Science and Technology Austria, Austria
[2] École Polytechnique Fédéral de Lausanne, Switzerland
[3] CNRS/Verimag, France
[4] Indian Institute of Technology Bombay, India

Abstract. We present the tool QUASY, a quantitative synthesis tool. QUASY takes qualitative and quantitative specifications and automatically constructs a system that satisfies the qualitative specification and optimizes the quantitative specification, if such a system exists. The user can choose between a system that satisfies and optimizes the specifications (a) under all possible environment behaviors or (b) under the most-likely environment behaviors given as a probability distribution on the possible input sequences. QUASY solves these two quantitative synthesis problems by reduction to instances of 2-player games and Markov Decision Processes (MDPs) with quantitative winning objectives. QUASY can also be seen as a game solver for quantitative games. Most notable, it can solve lexicographic mean-payoff games with 2 players, MDPs with mean-payoff objectives, and ergodic MDPs with mean-payoff parity objectives.

1 Introduction

Quantitative techniques have been successfully used to measure quantitative properties of systems, such as timing, performance, or reliability (cf. [1,9,2]). We believe that quantitative reasoning is also useful in the classically Boolean contexts of verification and synthesis because they allow the user to distinguish systems with respect to "soft constraints" like robustness [4] or default behavior [3]. This is particularly helpful in synthesis, where a system is automatically derived from a specification, because the designer can use soft constraints to guide the synthesis tool towards a desired implementation.

QUASY[1] is the first synthesis tool taking soft constraints into account. Soft constraints are specified using quantitative specifications, which are functions that map infinite words over atomic propositions to a set of values. Given a (classical) qualitative specification φ and a quantitative specification ψ over signals $\mathcal{I} \cup \mathcal{O}$, the tool constructs a reactive system with input signals $\mathcal{I}$ and output signals $\mathcal{O}$ that satisfies φ (if such a system exists) and optimizes ψ either under the worse-case behavior [3] or under the average-case behavior [7] of the environment. The average-case behavior of the environment can be specified by a probability distribution μ of the input sequences.

In summary, QUASY is the first tool for quantitative synthesis, both under adversarial environment as well as probabilistic environment. The underlying techniques to achieve

---

[1] http://pub.ist.ac.at/quasy/

P.A. Abdulla and K.R.M. Leino (Eds.): TACAS 2011, LNCS 6605, pp. 267–271, 2011.

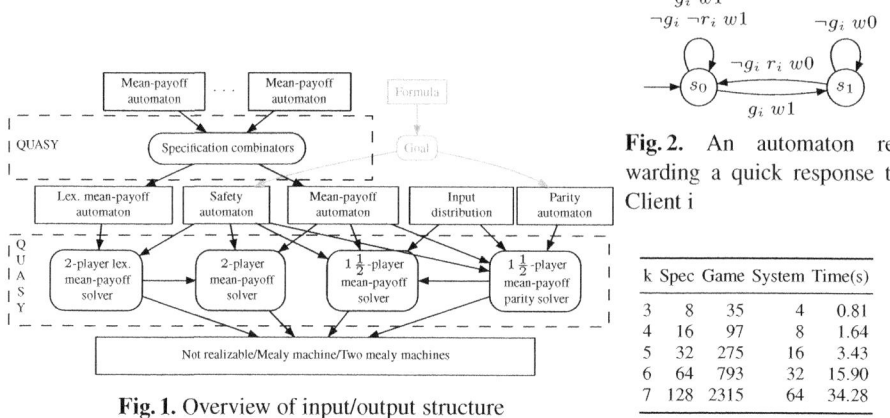

Fig. 2. An automaton rewarding a quick response to Client i

Fig. 1. Overview of input/output structure

k	Spec	Game	System	Time(s)
3	8	35	4	0.81
4	16	97	8	1.64
5	32	275	16	3.43
6	64	793	32	15.90
7	128	2315	64	34.28

Fig. 3. Results for MDPs

quantitative synthesis are algorithms to solve two-player games and MDPs with quantitative objectives. QUASY is the first tool that solves lexicographic mean-payoff games and ergodic MDPs with mean-payoff parity objectives.

2 Synthesis from Combined Specifications

QUASY handles several combinations of qualitative and quantitative specifications. We give a brief description of the input format the tool expects. Then, we summarize the implemented combinations (see Figure 1 for an overview) and give some results.

Specifications. QUASY accepts qualitative specifications given as deterministic safety or parity automaton in GOAL[2] format. LTL properties can be translated into the required format using GOAL. Quantitative properties are specified by (lexicographic) mean-payoff automata. A mean-payoff automaton A is a deterministic automaton with weights on edges that maps a word v to the average over the weights encountered along the run of A on v. Lexicographic mean-payoff automata [3] are a generalization of mean-payoff automata. They map edges to tuples of weights. Figure 2 shows a mean-payoff automaton with alphabet $2^{\{r_i, g_i\}}$ in the accepted format (GOAL). Labels of the form wk, for $k \in \mathbb{N}$, state that the edge has weight k. E.g., the edge from state s_0 to s_1 labeled $\neg g_i \neg r_i \ w1$ states that if r_i and g_i are false, then we can follow this edge and encoder a weight of 1. QUASY can combine a set of mean-payoff automata to (i) a lexicographic mean-payoff automaton or (ii) a mean-payoff automaton representing the sum of the weights.

Optimality. The user can choose between two modes: (1) the construction of a worst-case optimal system, or (2) the construction of an average-case optimal system. In the latter case, the user needs to provide a text file that assigns to each state of the

---

[2] GOAL is a graphical tool for manipulating Büchi automata and temporal formulae. It is available at http://goal.im.ntu.edu.tw/. We use it as graphical front-end for QUASY.

specification a probability distribution over the possible input values. For states that do not appear in the file, a uniform distribution over the input values is assumed.

Combinations and Results. The tool can construct worst-case optimal systems for mean-payoff and lexicographic mean-payoff specifications combined with safety specifications. For the average-case, QUASY accepts mean-payoff specifications combined with safety and parity specifications. Figure 3 shows (in Column 5) the time needed to construct a resource controller for k clients that (i) guarantees mutually exclusive access to the resource and (ii) optimizes the reaction time in a probabilistic environment, in which clients with lower id are more likely to send requests than clients with higher id. The quantitative specifications were built from k copies of the automaton shown in Figure 2. Column 2-4 in Figure 3 show the size of the specifications, the size of the corresponding game graphs, and the size of the final Mealy machines, respectively. These specifications were also used for preliminary experiments reported in [7]. QUASY significantly improves these runtimes, e.g., from 5 minutes to 16 seconds for game graphs with 800 states (6 clients). We provide more examples and results at http://pub.ist.ac.at/quasy/

3   Implementation Details

QUASY is implemented in the programming language SCALA[10] with an alternative mean-payoff MDP solver in C++. It transforms the input specifications into a game graph. States are represented explicitly. Labels are stored as two pairs of sets *pos* and *neg* corresponding to the positive and negative literals, respectively, for both input and output alphabets to facilitate efficient merging and splitting of edges during the game transformations. The games are solved using one of the game solvers described below. If a game is winning for the system, QUASY constructs a winning strategy, transforms it into a reactive system, and outputs it in GOAL format.

Two-Player Mean-Payoff Games. For two-player games with mean-payoff objectives, we have implemented the value iteration algorithm of [12]. The algorithm runs in steps. In every step k, the algorithm computes for each state s the minimal reward r_k player system can obtain within the next k steps starting from state s independent of the choice of the other player. The reward is the sum of the weights seen along the next k steps. The k-*step value* v_k, obtained by dividing the reward r_k by k, converges to the actual value of the game. After $n^3 \cdot W$ steps, where n is the size of the state space and W is the maximal weight, the k-step value uniquely identifies the value of the game [12]. Since the theoretical bound can be large and the value often converges before the theoretical bound is reached, we implemented the following *early stopping criteria*. The actual value of a state is a rational $\frac{e}{d}$ with $d \in \{1, \ldots, n\}$ and $e \in \{1, \ldots, d \cdot W\}$, and it is always in a $\frac{n \cdot W}{k}$-neighbourhood of the approximated value v_k [12]. So, we can fix the value of a state, if, for all d, there is only one integer in the interval $[d \cdot (v_k - \frac{n \cdot W}{k}), d \cdot (v_k + \frac{n \cdot W}{k})]$ and the integers obtained by varying d correspond to the same rational number. Fixing the value of these states, leads to a faster convergence. We implemented this criterion alternatively also by storing all possible values in an array and performing a binary search for a unique value. This method requires more memory but increases the speed of checking convergence.

Two-Player Lexicographic Mean-Payoff Games. For two-player lexicographic mean-payoff games, we have implemented three variants of value iterations. First, a straight forward adaption of the reduction described in [3]: given a lexicographic weight function $\overrightarrow{w}$ with two components $\overrightarrow{w}|_1$ and $\overrightarrow{w}|_2$, we construct a single weight function w defined by $w = c \cdot \overrightarrow{w}|_1 + \overrightarrow{w}|_2$, where the constant $c = n^2 \cdot W + 1$ depends on the maximal weight W in $\overrightarrow{w}|_2$ and the number n of states of the automaton. The other two variants keep the components separately and iterate over tuples. In the second version, we add the tuples component-wise and compare two tuples by evaluating them as a sum of successive division by powers of the base c. This avoids handling large integers but requires high precision. In the third version, we use the following addition modulo c on tuples to obtain a correct value iteration:

$$\begin{pmatrix} a_1 \\ a_2 \end{pmatrix} + \begin{pmatrix} b_1 \\ b_2 \end{pmatrix} = \begin{pmatrix} a_1 + b_1 + ((a_2 + b_2) \operatorname{div} c) \\ (a_2 + b_2) \operatorname{mod} c \end{pmatrix}.$$

We use lexicographic comparison because in many cases we do not need to compare all components to decide the ordering between two tuples. Furthermore, it allows us to handle large state spaces and large weights, which would lead to an overflow otherwise.

MDPs with Mean-Payoff and Mean-Payoff-Parity Objective. For ergodic MDPs with mean-payoff-parity objective, we implemented the algorithm described in [7]. QUASY produces two mealy machines A and B as output: (i) A is optimal wrt the mean-payoff objective and (ii) B almost-surely satisfies the parity objective. The actual system corresponds to a combination of the two mealy machines based on inputs from the environment switching over from one mealy machine to another based on a counter as explained in [7]. For MDPs with mean-payoff, QUASY implements the strategy improvement algorithm (cf. [8], Section 2.4) using two different methods to compute an improvement step of the algorithm: (i) Gaussian elimination that requires the complete probability matrix to be stored in memory (works well for dense and small game graphs) and (ii) GMRES iterative sparse matrix equation solver (works very well for sparse and large game graphs with optimizations as explained in [6]). The strategy for parity objective is computed using a reverse breadth first search from the set of least even-parity states ensuring that in every state we choose an action which shortens the distance to a least even-parity state.

4   Future Work

We will explore different directions to improve the performance of QUASY. In particular, a recent paper by Brim and Chaloupka [5] proposes a set of heuristics to solve mean-payoff games efficiently. It will be interesting to see if these heuristics can be extended to lexicographic mean-payoff games. Furthermore, Wimmer et al. [11] recently developed an efficient technique for solving MDP with mean-payoff objectives based on combining symbolic and explicit computation. We will investigate if symbolic and explicit computations can be combined for MDPs with mean-payoff parity objectives as well.

References

1. Baier, C., Haverkort, B.R., Hermanns, H., Katoen, J.-P.: Performance evaluation and model checking join forces. Commun. ACM 53(9) (2010)
2. Behrmann, G., Bengtsson, J., David, A., Larsen, K.G., Pettersson, P., Yi, W.: UPPAAL implementation secrets. In: Formal Techniques in Real-Time and Fault Tolerant Systems (2002)
3. Bloem, R., Chatterjee, K., Henzinger, T.A., Jobstmann, B.: Better quality in synthesis through quantitative objectives. In: Bouajjani, A., Maler, O. (eds.) CAV 2009. LNCS, vol. 5643, pp. 140–156. Springer, Heidelberg (2009)
4. Bloem, R., Greimel, K., Henzinger, T.A., Jobstmann, B.: Synthesizing robust systems. In: FMCAD (2009)
5. Brim, L., Chaloupka, J.: Using strategy improvement to stay alive. CoRR, 1006.1405 (2010)
6. Černý, P., Chatterjee, K., Henzinger, T., Radhakrishna, A., Singh, R.: Quantitative synthesis for concurrent programs. Technical Report IST-2010-0004, IST Austria (2010)
7. Chatterjee, K., Henzinger, T.A., Jobstmann, B., Singh, R.: Measuring and synthesizing systems in probabilistic environments. In: Touili, T., Cook, B., Jackson, P. (eds.) CAV 2010. LNCS, vol. 6174, pp. 380–395. Springer, Heidelberg (2010)
8. Feinberg, E.A., Shwartz, A.: Handbook of Markov Decision Processes: Methods and Applications. Springer, Heidelberg (2001)
9. Hinton, A., Kwiatkowska, M., Norman, G., Parker, D.: PRISM: A tool for automatic verification of probabilistic systems. In: Hermanns, H. (ed.) TACAS 2006. LNCS, vol. 3920, pp. 441–444. Springer, Heidelberg (2006)
10. The Scala programming language, http://www.scala-lang.org/
11. Wimmer, R., Braitling, B., Becker, B., Hahn, E.M., Crouzen, P., Hermanns, H., Dhama, A., Theel, O.: Symblicit calculation of long-run averages for concurrent probabilistic systems. In: QEST (2010)
12. Zwick, U., Paterson, M.: The complexity of mean payoff games on graphs. Theor. Comput. Sci. 158(1-2), 343–359 (1996)

Unbeast: Symbolic Bounded Synthesis[*]

Rüdiger Ehlers

Reactive Systems Group
Saarland University

Abstract. We present UNBEAST v.0.6, a tool for synthesising finite-state systems from specifications written in linear-time temporal logic (LTL). We combine bounded synthesis, specification splitting and symbolic game solving with binary decision diagrams (BDDs), which allows tackling specifications that previous tools were typically unable to handle. In case of realizability of a given specification, our tool computes a prototype implementation in a fully symbolic way, which is especially beneficial for settings with many input and output bits.

1 Introduction

Specification engineering is known to be a tedious and error-prone task. During the development of complex systems, the early versions of the specifications for the individual parts of the system often turn out to be incomplete or unrealisable. In order to effectively detect such problems early in the development cycle, specification debugging tools have been developed.

One particularly well-known representative of this class are synthesis tools for reactive systems. These take lists of input and output bits and a temporal logic formula in order to check whether there exists a reactive system reading from the specified input signals and writing to the given output signals that satisfies the specification. In case of a positive answer, they also generate a finite-state representation of such a system. Then, by simulating the resulting system and analysing its behaviour, missing constraints in the specification can often be found.

In this work, we report on the UNBEAST tool, which performs this task for specifications written in linear-time temporal logic (LTL). By combining the merits of the bounded synthesis approach [7] with using binary decision diagrams (BDDs) as the symbolic reasoning backbone and the idea of splitting the specification into safety and non-safety parts, we achieve competitive computation times in the synthesis process. Our approach extracts implementations for realisable specifications in a fully symbolic manner, which is especially fruitful for systems with many input and output bits. As a consequence, even in the development of comparably complex systems that typically fall into this class, our tool is applicable to at least the initial versions of a successively built specification, despite the fact that the synthesis problem is 2EXPTIME-complete.

[*] This work was partially supported by the German Research Foundation (DFG) as part of the Transregional Collaborative Research Center "Automatic Verification and Analysis of Complex Systems" (SFB/TR 14 AVACS).

P.A. Abdulla and K.R.M. Leino (Eds.): TACAS 2011, LNCS 6605, pp. 272–275, 2011.

2   Tool Description

Input language: The tool takes as input XML files that provide all information necessary for the synthesis process. We assume that the specification of the system has the form $(a_1 \wedge \ldots \wedge a_n) \rightarrow (g_1 \wedge \ldots \wedge g_m)$, where each of $a_1, \ldots, a_n, g_1, \ldots, g_m$ is an LTL formula. Such formulas are typical for cases in which a component of a larger system is to be synthesised: the *assumptions* $a_1, \ldots, a_n$ represent the behaviour of the environment that the system can assume, whereas the guarantees $g_1, \ldots, g_n$ reflect the obligations that the system to be synthesised needs to fulfil.

Tool output: In case of a realisable specification, a system implementation in form of a NuSMV [2] model is produced (if wanted). Alternatively, the user has the possibility to run a simulation of a system satisfying the specification, where the user provides the input to the system. If the specification is unrealisable, the roles in the simulation are swapped – the tool then demonstrates interactively which environment behaviour leads to a violation of the specification.

2.1   Technology

The UNBEAST V.0.6 tool implements the synthesis techniques presented in [7,3]. We use the library CUDD V.2.4.2 [8] for constructing and manipulating BDDs during the synthesis process. We constrain its automatic variable reordering feature by enforcing that predecessor and successor variables in the transition relation are pairwise coupled in the ordering.

The first step is to determine which of the given assumptions and guarantees are safety formulas. In order to detect also simple cases of *pathological safety* [6], this is done by computing an equivalent Büchi automaton using the external LTL-to-Büchi converter LTL2BA V.1.1 [5] and examining whether all maximal strongly connected components in the computed automaton do not have infinite non-accepting paths. We take special care of bounded look-ahead safety formulas: these are of the form $G(\psi)$ for the LTL globally operator G and some formula ψ in which the only temporal operator allowed is the next-time operator. They are immediately identified as being safety formulas.

In a second step, for the set of bounded look-ahead assumptions and the set of such guarantees, we build two safety automata for their respective conjunctions. Both of them are represented in a symbolic way, i.e., we allocate predecessor and successor state variables that encode the last few input/output bit valuations and compute a transition relation for this automaton in BDD form. For the remaining safety assumptions and guarantees, safety automata are built by taking the Büchi automata computed in the previous step and applying a subset construction for determinisation in a symbolic manner. For the remaining non-safety parts of the specification, a combined universal co-Büchi automaton is computed by calling LTL2BA again.

In the next phase, the given specification is checked for realisability. Here, for a successively increasing so-called *bound value*, the bounded synthesis approach [7] is performed by building a safety automaton from the co-Büchi automaton

Table 1. Running times (in seconds) of UNBEAST v.0.6 and ACACIA v.0.3.9.9 on the load balancing case study on a Sun XFire computer with 2.6 Ghz AMD Opteron processors running an x64-version of Linux. All tools considered are single-threaded. We restricted the memory usage to 2 GB and set a timeout of 3600 seconds.

Tool	Setting / # Clients	2	3	4	5	6	7	8	9
A		+0.4	+0.5	+0.7	+0.9	+1.4	+2.7	+5.5	+12.7
U+S	1	+0.0	+0.1	+0.0	+0.0	+0.1	+0.1	+0.1	+0.1
U−S		+0.0	+0.0	+0.0	+0.0	+0.0	+0.0	+0.1	+0.1
A		+0.4	+0.4	+0.4	+0.5	+0.6	+0.9	+1.6	+3.0
U+S	$1 \wedge 2$	+0.0	+0.1	+0.0	+0.0	+0.0	+0.0	+0.1	+0.1
U−S		+0.0	+0.0	+0.0	+0.0	+0.0	+0.0	+0.0	+0.1
A	$1 \wedge 2 \wedge 3$	-21.8	-484.3	timeout	timeout	timeout	memout	memout	timeout
U		-0.1	-0.1	-0.1	-0.1	-0.3	-1.0	-6.8	-73.2
A		+0.7	+1.4	+8.5	memout	memout	memout	memout	timeout
U+S	$1 \wedge 2 \wedge 4$	+0.2	+0.3	+1.0	+35.5	+214.1	timeout	timeout	timeout
U−S		+0.1	+0.2	+0.3	+2.5	+3.1	+11.2	+48.3	+386.4
A	$1 \wedge 2 \wedge 4 \wedge 5$	-148.7	timeout	timeout	timeout	memout	memout	memout	timeout
U		-0.2	-0.5	-909.4	timeout	timeout	timeout	timeout	timeout
A	$6 \rightarrow 1 \wedge 2 \wedge 4 \wedge 5$	-179.1	timeout	timeout	timeout	memout	memout	timeout	timeout
U		-0.1	-0.7	-585.5	timeout	timeout	timeout	timeout	timeout
A	$6 \wedge 7 \rightarrow 1 \wedge 2 \wedge 4 \wedge 5$	-182.7	memout	timeout	timeout	timeout	timeout	timeout	timeout
U		-0.2	-1.5	-787.9	timeout	timeout	timeout	timeout	timeout
A		+11.6	+68.8	+406.6	memout	timeout	timeout	timeout	timeout
U+S	$6 \wedge 7 \rightarrow 1 \wedge 2 \wedge 5 \wedge 8$	+0.1	+0.4	+1.4	+86.6	+1460.4	timeout	timeout	timeout
U−S		+0.1	+0.2	+0.4	+3.7	+3.9	+17.7	+84.1	+1414.3
A	$6 \wedge 7 \rightarrow 1 \wedge 2 \wedge 5 \wedge 8 \wedge 9$	-41.0	-1498.9	timeout	memout	timeout	timeout	timeout	timeout
U		-0.1	-0.1	-0.2	-0.9	-15.8	-427.9	timeout	timeout
A		+67.5	+660.1	memout	timeout	timeout	timeout	timeout	timeout
U+S	$6 \wedge 7 \wedge 10 \rightarrow 1 \wedge 2 \wedge 5 \wedge 8 \wedge 9$	+0.3	+2.2	+36.0	+899.3	timeout	timeout	timeout	timeout
U−S		+0.2	+0.6	+11.6	+16.9	+1222.2	timeout	timeout	timeout

for the non-safety part of the specification and solving the safety games induced by a special product of the automata involved [3].

Finally, if the specification is found to be realisable (i.e., the game computed in the previous phase is winning for the player representing the system to be synthesised), the symbolic representation of the winning states of the system is used to compute a prototype implementation satisfying the specification. Intuitively, this is done by constructing a circuit that keeps track of the current position in the game and computes a transition to another winning position whose input matches the one observed after each computation cycle. At the same time, the output labelling along the transition is used as the output for the system. For this computation, only the BDD representation of the winning positions is used.

3   Experimental Results

We use a load balancing system [3] as our case study. The specification is supposed to be successfully built by an engineer who applies a synthesis tool after each modification of the specification. The setting is parametrised by the number of servers the balancer is meant to work for. For some number of servers $n \in \mathbb{N}$, the load balancer has $n + 1$ input bits and n output bits.

Table 1 surveys the results. The individual assumptions and guarantees in the specification are numbered as in [3]. For example, the setting $6 \rightarrow 1 \wedge 2 \wedge 4 \wedge 5$

corresponds to a specification with the assumption no. 5 and the guarantees no. 2, 4, and 6. For every setting, the table describes whether the specification was found to be realisable ("+") or not ("−") and the running times for the ACACIA v.0.3.9.9 [4] tool (abbreviated by "A") and UNBEAST (abbreviated by "U"). Both tools can only check for either realisability or unrealisabilty at a time. Thus, we ran them for both cases concurrently and only report the running times of the instances that terminated with a positive answer. For realisable specifications, for our tool, we distinguish between the cases that a system is to be synthesised ("+S") or just realisability checking is to be performed ("−S"). We did not configure ACACIA to construct an implementation.

4   Conclusion

We presented UNBEAST, a tool for the synthesis of reactive systems from LTL specifications. In the experimental evaluation, we compared our tool against ACACIA on a case study from [3] and found it to be faster in all cases, sometimes even orders of magnitude. For academic purposes, the tool can freely be downloaded from `http://react.cs.uni-saarland.de/tools/unbeast`.

Especially when only realisability of a specification is to be checked, the BDD-based bounded synthesis approach turns out to work well. However, it can be observed that extracting a prototype implementation significantly increases the computation time. This is in line with the findings in [1], where the same observation is made in the context of synthesis from a subset of LTL. We see this as a strong indication that the problem of extracting winning strategies from symbolically represented games requires further research.

References

1. Bloem, R., Galler, S., Jobstmann, B., Piterman, N., Pnueli, A., Weiglhofer, M.: Specify, compile, run: Hardware from PSL. ENTCS 190(4), 3–16 (2007)
2. Cimatti, A., Clarke, E.M., Giunchiglia, E., Giunchiglia, F., Pistore, M., Roveri, M., Sebastiani, R., Tacchella, A.: NuSMV 2: An opensource tool for symbolic model checking. In: Brinksma, E., Larsen, K.G. (eds.) CAV 2002. LNCS, vol. 2404, pp. 359–364. Springer, Heidelberg (2002)
3. Ehlers, R.: Symbolic bounded synthesis. In: Touili, T., Cook, B., Jackson, P. (eds.) CAV 2010. LNCS, vol. 6174, pp. 365–379. Springer, Heidelberg (2010)
4. Filiot, E., Jin, N., Raskin, J.F.: An antichain algorithm for LTL realizability. In: Bouajjani, A., Maler, O. (eds.) CAV 2009. LNCS, vol. 5643, pp. 263–277. Springer, Heidelberg (2009)
5. Gastin, P., Oddoux, D.: Fast LTL to Büchi automata translation. In: Berry, G., Comon, H., Finkel, A. (eds.) CAV 2001. LNCS, vol. 2102, pp. 53–65. Springer, Heidelberg (2001)
6. Kupferman, O., Vardi, M.Y.: Model checking of safety properties. Formal Methods in System Design 19(3), 291–314 (2001)
7. Schewe, S., Finkbeiner, B.: Bounded synthesis. In: Namjoshi, K.S., Yoneda, T., Higashino, T., Okamura, Y. (eds.) ATVA 2007. LNCS, vol. 4762, pp. 474–488. Springer, Heidelberg (2007)
8. Somenzi, F.: CUDD: CU Decision Diagram package release 2.4.2 (2009)

Abstractions and Pattern Databases:
The Quest for Succinctness and Accuracy

Sebastian Kupferschmid and Martin Wehrle

University of Freiburg
Department of Computer Science
Freiburg, Germany
{kupfersc,mwehrle}@informatik.uni-freiburg.de

Abstract. Directed model checking is a well-established technique for detecting error states in concurrent systems efficiently. As error traces are important for debugging purposes, it is preferable to find as short error traces as possible. A wide spread method to find provably *shortest* error traces is to apply the A^* search algorithm with distance heuristics that never overestimate the real error distance. An important class of such distance estimators is the class of *pattern database* heuristics, which are built on abstractions of the system under consideration. In this paper, we propose a systematic approach for the construction of pattern database heuristics. We formally define a concept to measure the accuracy of abstractions. Based on this technique, we address the challenge of finding abstractions that are succinct on the one hand, and accurate to produce informed pattern databases on the other hand. We evaluate our approach on large and complex industrial problems. The experiments show that the resulting distance heuristic impressively advances the state of the art.

1 Introduction

When model checking safety properties of large systems, the ultimate goal is to prove the system correct with respect to a given property. However, for practically relevant systems, this is often not possible because of the state explosion problem. Complementary to proving a system correct, model checking is often used to detect reachable *error states*. To be able to debug a system effectively, it is important to have *short* or preferably *shortest possible* error traces in such cases. Directed model checking (DMC) is a well established technique to find reachable error states in concurrent systems and has recently found much attention[4, 6, 7, 10, 14, 15, 19–23]. The main idea of DMC is to focus on those parts of the state space that show promise to contain reachable error states. Thus, error states can often be found by only exploring a small fraction of the entire reachable state space. DMC achieves this by applying a *distance heuristic* that estimates for each state encountered during the search the distance to a nearest error state. These heuristics are usually based on abstractions and computed fully automatically. One important discipline in DMC is to find *optimal*, i. e., *shortest* possible error traces. This can be achieved by applying an *admissible* distance heuristic, i. e., a heuristic that never overestimates the real error distance, with the A^* search algorithm [8, 9].

An important class of admissible distance heuristics is based on *pattern databases* (PDB). They have originally been introduced in the area of Artificial Intelligence [2, 5].

P.A. Abdulla and K.R.M. Leino (Eds.): TACAS 2011, LNCS 6605, pp. 276–290, 2011.

A *pattern* in this context is a subset of the variables or the predicates that describe the original system $\mathcal{M}$. A PDB is essentially the state space of an abstraction of $\mathcal{M}$ based on the selected pattern. The heuristic estimate for a state encountered during the model checking process is the error distance of the corresponding abstract state. The most crucial part in the design of a pattern database heuristic is the choice of the pattern, which determines the heuristic's behavior and therefore the overall quality of the resulting heuristic. Ultimately, one seeks for patterns that are as small as possible to be able to handle large systems on the one hand. On the other hand, the corresponding abstract system that is determined by the pattern should be as similar to the original system as possible to appropriately reflect the original system behavior. This is equivalent to maximizing the distance estimates of the pattern database, as higher distance values lead to more informed admissible heuristics. More precisely, for admissible heuristics h_1 and h_2, it is known that A* with h_1 performs better than with h_2 if h_1 delivers greater distance values than h_2 [18]. Therefore, it is desirable to have admissible heuristics that deliver as high distance values as possible.

In this paper, we present *downward pattern refinement*, a systematic approach to the pattern selection problem. Complementary to other approaches, we successively abstract the original system as long as only little spurious behavior is introduced. For this purpose we developed a suitable notion to measure the similarity of systems which corresponds to the accuracy of abstractions. This often results in small patterns that still lead to very informed pattern database heuristics. We demonstrate that downward pattern refinement is a powerful approach, and we are able to handle large problems that could not be solved optimally before. In particular, we show that the resulting pattern database heuristic recognizes many dead end states. Correctly identifying dead end states is useful to reduce the search effort significantly, since such states can be excluded from the search process without losing completeness. This even allows us to efficiently verify *correct* systems with directed model checking techniques.

The remainder of this paper is structured as follows. Section 2 provides notations and the necessary background in directed model checking. In Sec. 3 we detail the main part of our contribution. This is followed by a discussion of related work. Afterwards, in Sec. 5 we empirically evaluate our approach by comparing it to other state-of-the-art PDB heuristics. Section 6 concludes the paper.

2 Preliminaries

In this section we introduce the notations used throughout this paper. This is followed by a short introduction to directed model checking and pattern database heuristics.

2.1 Notation

The approach we are presenting here is applicable to a broad class of transition systems, including systems featuring parallelism and interleaving, shared variables or binary synchronization. For the sake of presentation, we decided to define systems rather generally. The systems considered here consist of parallel processes using a global synchronization mechanism. Throughout this paper, Σ denotes a finite set of synchronization labels.

Definition 1 (Process). *A process* $p = \langle L, L^*, T \rangle$ *is a labeled directed graph, where* $L \neq \emptyset$ *is a finite set of locations,* $L^* \subseteq L$ *is a set of error locations, and* $T \subseteq L \times \Sigma \times L$ *is a set of local transitions.*

For a local transition $(l, a, l') \in T$ we also write $l \xrightarrow{a} l'$. The systems we are dealing with are running in lockstep. This means that a process can only perform a local transition, if all other processes of the system simultaneously perform a local transition with the same label. In this paper, a system $\mathcal{M}$ is just the parallel composition of a finite number of processes $p_1, \ldots, p_n$, the components of $\mathcal{M}$.

Definition 2 (Parallel composition). *Let* $p_1, \ldots, p_n$ *be a finite number of processes with* $p_i = \langle L_i, L_i^*, T_i \rangle$ *for* $i \in \{1, \ldots, n\}$. *The parallel composition* $p_1 \parallel \ldots \parallel p_n$ *of these processes is the process* $\langle S, S^*, \Delta \rangle$, *where* $S = L_1 \times \cdots \times L_n$, $S^* = L_1^* \times \cdots \times L_n^*$, *and* $\Delta \subseteq S \times \Sigma \times S$ *is a transition relation. There is a transition* $(l_1, \ldots, l_n) \xrightarrow{a} (l'_1, \ldots, l'_n) \in \Delta$, *if and only if* $l_i \xrightarrow{a} l'_i \in T_i$ *for each* $i \in \{1, \ldots, n\}$.

Note that parallel composition as defined above is an associative and commutative operation. This gives rise to the following definition of systems.

Definition 3 (System). *A system* $\mathcal{M} = \{p_1, \ldots, p_n\}$ *is a set of processes, the components of* $\mathcal{M}$. *The semantics of* $\mathcal{M}$ *is given as the composite process* $\langle S, S^*, \Delta \rangle = p_1 \parallel \ldots \parallel p_n$. *We use* $S(\mathcal{M})$ *and* $\Delta(\mathcal{M})$ *to denote the set of system states and global transitions, respectively. We denote the set of error states with* $S^*(\mathcal{M})$.

To distinguish between local states of an atomic process and the global state of the overall system, we use the term *location* for the former and *state* for the latter. The problem we address in this paper is the detection of reachable error states $s \in S^*(\mathcal{M})$ for a given system $\mathcal{M}$. Formally, we define model checking tasks as follows.

Definition 4 (Model checking task). *A model checking task is a tuple* $\langle \mathcal{M}, s_0 \rangle$, *where* $\mathcal{M}$ *is a system and* $s_0 \in S(\mathcal{M})$ *is the initial state. The objective is to find a sequence* $\pi = t_1, \ldots, t_n$, *of transitions, so that* $t_i = s_{i-1} \xrightarrow{a_i} s_i \in \Delta(\mathcal{M})$ *for* $i \in \{1, \ldots, n\}$ *and* $s_n \in S^*(\mathcal{M})$.

We call a sequence π of successively applicable transitions leading from $s \in S(\mathcal{M})$ to $s' \in S(\mathcal{M})$ a trace. If $s' \in S^*(\mathcal{M})$, then π is an *error trace*. The length of a trace, denoted with $\|\pi\|$, is the number of its transitions, e. g., the length of π from the last definition is n.

We conclude this section with a short remark on the close correspondence between solving model checking tasks and the nonemptiness problem for intersections of regular automata. From this perspective, a system component corresponds to a regular automaton and the error locations correspond to accepting states of the automaton. Parallel composition corresponds to language intersection. This view is not necessarily useful for efficiently solving model checking tasks, but it shows that deciding existence of error traces is PSPACE-complete [11].

2.2 Directed Model Checking

Directed model checking (DMC) is the application of heuristic search [18] to model checking. DMC is an explicit search strategy which is especially tailored to the fast detection of reachable error states. This is achieved by focusing the search on those parts of the state space that show promise to contain error states. More precisely, DMC applies a *distance heuristic* to influence the order in which states are explored. The most successful distance heuristics are fully automatically generated based on a declarative description of the given model checking task. A distance heuristic for a system $\mathcal{M}$ is a function $h : S(\mathcal{M}) \to \mathbb{N}_0 \cup \{\infty\}$ which maps each state $s \in S(\mathcal{M})$ to an integer, estimating $d(s)$, the distance from s to a nearest error state in $S^*(\mathcal{M})$. When we want to stress that $h(s)$ is a heuristic estimate for $s \in S(\mathcal{M})$, we write $h(s, \mathcal{M})$. Typically, heuristics in DMC are based on abstractions, i.e., the heuristic estimate for a states s is the length of a corresponding error trace in an abstraction of $\mathcal{M}$. During search, such a heuristic is used to determine which state to explore next. There are many different ways how to prioritize states, e.g., the wide-spread methods A^* [8, 9] and *greedy search* [18]. In the former, states are explored by increasing value of $c(s) + h(s)$, where $c(s)$ is the length of the trace on that state s was reached. If h is *admissible*, i.e., if it never overestimates the real error distance, then A^* is guaranteed to return a shortest possible error trace. In greedy search, states are explored by increasing value of $h(s)$. This gives no guarantee on the length of the detected error trace, but tends to explore fewer states in practice. Figure 1 shows a basic directed model checking algorithm.

```
1  function dmc(M, h):
2      open = empty priority queue
3      closed = ∅
4      open.insert(s₀, priority(s₀))
5      while open ≠ ∅ do:
6          s = open.getMinimum()
7          if s ∈ S*(M) then:
8              return False
9          closed = closed ∪ {s}
10         for each transition t = s →ᵃ s' ∈ Δ(M) do:
11             if s' ∉ closed then:
12                 open.insert(s', priority(s'))
13     return True
```

Fig. 1. A basic directed model checking algorithm

The algorithm takes a system $\mathcal{M}$ and a heuristic h as input. It returns *False* if there is a reachable error state, otherwise it returns *True*. The state s_0 is the initial state of $\mathcal{M}$. The algorithm maintains a priority queue *open* which contains visited but not yet explored states. When *open.getMinimum* is called, *open* returns a minimum element, i.e., a state with minimal priority value. States that have been explored are stored in *closed*. Every encountered state is first checked if it is an error state. If this is not the case, its successors are computed. Every successor that has not been visited before is inserted into *open* according to its priority value. The *priority* function depends on the

applied version of directed model checking, i.e., if applied with A* or greedy search
(cf. [8, 18]). As already mentioned, for A*, $priority(s)$ returns $h(s) + c(s)$, where $c(s)$
is the length of the path on which s was reached for the first time. For greedy search, it
simply evaluates to $h(s)$. When every successor has been computed and prioritized, the
process continues with the next state from *open* with lowest priority value.

2.3 Pattern Database Heuristics

Pattern database (PDB) heuristics [2] are a family of abstraction-based heuristics. Originally, they were proposed for solving single-agent problems. Today they are one of the
most successful approaches for creating admissible heuristics.

For a given system $\mathcal{M} = \{p_1, \ldots, p_n\}$ a PDB heuristic is defined by a *pattern*
$\mathcal{P} \subseteq \mathcal{M}$, i.e., a subset of the components of $\mathcal{M}$. The pattern $\mathcal{P}$ can be interpreted as an
abstraction of $\mathcal{M}$. To stress that $\mathcal{P}$ is an abstraction of $\mathcal{M}$, we will denote this system
with $\mathcal{M}|_\mathcal{P}$. It is not difficult to see that this kind of abstraction is a projection, and the
abstract system is an overapproximation of $\mathcal{M}$. A PDB is built prior to solving the actual
model checking task. For this, the entire reachable state space of the abstract system is
enumerated. Every reachable abstract state is then stored together with its abstract error
distance in some kind of lookup table, the so-called *pattern database*. Typically, these
distances are computed by a breadth-first search. Note that abstract systems have to
be chosen so that they are much smaller than their original counterparts and hence are
much easier to solve.

When solving the original model checking task with such a PDB, distances of concrete states are estimated as follows. A concrete state s is mapped to its corresponding
abstract state $s|_\mathcal{P}$; the heuristic estimate for s is then looked up in the PDB. This is
given formally in the definition of PDB heuristics.

Definition 5 (Pattern database heuristic). *Let $\mathcal{M}$ be a system and $\mathcal{P}$ be a pattern for
$\mathcal{M}$. The heuristic value for a state $s \in S(\mathcal{M})$ induced by $\mathcal{P}$ is defined as follows:*

$$h^\mathcal{P}(s) = \min\{\|\pi\| \mid \pi \text{ is error trace of model checking task } \langle \mathcal{M}|_\mathcal{P}, s|_\mathcal{P} \rangle\},$$

where $s|_\mathcal{P}$ is the projection of s onto $\mathcal{P}$.

The main problem with PDB heuristics is the *pattern selection problem*. The informed-
ness of a PDB heuristic crucially depends on the selected pattern. For instance, if the
selected pattern $\mathcal{P}$ contains all processes, i.e., $\mathcal{M} = \mathcal{M}|_\mathcal{P}$, then we obtain a *perfect
heuristic*, i.e., a heuristic that returns the real error distance for all states of the sys-
tem. On the other hand, since we have to enumerate the reachable abstract state space
exhaustively, which coincides with the concrete one, this exactly performs like blind
breadth-first search. On the other end of an extreme spectrum, the PDB heuristic induced by the empty pattern can be computed very efficiently, but on the negative side,
the resulting heuristic constantly evaluates to zero and thus also behaves like unin-
formed search. Good patterns are somewhere in between and how to find good patterns
is the topic of this paper.

3 Pattern Selection Based on Downward Refinement

In this section, we describe our approach for the pattern selection that underlies the pattern database heuristic. On the one hand, as the abstract state space has to be searched exhaustively to build the pattern database, the ultimate goal is to find patterns that lead to *small* abstractions to be able to handle large systems. On the other hand, the patterns should yield abstractions that are as *similar* to the original system as possible to retain as much of the original system behavior as possible. An obvious question in this context is the question about similarity: what does it mean for a system to be "similar" to an abstract system? In Sec. 3.1 and Sec. 3.2, we derive precise, but computationally hard properties of similarity of abstract systems. Furthermore, based on these considerations, we provide ways to efficiently approximate these notions in practice. Based on these techniques, we finally describe an algorithm for the pattern selection based on *downward pattern refinement* in Sec. 3.3.

3.1 Sufficiently Accurate Distance Heuristics

In this section, we derive a precise measure for abstractions to obtain informed pattern database heuristics. As already outlined above, the most important question in this context is the question about similarity. At the extreme end of the spectrum of possible abstractions, one could choose a pattern that leads to bisimilar abstractions to the original system. This yields a pattern database heuristic $h^{\mathcal{P}}$ that is *perfect*, i.e., $h^{\mathcal{P}}(s) = d(s)$ for all states s, where d is the real error distance function. However, apart from being not feasible in practice, we will see that this condition is stricter than needed for obtaining perfect search behavior. It suffices to require $h^{\mathcal{P}}(s) = d(s)$ only for states s that are possibly explored by A*. In this context, Pearl [18] gives a necessary and sufficient condition for a state to be explored by A*. Consider a model checking task $\langle \mathcal{M}, s_0 \rangle$ and let $d(s_0)$ denote the length of a shortest error trace of $\mathcal{M}$. Recall that the priority function of A* is $priority(s) = h(s) + c(s)$, where $c(s)$ is the length of a shortest trace from s_0 to s. Pearl shows that if $priority(s) < d(s_0)$, then s is necessarily explored by A*, whereas exploring s implies that $priority(s) \leq d(s_0)$. This gives rise to the following definition for a distance heuristic to be *sufficiently accurate*.

Definition 6 (Sufficiently accurate). *Let $\mathcal{M}$ be a system with shortest error trace length $d(s_0)$, $\mathcal{P} \subseteq \mathcal{M}$ be a pattern of $\mathcal{M}$, and $h^{\mathcal{P}}$ be the pattern database heuristic for $\mathcal{P}$. If $h^{\mathcal{P}}(s) = d(s)$ for all states s with $h^{\mathcal{P}}(s) + c(s) \leq d(s_0)$, then $\mathcal{M}|_{\mathcal{P}}$ is called a* sufficiently accurate abstraction *of $\mathcal{M}$, and $h^{\mathcal{P}}$ is called* sufficiently accurate distance heuristic *for $\mathcal{M}$.*

Obviously, the requirement for a distance heuristic $h^{\mathcal{P}}$ to be sufficiently accurate is weaker than the requirement $h^{\mathcal{P}}(s) = d(s)$ for *all* possible states. However, with the results given by Pearl, we still know that A* with a sufficiently accurate distance heuristic delivers perfect search behavior, i.e., the same search behavior as that of A* with d. This justifies Def. 6 and is stated formally in the following proposition.

Proposition 1. *Let $\mathcal{M}$ be a system, $h^{\mathcal{P}}$ be a distance heuristic that is sufficiently accurate for $\mathcal{M}$. Then the set of explored states with A* applied with d is equal to the set of explored states of A* applied with $h^{\mathcal{P}}$.*

Proof. The claim follows immediately from the results given by Pearl [18]. As $h^{\mathcal{P}}$ is sufficiently accurate, we know that for every state s that is possibly explored by A* applied with $h^{\mathcal{P}}$ it holds $h^{\mathcal{P}}(s) = d(s)$. Therefore, the behavior of A* with d and $h^{\mathcal{P}}$ is identical.

As a first and immediate result of the above considerations, it suffices to have patterns that lead to sufficiently accurate distance heuristics to obtain perfect search behavior with A*. On the one hand, this notion is intuitive and reasonable. On the other hand, it is still of rather theoretical nature. It should be obvious that a sufficiently accurate heuristic is hard to compute as it relies on *exact* error distances d; as a side remark, if d was given, the overall model checking problem would be already solved, and there would be no need to compute a pattern database heuristic. However, Def. 6 also provides a first intuitive way for approximating this property, an approximation which is described next.

According to Def. 6, an abstraction $\mathcal{M}|_{\mathcal{P}}$ is sufficiently accurate if $h^{\mathcal{P}}(s) = d(s)$ for all states s that are possibly explored by A*. In this case, the pattern database heuristic based on $\mathcal{M}|_{\mathcal{P}}$ is sufficiently accurate for $\mathcal{M}$. For the following considerations, note that $h^{\mathcal{P}}(s) = d(s|_{\mathcal{P}}, \mathcal{M}|_{\mathcal{P}})$, and therefore, a direct way to approximate this test is to use a *distance heuristic* h instead of d. This is reasonable as distance heuristics are designed exactly for the purpose of approximating d, and various distance heuristics have been proposed in the directed model checking literature. Furthermore, as checking *all* states that are possibly explored by A* is not feasible either, we check this property only for the *initial* system state. This is the only state for which we know a priori that it is explored by A*. Overall, this gives rise to the following definition of *relatively accurate* abstractions.

Definition 7 (Relatively accurate). *Let $\langle \mathcal{M}, s_0 \rangle$ be a model checking task with system $\mathcal{M}$ and initial state s_0. Further, let $\mathcal{P} \subseteq \mathcal{M}$ be a pattern of $\mathcal{M}$, and let $\mathcal{M}|_{\mathcal{P}}$ be the corresponding abstraction to $\mathcal{P}$ with abstract initial state $s_0|_{\mathcal{P}}$. Furthermore, let h be a distance heuristic, $h(s_0, \mathcal{M})$ the distance estimate of $s_0 \in S(\mathcal{M})$, and $h(s_0|_{\mathcal{P}}, \mathcal{M}|_{\mathcal{P}})$ the distance estimate of $s_0|_{\mathcal{P}} \in S(\mathcal{M}|_{\mathcal{P}})$. If*

$$h(s_0, \mathcal{M}) = h(s_0|_{\mathcal{P}}, \mathcal{M}|_{\mathcal{P}}),$$

then $\mathcal{M}|_{\mathcal{P}}$ is called the relatively accurate abstraction *of $\mathcal{M}$ induced by h and $\mathcal{P}$.*

Obviously, the quality of this approximation strongly depends on the quality of the applied distance heuristic h. We want to emphasize that, according to the above definition, we apply a *second* distance heuristic h to determine our pattern database heuristic $h^{\mathcal{P}}$. In the experimental section, we will see that even this rather simple approximation of sufficient accurateness yields very informed abstract systems for sophisticated h.

3.2 Concretizable Traces and Safe Abstractions

In addition to the criterion from the last section, we derive a sufficient criterion for a distance heuristic to be sufficiently accurate that is still weaker than the requirement $h^{\mathcal{P}}(s) = d(s)$ for *all* states s. It is based on the observation that abstract systems where every spurious error trace is longer than $d(s_0)$ are not harmful. This is stated formally in the following proposition.

Proposition 2. *Let $\mathcal{M}$ be a system, $\mathcal{P} \subseteq \mathcal{M}$ be a pattern such that every spurious error trace π in the corresponding abstraction $\mathcal{M}|_{\mathcal{P}}$ is longer than a shortest possible error trace in $\mathcal{M}$, i.e., $\|\pi\| > d(s_0)$. Then $h^{\mathcal{P}}$ is sufficiently accurate for $\mathcal{M}$.*

Proof. First, recall that $h^{\mathcal{P}}(s) \leq d(s)$ for all states $s \in S(\mathcal{M})$ because $\mathcal{M}|_{\mathcal{P}}$ is an overapproximation of $\mathcal{M}$. We show that $h^{\mathcal{P}}(s) + c(s) > d(s_0)$ for all states $s \in S(\mathcal{M})$ with $h^{\mathcal{P}}(s) < d(s)$. Assume $h^{\mathcal{P}}(s) < d(s)$ for a state $s \in S(\mathcal{M})$. Let $s|_{\mathcal{P}} \in S(\mathcal{M}|_{\mathcal{P}})$ be the corresponding abstract state to s. As $h^{\mathcal{P}}(s) < d(s)$, there is an abstract trace $\pi_{\mathcal{P}}$ that is spurious and contains $s|_{\mathcal{P}}$. As all spurious error traces are longer than $d(s_0)$ by assumption, we have $\|\pi_{\mathcal{P}}\| > d(s_0)$. Therefore, $\|\pi_{\mathcal{P}}\| = c^{\mathcal{P}}(s|_{\mathcal{P}}) + d^{\mathcal{P}}(s|_{\mathcal{P}}) > d(s_0)$, where $c^{\mathcal{P}}(s|_{\mathcal{P}})$ denotes the length of a shortest abstract trace from the initial abstract state to $s|_{\mathcal{P}}$, and $d^{\mathcal{P}}(s|_{\mathcal{P}})$ denotes the abstract error distance of $s|_{\mathcal{P}} \in S(\mathcal{M}|_{\mathcal{P}})$. As $d^{\mathcal{P}}(s|_{\mathcal{P}}) = h^{\mathcal{P}}(s)$ and $c(s) \geq c^{\mathcal{P}}(s|_{\mathcal{P}})$, we have $c(s) + h^{\mathcal{P}}(s) > d(s_0)$.

Again, identifying abstractions with the property given by the above proposition is computationally hard as it relies on checking *all* possible spurious error traces. In the following, we show that a subclass of abstractions for a slightly stronger condition can be identified efficiently. To be more precise, we focus on abstractions that only introduce spurious error traces that can be *concretized* in the following sense.

Definition 8 (Concretizable Trace). *Let $\mathcal{M}$ be a system, $\mathcal{P} \subseteq \mathcal{M}$ be a pattern, and $\mathcal{M}|_{\mathcal{P}}$ be the corresponding abstraction of $\mathcal{M}$. Let $\pi_{\mathcal{P}} = t_1^{\#}, \ldots, t_n^{\#}$ be an abstract error trace of $\mathcal{M}|_{\mathcal{P}}$ with corresponding concrete transitions $t_1, \ldots, t_n$ of $\mathcal{M}$. Let $\pi_{\mathcal{P}}$ be spurious, i.e., $t_1, \ldots, t_n$ is not a concrete error trace of $\mathcal{M}$. The error trace $\pi_{\mathcal{P}}$ is concretizable in $\mathcal{M}$ if and only if there is a concrete error trace*

$$\pi = \pi_0, t_1, \pi_1, t_2, \pi_2, \ldots, \pi_{n-1}, t_n, \pi_n$$

in $\mathcal{M}$ that embeds $t_1, \ldots, t_n$. The π_i are traces in $\mathcal{M}$ with $\|\pi_i\| \geq 0$, for $i \in \{0, \ldots, n\}$.

Informally speaking, an abstract trace $\pi_{\mathcal{P}}$ in $\mathcal{M}|_{\mathcal{P}}$ is concretizable in $\mathcal{M}$ if there is a concrete trace in $\mathcal{M}$ so that the corresponding abstract trace in $\mathcal{M}|_{\mathcal{P}}$ is equal to $\pi_{\mathcal{P}}$. Note that from the above definition, concretizable error traces are a subclass of spurious error traces; as a side remark, these are exactly those error traces that preserve dead-ends in $\mathcal{M}$, i.e., states from which no error state is reachable. In the following, we focus on finding abstractions that do not introduce error traces that are not concretizable. We observe that *safe abstraction* is an effective technique for this purpose.

Safe abstraction for directed model checking has been introduced by Wehrle and Helmert [21]. Essentially, processes identified by safe abstraction can change their locations independently of and without affecting any other process, and every possible location is reachable. We briefly give the formal definitions in the following, starting with the notion of *independent* processes.

Definition 9 (Independent process). *Let $\mathcal{M}$ be a system and let $p \in \mathcal{M}$ be a process. Process $p = \langle L, L^*, T \rangle$ is independent in $\mathcal{M}$ if for every $(l_1, a, l_2) \in T$ with $l_1 \neq l_2$ and every process $\langle L', L^{*'}, T' \rangle = p' \in \mathcal{M} \setminus \{p\}$, the following two conditions hold: For every $l' \in L'$, there is $(l', a, l'') \in T'$, and for every $(l', a, l'') \in T' \colon l' = l''$.*

According to the above definition, independent processes p can change their locations independently of the current locations of other processes, and changing locations in p has no side effect on other processes either. Based on this notion, we define safe processes as follows.

Definition 10 (Safe process). *Let $\mathcal{M}$ be a system and let $\langle L, L^*, T \rangle = p \in \mathcal{M}$ be a process. Process p is safe in $\mathcal{M}$ if p is independent in $\mathcal{M}$, and for all locations $l, l' \in L$ there is a sequence of local transitions in T that leads from l to l', i.e., p is strongly connected.*

Safe processes can be efficiently identified by an analysis of the system processes' causal dependencies. Wehrle and Helmert exploited this property by performing directed model checking directly on the safe abstracted system. Corresponding abstract error traces in $\mathcal{M}|_{\mathcal{P}}$ have been finally extended to a concrete error trace in $\mathcal{M}$. Doing so, however, is not optimality preserving: shortest abstract error traces in $\mathcal{M}|_{\mathcal{P}}$ may not correspond to *any* shortest error trace in $\mathcal{M}$.

In this work, we use safe abstraction in a different context, namely to select patterns for a pattern database. For the following proposition, we assume without loss of generality that for every process p the target location set for p is not empty, and each label a that occurs in a transition of any process also occurs in a transition of p. Under these assumptions, every abstract error trace can be concretized. This is summarized in the following proposition. A proof is given by Wehrle and Helmert [21].

Proposition 3. *Let $\mathcal{M}$ be a system and let $p \in \mathcal{M}$ be a safe process of $\mathcal{M}$. Let $\mathcal{P} = \mathcal{M} \backslash \{p\}$ be the pattern obtained by removing p from $\mathcal{M}$, and $\mathcal{M}|_{\mathcal{P}}$ be the corresponding abstract system. Then every abstract error trace in $\mathcal{M}|_{\mathcal{P}}$ is concretizable.*

We observe that under the assumptions of Prop. 3, the set of unconcretizable abstract error traces is empty, and of course, so is the set of shorter or equally long abstract traces $\{\pi_{\mathcal{P}} \mid \pi_{\mathcal{P}}$ is not concretizable and $\|\pi_{\mathcal{P}}\| \leq d(s_0)\}$. In other words, abstracting safe variables does not introduce error traces that are longer than $d(s_0)$ and are not concretizable. We observe that safe abstraction provides an effective technique to approximate Prop. 2, where the condition of spuriousness is strengthened to concretizablility. The causal analysis required for safe abstraction can be done statically, is cheap to compute, and identifies processes that have the property that corresponding abstract systems approximate the conditions of Prop. 2. At this point, we emphasize again that we do not claim to introduce safe abstraction; however, we rather *apply* this technique for a different purpose than it was originally introduced.

Overall, based on the computationally hard notion of sufficiently accurate distance heuristics and Prop. 2, we have introduced ways to find abstract systems that are similar to the original system. Based on these techniques, we propose an algorithm for the pattern selection in the next section.

3.3   An Algorithm for Pattern Selection Based on Downward Refinement

In this section, we put the pieces together. So far, we have identified the notion of sufficiently accurate abstractions, and proposed techniques for approximating these concepts. Based on these techniques, we introduce an algorithm for the pattern selection

```
1 function dpr(M, s₀, h):
2     P := M \ {p | p safe process in M}
3     for each p ∈ P do:
4         if h(s₀, M) = h(s₀|_{P\{p}}, M|_{P\{p}}) then:
5             P := P \ {p}
6             goto 3
7 return P
```

Fig. 2. The downward pattern refinement algorithm

which we call *downward pattern refinement*. It starts with the full pattern, and iteratively refines it as long as the confidence is high enough that the resulting abstraction
yields an informed pattern database. The algorithm is shown in Figure 2.

Roughly speaking, the overall approach works as follows. We start with the pattern
that contains all system processes. In this case, the resulting pattern database heuristic
would deliver perfect search behavior. However, as we have already discussed, such
systems usually become too huge and cannot be handled in general due to the state
explosion problem. Therefore, we iteratively remove processes such that the resulting
abstraction is still similar to the original system.

We start with identifying all processes that do not introduce error traces that are not
concretizable. Therefore, we remove all processes that are safe according to the safe
abstraction approach (line 2). From the resulting abstract system $M|_P$, we iteratively
remove processes p that lead to relatively accurate abstractions for the given distance
heuristic h, i.e., for which the distance estimate of the initial abstract state does not
decrease (lines 3–6). In particular, in line 4, we check for the current abstraction $M|_P$
if it can be further abstracted without reducing the distance estimation provided by h.
The search stops when no more processes can be removed without decreasing h, i.e.,
when a fixpoint is reached. Termination is guaranteed after at most $|P|$ iterations as we
remove one process from the pattern in each iteration. We finally return the obtained
pattern (line 7). We remark that the order in which the processes are considered may influence the resulting pattern. However, in our experiments, we observed that the pattern
is invariant with respect to this order.

4   Related Work

Directed model checking has recently found much attention in different versions to
efficiently detect error states in concurrent systems [4, 6, 7, 10, 14, 15, 19–23]. In the
following, we give a very brief comparison of downward pattern refinement with other
PDB heuristics. Overall, they mainly differ from our approach in the pattern selection
scheme. The h^{rd} heuristic [15] uses a counterexample-guided pattern selection scheme,
where those variables that occur in a certain abstract error trace are selected. The pattern
selection mechanism of the h^{coi} heuristic [19] is based on a cone of influence analysis.
It is based on the idea that variables that occur "closer" to those variables of the property
are more important than other ones. The h^{pa} heuristic [10] splits a system into several
parts and uses predicate abstraction to build a PDB heuristic for each of the parts. The
resulting heuristic is the sum of all these heuristics.

Further admissible non-PDB heuristics are the h^L and h^{aa} heuristics. The underlying abstraction of the h^{aa} heuristic [4] is obtained by iteratively replacing two components of a system by an overapproximation of their cross product. The h^L heuristic is based on the monotonicity abstraction [14]. The main idea of this abstraction is that variables are set-valued and these sets grow monotonically over transition application. In the following section, we will provide an experimental comparison of these heuristics with our approach.

5   Evaluation

We have implemented downward pattern refinement into our model checker MCTA [16] and empirically evaluated its potential on a range of practically relevant systems coming from an industrial case study. We call the resulting heuristic h^{dpr}. We compare it with various other admissible distance heuristics as implemented in the directed model checking tools UPPAAL/DMC [13] and MCTA.

5.1   Implementation Details

As outlined in the preliminaries, we chose our formalism mainly to ease presentation. Actually, our benchmarks are modeled as timed automata consisting of finitely many parallel automata with clocks and bounded integer variables. The downward pattern refinement algorithm works directly on timed automata. In a nutshell, automata and integer variables correspond to processes in our formalism. As it is not overly important how abstractions of timed automata systems are built, we omit a more detailed description here for lack of space. We remark that this formalism is handled as in Kupferschmid et al.'s Russian doll approach [15].

To identify relatively accurate abstractions, we use the (inadmissible) h^U heuristic [14] for the following reasons. First, it is one of MCTA's fastest to compute heuristics for this purpose. This is an important property since the heuristic is often called for different patterns during the downward pattern refinement procedure. Second, among MCTA's fastest heuristics, the h^U heuristic is the most informed one. The more informed a heuristic is the better it is suited for the evaluation of patterns. As in the computation of the h^U heuristic clocks are ignored, we always include all clocks from the original system in the selected pattern. By doing so, the resulting h^{dpr} is able to reason about clocks. We will come back to this in Sec. 6.

To identify safe variables, each automaton and each bounded integer variable corresponds essentially to a process p in the sense of Def. 1. Both kinds of processes can be subject to safe abstraction as described in Sec. 3.2.

5.2   Experimental Setup

We evaluate the h^{dpr} distance heuristic with A^* search by comparing it with other distance heuristics implemented in MCTA or UPPAAL/DMC. In more details, we compare to h^{rd}, h^{coi}, h^{pa}, h^{aa} and h^L heuristics as described in the related work section. Furthermore, we compare to UPPAAL's[1] breadth-first search (*BFS*) as implemented in the

---

[1] http://www.uppaal.com/

current version (4.0.13). Note that we do not compare our method with inadmissible heuristics like the h^U heuristic, as we do not necessarily get shortest error traces when applied with A*. All experiments have been performed on an AMD Opteron 2.3 GHz system with 4 GByte of memory.

As benchmarks, we use the *Single-tracked Line Segment* case study, which comes from an industrial project partner of the UniForM-project [12]. The case study models a distributed real-time controller for a segment of tracks where trams share a piece of track. A distributed controller has to ensure that never two trams are simultaneously in the critical section driving in different directions. The controller was modeled in terms of PLC automata [3], which is an automata-like notation for real-time programs. With the tool MOBY/RT [17], we transformed the PLC automata system into abstractions of its semantics in terms of timed automata [1]. For the evaluation of our approach we chose the property that never both directions are given permission to enter the shared segment simultaneously. We use three problem families to evaluate our approach, denoted with C, D, and E. They have been obtained by applying different abstractions to the case study. For each of them, we constructed nine models of increasing size by decreasing the number of abstracted variables. Note that all these problems are very large. The number of variables in the C instances ranges from 15 to 28, the number of automata ranges from 5 to 10. The corresponding numbers in the D problems range from 29 to 54 (variables) and from 7 to 13 (automata). The E instances have 44 to 54 variables and 9 to 13 automata. We injected an error into the C and D examples by manipulating an upper time bound. The E instances are correct with respect to the chosen property.

5.3 Experimental Results

Our experimental results are presented in Table 1. We compare h^{dpr} with the other heuristics and UPPAAL's breadth-first search (*BFS*) in terms of total runtime (including the preprocessing to build the pattern database for the PDB heuristics) and in terms of number of explored concrete states during the actual model checking process. The results are impressive. Most strikingly, h^{dpr} is the only heuristic that is able to solve every (erroneous and correct) problem instance. Looking a bit more closely, we also observe that h^{dpr} is always among the fastest approaches. In the C instances, only h^{rd} is faster, whereas in the smaller D instances, h^{dpr} outperforms the other approaches except for D_1. The larger D instances cannot be handled by any of the other heuristics at all. Moreover, we observe that the pruning power of h^{dpr} is high, and hence, we are able to verify correct systems that are even out of scope for the current version of UPPAAL. In many cases, the initial system state s_0 is already evaluated to infinity; this means that there is provably no concrete error trace from s_0 and there is no need to search in the concrete system at all. In particular, this results in a total number of explored states of zero. We will discuss these points in more details in the next section.

5.4 Directed Model Checking for Correct Systems?

As outlined in the introduction, directed model checking is tailored to the fast detection of reachable error states. The approach is sound and complete as only the order is

Table 1. Experimental results for A^* search. Abbreviations: "runtime": overall runtime including any preprocessing in seconds, "explored states": number of explored states before an error state was encountered or the instance was proven correct, dashes indicate out of memory ($> 4\,\mathrm{GByte}$)

Inst.	\multicolumn{7}{c}{runtime in s}							\multicolumn{7}{c}{explored states}							trace length
	h^{dpr}	h^{rd}	h^{coi}	h^{pa}	h^{aa}	h^L	BFS	h^{dpr}	h^{rd}	h^{coi}	h^{pa}	h^{aa}	h^L	BFS	length
\multicolumn{16}{c}{erroneous instances}
C_1	1.8	0.7	0.6	1.1	0.1	0.1	0.2	55	130	130	7088	8649	8053	21008	54
C_2	2.3	1.0	1.6	1.2	0.4	0.2	0.4	55	89813	187	15742	21719	21956	55544	54
C_3	2.2	0.6	2.6	1.2	0.4	0.4	0.6	55	197	197	15586	28753	24951	74791	54
C_4	2.5	0.7	23.4	2.2	1.9	2.3	5.3	253	1140	466	108603	328415	170325	553265	55
C_5	2.6	0.9	223.9	6.8	12.6	18.2	46.7	1083	7530	2147	733761	2.5e+6	1.2e+6	4.0e+6	56
C_6	4.1	0.8	227.4	53.6	176.2	165.2	464.7	2380	39436	6229	7.4e+6	2.5e+6	1.0e+7	3.4e+7	56
C_7	3.7	1.3	227.7	–	–	–	–	3879	149993	16357	–	–	–	–	56
C_8	3.7	1.3	182.3	–	–	–	–	5048	158361	16353	–	–	–	–	56
C_9	3.3	1.3	–	–	–	–	–	12651	127895	–	–	–	–	–	57
D_1	5.9	81.1	217.6	9.0	2.3	28.0	76.6	2450	4.6e+6	414	475354	2.6e+6	888779	4.1e+6	78
D_2	6.7	218.5	213.5	35.6	11.5	134.0	458.4	4401	4223	4223	2.5e+6	1.4e+6	4.0e+6	2.2e+7	79
D_3	6.7	222.7	215.0	36.3	11.8	152.3	466.5	4713	2993	2993	2.5e+6	1.4e+6	4.6e+6	2.2e+7	79
D_4	7.1	218.7	216.3	27.9	9.9	79.7	404.4	979	2031	2031	2.0e+6	1.3e+6	2.4e+6	1.8e+7	79
D_5	48.9	–	–	–	–	–	–	75631	–	–	–	–	–	–	102
D_6	52.6	–	–	–	–	–	–	255486	–	–	–	–	–	–	103
D_7	55.5	–	–	–	–	–	–	131275	–	–	–	–	–	–	104
D_8	52.6	–	–	–	–	–	–	22267	–	–	–	–	–	–	104
D_9	55.3	–	–	–	–	–	–	11960	–	–	–	–	–	–	105
\multicolumn{16}{c}{error-free instances}
E_1	5.9	–	1.5	1.6	0.2	0.3	0.3	0	–	59210	22571	24842	18533	43108	n/a
E_2	23.4	–	–	91.6	65.7	140.1	157.0	0	–	–	6.1e+6	6.4e+6	4.6e+6	1.1e+7	n/a
E_3	53.1	–	–	–	–	–	–	1	–	–	–	–	–	–	n/a
E_4	156.1	–	–	–	–	–	–	1	–	–	–	–	–	–	n/a
E_5	158.0	–	–	–	–	–	–	0	–	–	–	–	–	–	n/a
E_6	161.9	–	–	–	–	–	–	0	–	–	–	–	–	–	n/a
E_7	168.1	–	–	–	–	–	–	0	–	–	–	–	–	–	n/a
E_8	172.8	–	–	–	–	–	–	13	–	–	–	–	–	–	n/a
E_9	180.1	–	–	–	–	–	–	0	–	–	–	–	–	–	n/a

influenced in which the states are explored. However, one may wonder why a technique that influences the order of explored states is also capable of efficiently proving a system correct. The answer is that admissible distance heuristics like h^{dpr}, i.e., heuristics h with $h(s) \le d(s)$ for all states s and the real error distance function d, also admit pruning power in the following sense. If $h(s) = \infty$ for an admissible heuristic h and a state s, then there is no abstract error trace that starts from the corresponding abstract state of s. Therefore, s can be pruned without losing completeness because $d(s) = \infty$ as well, as there is no concrete error trace starting from s either. Therefore, the absence of error states might be shown without actually exploring the entire reachable state space. In our experiments, we observe that h^{dpr} is very successful for this purpose as well. This is caused by the suitable abstraction found by our downward refinement algorithm that preserves much of the original system behavior. The other distance heuristics do not perform as well in this respect. This is either because the underlying abstraction is too coarse (and hence, not many states are recognized that can be pruned), or it is too large such that no pattern database could be built because of lack of memory. Obviously, the abstractions of h^{dpr} identify a sweet spot of the trade-off to be as succinct as possible on the one hand, and as accurate as possible on the other hand.

6   Conclusions

We have introduced an approach to find abstractions and to build pattern database heuristics by systematically exploiting a tractable notion of system similarity. Based on these techniques, we presented a powerful algorithm for selecting patterns based on downward refinement. The experimental evaluation shows impressive performance improvements compared to previously proposed, state-of-the-art distance heuristics on a range of large and complex real world problems. In particular, we have learned that directed model checking with admissible distance heuristics can also be successfully applied to verify correct systems. For both erroneous and correct systems, we are able to solve very large problems that could not be optimally solved before. Overall, we observe that directed model checking with abstraction based distance heuristics faces similar problems as other (abstraction based) approaches to solve model checking tasks. In all these areas, the common problem is to find abstractions that are both succinct and accurate. This is also reflected in the future work, where it will be interesting to further investigate the class of pattern database heuristics and, in particular, to find suitable abstractions for pattern databases. In this context, counterexample-guided abstraction refinement could serve as a technique to further push our approach. Moreover, for the class of timed automata, we expect that the development of heuristics that consider clocks in the computation of heuristic values (rather than ignoring them) will improve our approach as such heuristics are better suited for the evaluation of patterns.

Acknowledgments

This work was partly supported by the German Research Foundation (DFG) as part of the Transregional Collaborative Research Center "Automatic Verification and Analysis of Complex Systems" (SFB/TR 14 AVACS, http://www.avacs.org/).

References

1. Alur, R., Dill, D.L.: A theory of timed automata. Theor. Comput. Sci. 126(2), 183–235 (1994)
2. Culberson, J.C., Schaeffer, J.: Pattern databases. Comp. Int. 14(3), 318–334 (1998)
3. Dierks, H.: Time, Abstraction and Heuristics – Automatic Verification and Planning of Timed Systems using Abstraction and Heuristics. Habilitation Thesis, University of Oldenburg, Germany (2005)
4. Dräger, K., Finkbeiner, B., Podelski, A.: Directed model checking with distance-preserving abstractions. STTT 11(1), 27–37 (2009)
5. Edelkamp, S.: Planning with pattern databases. In: Proc. ECP, pp. 13–24 (2001)
6. Edelkamp, S., Leue, S., Lluch-Lafuente, A.: Directed explicit-state model checking in the validation of communication protocols. STTT 5(2), 247–267 (2004)

7. Edelkamp, S., Schuppan, V., Bošnački, D., Wijs, A., Fehnker, A., Aljazzar, H.: Survey on directed model checking. In: Peled, D.A., Wooldridge, M.J. (eds.) MoChArt 2008. LNCS, vol. 5348, pp. 65–89. Springer, Heidelberg (2009)
8. Hart, P.E., Nilsson, N.J., Raphael, B.: A formal basis for the heuristic determination of minimum cost paths. IEEE Trans. Systems Science and Cybernetics 4(2), 100–107 (1968)
9. Hart, P.E., Nilsson, N.J., Raphael, B.: Correction to a formal basis for the heuristic determination of minimum cost paths. SIGART Newsletter 37, 28–29 (1972)
10. Hoffmann, J., Smaus, J.-G., Rybalchenko, A., Kupferschmid, S., Podelski, A.: Using predicate abstraction to generate heuristic functions in UPPAAL. In: Edelkamp, S., Lomuscio, A. (eds.) MoChArt IV. LNCS (LNAI), vol. 4428, pp. 51–66. Springer, Heidelberg (2007)
11. Kozen, D.: Lower bounds for natural proof systems. In: Proc. FOCS, pp. 254–266. IEEE Computer Society, Los Alamitos (1977)
12. Krieg-Brückner, B., Peleska, J., Olderog, E.R., Baer, A.: The UniForM workbench, a universal development environment for formal methods. In: Woodcock, J.C.P., Davies, J. (eds.) FM 1999. LNCS, vol. 1709, pp. 1186–1205. Springer, Heidelberg (1999)
13. Kupferschmid, S., Dräger, K., Hoffmann, J., Finkbeiner, B., Dierks, H., Podelski, A., Behrmann, G.: UPPAAL/DMC – abstraction-based heuristics for directed model checking. In: Grumberg, O., Huth, M. (eds.) TACAS 2007. LNCS, vol. 4424, pp. 679–682. Springer, Heidelberg (2007)
14. Kupferschmid, S., Hoffmann, J., Dierks, H., Behrmann, G.: Adapting an AI planning heuristic for directed model checking. In: Valmari, A. (ed.) SPIN 2006. LNCS, vol. 3925, pp. 35–52. Springer, Heidelberg (2006)
15. Kupferschmid, S., Hoffmann, J., Larsen, K.G.: Fast directed model checking via russian doll abstraction. In: Ramakrishnan, C.R., Rehof, J. (eds.) TACAS 2008. LNCS, vol. 4963, pp. 203–217. Springer, Heidelberg (2008)
16. Kupferschmid, S., Wehrle, M., Nebel, B., Podelski, A.: Faster than UPPAAL? In: Gupta, A., Malik, S. (eds.) CAV 2008. LNCS, vol. 5123, pp. 552–555. Springer, Heidelberg (2008)
17. Olderog, E.R., Dierks, H.: Moby/RT: A tool for specification and verification of real-time systems. J. UCS 9(2), 88–105 (2003)
18. Pearl, J.: Heuristics: Intelligent Search Strategies for Computer Problem Solving. Addison-Wesley, Reading (1984)
19. Qian, K., Nymeyer, A.: Guided invariant model checking based on abstraction and symbolic pattern databases. In: Jensen, K., Podelski, A. (eds.) TACAS 2004. LNCS, vol. 2988, pp. 497–511. Springer, Heidelberg (2004)
20. Smaus, J.-G., Hoffmann, J.: Relaxation refinement: A new method to generate heuristic functions. In: Peled, D.A., Wooldridge, M.J. (eds.) MoChArt 2008. LNCS, vol. 5348, pp. 147–165. Springer, Heidelberg (2009)
21. Wehrle, M., Helmert, M.: The causal graph revisited for directed model checking. In: Palsberg, J., Su, Z. (eds.) SAS 2009. LNCS, vol. 5673, pp. 86–101. Springer, Heidelberg (2009)
22. Wehrle, M., Kupferschmid, S.: Context-enhanced directed model checking. In: van de Pol, J., Weber, M. (eds.) Model Checking Software. LNCS, vol. 6349, pp. 88–105. Springer, Heidelberg (2010)
23. Wehrle, M., Kupferschmid, S., Podelski, A.: Transition-based directed model checking. In: Kowalewski, S., Philippou, A. (eds.) TACAS 2009. LNCS, vol. 5505, pp. 186–200. Springer, Heidelberg (2009)

The ACL2 Sedan Theorem Proving System

Harsh Raju Chamarthi, Peter Dillinger, Panagiotis Manolios, and Daron Vroon

College of Computer and Information Science
Northeastern University
360 Huntington Ave., Boston MA 02115, USA
{harshrc,pcd,pete}@ccs.neu.edu, daron.vroon@gmail.com

Abstract. The ACL2 Sedan theorem prover (ACL2s) is an Eclipse plug-in that provides a modern integrated development environment, supports several modes of interaction, provides a powerful termination analysis engine, and includes fully automatic bug-finding methods based on a synergistic combination of theorem proving and random testing. ACL2s is publicly available and open source. It has also been used in several sections of a required freshman course at Northeastern University to teach over 200 undergraduate students how to reason about programs.

1   Introduction

ACL2 is a powerful system for integrated modeling, simulation, and theorem proving [5,4,6]. Think of ACL2 as a finely-tuned racecar. In the hands of experts, it has been used to prove some of the most the complex theorems ever proved about commercially designed systems. Novices, however, tend to have a different experience: they crash and burn. Our motivation in developing ACL2s, the ACL2 Sedan, was to bring computer-aided reasoning to the masses by developing a user-friendly system that retained the power of ACL2, but made it possible for new users to quickly, easily learn how to develop and reason about programs.

Usability is one of the major factors contributing to ACL2's steep learning curve. To address the usability problem, ACL2s provides a modern graphical integrated development environment. It is an Eclipse plug-in that includes syntax highlighting, character pair matching, input command demarcation and classification, automatic indentation, auto-completion, a powerful undo facility, various script management capabilities, a clickable proof-tree viewer, clickable icons and keybindings for common actions, tracing support, support for graphics development, and a collection of session modes ranging from beginner modes to advanced user modes. ACL2s also provides GUI support for the "method," an approach to developing programs and theorems advocated in the ACL2 book [5]. Most of these features have been described previously, so we will not dwell on them any further [3].

The other major challenge new users are confronted with is formal reasoning. A major advantage of ACL2 is that it is based on a simple applicative programming language, which is easy to teach. What students find more challenging is the ACL2 logic. The first issue they confront is that functions must be shown

P.A. Abdulla and K.R.M. Leino (Eds.): TACAS 2011, LNCS 6605, pp. 291–295, 2011.

to terminate. Termination is used to both guarantee soundness and to introduce induction schemes. We have developed and implemented Calling-Context Graph termination analysis (CCG), which is able to automatically prove termination of the kinds of functions arising in undergraduate classes [7]. However, beginners often define non-terminating functions. A new feature of ACL2s is that it provides support for the interactive use of CCG analysis. In particular, we provide termination counterexamples and a powerful interface for users to direct the CCG analysis. This is described in Section 2.

Once their function definitions are admitted, new users next learn how to reason about such functions, which first requires learning how to specify properties. We have seen that beginners often make specification errors. ACL2s provides a new lightweight and fully automatic synergistic integration of testing and theorem proving that often generates counterexamples to false conjectures. The counterexamples allow users to quickly fix specification errors and to learn the valuable skill of generating correct specifications. This works well pedagogically because students know how to program, so they understand evaluation. Invalidating a conjecture simply involves finding inputs for which their conjecture evaluates to false. This is similar to the unit testing they do when they develop programs, except that it is automated. An overview of our synergistic integration of testing and theorem proving is given in Section 3.

ACL2s has been successfully used to teach novices. We have used ACL2s at Northeastern University to teach eight sections of a required second-semester freshman course entitled "Logic and Computation." The goal of the class is to teach fundamental techniques for describing and reasoning about computation. Students learn that they can gain predictive power over the programs they write by using logic and automated theorem proving. They learn to use ACL2s to model systems, to specify correctness, to validate their designs using lightweight methods, and to ultimately prove theorems that are mechanically checked. For example, students reason about data structures, circuits, and algorithms; they prove that a simple compiler is correct; they prove equivalence between various programs; they show that library routines are observationally equivalent; and they develop and reason about video games.

ACL2s is freely available, open-source, and well supported [1]. Installation is simple, *e.g.*, we provide prepackaged images for Mac, Linux, and Windows platforms. In addition, everything described in this paper is implemented and available in the current version of ACL2s.

2   Termination Analysis Using Calling Context Graphs

Consider the function definitions in Figure 1, where zp is false iff its argument is a positive integer and expt is exponentiation.

This program was generated by applying weakest precondition analysis to a triply-nested loop. An expert with over a decade of theorem proving experience

```
(defun f1 (w r z s x y a b zs)        (defun f2 (w r z s x y a b zs)
  (if (not (zp z))                      (if (not (zp x))
      (f2 w r z 0 r w 0 0 zs)               (f3 w r z s x y y s zs)
      (= w (expt r zs))))                   (f1 s r (- z 1) 0 0 0 0 zs))))

              (defun f3 (w r z s x y a b zs)
                (if (not (zp a))
                    (f3 w r z s x y (- a 1) (+ b 1) zs)
                    (f2 w r z b (- x 1) y 0 0 zs)))
```

Fig. 1. An interesting termination problem

spent 4–6 hours attempting to construct a measure function that could be used to prove termination, before giving up. Readers are encouraged to construct a measure and to mechanically verify it. (It took us about 20 minutes.) We also tried our CCG termination analysis, as implemented in ACL2s: it proved termination in under 2 seconds, fully automatically with no user guidance.

We have found that if beginners write a terminating function, our CCG analysis will almost certainly prove termination automatically. Unfortunately, beginners often write non-terminating programs, and then want to know why termination analysis failed. This lead us to develop an algorithm that generates a simplified version of the user's program that highlights the reason for the failure. We call the simplified program that is generated a *termination core*, and it corresponds to a single simple cycle of the original program which the CCG analysis was unable to prove terminating.

The termination core can be seen as an explanation of why the CCG analysis failed. After examining the termination core, a user has three options. First, the user can change the function definition. This is the logical course of action if the loop returned by CCG reveals that the program as defined is really not terminating. Second, the user can guide the CCG analysis by providing hints that tell the CCG analysis what local measures to consider. We provide a hint mechanism for doing this. The user can provide either the CONSIDER hint or the CONSIDER-ONLY hint. The former tells CCG to add the user-provided local measures to the local measures it heuristically guesses, while the latter tells CCG to use only the user-provided local measures. This is an effective means of guiding CCG if its heuristics fail to guess the appropriate local measures, and is much simpler than the previous alternative which was to construct a global measure. Finally, it may be the case that the CCG analysis guessed the appropriate local measures but was unable to prove the necessary theorems to show that those measures decrease from one step to the next. In this case, the user can prove the appropriate lemmas.

The result of integrating CCG with ACL2s is a highly automated, intuitive, and interactive termination analysis that eases the steep learning curve for new users of ACL2 and streamlines the ACL2 development process for expert users. The ACL2 Sedan includes extensive documentation of CCG analysis.

3   Random Testing and Proving: Synergistic Combination

Users of ACL2 spend much of their time and effort steering the theorem prover towards proofs of conjectures. During this process users invariably consider conjectures that are in fact false. Often, it is difficult even for experts to determine whether the theorem prover failed because the conjecture is not true or because the theorem prover needs further user guidance.

ACL2s provides a lightweight method based on the synergistic combination of random testing [2] and theorem proving, for debugging and understanding conjectures. This has turned out to be invaluable in helping beginners become effective users of formal methods. We have integrated random testing into ACL2s in a deep way: it is enabled by default and requires no special syntax so that users get the benefit of random testing without any effort on their part.

Since ACL2 formulas are executable, random testing in ACL2 involves randomly instantiating the free variables in a formula and then evaluating the result. This is a small part of the picture because this naive approach is unlikely to find counterexamples in all but the simplest of cases. This is especially true in a theorem prover for an untyped logic, like ACL2, where every variable can take on any value. As might be expected, conjectures typically contain hypotheses that constrain variables. Therefore, we randomly instantiate variables subject to these constraints. We do this by introducing a flexible and powerful data definition framework in ACL2s which provides support for defining union types, product types, list types, record types, enumeration types, and mutually-recursive data definitions. It allows the use of macros inside definitions and supports custom data definitions (*e.g.*, primes). The data definition framework is integrated with our random testing framework in several important ways. For example, we guarantee that random testing will automatically generate examples that satisfy any hypothesis restricting the type of a variable.

Complex conjectures often involve many variables with many hypotheses and intricate propositional structure involving complex hierarchies of user-defined functions. Testing such conjectures directly is unlikely to yield counterexamples. We address this by integrating our testing framework with the core theorem proving engine in a synergistic fashion, using the full power of ACL2 to simplify conjectures for better testing. The main idea is to let ACL2 use all of the proof techniques at its disposal to simplify conjectures into subgoals, and to then test the "interesting" subgoals. This winds up requiring lots of care. For example, ACL2 employs proof techniques that can generate radically transformed subgoals, where variables disappear or are replaced with new variables that are related to the original variables via certain constraints. Finally, our analysis is *sound*, *i.e.*, any counterexamples generated truly are counterexamples to the original conjecture.

References

1. Chamarthi, H.R., Dillinger, P.C., Manolios, P., Vroon, D.: ACL2 Sedan homepage, http://acl2s.ccs.neu.edu/
2. Claessen, K., Hughes, J.: QuickCheck: a lightweight tool for random testing of Haskell programs. In: ICFP, pp. 268–279 (2000)
3. Dillinger, P.C., Manolios, P., Vroon, D., Strother Moore, J.: ACL2s: "The ACL2 Sedan". Electr. Notes Theor. Comput. Sci. 174(2), 3–18 (2007)
4. Kaufmann, M., Manolios, P., Strother Moore, J. (eds.): Computer-Aided Reasoning: ACL2 Case Studies. Kluwer Academic Publishers, Dordrecht (2000)
5. Kaufmann, M., Manolios, P., Strother Moore, J.: Computer-Aided Reasoning: An Approach. Kluwer Academic Publishers, Dordrecht (2000)
6. Kaufmann, M., Strother Moore, J.: ACL2 homepage, http://www.cs.utexas.edu/users/moore/acl2
7. Manolios, P., Vroon, D.: Termination analysis with calling context graphs. In: Ball, T., Jones, R.B. (eds.) CAV 2006. LNCS, vol. 4144, pp. 401–414. Springer, Heidelberg (2006)

On Probabilistic Parallel Programs with Process Creation and Synchronisation*

Stefan Kiefer and Dominik Wojtczak

Oxford University Computing Laboratory, UK
{stefan.kiefer,dominik.wojtczak}@comlab.ox.ac.uk

Abstract. We initiate the study of probabilistic parallel programs with dynamic process creation and synchronisation. To this end, we introduce *probabilistic split-join systems (pSJSs)*, a model for parallel programs, generalising both probabilistic pushdown systems (a model for sequential probabilistic procedural programs which is equivalent to recursive Markov chains) and stochastic branching processes (a classical mathematical model with applications in various areas such as biology, physics, and language processing). Our pSJS model allows for a possibly recursive spawning of parallel processes; the spawned processes can synchronise and return values. We study the basic performance measures of pSJSs, especially the distribution and expectation of space, work and time. Our results extend and improve previously known results on the subsumed models. We also show how to do performance analysis in practice, and present two case studies illustrating the modelling power of pSJSs.

1 Introduction

The verification of probabilistic programs with possibly recursive procedures has been intensely studied in the last years. The Markov chains or Markov Decision Processes underlying these systems may have infinitely many states. Despite this fact, which prevents the direct application of the rich theory of finite Markov chains, many positive results have been obtained. Model-checking algorithms have been proposed for both linear and branching temporal logics [11,15,23], algorithms deciding properties of several kinds of games have been described (see e.g. [14]), and distributions and expectations of performance measures such as run-time and memory consumption have been investigated [12,4,5].

In all these papers programs are modelled as *probabilistic pushdown systems (pPDSs)* or, equivalently [9], as recursive Markov chains. Loosely speaking, a pPDS is a pushdown automaton whose transitions carry probabilities. The *configurations* of a pPDS are pairs containing the current control state and the current stack content. In each *step*, a new configuration is obtained from its predecessor by applying a transition rule, which may modify the control state and the top of the stack.

The programs modelled by pPDSs are necessarily sequential: at each point in time, only the procedure represented by the topmost stack symbol is active. Recursion, however, is a useful language feature also for multithreaded and other parallel programming

---

* The first author is supported by a postdoctoral fellowship of the German Academic Exchange Service (DAAD). The second author is supported by EPSRC grant EP/G050112/1.

P.A. Abdulla and K.R.M. Leino (Eds.): TACAS 2011, LNCS 6605, pp. 296–310, 2011.

languages, such as Cilk and JCilk, which allow, e.g., for a natural parallelisation of divide-and-conquer algorithms [7,8]. To model parallel programs in probabilistic scenarios, one may be tempted to use *stochastic multitype branching processes*, a classical mathematical model with applications in numerous fields including biology, physics and natural language processing [17,2]. In this model, each process has a type, and each type is associated with a probability distribution on transition rules. For instance, a branching process with the transition rules $X \xrightarrow{2/3} \{\}, X \xrightarrow{1/3} \{X,Y\}, Y \xrightarrow{1} \{X\}$ can be thought of describing a parallel program with two types of processes, X and Y. A process of type X terminates with probability $2/3$, and with probability $1/3$ stays active and spawns a new process of type Y. A process of type Y changes its type to X. A *configuration* of a branching process consists of a pool of currently active processes. In each *step*, all active processes develop in parallel, each one according to a rule which is chosen probabilistically. For instance, a step transforms the configuration $\langle XY \rangle$ into $\langle XYX \rangle$ with probability $\frac{1}{3} \cdot 1$, by applying the second X-rule to the X-process and, in parallel, the Y-rule to the Y-process.

Branching processes do not satisfactorily model parallel programs, because they lack two key features: synchronisation and returning values. In this paper we introduce *probabilistic split-join systems (pSJSs)*, a model which offers these features. Parallel spawns are modelled by rules of the form $X \hookrightarrow \langle YZ \rangle$. The spawned processes Y and Z develop independently; e.g., a rule $Y \hookrightarrow Y'$ may be applied to the Y-process, replacing Y by Y'. When terminating, a process enters a *synchronisation state*, e.g. with rules $Y' \hookrightarrow q$ and $Z \hookrightarrow r$ (where q and r are synchronisation states). Once a process terminates in a synchronisation state, it waits for its *sibling* to terminate in a synchronisation state as well. In the above example, the spawned processes wait for each other, until they terminate in q and r. At that point, they may *join* to form a single process, e.g. with a rule $\langle qr \rangle \hookrightarrow W$. So, synchronisation is achieved by the siblings waiting for each other to terminate. All rules could be probabilistic. Notice that synchronisation states can be used to return values; e.g., if the Y-process returns q' instead of q, this can be recorded by the existence of a rule $\langle q'r \rangle \hookrightarrow W'$, so that the resulting process (i.e., W or W') depends on the values computed by the joined processes. For the notion of siblings to make sense, a *configuration* of a pSJS is not a set, but a binary tree whose leaves are *process symbols* (such as X, Y, Z) or synchronisation states (such as q, r). A *step* transforms the leaves of the binary tree in parallel by applying rules; if a leaf is not a process symbol but a synchronisation state, it remains unchanged unless its sibling is also a synchronisation state and a joining rule (such as $\langle qr \rangle \hookrightarrow W$) exists, which removes the siblings and replaces their parent node with the right hand side.

Related work. The probabilistic models closest to ours are pPDSs, recursive Markov chains, and stochastic branching processes, as described above. The non-probabilistic (i.e., nondeterministic) version of pSJSs (SJSs, say) can be regarded as a special case of *ground tree rewriting systems*, see [19] and the references therein. A configuration of a ground tree rewriting system is a node-labelled tree, and a rewrite rule replaces a subtree. The *process rewrite system* (PRS) hierarchy of [21] features sequential and parallel process composition. Due to its syntactic differences, it is not obvious whether SJSs are in that hierarchy. They would be above pushdown systems (which is the sequential fragment of PRSs), because SJSs subsume pushdown systems, as we show in

Section 3.1 for the probabilistic models. *Dynamic pushdown networks* (DPNs) [3] are a parallel extension of pushdown systems. A configuration of a DPN is a list of configurations of pushdown systems running in parallel. DPNs feature the spawning of parallel threads, and an extension of DPNs, called *constrained DPNs*, can also model joins via regular expressions on spawned children. The DPN model is more powerful and more complicated than SJSs. All those models are non-probabilistic.

Organisation of the paper. In Section 2 we formally define our model and provide further preliminaries. Section 3 contains our main results: we study the relationship between pSJSs and pPDSs (Section 3.1), we show how to compute the probabilities for termination and finite space, respectively (Sections 3.2 and 3.3), and investigate the distribution and expectation of work and time (Section 3.4). In Section 4 we present two case studies illustrating the modelling power of pSJSs. We conclude in Section 5. All proofs are provided in a technical report [18].

2   Preliminaries

For a finite or infinite word w, we write $w(0), w(1), \ldots$ to refer to its individual letters. We assume throughout the paper that $\mathcal{B}$ is a fixed infinite set of *basic process symbols*. We use the symbols '$\langle$' and '$\rangle$' as special letters not contained in $\mathcal{B}$. For an alphabet Σ, we write $\langle \Sigma \Sigma \rangle$ to denote the language $\{\langle \sigma_1 \sigma_2 \rangle \mid \sigma_1, \sigma_2 \in \Sigma\}$ and $\Sigma^{1,2}$ to denote $\Sigma \cup \langle \Sigma \Sigma \rangle$. To a set Σ we associate a set $T(\Sigma)$ of binary trees whose leaves are labelled with elements of Σ. Formally, $T(\Sigma)$ is the smallest language that contains Σ and $\langle T(\Sigma) T(\Sigma) \rangle$. For instance, $\langle \langle \sigma \sigma \rangle \sigma \rangle \in T(\{\sigma\})$.

Definition 1 (pSJS). *Let Q be a finite set of* synchronisation states *disjoint from $\mathcal{B}$ and not containing '$\langle$' or '$\rangle$'. Let Γ be a finite set of* process symbols, *such that $\Gamma \subset \mathcal{B} \cup \langle QQ \rangle$. Define the alphabet $\Sigma := \Gamma \cup Q$. Let $\delta \subseteq \Gamma \times \Sigma^{1,2}$ be a transition relation. Let $Prob : \delta \to (0,1]$ be a function so that for all $a \in \Gamma$ we have $\sum_{a \hookrightarrow \alpha \in \delta} Prob(a \hookrightarrow \alpha) = 1$. Then the tuple $S = (\Gamma, Q, \delta, Prob)$ is a* probabilistic split-join system *(pSJS). A pSJS with $\Gamma \cap \langle QQ \rangle = \emptyset$ is called* branching process.

We usually write $a \overset{p}{\hookrightarrow} \alpha$ instead of $Prob(a \hookrightarrow \alpha) = p$. For technical reasons we allow branching processes of "degree 3", i.e., branching processes where $\Sigma^{1,2}$ may be extended to $\Sigma^{1,2,3} := \Sigma^{1,2} \cup \{\langle \sigma_1 \sigma_2 \sigma_3 \rangle \mid \sigma_1, \sigma_2, \sigma_3 \in \Sigma\}$. In branching processes, it is usually sufficient to have $|Q| = 1$.

A *Markov chain* is a stochastic process that can be described by a triple $M = (D, \to, Prob)$ where D is a finite or countably infinite set of *states*, $\to \subseteq D \times D$ is a *transition relation*, and $Prob$ is a function which to each transition $s \to t$ of M assigns its probability $Prob(s \to t) > 0$ so that for every $s \in D$ we have $\sum_{s \to t} Prob(s \to t) = 1$ (as usual, we write $s \overset{x}{\to} t$ instead of $Prob(s \to t) = x$). A *path* (or *run*) in M is a finite (or infinite, resp.) word $u \in D^+ \cup D^\omega$, such that $u(i-1) \to u(i)$ for every $1 \leq i < |u|$. The set of all runs that start with a given path u is denoted by $Run[M](u)$ (or $Run(u)$, if M is understood). To every $s \in D$ we associate the probability space $(Run(s), \mathcal{F}, \mathcal{P})$ where $\mathcal{F}$ is the σ-field generated by all *basic cylinders* $Run(u)$ where u is a path starting with s, and $\mathcal{P} : \mathcal{F} \to [0,1]$ is the unique probability measure such that $\mathcal{P}(Run(u)) = \Pi_{i=1}^{|u|-1} x_i$ where $u(i-1) \overset{x_i}{\to} u(i)$ for every

$1 \leq i < |u|$. Only certain subsets of $Run(s)$ are $\mathcal{P}$-measurable, but in this paper we only deal with "safe" subsets that are guaranteed to be in $\mathcal{F}$. If $\mathbf{X}_s$ is a random variable over $Run(s)$, we write $\mathbb{E}[\mathbf{X}_s]$ for its expectation. For $s, t \in D$, we define $Run(s{\downarrow}t) := \{w \in Run(s) \mid \exists i \geq 0 : w(i) = t\}$ and $[s{\downarrow}t] := \mathcal{P}(Run(s{\downarrow}t))$.

To a pSJS $S = (\Gamma, Q, \delta, Prob)$ with alphabet $\Sigma = \Gamma \cup Q$ we associate a Markov chain M_S with $T(\Sigma)$ as set of states. For $t \in T(\Sigma)$, we define $Front(t) = a_1, \ldots, a_k$ as the unique finite sequence of subwords of t (read from left to right) with $a_i \in \Gamma$ for all $1 \leq i \leq k$. We write $|Front(t)| = k$. If $k = 0$, then t is called *terminal*. The Markov chain M_S has a transition $t \xrightarrow{p} t'$, if: $Front(t) = a_1, \ldots, a_k$; $a_i \xhookrightarrow{p_i} \alpha_i$ are transitions in S for all i; t' is obtained from t by replacing a_i with α_i for all i; and $p = \prod_{i=1}^{k} p_i$. Note that $t \xrightarrow{1} t$, if t is terminal. For branching processes of degree 3, the set $T(\Sigma)$ is extended in the obvious way to trees whose nodes may have two or three children.

Denote by $\mathbf{T}_\sigma$ a random variable over $Run(\sigma)$ where $\mathbf{T}_\sigma(w)$ is either the least $i \in \mathbb{N}$ such that $w(i)$ is terminal, or ∞, if no such i exists. Intuitively, $\mathbf{T}_\sigma(w)$ is the number of steps in which w terminates, i.e., the *termination time*. Denote by $\mathbf{W}_\sigma$ a random variable over $Run(\sigma)$ where $\mathbf{W}_\sigma(w) := \sum_{i=0}^{\infty} |Front(w(i))|$. Intuitively, $\mathbf{W}_\sigma(w)$ is the total *work* in w. Denote by $\mathbf{S}_\sigma$ a random variable over $Run(\sigma)$ where $\mathbf{S}_\sigma(w) := \sup_{i=0}^{\infty} |w(i)|$, and $|w(i)|$ is the length of $w(i)$ not counting the symbols '$\langle$' and '$\rangle$'. Intuitively, $\mathbf{S}_\sigma(w)$ is the maximal number of processes during the computation, or, short, the *space* of w.

Example 2. Consider the pSJS with $\Gamma = \{X, \langle qr \rangle\}$ and $Q = \{q, r\}$ and the transitions $X \xhookrightarrow{0.5} \langle XX \rangle$, $X \xhookrightarrow{0.3} q$, $X \xhookrightarrow{0.2} r$, $\langle qr \rangle \xhookrightarrow{1} X$. Let $u = X \langle XX \rangle \langle qr \rangle X q q$. Then u is a path, because we have $X \xrightarrow{0.5} \langle XX \rangle \xrightarrow{0.06} \langle qr \rangle \xrightarrow{1} X \xrightarrow{0.3} q \xrightarrow{1} q$. Note that q is terminal. The set $Run(u)$ contains only one run, namely $w := u(0)u(1)u(2)u(3)u(4) u(4) \cdots$. We have $\mathcal{P}(Run(u)) = 0.5 \cdot 0.06 \cdot 0.3$, and $\mathbf{T}_X(w) = 4$, $\mathbf{W}_X(w) = 5$, and $\mathbf{S}_X(w) = 2$. The dags in Figure 1 graphically represent this run (on the left), and another example run (on the right) with $\mathbf{T}_X = 3$, $\mathbf{W}_X = 5$, and $\mathbf{S}_X = 3$.

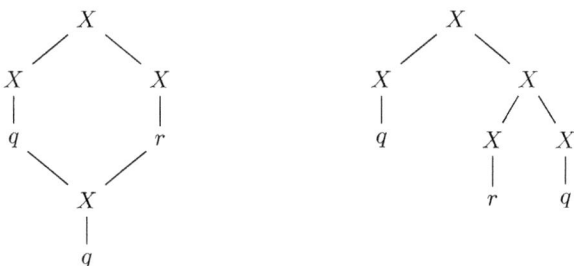

Fig. 1. Two terminating runs

Remark 3. Our definition of pSJSs may be more general than needed from a modelling perspective: e.g., our rules allow for both synchronisation and splitting in a single step. We choose this definition for technical convenience and to allow for easy comparisons with pPDSs (Section 3.1).

The complexity-theoretic statements in this paper are with respect to the *size* of the given pSJS $S = (\Gamma, Q, \delta, Prob)$, which is defined as $|\Gamma| + |Q| + |\delta| + |Prob|$, where $|Prob|$ equals the sum of the sizes of the binary representations of the values of $Prob$. A formula of $ExTh(\mathbb{R})$, the existential fragment of the first-order theory of the reals, is of the form $\exists x_1 \ldots \exists x_m R(x_1, \ldots, x_n)$, where $R(x_1, \ldots, x_n)$ is a boolean combination of comparisons of the form $p(x_1, \ldots, x_n) \sim 0$, where $p(x_1, \ldots, x_n)$ is a multivariate polynomial and $\sim \in \{<, >, \leq, \geq, =, \neq\}$. The validity of closed formulas ($m = n$) is decidable in PSPACE [6,22]. We say that one can *efficiently express* a value $c \in \mathbb{R}$ associated with a pSJS, if one can, in polynomial space, construct a formula $\phi(x)$ in $ExTh(\mathbb{R})$ of polynomial length such that x is the only free variable in $\phi(x)$, and $\phi(x)$ is true if and only if $x = c$. Notice that if c is efficiently expressible, then $c \sim \tau$ for $\tau \in \mathbb{Q}$ is decidable in PSPACE for $\sim \in \{<, >, \leq, \geq, =, \neq\}$.

For some lower bounds, we prove hardness (with respect to P-time many-one reductions) in terms of the PosSLP decision problem. The PosSLP (Positive Straight-Line Program) problem asks whether a given straight-line program or, equivalently, arithmetic circuit with operations $+$, $-$, $\cdot$, and inputs 0 and 1, and a designated output gate, outputs a positive integer or not. PosSLP is in PSPACE. More precisely, it is known to be on the 4th level of the Counting Hierarchy [1]; it is not known to be in NP. The PosSLP problem is a fundamental problem for numerical computation; it is complete for the class of decision problems that can be solved in polynomial time on models with unit-cost exact rational arithmetic, see [1,15] for more details.

3   Results

3.1   Relationship with Probabilistic Pushdown Systems (pPDSs)

We show that pSJSs subsume pPDSs. A *probabilistic pushdown system (pPDS)* [11,12,4,5] is a tuple $S = (\Gamma, Q, \delta, Prob)$, where Γ is a finite *stack alphabet*, Q is a finite set of *control states*, $\delta \subseteq Q \times \Gamma \times Q \times \Gamma^{\leq 2}$ (where $\Gamma^{\leq 2} = \{\alpha \in \Gamma^*, |\alpha| \leq 2\}$) is a *transition relation*, and $Prob : \delta \to (0, 1]$ is a function so that for all $q \in Q$ and $a \in \Gamma$ we have $\sum_{qa \hookrightarrow r\alpha} Prob(qa \hookrightarrow r\alpha) = 1$. One usually writes $qa \xrightarrow{p} r\alpha$ instead of $Prob(qa \to r\alpha) = p$. To a pPDS $S = (\Gamma, Q, \delta, Prob)$ one associates a Markov chain M_S with $Q \times \Gamma^*$ as set of states, and transitions $q \xrightarrow{1} q$ for all $q \in Q$, and $qa\beta \xrightarrow{p} r\alpha\beta$ for all $qa \xrightarrow{p} r\alpha$ and all $\beta \in \Gamma^*$.

A pPDS S_P with Γ_P as stack alphabet, Q_P as set of control states, and transitions $\xrightarrow{p}_P$ can be transformed to an equivalent pSJS S: Take $Q := Q_P \cup \Gamma_P$ as synchronisation states; $\Gamma := \{\langle qa \rangle \mid q \in Q_P, \ a \in \Gamma_P\}$ as process symbols; and transitions $\langle qa \rangle \xrightarrow{p} \langle \langle rb \rangle c \rangle$ for all $qa \xrightarrow{p}_P rbc$, $\langle qa \rangle \xrightarrow{p} \langle rb \rangle$ for all $qa \xrightarrow{p}_P rb$, and $\langle qa \rangle \xrightarrow{p} r$ for all $qa \xrightarrow{p}_P r$. The Markov chains M_{S_P} and M_S are isomorphic. Therefore, we occasionally say that a pSJS *is* a pPDS, if it can be obtained from a pPDS by this transformation. Observe that in pPDSs, we have $\mathbf{T} = \mathbf{W}$, because there is no parallelism.

Conversely, a pSJS S with alphabet $\Sigma = \Gamma \cup Q$ can be transformed into a pPDS S_P by "serialising" S: Take $Q_P := \{\Box\} \cup \{\overline{q} \mid q \in Q\}$ as control states; $\Gamma_P := \Gamma \cup Q \cup \{\widetilde{q} \mid q \in Q\}$ as stack alphabet; and transitions $\Box a \xrightarrow{p}_P \Box\sigma_1\sigma_2$ for all $a \xrightarrow{p} \langle\sigma_1\sigma_2\rangle$,

$\Box a \xrightarrow{p}_P \Box\sigma$ for all $a \xrightarrow{p} \sigma$ with $\sigma \in \Sigma \setminus \langle QQ \rangle$, and $\Box q \xrightarrow{1}_P \bar{q}$ for all $q \in Q$, and $\bar{q}\sigma \xrightarrow{1}_P \Box\sigma\tilde{q}$ for all $q \in Q$ and $\sigma \in \Sigma$, and $\bar{r}\tilde{q} \xrightarrow{1}_P \Box\langle qr \rangle$ for all $q, r \in Q$. The Markov chains M_S and M_{S_P} are *not* isomorphic. However, we have:

Proposition 4. *There is a probability-preserving bijection between the runs $Run(\sigma\downarrow q)$ in M_S and the runs $Run(\Box\sigma\downarrow\bar{q})$ in M_{S_P}. In particular, we have $[\sigma\downarrow q] = [\Box\sigma\downarrow\bar{q}]$.*

For example, the pSJS run on the left side of Figure 1 corresponds to the pPDS run
$$\Box X \xrightarrow{0.5} \Box XX \xrightarrow{0.3} \Box qX \xrightarrow{1} \bar{q}X \xrightarrow{1} \Box X\tilde{q} \xrightarrow{0.2} \Box r\tilde{q} \xrightarrow{1} \bar{r}\tilde{q} \xrightarrow{1} \Box\langle qr \rangle \xrightarrow{1} \Box X \xrightarrow{0.3}$$
$$\Box q \xrightarrow{1} \bar{q} \xrightarrow{1} \bar{q} \xrightarrow{1} \ldots$$

3.2 Probability of Termination

We call a run *terminating*, if it reaches a terminal tree. Such a tree can be a single synchronisation state (e.g., q on the left of Figure 1), or another terminal tree (e.g., $\langle q \langle rq \rangle \rangle$ on the right of Figure 1). For any $\sigma \in \Sigma$, we denote by $[\sigma\downarrow]$ the termination probability when starting in σ; i.e., $[\sigma\downarrow] = \sum_{t \text{ is terminal}} [\sigma\downarrow t]$. One can transform any pSJS S into a pSJS S' such that whenever a run in S terminates, then a corresponding run in S' terminates in a synchronisation state. This transformation is by adding a fresh state $\check{q}$, and transitions $\langle rs \rangle \xrightarrow{1} \check{q}$ for all $r, s \in Q$ with $\langle rs \rangle \notin \Gamma$, and $\langle \check{q}r \rangle \xrightarrow{1} \check{q}$ and $\langle r\check{q} \rangle \xrightarrow{1} \check{q}$ for all $r \in Q$. It is easy to see that this keeps the probability of termination unchanged, and modifies the random variables $\mathbf{T}_\sigma$ and $\mathbf{W}_\sigma$ by at most a factor 2. Notice that the transformation can be performed in polynomial time. After the transformation we have $[\sigma\downarrow] = \sum_{q \in Q}[\sigma\downarrow q]$. A pSJS which satisfies this equality will be called *normalised* in the following. From a modelling point of view, pSJSs may be expected to be normalised in the first place: a terminating program should terminate all its processes.

We set up an equation system for the probabilities $[\sigma\downarrow q]$. For each $\sigma \in \Sigma$ and $q \in Q$, the equation system has a variable of the form $[\![\sigma\downarrow q]\!]$ and an equation of the form $[\![\sigma\downarrow q]\!] = f_{[\![\sigma\downarrow q]\!]}$, where $f_{[\![\sigma\downarrow q]\!]}$ is a multivariate polynomial with nonnegative coefficients. More concretely: If $q \in Q$, then we set $[\![q\downarrow q]\!] = 1$; if $r \in Q \setminus \{q\}$, then we set $[\![r\downarrow q]\!] = 0$; if $a \in \Gamma$, then we set

$$[\![a\downarrow q]\!] = \sum_{\substack{a \xrightarrow{p} \langle \sigma_1\sigma_2 \rangle \\ \langle q_1 q_2 \rangle \in \Gamma \cap \langle QQ \rangle}} p \cdot [\![\sigma_1\downarrow q_1]\!] \cdot [\![\sigma_2\downarrow q_2]\!] \cdot [\![\langle q_1q_2 \rangle\downarrow q]\!] \ + \sum_{\substack{a \xrightarrow{p} \sigma' \\ \sigma' \in \Sigma \setminus \langle QQ \rangle}} p \cdot [\![\sigma'\downarrow q]\!].$$

Proposition 5. *Let $\sigma \in \Sigma$ and $q \in Q$. Then $[\sigma\downarrow q]$ is the value for $[\![\sigma\downarrow q]\!]$ in the least (w.r.t. componentwise ordering) nonnegative solution of the above equation system.*

One can efficiently approximate $[\sigma\downarrow q]$ by applying Newton's method to the fixed-point equation system from Proposition 5, cf. [15]. The convergence speed of Newton's method for such equation systems was recently studied in detail [10]. The simpler "Kleene" method (sometimes called "fixed-point iteration") often suffices, but can be much slower. In the case studies of Section 4, using Kleene for computing the termination probabilities up to machine accuracy was not a bottleneck. The following theorem essentially follows from similar results for pPDSs:

Theorem 6 (cf. [13,15]). *Consider a pSJS with alphabet $\Sigma = \Gamma \cup Q$. Let $\sigma \in \Sigma$ and $q \in Q$. Then (1) one can efficiently express (in the sense defined in Section 2) the value of $[\sigma \downarrow q]$, (2) deciding whether $[\sigma \downarrow q] = 0$ is in P, and (3) deciding whether $[\sigma \downarrow q] < 1$ is PosSLP-hard even for pPDSs.*

3.3 Probability of Finite Space

A run $w \in Run(\sigma)$ is either (i) terminating, or (ii) nonterminating with $\mathbf{S}_\sigma < \infty$, or (iii) nonterminating with $\mathbf{S}_\sigma = \infty$. From a modelling point of view, some programs may be considered incorrect, if they do not terminate with probability 1. As is well-known, this does not apply to programs like operating systems, network servers, system daemons, etc., where nontermination may be tolerated or desirable. Such programs may be expected not to need an infinite amount of space; i.e., $\mathbf{S}_\sigma$ should be finite.

Given a pSJS S with alphabet $\Sigma = \Gamma \cup Q$, we show how to construct, in polynomial time, a normalised pSJS $\overline{S}$ with alphabet $\overline{\Sigma} = \overline{\Gamma} \cup \overline{Q} \supseteq \Sigma$ where $\overline{Q} = Q \cup \{\overline{q}\}$ for a fresh synchronisation state $\overline{q}$, and $\mathcal{P}\left(\mathbf{S}_a < \infty = \mathbf{T}_a \mid Run(a)\right) = [a \downarrow \overline{q}]$ for all $a \in \Gamma$. Having done that, we can compute this probability according to Section 3.2.

For the construction, we can assume w.l.o.g. that S has been normalised using the procedure of Section 3.2. Let $U := \{a \in \Gamma \mid \forall n \in \mathbb{N} : \mathcal{P}\left(\mathbf{S}_a > n\right) > 0\}$.

Lemma 7. *The set U can be computed in polynomial time.*

Let $B := \{a \in \Gamma \setminus U \mid \forall q \in Q : [a \downarrow q] = 0\}$, so B is the set of process symbols a that are both "bounded above" (because $a \notin U$) and "bounded below" (because a cannot terminate). By Theorem 6 (2) and Lemma 7 we can compute B in polynomial time. Now we construct $\overline{S}$ by modifying S as follows: we set $\overline{Q} := Q \cup \{\overline{q}\}$ for a fresh synchronisation state $\overline{q}$; we remove all transitions with symbols $b \in B$ on the left hand side and replace them with a new transition $b \overset{1}{\hookrightarrow} \overline{q}$; we add transitions $\langle q_1 q_2 \rangle \overset{1}{\hookrightarrow} \overline{q}$ for all $q_1, q_2 \in \overline{Q}$ with $\overline{q} \in \{q_1, q_2\}$. We have the following proposition.

Proposition 8. *(1) The pSJS $\overline{S}$ is normalised; (2) the value $[a \downarrow q]$ for $a \in \Gamma$ and $q \in Q$ is the same in S and $\overline{S}$; (3) we have $\mathcal{P}\left(\mathbf{S}_a < \infty = \mathbf{T}_a \mid Run(a)\right) = [a \downarrow \overline{q}]$ for all $a \in \Gamma$.*

Proposition 8 allows for the following theorem.

Theorem 9. *Consider a pSJS with alphabet $\Sigma = \Gamma \cup Q$ and $a \in \Gamma$. Let $s := \mathcal{P}\left(\mathbf{S}_a < \infty\right)$. Then (1) one can efficiently express s, (2) deciding whether $s = 0$ is in P, and (3) deciding whether $s < 1$ is PosSLP-hard even for pPDSs.*

Theorem 9, applied to pPDSs, improves Corollary 6.3 of [12]. There it is shown for pPDSs that comparing $\mathcal{P}\left(\mathbf{S}_a < \infty\right)$ with $\tau \in \mathbb{Q}$ is in EXPTIME, and in PSPACE if $\tau \in \{0, 1\}$. With Theorem 9 we get PSPACE for $\tau \in \mathbb{Q}$, and P for $\tau = 0$.

3.4 Work and Time

We show how to compute the distribution and expectation of work and time of a given pSJS S with alphabet $\Sigma = \Gamma \cup Q$.

Distribution. For $\sigma \in \Sigma$ and $q \in Q$, let $T_{\sigma \downarrow q}(k) := \mathcal{P}\left(Run(\sigma \downarrow q), \mathbf{T}_\sigma = k \mid Run(\sigma)\right)$. It is easy to see that, for $k \geq 1$ and $a \in \Gamma$ and $q \in Q$, we have

$$T_{a\downarrow q}(k) = \sum_{\substack{a \xrightarrow{p} \langle \sigma_1 \sigma_2 \rangle \\ \langle q_1 q_2 \rangle \in \Gamma \cap \langle QQ \rangle}} p \cdot \sum_{\substack{\ell_1, \ell_2, \ell_3 \geq 0 \\ \max\{\ell_1, \ell_2\} + \ell_3 = k-1}} T_{\sigma_1 \downarrow q_1}(\ell_1) \cdot T_{\sigma_2 \downarrow q_2}(\ell_2) \cdot T_{\langle q_1 q_2 \rangle \downarrow q}(\ell_3) +$$

$$\sum_{\substack{a \xrightarrow{p} \sigma' \\ \sigma' \in \Sigma \setminus \langle QQ \rangle}} p \cdot T_{\sigma' \downarrow q}(k - 1) .$$

This allows to compute the distribution of time (and, similarly, work) using dynamic programming. In particular, for any k, one can compute $\overrightarrow{T}_{\sigma \downarrow q}(k) := \mathcal{P}(\mathbf{T}_\sigma > k \mid Run(\sigma \downarrow q)) = 1 - \frac{1}{[\sigma \downarrow q]} \sum_{i=0}^{k} T_{\sigma \downarrow q}(k)$.

Expectation. For any random variable Z taking positive integers as value, it holds $\mathbb{E}Z = \sum_{k=0}^{\infty} \mathcal{P}(Z > k)$. Hence, one can approximate $\mathbb{E}[\mathbf{T}_\sigma \mid Run(\sigma \downarrow q)] = \sum_{k=0}^{\infty} \overrightarrow{T}_{\sigma \downarrow q}(k)$ by computing $\sum_{k=0}^{\ell} \overrightarrow{T}_{\sigma \downarrow q}(k)$ for large ℓ. In the rest of the section we show how to decide on the finiteness of expected work and time. It follows from Propositions 10 and 11 below that the expected work $\mathbb{E}[\mathbf{W}_\sigma \mid Run(\sigma \downarrow q)]$ is easier to compute: it is the solution of a linear equation system.

We construct a branching process $\overline{S}$ with process symbols $\overline{\Gamma} = \{(aq) \mid a \in \Gamma, q \in Q, [a \downarrow q] > 0\}$, synchronisation states $\overline{Q} = \{\bot\}$ and transitions as follows. For notational convenience, we identify $\bot$ and (qq) for all $q \in Q$. For $(aq) \in \overline{\Gamma}$, we set

- $(aq) \xrightarrow{y/[a\downarrow q]} \langle (\sigma_1 q_1)(\sigma_2 q_2)((q_1 q_2)q) \rangle$ for all $a \xrightarrow{p} \langle \sigma_1 \sigma_2 \rangle$ and $\langle q_1 q_2 \rangle \in \Gamma \cap \langle QQ \rangle$, where $y := p \cdot [\sigma_1 \downarrow q_1] \cdot [\sigma_2 \downarrow q_2] \cdot [\langle q_1 q_2 \rangle \downarrow q] > 0$;
- $(aq) \xrightarrow{y/[a\downarrow q]} (\sigma' q)$ for all $a \xrightarrow{p} \sigma'$ with $\sigma' \in \Sigma \setminus \langle QQ \rangle$, where $y := p \cdot [\sigma' \downarrow q] > 0$.

The following proposition (inspired by a statement on pPDSs [5]) links the distributions of $\mathbf{W}_\sigma$ and $\mathbf{T}_\sigma$ conditioned under termination in q with the distributions of $\mathbf{W}_{(\sigma q)}$ and $\mathbf{T}_{(\sigma q)}$.

Proposition 10. *Let* $\sigma \in \Sigma$ *and* $q \in Q$ *with* $[\sigma \downarrow q] > 0$. *Then*

$$\mathcal{P}(\mathbf{W}_\sigma = n \mid Run(\sigma \downarrow q)) = \mathcal{P}(\mathbf{W}_{(\sigma q)} = n \mid Run((\sigma q))) \quad \text{for all } n \geq 0 \quad \text{and}$$
$$\mathcal{P}(\mathbf{T}_\sigma \leq n \mid Run(\sigma \downarrow q)) \leq \mathcal{P}(\mathbf{T}_{(\sigma q)} \leq n \mid Run((\sigma q))) \quad \text{for all } n \geq 0.$$

In particular, we have $[(\sigma q) \downarrow] = 1$.

Proposition 10 allows us to focus on branching processes. For $X \in \Gamma$ and a finite sequence $\sigma_1, \ldots, \sigma_k$ with $\sigma_i \in \Sigma$, define $|\sigma_1, \ldots, \sigma_k|_X := |\{i \mid 1 \leq i \leq k, \sigma_i = X\}|$, i.e., the number of X-symbols in the sequence. We define the *characteristic matrix* $A \in \mathbb{R}^{\Gamma \times \Gamma}$ of a branching process by setting

$$A_{X,Y} := \sum_{X \xrightarrow{p} \langle \sigma_1 \sigma_2 \sigma_3 \rangle} p \cdot |\sigma_1, \sigma_2, \sigma_3|_Y + \sum_{X \xrightarrow{p} \langle \sigma_1 \sigma_2 \rangle} p \cdot |\sigma_1, \sigma_2|_Y + \sum_{X \xrightarrow{p} \sigma_1} p \cdot |\sigma_1|_Y .$$

It is easy to see that the (X, Y)-entry of A is the expected number of Y-processes after the first step, if starting in a single X-process. If S is a branching process and $X_0 \in \Gamma$, we call the pair (S, X_0) a *reduced branching process*, if for all $X \in \Gamma$ there is $i \in \mathbb{N}$ such that $(A^i)_{X_0, X} > 0$. Intuitively, (S, X_0) is reduced, if, starting in X_0, all process symbols can be reached with positive probability. If (S, X_0) is not reduced, it is easy to reduce it in polynomial time by eliminating all non-reachable process symbols.

The following proposition characterises the finiteness of both expected work and expected time in terms of the spectral radius $\rho(A)$ of A. (Recall that $\rho(A)$ is the largest absolute value of the eigenvalues of A.)

Proposition 11. *Let (S, X_0) be a reduced branching process. Let A be the associated characteristic matrix. Then the following statements are equivalent:*

(1) $\mathbb{E}\mathbf{W}_{X_0}$ is finite;   *(2) $\mathbb{E}\mathbf{T}_{X_0}$ is finite;*   *(3) $\rho(A) < 1$.*

Further, if $\mathbb{E}\mathbf{W}_{X_0}$ is finite, then it equals the X_0-component of $(I - A)^{-1} \cdot \mathbf{1}$, where I is the identity matrix, and $\mathbf{1}$ is the column vector with all ones.

Statements similar to Proposition 11 do appear in the standard branching process literature [17,2], however, not explicitly enough to cite directly or with stronger assumptions[1]. Our proof adapts a technique which was developed in [4] for a different purpose. It uses only basic tools and Perron-Frobenius theory, the spectral theory of nonnegative matrices. Proposition 11 has the following consequence:

Corollary 12. *Consider a branching process with process symbols Γ and $X_0 \in \Gamma$. Then $\mathbb{E}\mathbf{W}_{X_0}$ and $\mathbb{E}\mathbf{T}_{X_0}$ are both finite or both infinite. Distinguishing between those cases is in* P.

By combining the previous results we obtain the following theorem.

Theorem 13. *Consider a pSJS S with alphabet $\Sigma = \Gamma \cup Q$. Let $a \in \Gamma$. Then $\mathbb{E}\mathbf{W}_a$ and $\mathbb{E}\mathbf{T}_a$ are both finite or both infinite. Distinguishing between those cases is in* PSPACE, *and* PosSLP-*hard even for pPDSs. Further, if S is normalised and $\mathbb{E}\mathbf{W}_a$ is finite, one can efficiently express $\mathbb{E}\mathbf{W}_a$.*

Theorem 13 can be interpreted as saying that, although the pSJS model does not impose a bound on the number of active processes at a time, its parallelism *cannot* be used to do an infinite expected amount of work in a finite expected time. However, the "speedup" $\mathbb{E}[\mathbf{W}]/\mathbb{E}[\mathbf{T}]$ may be unbounded:

Proposition 14. *Consider the family of branching processes with transitions $X \xrightarrow{p} \langle XX \rangle$ and $X \xrightarrow{1-p} \bot$, where $0 < p < 1/2$. Then the ratio $\mathbb{E}[\mathbf{W}_X]/\mathbb{E}[\mathbf{T}_X]$ is unbounded for $p \to 1/2$.*

4 Case Studies

We have implemented a prototype tool in the form of a Maple worksheet, which allows to compute some of the quantities from the previous section: the termination probabilities, and distributions and expectations of work and time. In this section, we use our tool for two case studies[2], which also illustrate how probabilistic parallel programs can be modelled with pSJSs. We only deal with normalised pSJSs in this section.

---

[1] For example, [2] assumes that there is $n \in \mathbb{N}$ such that A^n is positive in all entries, a restriction which is not natural for our setting.

[2] Available at http://www.comlab.ox.ac.uk/people/Stefan.Kiefer/case-studies.mws

4.1 Divide and Conquer

The pSJS model lends itself to analyse parallel divide-and-conquer programs. For simplicity, we assume that the problem is already given as a binary tree, and solving it means traversing the tree and combining the results of the children. Figure 2 shows generic parallel code for such a problem.

```
function divCon(node)
    if node.leaf() then return node.val()
    else parallel ⟨ val1 := divCon(node.c1), val2 := divCon(node.c2) ⟩
        return combine(val1, val2)
```

Fig. 2. A generic parallel divide-and-conquer program

For an example, think of a routine for numerically approximating an integral $\int_0^1 f(x)\, dx$. Given the integrand f and a subinterval $I \subseteq [0,1]$, we assume that there is a function which computes $osc_f(I) \in \mathbb{N}$, the "oscillation" of f in the interval I, a measure for the need for further refinement. If $osc_f(I) = 0$, then the integration routine returns the approximation $1 \cdot f(1/2)$, otherwise it returns $I_1 + I_2$, where I_1 and I_2 are recursive approximations of $\int_0^{1/2} f(x)\, dx$ and $\int_{1/2}^1 f(x)\, dx$, respectively.[3]

We analyse such a routine using probabilistic assumptions on the integrand: Let n, n_1, n_2 be nonnegative integers such that $0 \le n_1 + n_2 \le n$. If $osc_f([a,b]) = n$, then $osc_f([a, (a+b)/2]) = n_1$ and $osc_f([(a+b)/2, b]) = n_2$ with probability $x(n, n_1, n_2) :=$ $\binom{n}{n_1} \cdot \binom{n-n_1}{n_2} \cdot \left(\frac{p}{2}\right)^{n_1} \cdot \left(\frac{p}{2}\right)^{n_2} \cdot (1-p)^{n-n_1-n_2}$, where $0 < p < 1$ is some parameter.[4] Of course, other distributions could be used as well. The integration routine can then be modelled by the pSJS with $Q = \{q\}$ and $\Gamma = \{\langle qq \rangle, 0, \ldots, n_{\max}\}$ and the following rules:

$$0 \xrightarrow{1} q \quad \text{and} \quad \langle q\, q \rangle \xrightarrow{1} q \quad \text{and} \quad n \xrightarrow{x(n,n_1,n_2)} \langle n_1\, n_2 \rangle \text{ for all } 1 \le n \le n_{\max} ,$$

where $0 \le n_1 + n_2 \le n$. (Since we are merely interested in the performance of the algorithm, we can identify all return values with a single synchronisation state q.)

Using our prototype, we computed $\mathbb{E}[\mathbf{W}_n]$ and $\mathbb{E}[\mathbf{T}_n]$ for $p = 0.8$ and $n = 0, 1, \ldots,$ 10. Figure 3 shows that $\mathbb{E}[\mathbf{W}_n]$ increases faster with n than $\mathbb{E}[\mathbf{T}_n]$; i.e., the parallelism increases.

4.2 Evaluation of Game Trees

The evaluation of game trees is a central task of programs that are equipped with "artificial intelligence" to play games such as chess. These game trees are min-max trees (see Figure 4): each node corresponds to a position of the game, and each edge from a parent to a child corresponds to a move that transforms the position represented by the

---

[3] Such an adaptive approximation scheme is called "local" in [20].

[4] That means, the oscillation n in the interval $[a,b]$ can be thought of as distributed between $[a, (a+b)/2]$ and $[(a+b)/2, b]$ according to a ball-and-urn experiment, where each of the n balls is placed in the $[a, (a+b)/2]$-urn and the $[(a+b)/2, b]$-urn with probability $p/2$, respectively, and in a trash urn with probability $1-p$.

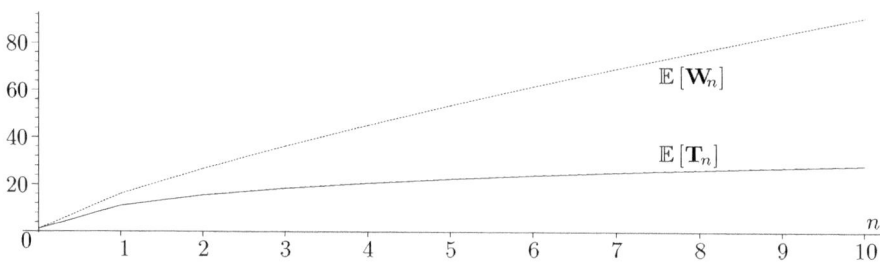

Fig. 3. Expectations of time and work

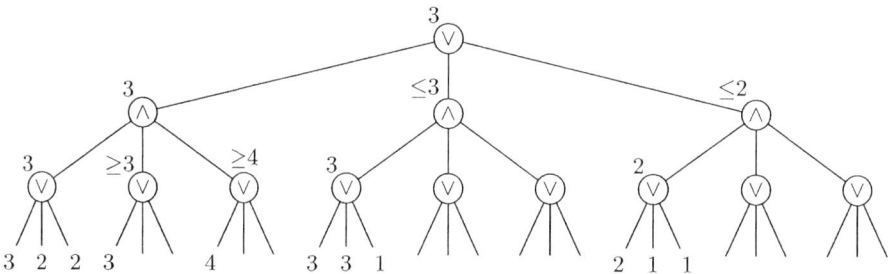

Fig. 4. A game tree with value 3

parent to a child position. Since the players have opposing objectives, the nodes alternate between max-nodes and min-nodes (denoted $\vee$ and $\wedge$, respectively). A leaf of a game tree corresponds either to a final game position or to a position which is evaluated heuristically by the game-playing program; in both cases, the leaf is assigned a number. Given such a leaf labelling, a number can be assigned to each node in the tree in the straightforward way; in particular, *evaluating a tree* means computing the root value.

In the following, we assume for simplicity that each node is either a leaf or has exactly three children. Figure 5 shows a straightforward recursive parallel procedure for evaluating a max-node of a game tree. (Of course, there is a symmetrical procedure for min-nodes.)

```
function parMax(node)
    if node.leaf() then return node.val()
    else parallel ⟨ val1 := parMin(node.c1), val2 := parMin(node.c2), val3 := parMin(node.c3) ⟩
        return max{val1, val2, val3}
```

Fig. 5. A simple parallel program for evaluating a game tree

Notice that in Figure 4 the value of the root is 3, independent of some missing leaf values. Game-playing programs aim at evaluating a tree as fast as possible, possibly by not evaluating nodes which are irrelevant for the root value. The classic technique is called *alpha-beta pruning*: it maintains an interval $[\alpha, \beta]$ in which the value of the current node is to be determined exactly. If the value turns out to be below α or above β, it is safe to return α or β, respectively. This may be the case even before all children have been evaluated (a so-called *cut-off*). Figure 6 shows a sequential program for alpha-beta pruning, initially to be called "seqMax(root, $-\infty$, $+\infty$)". Applying seqMax to the tree from Figure 4 results in several cut-offs: all non-labelled leaves are pruned.

```
function seqMax(node, α, β)
    if node.leaf() then if node.val() ≤ α then return α
                         elsif node.val() ≥ β then return β
                         else return node.val()
    else val1 := seqMin(node.c1, α, β)
         if val1 = β then return β
         else val2 := seqMin(node.c2, val1, β)
              if val2 = β then return β
              else return seqMin(node.c3, val2, β)
```

Fig. 6. A sequential program for evaluating a game tree using alpha-beta pruning

Although alpha-beta pruning may seem inherently sequential, parallel versions have been developed, often involving the *Young Brothers Wait (YBW)* strategy [16]. It relies on a good ordering heuristic, i.e., a method that sorts the children of a max-node (resp. min-node) in increasing (resp. decreasing) order, without actually evaluating the children. Such an ordering heuristic is often available, but usually not perfect. The tree in Figure 4 is ordered in this way. If alpha-beta pruning is performed on such an ordered tree, then either all children of a node are evaluated or only the first one. The YBW method first evaluates the first child only and hopes that this creates a cut-off or, at least, decreases the interval $[\alpha, \beta]$. If the first child fails to cause a cut-off, YBW *speculates* that both "younger brothers" need to be evaluated, which can be done in parallel without wasting work. A wrong speculation may affect the performance, but not the correctness. Figure 7 shows a YBW-based program. Similar code is given in [7] using *Cilk*, a C-based parallel programming language.

```
function YBWMax(node, α, β)
    if node.leaf() then if node.val() ≤ α then return α
                         elsif node.val() ≥ β then return β
                         else return node.val()
    else val1 := YBWMin(node.c1, α, β)
         if val1 = β then return β
         else parallel ⟨ val2 := YBWMin(node.c2, val1, β), val3 := YBWMin(node.c3, val1, β) ⟩
              return max{val2, val3}
```

Fig. 7. A parallel program based on YBW for evaluating a game tree

We evaluate the performance of these three (deterministic) programs using probabilistic assumptions about the game trees. More precisely, we assume the following: Each node has exactly three children with probability p, and is a leaf with probability $1-p$. A leaf (and hence any node) takes as value a number from $\mathbb{N}_4 := \{0, 1, 2, 3, 4\}$, according to a distribution described below. In order to model an ordering heuristic on the children, each node carries a parameter $e \in \mathbb{N}_4$ which intuitively corresponds to its expected value. If a max-node with parameter e has children, then they are min-nodes with parameters e, $e\ominus1$, $e\ominus2$, respectively, where $a\ominus b := \max\{a-b, 0\}$; similarly, the children of a min-node with parameter e are max-nodes with parameters e, $e\oplus1$, $e\oplus2$, where $a\oplus b := \min\{a+b, 4\}$. A leaf-node with parameter e takes value k with probability $\binom{4}{k} \cdot (e/4)^k \cdot (1-e/4)^{4-k}$; i.e., a leaf value is binomially distributed with expectation e. One could think of a game tree as the terminal tree of a branching process with $\Gamma = \{Max(e), Min(e) \mid e \in \{0, \ldots, 4\}\}$ and $Q = \mathbb{N}_4$ and the rules $Max(e) \xrightarrow{p} \langle Min(e)\ Min(e\ominus1)\ Min(e\ominus2)\rangle$ and $Max(e) \xrightarrow{x(k)} k$, with $x(k) := (1-p) \cdot \binom{4}{k} \cdot (e/4)^k \cdot (1-e/4)^{4-k}$ for all $e, k \in \mathbb{N}_4$, and similar rules for $Min(e)$.

We model the YBW-program from Figure 7 running on such random game trees by the pSJS with $Q = \{0, 1, 2, 3, 4, q(\vee), q(\wedge)\} \cup \{q(\alpha, \beta, \vee, e), q(\alpha, \beta, \wedge, e) \mid 0 \le \alpha < \beta \le 4,\ 0 \le e \le 4\} \cup \{q(a, b) \mid 0 \le a, b \le 4\}$ and the following rules:

$$Max(\alpha, \beta, e) \xrightarrow{x(0)+\cdots+x(\alpha)} \alpha, \quad Max(\alpha, \beta, e) \xrightarrow{x(\beta)+\cdots+x(4)} \beta, \quad Max(\alpha, \beta, e) \xrightarrow{x(k)} k$$

$$Max(\alpha, \beta, e) \xrightarrow{p} \langle Min(\alpha, \beta, e)\ q(\alpha, \beta, \vee, e\ominus1)\rangle$$

$$\langle \beta\ q(\alpha, \beta, \vee, e)\rangle \xrightarrow{1} \beta, \qquad \langle \gamma\ q(\alpha, \beta, \vee, e)\rangle \xrightarrow{1} \langle Max2(\gamma, \beta, e)\ q(\vee)\rangle$$

$$Max2(\alpha, \beta, e) \xrightarrow{1} \langle Min(\alpha, \beta, e)\ Min(\alpha, \beta, e\ominus1)\rangle$$

$$\langle a\ b\rangle \xrightarrow{1} q(a, b), \qquad \langle q(a, b)\ q(\vee)\rangle \xrightarrow{1} \max\{a, b\},$$

where $0 \le \alpha \le \gamma < \beta \le 4$ and $\alpha < k < \beta$ and $0 \le e \le 4$ and $0 \le a, b \le 4$. There are analogous rules with Min and Max exchanged. Notice that the rules closely follow the program from Figure 7. The programs parMax and seqMax from Figures 5 and 6 can be modelled similarly.

Let $T(\text{YBW}, p) := \mathbb{E}\left[\mathbf{T}_{Max(0,4,2)} \mid Run(Max(0, 4, 2)\!\downarrow\!2)\right]$; i.e., $T(\text{YBW}, p)$ is the expected time of the YBW-program called with a tree with value 2 and whose root is a max-node with parameter 2. (Recall that p is the probability that a node has children.) Let $W(\text{YBW}, p)$ defined similarly for the expected work, and define these numbers also for par and seq instead of YBW, i.e., for the programs from Figures 5 and 6. Using our prototype we computed $W(\text{seq}, p) = 1.00, 1.43, 1.96, 2.63, 3.50, 4.68, 6.33$ for $p = 0.00, 0.05, 0.10, 0.15, 0.20, 0.25, 0.30$. Since the program seq is sequential, we have the same sequence for $T(\text{seq}, p)$. To assess the speed of the parallel programs par and YBW, we also computed the percentaged increase of their runtime relative to seq, i.e., $100 \cdot (T(\text{par}, p)/T(\text{seq}, p) - 1)$, and similarly for YBW. Figure 8 shows the results. One can observe that for small values of p (i.e., small trees), the program par is slightly faster than seq because of its parallelism. For larger values of p, par still evaluates all nodes in the tree, whereas seq increasingly benefits from cut-offs of potentially deep branches. Using Proposition 11, one can prove $W(\text{par}, \frac{1}{3}) = T(\text{par}, \frac{1}{3}) = \infty > W(\text{seq}, \frac{1}{3})$.[5]

---

[5] In fact, $W(\text{seq}, p)$ is finite even for values of p which are slightly larger than $\frac{1}{3}$; in other words, seq cuts off infinite branches.

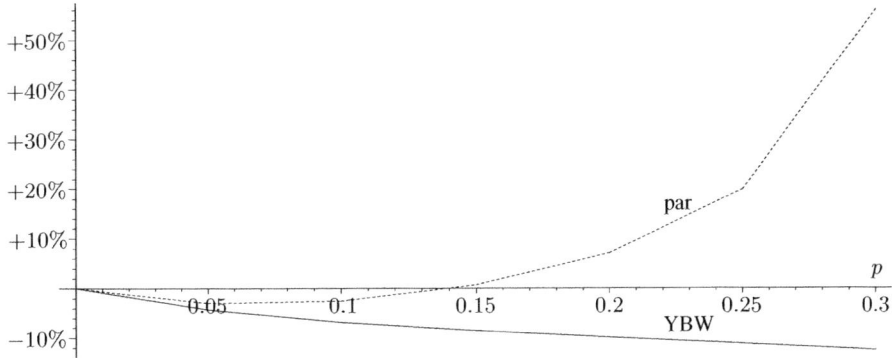

Fig. 8. Percentaged runtime increase of par and YBW relative to seq

The figure also shows that the YBW-program is faster than seq: the advantage of YBW increases with p up to about 10%.

We also compared the *work* of YBW with seq, and found that the percentaged increase ranges from 0 to about $+0.4\%$ for p between 0 and 0.3. This means that YBW wastes almost no work; in other words, a sequential version of YBW would be almost as fast as seq. An interpretation is that the second child rarely causes large cut-offs. Of course, all of these findings could depend on the exact probabilistic assumptions on the game trees.

5   Conclusions and Future Work

We have introduced pSJSs, a model for probabilistic parallel programs with process spawning and synchronisation. We have studied the basic performance measures of termination probability, space, work, and time. In our results the upper complexity bounds coincide with the best ones known for pPDSs, and the lower bounds also hold for pPDSs. This suggests that analysing pSJSs is no more expensive than analysing pPDSs. The pSJS model is amenable to a practical performance analysis. Our two case studies have demonstrated the modelling power of pSJSs: one can use pSJSs to model, analyse, and compare the performance of parallel programs under probabilistic assumptions.

We intend to develop model-checking algorithms for pSJSs. It seems to us that a meaningful functional analysis should not only model-check the Markov chain induced by the pSJS, but rather take the individual process "histories" into account.

Acknowledgements. We thank Javier Esparza, Alastair Donaldson, Barbara König, Markus Müller-Olm, Luke Ong and Thomas Wahl for helpful discussions on the non-probabilistic version of pSJSs. We also thank the anonymous referees for valuable comments.

References

1. Allender, E., Bürgisser, P., Kjeldgaard-Pedersen, J., Miltersen, P.B.: On the complexity of numerical analysis. In: IEEE Conf. on Computational Complexity, pp. 331–339 (2006)
2. Athreya, K.B., Ney, P.E.: Branching Processes. Springer, Heidelberg (1972)
3. Bouajjani, A., Müller-Olm, M., Touili, T.: Regular symbolic analysis of dynamic networks of pushdown systems. In: Jayaraman, K., de Alfaro, L. (eds.) CONCUR 2005. LNCS, vol. 3653, pp. 473–487. Springer, Heidelberg (2005)
4. Brázdil, T., Esparza, J., Kiefer, S.: On the memory consumption of probabilistic pushdown automata. In: Proceedings of FSTTCS, pp. 49–60 (2009)
5. Brázdil, T., Kiefer, S., Kučera, A., Vařeková, I.H.: Runtime analysis of probabilistic programs with unbounded recursion (2010) (submitted for publication), http://arxiv.org/abs/1007.1710
6. Canny, J.: Some algebraic and geometric computations in PSPACE. In: STOC 1988, pp. 460–467 (1988)
7. Dailey, D., Leiserson, C.E.: Using Cilk to write multiprocessor chess programs. The Journal of the International Computer Chess Association (2002)
8. Danaher, J.S., Lee, I.A., Leiserson, C.E.: Programming with exceptions in JCilk. Science of Computer Programming (SCP) 63(2), 147–171 (2006)
9. Esparza, J., Etessami, K.: Verifying probabilistic procedural programs. In: Lodaya, K., Mahajan, M. (eds.) FSTTCS 2004. LNCS, vol. 3328, pp. 16–31. Springer, Heidelberg (2004)
10. Esparza, J., Kiefer, S., Luttenberger, M.: Computing the least fixed point of positive polynomial systems. SIAM Journal on Computing 39(6), 2282–2335 (2010)
11. Esparza, J., Kučera, A., Mayr, R.: Model checking probabilistic pushdown automata. In: LICS 2004, pp. 12–21. IEEE, Los Alamitos (2004)
12. Esparza, J., Kučera, A., Mayr, R.: Quantitative analysis of probabilistic pushdown automata: Expectations and variances. In: LICS 2005, pp. 117–126. IEEE, Los Alamitos (2005)
13. Etessami, K., Yannakakis, M.: Algorithmic verification of recursive probabilistic state machines. In: Halbwachs, N., Zuck, L.D. (eds.) TACAS 2005. LNCS, vol. 3440, pp. 253–270. Springer, Heidelberg (2005)
14. Etessami, K., Yannakakis, M.: Recursive concurrent stochastic games. Logical Methods in Computer Science 4(4) (2008)
15. Etessami, K., Yannakakis, M.: Recursive markov chains, stochastic grammars, and monotone systems of nonlinear equations. Journal of the ACM 56(1), 1–66 (2009)
16. Feldmann, R., Monien, B., Mysliwietz, P., Vornberger, O.: Distributed game-tree search. ICCA Journal 12(2), 65–73 (1989)
17. Harris, T.E.: The Theory of Branching Processes. Springer, Heidelberg (1963)
18. Kiefer, S., Wojtczak, D.: On probabilistic parallel programs with process creation and synchronisation. Technical report, arxiv.org (2010), http://arxiv.org/abs/1012.2998
19. Löding, C.: Reachability problems on regular ground tree rewriting graphs. Theory of Computing Systems 39, 347–383 (2006)
20. Malcolm, M.A., Simpson, R.B.: Local versus global strategies for adaptive quadrature. ACM Transactions on Mathematical Software 1(2), 129–146 (1975)
21. Mayr, R.: Process rewrite systems. Information and Computation 156(1-2), 264–286 (2000)
22. Renegar, J.: On the computational complexity and geometry of the first-order theory of the reals. Parts I–III. Journal of Symbolic Computation 13(3), 255–352 (1992)
23. Yannakakis, M., Etessami, K.: Checking LTL properties of recursive Markov chains. In: QEST 2005, pp. 155–165 (2005)

Confluence Reduction for Probabilistic Systems

Mark Timmer, Mariëlle Stoelinga, and Jaco van de Pol*

Formal Methods and Tools, Faculty of EEMCS
University of Twente, The Netherlands
{timmer,marielle,vdpol}@cs.utwente.nl

Abstract. This paper presents a novel technique for state space reduction of probabilistic specifications, based on a newly developed notion of confluence for probabilistic automata. We prove that this reduction preserves branching probabilistic bisimulation and can be applied on-the-fly. To support the technique, we introduce a method for detecting confluent transitions in the context of a probabilistic process algebra with data, facilitated by an earlier defined linear format. A case study demonstrates that significant reductions can be obtained.

1 Introduction

Model checking of probabilistic systems is getting more and more attention, but there still is a large gap between the number of techniques supporting traditional model checking and those supporting probabilistic model checking. Especially methods aimed at reducing state spaces are greatly needed to battle the omnipresent state space explosion.

In this paper, we generalise the notion of confluence [8] from labelled transition systems (LTSs) to probabilistic automata (PAs) [14]. Basically, we define under which conditions unobservable transitions (often called τ-transitions) do not influence a PA's behaviour (i.e., they commute with all other transitions). Using this new notion of probabilistic confluence, we introduce a symbolic technique that reduces PAs while preserving branching probabilistic bisimulation.

The non-probabilistic case. Our methodology follows the approach for LTSs from [4]. It consists of the following steps: (i) a system is specified as the parallel composition of several processes with data; (ii) the specification is linearised to a canonical form that facilitates symbolic manipulations; (iii) first-order logic formulas are generated to check symbolically which τ-transitions are confluent; (iv) an LTS is generated in such a way that confluent τ-transitions are given priority, leading to an on-the-fly (potentially exponential) state space reduction. Refinements by [12] make it even possible to perform confluence detection on-the-fly by means of boolean equation systems.

The probabilistic case. After recalling some basic concepts from probability theory and probabilistic automata, we introduce three novel notions of probabilistic

---

* This research has been partially funded by NWO under grant 612.063.817 (SYRUP) and grant Dn 63-257 (ROCKS), and by the European Union under FP7-ICT-2007-1 grant 214755 (QUASIMODO).

P.A. Abdulla and K.R.M. Leino (Eds.): TACAS 2011, LNCS 6605, pp. 311–325, 2011.
© Springer-Verlag Berlin Heidelberg 2011

confluence. Inspired by [3], these are *weak probabilistic confluence, probabilistic confluence* and *strong probabilistic confluence* (in decreasing order of reduction power, but in increasing order of detection efficiency).

We prove that the stronger notions imply the weaker ones, and that τ-transitions that are confluent according to any of these notions always connect branching probabilistically bisimilar states. Basically, this means that they can be given priority without losing any behaviour. Based on this idea, we propose a reduction technique that can be applied using the two stronger notions of confluence. For each set of states that can reach each other by traversing only confluent transitions, it chooses a representative state that has all relevant behaviour. We prove that this reduction technique yields a branching probabilistically bisimilar PA. Therefore, it preserves virtually all interesting temporal properties.

As we want to analyse systems that would normally be too large, we need to detect confluence symbolically and use it to reduce on-the-fly during state space generation. That way, the unreduced PA never needs to be generated. Since it is not clear how not to detect (weak) probabilistic confluence efficiently, we only provide a detection method for strong probabilistic confluence. Here, we exploit a previously defined probabilistic process-algebraic linear format, which is capable of modelling any system consisting of parallel components with data [10]. In this paper, we show how symbolic τ-transitions can be proven confluent by solving formulas in first-order logic over this format. As a result, confluence can be detected symbolically, and the reduced PA can be generated on-the-fly. We present a case study of leader election protocols, showing significant reductions.

Proofs for all our propositions and theorems can be found in an extended version of this paper [17].

Related work. As mentioned before, we basically generalise the techniques presented in [4] to PAs.

In the probabilistic setting, several reduction techniques similar to ours exist. Most of these are generalisations of the well-known concept of partial-order reduction (POR) [13]. In [2] and [5], the concept of POR was lifted to Markov decision processes, providing reductions that preserve quantitative LTL\X. This was refined in [1] to probabilistic CTL, a branching logic. Recently, a revision of POR for distributed schedulers was introduced and implemented in PRISM [7].

Our confluence reduction differs from these techniques on several accounts. First, POR is applicable on state-based systems, whereas our confluence reduction is the first technique that can be used for action-based systems. As the transformation between action- and state-based blows up the state space [11], having confluence reduction really provides new possibilities. Second, the definition of confluence is quite elegant, and (strong) confluence seems to be of a more local nature (which makes the correctness proofs easier). Third, the detection of POR requires language-specific heuristics, whereas confluence reduction acts at a more semantic level and can be implemented by a generic theorem prover. (Alternatively, decision procedures for a fixed set of data types could be devised.)

Our case study shows that the reductions obtained using probabilistic confluence exceed the reductions obtained by probabilistic POR [9].

2   Preliminaries

Given a set S, an element $s \in S$ and an equivalence relation $R \subseteq S \times S$, we write $[s]_R$ for the *equivalence class* of s under R, i.e., $[s]_R = \{s' \in S \mid (s, s') \in R\}$. We write $S/R = \{[s]_R \mid s \in S\}$ for the set of all equivalence classes in S.

2.1   Probability Theory and Probabilistic Automata

Definition 1 (Probability distributions). *A* probability distribution *over a countable set S is a function $\mu \colon S \to [0, 1]$ such that $\sum_{s \in S} \mu(s) = 1$. Given $S' \subseteq S$, we write $\mu(S')$ to denote $\sum_{s' \in S'} \mu(s')$. We use $\mathrm{Distr}(S)$ to denote the set of all probability distributions over S, and $\mathrm{Distr}^*(S)$ for the set of all substochastic probability distributions over S, i.e., where $0 \leq \sum_{s \in S} \mu(s) \leq 1$.*

Given a probability distribution μ with $\mu(s_1) = p_1$, $\mu(s_2) = p_2, \ldots$ $(p_i \neq 0)$, we write $\mu = \{s_1 \mapsto p_1, s_2 \mapsto p_2, \ldots\}$ and let $\mathrm{spt}(\mu) = \{s_1, s_2, \ldots\}$ denote its support. *For the deterministic distribution μ determined by $\mu(t) = 1$ we write $\mathbb{1}_t$.*

Given an equivalence relation R over S and two probability distributions μ, μ' over S, we say that $\mu \equiv_R \mu'$ if and only if $\mu(C) = \mu'(C)$ for all $C \in S/R$.

Probabilistic automata (PAs) are similar to labelled transition systems, except that transitions do not have a fixed successor state anymore. Instead, the state reached after taking a certain transition is determined by a probability distribution [14]. The transitions themselves can be chosen nondeterministically.

Definition 2 (Probabilistic automata). *A* probabilistic automaton (PA) *is a tuple $\mathcal{A} = \langle S, s^0, L, \Delta \rangle$, where S is a countable set of states of which $s^0 \in S$ is initial, L is a countable set of actions, and $\Delta \subseteq S \times L \times \mathrm{Distr}(S)$ is a countable transition relation. We assume that every PA contains an unobservable action $\tau \in L$. If $(s, a, \mu) \in \Delta$, we write $s \xrightarrow{a} \mu$, meaning that state s enables action a, after which the probability to go to $s' \in S$ is $\mu(s')$. If $\mu = \mathbb{1}_t$, we write $s \xrightarrow{a} t$.*

Definition 3 (Paths and traces). *Given a PA $\mathcal{A} = \langle S, s^0, L, \Delta \rangle$, we define a* path *of $\mathcal{A}$ to be either a finite sequence $\pi = s_0 \overset{a_1, \mu_1}{\rightsquigarrow} s_1 \overset{a_2, \mu_2}{\rightsquigarrow} s_2 \overset{a_3, \mu_3}{\rightsquigarrow} \ldots \overset{a_n, \mu_n}{\rightsquigarrow} s_n$, or an infinite sequence $\pi' = s_0 \overset{a_1, \mu_1}{\rightsquigarrow} s_1 \overset{a_2, \mu_2}{\rightsquigarrow} s_2 \overset{a_3, \mu_3}{\rightsquigarrow} \ldots$, where for finite paths we require $s_i \in S$ for all $0 \leq i \leq n$, and $s_i \xrightarrow{a_{i+1}} \mu_{i+1}$ as well as $\mu_{i+1}(s_{i+1}) > 0$ for all $0 \leq i < n$. For infinite paths these properties should hold for all $i \geq 0$. A fragment $s \overset{a, \mu}{\rightsquigarrow} s'$ denotes that the transition $s \xrightarrow{a} \mu$ was chosen from state s, after which the successor s' was selected by chance (so $\mu(s') > 0$).*

- *If $\pi = s_0 \overset{a, \mathbb{1}_{s_1}}{\rightsquigarrow} s_1 \overset{a, \mathbb{1}_{s_2}}{\rightsquigarrow} \ldots \overset{a, \mathbb{1}_{s_n}}{\rightsquigarrow} s_n$ is a path of $\mathcal{A}$ $(n \geq 0)$, we write $s_0 \xrightarrow{a} s_n$. In case we also allow steps of the form $s_i \overset{a, \mathbb{1}_{s_{i+1}}}{\longleftrightarrow} s_{i+1}$, we write $s_0 \overset{a}{\longleftrightarrow} s_n$. If there exists a state t such that $s \xrightarrow{a} t$ and $s' \xrightarrow{a} t$, we write $s \xrightarrow{a} \overset{a}{\leftarrow} s'$.*
- *We use $\mathrm{prefix}(\pi, i)$ to denote $s_0 \overset{a_1, \mu_1}{\rightsquigarrow} \ldots \overset{a_i, \mu_i}{\rightsquigarrow} s_i$, and $\mathrm{step}(\pi, i)$ to denote the transition (s_{i-1}, a_i, μ_i). When π is finite we define $|\pi| = n$ and $\mathrm{last}(\pi) = s_n$.*
- *We use $\mathrm{finpaths}_{\mathcal{A}}$ to denote the set of all finite paths of $\mathcal{A}$, and $\mathrm{finpaths}_{\mathcal{A}}(s)$ for all finite paths where $s_0 = s$.*
- *A path's* trace *is the sequence of actions obtained by omitting all its states, distributions and τ-steps; given $\pi = s_0 \overset{a_1, \mu_1}{\rightsquigarrow} s_1 \overset{\tau, \mu_2}{\rightsquigarrow} s_2 \overset{a_3, \mu_3}{\rightsquigarrow} \ldots \overset{a_n, \mu_n}{\rightsquigarrow} s_n$, we denote the sequence $a_1 a_3 \ldots a_n$ by $\mathrm{trace}(\pi)$.*

2.2 Schedulers

To resolve the nondeterminism in PAs, schedulers are used [16]. Basically, a scheduler is a function defining for each finite path which transition to take next. The decisions of schedulers are allowed to be *randomised*, i.e., instead of choosing a single transition a scheduler might resolve a nondeterministic choice by a probabilistic choice. Schedulers can be *partial*, i.e., they might assign some probability to the decision of not choosing any next transition.

Definition 4 (Schedulers). *A scheduler for a PA $\mathcal{A} = \langle S, s^0, L, \Delta \rangle$ is a function*

$$\mathcal{S} : \mathit{finpaths}_\mathcal{A} \to \mathrm{Distr}(\{\bot\} \cup \Delta),$$

such that for every $\pi \in \mathit{finpaths}_\mathcal{A}$ the transitions (s, a, μ) that are scheduled by $\mathcal{S}$ after π are indeed possible after π, i.e., $\mathcal{S}(\pi)(s, a, \mu) > 0$ implies $s = \mathit{last}(\pi)$. The decision of not choosing any transition is represented by $\bot$.

We now define the notions of finite and maximal paths of a PA given a scheduler.

Definition 5 (Finite and maximal paths). *Let $\mathcal{A}$ be a PA and $\mathcal{S}$ a scheduler for $\mathcal{A}$. Then, the set of finite paths of $\mathcal{A}$ under $\mathcal{S}$ is given by*

$$\mathit{finpaths}^{\mathcal{S}}_\mathcal{A} = \{\pi \in \mathit{finpaths}_\mathcal{A} \mid \forall 0 \le i < |\pi| \ . \ \mathcal{S}(\mathit{prefix}(\pi, i))(\mathit{step}(\pi, i+1)) > 0\}.$$

We define $\mathit{finpaths}^{\mathcal{S}}_\mathcal{A}(s) \subseteq \mathit{finpaths}^{\mathcal{S}}_\mathcal{A}$ as the set of all such paths starting in s. The set of maximal paths of $\mathcal{A}$ under $\mathcal{S}$ is given by

$$\mathit{maxpaths}^{\mathcal{S}}_\mathcal{A} = \{\pi \in \mathit{finpaths}^{\mathcal{S}}_\mathcal{A} \mid \mathcal{S}(\pi)(\bot) > 0\}.$$

Similarly, $\mathit{maxpaths}^{\mathcal{S}}_\mathcal{A}(s)$ is the set of maximal paths of $\mathcal{A}$ under $\mathcal{S}$ starting in s.

We now define the behaviour of a PA $\mathcal{A}$ under a scheduler $\mathcal{S}$. As schedulers resolve all nondeterministic choices, this behaviour is fully probabilistic. We can therefore compute the probability that, starting from a given state s, the path generated by $\mathcal{S}$ has some finite prefix π. This probability is denoted by $P^{\mathcal{S}}_{\mathcal{A},s}(\pi)$.

Definition 6 (Path probabilities). *Let $\mathcal{A}$ be a PA, $\mathcal{S}$ a scheduler for $\mathcal{A}$, and s a state of $\mathcal{A}$. Then, we define the function $P^{\mathcal{S}}_{\mathcal{A},s} : \mathit{finpaths}_\mathcal{A}(s) \to [0, 1]$ by*

$$P^{\mathcal{S}}_{\mathcal{A},s}(s) = 1; \qquad P^{\mathcal{S}}_{\mathcal{A},s}(\pi \xrightarrow{a,\mu} t) = P^{\mathcal{S}}_{\mathcal{A},s}(\pi) \cdot \mathcal{S}(\pi)(\mathit{last}(\pi), a, \mu) \cdot \mu(t).$$

Based on these probabilities we can compute the probability distribution $F^{\mathcal{S}}_\mathcal{A}(s)$ over the states where a PA $\mathcal{A}$ under a scheduler $\mathcal{S}$ terminates, when starting in state s. Note that $F^{\mathcal{S}}_\mathcal{A}(s)$ is potentially substochastic (i.e., the probabilities do not add up to 1) if $\mathcal{S}$ allows infinite behaviour.

Definition 7 (Final state probabilities). *Let $\mathcal{A}$ be a PA and $\mathcal{S}$ a scheduler for $\mathcal{A}$. Then, we define the function $F^{\mathcal{S}}_\mathcal{A} : S \to \mathrm{Distr}^*(S)$ by*

$$F^{\mathcal{S}}_\mathcal{A}(s) = \left\{ s' \mapsto \sum_{\substack{\pi \in \mathit{maxpaths}^{\mathcal{S}}_\mathcal{A}(s) \\ \mathit{last}(\pi) = s'}} P^{\mathcal{S}}_{\mathcal{A},s}(\pi) \cdot \mathcal{S}(\pi)(\bot) \mid s' \in S \right\} \qquad \forall s \in S.$$

3   Branching Probabilistic Bisimulation

The notion of branching bisimulation for non-probabilistic systems was first introduced in [19]. Basically, it relates states that have an identical branching structure in the presence of τ-actions. Segala defined a generalisation of branching bisimulation for PAs [15], which we present here using the simplified definitions of [16]. First, we intuitively explain weak steps for PAs. Based on these ideas, we then formally introduce branching probabilistic bisimulation.

3.1   Weak Steps for Probabilistic Automata

As τ-steps cannot be observed, we want to abstract from them. Non-probabilistically, this is done via the weak step. A state s can do a weak step to s' under an action a, denoted by $s \xrightarrow{a} s'$, if there exists a path $s \xrightarrow{\tau} s_1 \xrightarrow{\tau} \ldots \xrightarrow{\tau} s_n \xrightarrow{a} s'$ with $n \geq 0$ (often, also τ-steps after the a-action are allowed, but this will not concern us). Traditionally, $s \xRightarrow{a} s'$ is thus satisfied by an *appropriate path*.

In the probabilistic setting, $s \xRightarrow{a} \mu$ is satisfied by an *appropriate scheduler*. A scheduler $\mathcal{S}$ is appropriate if for every maximal path π that is scheduled from s with non-zero probability, $trace(\pi) = a$ and the a-transition is the last transition of the path. Also, the final state distribution $F_{\mathcal{A}}^{\mathcal{S}}(s)$ must be equal to μ.

Example 8. Consider the PA shown in Figure 1(a). We demonstrate that $s \xRightarrow{a} \mu$, with $\mu = \{s_1 \mapsto \frac{8}{24}, s_2 \mapsto \frac{7}{24}, s_3 \mapsto \frac{1}{24}, s_4 \mapsto \frac{4}{24}, s_5 \mapsto \frac{4}{24}\}$. Take the scheduler $\mathcal{S}$:

$$\mathcal{S}(s) = \{(s, \tau, \mathbb{1}_{t_2}) \mapsto 2/3, (s, \tau, \mathbb{1}_{t_3}) \mapsto 1/3\}$$
$$\mathcal{S}(t_2) = \{(t_2, a, \mathbb{1}_{s_1}) \mapsto 1/2, (t_2, \tau, \mathbb{1}_{t_4}) \mapsto 1/2\}$$
$$\mathcal{S}(t_3) = \{(t_3, a, \{s_4 \mapsto 1/2, s_5 \mapsto 1/2\}) \mapsto 1\}$$
$$\mathcal{S}(t_4) = \{(t_4, a, \mathbb{1}_{s_2}) \mapsto 3/4, (t_4, a, \{s_2 \mapsto 1/2, s_3 \mapsto 1/2\}) \mapsto 1/4\}$$
$$\mathcal{S}(t_1) = \mathcal{S}(s_1) = \mathcal{S}(s_2) = \mathcal{S}(s_3) = \mathcal{S}(s_4) = \mathcal{S}(s_5) = \mathbb{1}_{\perp}$$

Here we used $\mathcal{S}(s)$ to denote the choice made for every possible path ending in s.

The scheduler is depicted in Figure 1(b). Where it chooses probabilistically between two transitions with the same label, this is represented as a *combined transition*. For instance, from t_4 the transition $(t_4, a, \{s_2 \mapsto 1\})$ is selected with probability 3/4, and $(t_4, a, \{s_2 \mapsto 1/2, s_3 \mapsto 1/2\})$ with probability 1/4. This corresponds to the combined transition $(t_4, a, \{s_2 \mapsto 7/8, s_3 \mapsto 1/8\})$.

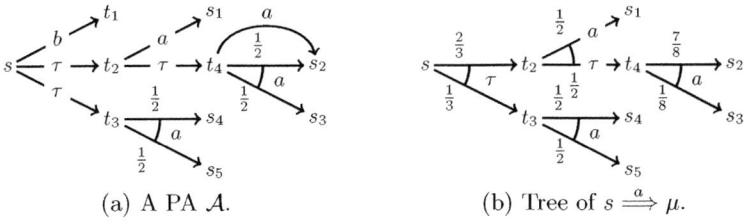

(a) A PA $\mathcal{A}$.   (b) Tree of $s \xRightarrow{a} \mu$.

Fig. 1. Weak steps

Clearly, all maximal paths enabled from s have trace a and end directly after their a-transition. The path probabilities can also be calculated. For instance,

$$P_{\mathcal{A},s}^{\mathcal{S}}(s \overset{\tau,\{t_2 \mapsto 1\}}{\leadsto} t_2 \overset{\tau,\{t_4 \mapsto 1\}}{\leadsto} t_4 \overset{a,\{s_2 \mapsto 1\}}{\leadsto} s_2) = \left(\tfrac{2}{3} \cdot 1\right) \cdot \left(\tfrac{1}{2} \cdot 1\right) \cdot \left(\tfrac{3}{4} \cdot 1\right) = \tfrac{6}{24}$$

$$P_{\mathcal{A},s}^{\mathcal{S}}(s \overset{\tau,\{t_2 \mapsto 1\}}{\leadsto} t_2 \overset{\tau,\{t_4 \mapsto 1\}}{\leadsto} t_4 \overset{a,\{s_2 \mapsto 1/2, s_3 \mapsto 1/2\}}{\leadsto} s_2) = \left(\tfrac{2}{3} \cdot 1\right) \cdot \left(\tfrac{1}{2} \cdot 1\right) \cdot \left(\tfrac{1}{4} \cdot \tfrac{1}{2}\right) = \tfrac{1}{24}$$

As no other maximal paths from s go to s_2, $F_{\mathcal{A}}^{\mathcal{S}}(s)(s_2) = \tfrac{6}{24} + \tfrac{1}{24} = \tfrac{7}{24} = \mu(s_2)$. Similarly, it can be shown that $F_{\mathcal{A}}^{\mathcal{S}}(s)(s_i) = \mu(s_i)$ for every $i \in \{1, 3, 4, 5\}$, so indeed $F_{\mathcal{A}}^{\mathcal{S}}(s) = \mu$.   □

3.2   Branching Probabilistic Bisimulation

Before introducing branching probabilistic bisimulation, we need a restriction on weak steps. Given an equivalence relation R, we let $s \overset{a}{\Longrightarrow}_R \mu$ denote that $(s, t) \in R$ for every state t before the a-step in the tree corresponding to $s \overset{a}{\Longrightarrow} \mu$.

Definition 9 (Branching steps). *Let $\mathcal{A} = \langle S, s^0, L, \Delta \rangle$ be a PA, $s \in S$, and R an equivalence relation over S. Then, $s \overset{a}{\Longrightarrow}_R \mu$ if either (1) $a = \tau$ and $\mu = \mathbb{1}_s$, or (2) there exists a scheduler $\mathcal{S}$ such that $F_{\mathcal{A}}^{\mathcal{S}}(s) = \mu$ and for every maximal path $s \overset{a_1,\mu_1}{\leadsto} s_1 \overset{a_2,\mu_2}{\leadsto} s_2 \overset{a_3,\mu_3}{\leadsto} \ldots \overset{a_n,\mu_n}{\leadsto} s_n \in maxpaths_{\mathcal{A}}^{\mathcal{S}}(s)$ it holds that $a_n = a$, as well as $a_i = \tau$ and $(s, s_i) \in R$ for all $1 \leq i < n$.*

Definition 10 (Branching probabilistic bisimulation). *Let $\mathcal{A} = \langle S, s^0, L, \Delta \rangle$ be a PA, then an equivalence relation $R \subseteq S \times S$ is a* branching probabilistic *bisimulation for $\mathcal{A}$ if for all $(s, t) \in R$*

$$s \overset{a}{\rightarrow} \mu \text{ implies } \exists \mu' \in \text{Distr}(S) \,.\, t \overset{a}{\Longrightarrow}_R \mu' \wedge \mu \equiv_R \mu'.$$

We say that $p, q \in S$ are branching probabilistically bisimilar*, denoted $p \underset{\text{bp}}{\leftrightarrow} q$, if there exists a branching probabilistic bisimulation R for $\mathcal{A}$ such that $(p, q) \in R$.*

Two PAs are branching probabilistically bisimilar if their initial states are (in the disjoint union of the two systems; see Remark 5.3.4 of [16] for the details).

This notion has some appealing properties. First, the definition is robust in the sense that it can be adapted to using $s \overset{a}{\Longrightarrow}_R \mu$ instead of $s \overset{a}{\rightarrow} \mu$ in its condition. Although this might seem to strengthen the concept, it does not. Second, the relation $\underset{\text{bp}}{\leftrightarrow}$ induced by the definition is an equivalence relation.

Proposition 11. *Let $\mathcal{A} = \langle S, s^0, L, \Delta \rangle$ be a PA. Then, an equivalence relation $R \subseteq S \times S$ is a branching probabilistic bisimulation for $\mathcal{A}$ if and only if for all $(s, t) \in R$*

$$s \overset{a}{\Longrightarrow}_R \mu \text{ implies } \exists \mu' \in \text{Distr}(S) \,.\, t \overset{a}{\Longrightarrow}_R \mu' \wedge \mu \equiv_R \mu'.$$

Proposition 12. *The relation $\underset{\text{bp}}{\leftrightarrow}$ is an equivalence relation.*

Moreover, Segala showed that branching bisimulation preserves all properties that can be expressed in the probabilistic temporal logic WPCTL (provided that no infinite path of τ-actions can be scheduled with non-zero probability) [15].

4   Confluence for Probabilistic Automata

As branching probabilistic bisimulation minimisation cannot easily be performed on-the-fly, we introduce a reduction technique based on sets of *confluent* τ-transitions. Basically, such transitions do not influence a system's behaviour, i.e., a confluent step $s \xrightarrow{\tau} s'$ implies that $s \leftrightarrow_{bp} s'$. Confluence therefore paves the way for state space reductions modulo branching probabilistic bisimulation (e.g., by giving confluent τ-transitions priority). Not all τ-transitions connect bisimilar states; even though their actions are unobservable, τ-steps might disable behaviour. The aim of our analysis is to efficiently underapproximate which τ-transitions are confluent.

For non-probabilistic systems, several notions of confluence already exist [3]. Basically, they all require that if an action a is enabled from a state that also enables a confluent τ-transition, then (1) a will still be enabled after taking that τ-transition (possibly requiring some additional confluent τ-transitions first), and (2) we can always end up in the same state traversing only confluent τ-steps and the a-step, no matter whether we started by the τ- or the a-transition.

Figure 2 depicts the three notions of confluence we will generalise [3]. Here, the notation τ_c is used for confluent τ-transitions. The diagrams should be interpreted as follows: for any state from which the solid transitions are enabled (universally quantified), there should be a matching for the dashed transitions (existentially quantified). A double-headed arrow denotes a path of zero of more transitions with the corresponding label, and an arrow with label $\bar{a}$ denotes a step that is optional in case $a = \tau$ (i.e., its source and target state may then coincide). The weaker the notion, the more reduction potentially can be achieved (although detection is harder). Note that we first need to find a subset of τ-transitions that we believe are confluent; then, the diagrams are checked.

For probabilistic systems, no similar notions of confluence have been defined before. The situation is indeed more difficult, as transitions do not have a single target state anymore. To still enable reductions based on confluence, only τ-transitions with a unique target state might be considered confluent. The next example shows what goes wrong without this precaution. For brevity, from now on we use *bisimilar* as an abbreviation for *branching probabilistically bisimilar*.

Example 13. Consider two people each throwing a die. The PA in Figure 3(a) models this behaviour given that it is unknown who throws first. The first

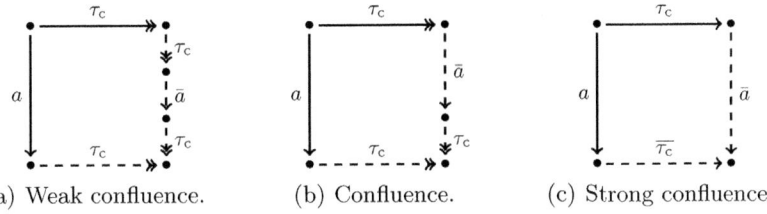

(a) Weak confluence.   (b) Confluence.   (c) Strong confluence.

Fig. 2. Three variants of confluence

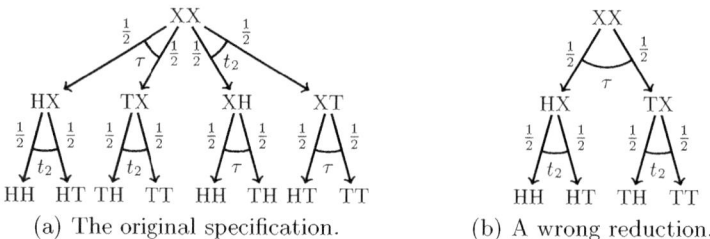

(a) The original specification.   (b) A wrong reduction.

Fig. 3. Two people throwing dice

character of each state name indicates whether the first player has not thrown yet (X), or threw heads (H) or tails (T), and the second character indicates the same for the second player. For lay-out purposes, some states were drawn twice.

We hid the first player's throw action, and kept the other one visible. Now, it might appear that the order in which the t_2- and the τ-transition occur does not influence the behaviour. However, the τ-step does not connect bisimilar states (assuming HH, HT, TH, and TT to be distinct). After all, from state XX it is possible to reach a state (XH) from where HH is reached with probability 0.5 and TH with probability 0.5. From HX and TX no such state is reachable anymore. Giving the τ-transition priority, as depicted in Figure 3(b), therefore yields a reduced system that is *not* bisimilar to the original system anymore.   □

Another difficulty arises in the probabilistic setting. Although for LTSs it is clear that a path $a\tau$ should reach the same state as τa, for PAs this is more involved as the a-step leads us to a distribution over states. So, how should the

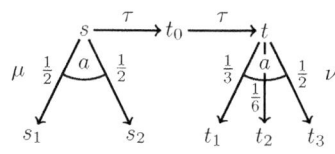

model shown here on the right be completed for the τ-steps to be confluent?

Since we want confluent τ-transitions to connect bisimilar states, we must assure that s, t_0, and t are bisimilar. Therefore, μ and ν must assign equal probabilities to each *class* of bisimilar states. Basically, given the assumption that the other confluent τ-transitions already connect bisimilar states, this is the case if $\mu \equiv_R \nu$ for $R = \{(s, s') \mid s \xrightarrow{\tau}\!\!\!\twoheadrightarrow \twoheadleftarrow\!\!\!\xleftarrow{\tau} s' \text{ using only confluent } \tau\text{-steps}\}$. The following definition formalises these observations. Here we use the notation $s \xrightarrow{\tau_c} s'$, given a set of τ-transitions c, to denote that $s \xrightarrow{\tau} s'$ and $(s, \tau, s') \in c$.

We define three notions of probabilistic confluence, all requiring the target state of a confluent step to be able to mimic the behaviour of its source state. In the weak version, mimicking may be postponed and is based on joinability (Definition 14a). In the default version, mimicking must happen immediately, but is still based on joinability (Definition 14b). Finally, the strong version requires immediate mimicking by directed steps (Definition 16).

Definition 14 ((Weak) probabilistic confluence). *Let $\mathcal{A} = \langle S, s^0, L, \Delta \rangle$ be a PA and $c \subseteq \{(s, a, \mu) \in \Delta \mid a = \tau, \mu \text{ is deterministic}\}$ a set of τ-transitions. (a) Then, c is weakly probabilistically confluent if $R = \{(s, s') \mid s \xrightarrow{\tau_c}\!\!\!\twoheadrightarrow \twoheadleftarrow\!\!\!\xleftarrow{\tau_c} s'\}$ is an equivalence relation, and for every path $s \xrightarrow{\tau_c} t$ and all $a \in L, \mu \in \text{Distr}(S)$*

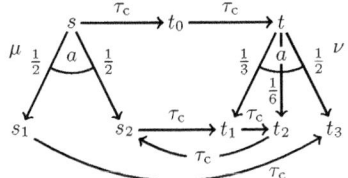 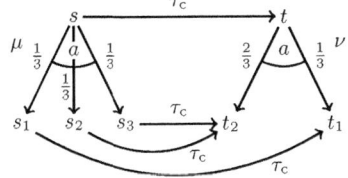

(a) Weak probabilistic confluence.   (b) Strong probabilistic confluence.

Fig. 4. Weak versus strong probabilistic confluence

$$s \xrightarrow{a} \mu \implies \exists t' \in S . t \xrightarrow{\tau_c} t' \wedge$$
$$\left(\left(\exists \nu \in \text{Distr}(S) . t' \xrightarrow{a} \nu \wedge \mu \equiv_R \nu \right) \vee \left(a = \tau \wedge \mu \equiv_R \mathbb{1}_{t'} \right) \right).$$

(b) If for every path $s \xrightarrow{\tau_c} t$ and every transition $s \xrightarrow{a} \mu$ the above implication can be satisfied by taking $t' = t$, then we say that c is probabilistically confluent.

For the strongest variant of confluence, moreover, we require the target states of μ to be connected by direct τ_c-transitions to the target states of ν:

Definition 15 (Equivalence up to τ_c-steps). *Let μ, ν be two probability distributions, and let $\nu = \{t_1 \mapsto p_1, t_2 \mapsto p_2, \dots\}$. Then, μ is equivalent to ν up to τ_c-steps, denoted by $\mu \overset{\tau_c}{\leadsto} \nu$, if there exists a partition $\text{spt}(\mu) = \biguplus_{i=1}^{n} S_i$ such that $n = |\text{spt}(\nu)|$ and $\forall 1 \le i \le n . \mu(S_i) = \nu(t_i) \wedge \forall s \in S_i . s \xrightarrow{\tau_c} t_i$.*

Definition 16 (Strong probabilistic confluence). *Let $\mathcal{A} = \langle S, s^0, L, \Delta \rangle$ be a PA and $c \subseteq \{(s, a, \mu) \in \Delta \mid a = \tau, \mu \text{ is deterministic}\}$ a set of τ-transitions, then c is* strongly probabilistically confluent *if for all $s \xrightarrow{\tau_c} t, a \in L, \mu \in \text{Distr}(S)$*

$$s \xrightarrow{a} \mu \implies \left(\left(\exists \nu \in \text{Distr}(S) . t \xrightarrow{a} \nu \wedge \mu \overset{\tau_c}{\leadsto} \nu \right) \vee \left(a = \tau \wedge \mu = \mathbb{1}_t \right) \right).$$

Proposition 17. *Strong probabilistic confluence implies probabilistic confluence, and probabilistic confluence implies weak probabilistic confluence.*

A transition $s \xrightarrow{\tau} t$ is called *(weakly, strongly) probabilistically confluent* if there exists a (weakly, strongly) probabilistically confluent set c such that $(s, \tau, t) \in c$.

Example 18. Observe the PAs in Figure 4. Assume that all transitions of s, t_0 and t are shown, and that all s_i, t_i, are potentially distinct. We marked all τ-transitions as being confluent, and will verify this for some of them.

In Figure 4(a), both the upper τ_c-steps are weakly probabilistically confluent, most interestingly $s \xrightarrow{\tau_c} t_0$. To verify this, first note that $t_0 \xrightarrow{\tau_c} t$ is (as t_0 has no other outgoing transitions), from where the a-transition of s can be mimicked. To see that indeed $\mu \equiv_R \nu$ (using R from Definition 14), observe that R yields three equivalence classes: $C_1 = \{s_2, t_1, t_2\}$, $C_2 = \{s_1, t_3\}$ and $C_3 = \{s, t_0, t\}$. As required, $\mu(C_1) = \frac{1}{2} = \nu(C_1)$ and $\mu(C_2) = \frac{1}{2} = \nu(C_2)$. Clearly $s \xrightarrow{\tau_c} t_0$ is not probabilistically confluent, as t_0 cannot immediately mimic the a-transition of s.

In Figure 4(b) the upper τ_c-transition is strongly probabilistically confluent (and therefore also (weakly) probabilistically confluent), as t is able to directly mimic the a-transition from s via $t \xrightarrow{a} \nu$. As required, $\mu \overset{\tau_c}{\rightsquigarrow} \nu$ also holds, which is easily seen by taking the partition $S_1 = \{s_1\}, S_2 = \{s_2, s_3\}$.   □

The following theorem shows that weakly probabilistically confluent τ-transitions indeed connect bisimilar states. With Proposition 17 in mind, this also holds for (strong) probabilistic confluence. Additionally, we show that confluent sets can be joined (so there is a unique maximal confluent set of τ-transitions).

Theorem 19. *Let* $\mathcal{A} = \langle S, s^0, L, \Delta \rangle$ *be a PA,* $s, s' \in S$ *two of its states, and* c *a weakly probabilistically confluent subset of its* τ*-transitions. Then,*

$$s \xleftarrow{\tau_c} s' \text{ implies } s \leftrightarrow_{bp} s'.$$

Proposition 20. *Let* c, c' *be (weakly, strongly) probabilistically confluent sets of* τ*-transitions. Then,* $c \cup c'$ *is also (weakly, strongly) probabilistically confluent.*

5   State Space Reduction Using Probabilistic Confluence

As confluent τ-transitions connect branching probabilistic bisimilar states, all states that can reach each other via such transitions can be merged. That is, we can take the original PA modulo the equivalence relation $\xleftrightarrow{\tau_c}$ and obtain a reduced and bisimilar system. The downside of this method is that, in general, it is hard to compute the equivalence classes according to $\xleftrightarrow{\tau_c}$. Therefore, a slightly adapted reduction technique was proposed in [3], and later used in [4]. It chooses a representative state s for each equivalence class, such that all transitions leaving the equivalence class are directly enabled from s. This method relies on (strong) probabilistic confluence, and does not work for the weak variant.

To find a valid representative, we first look at the directed (unlabeled) graph $G = (S, \xrightarrow{\tau_c})$. It contains all states of the original system, and denotes precisely which states can reach each other by taking only τ_c-transitions. Because of the restrictions on τ_c-transitions, the subgraph of G corresponding to each equivalence class $[s]_{\xleftrightarrow{\tau_c}}$ has exactly one terminal strongly connected component (TSCC), from which the representative state for that equivalence class should be chosen. Intuitively, this follows from the fact that τ_c-transitions always lead to a state with at least the same observable transitions as the previous state, and maybe more. (This is not the case for weak probabilistic confluence, therefore the reduction using representatives does not work for that variant of confluence.) The next definition formalises these observations.

Definition 21 (Representation maps). *Let* $\mathcal{A} = \langle S, s^0, L, \Delta \rangle$ *be a PA and* c *a subset of its* τ*-transitions. Then, a function* $\phi_c \colon S \to S$ *is a representation* map *for* $\mathcal{A}$ *under* c *if*

- $\forall s, s' \in S . s \xrightarrow{\tau_c} s' \implies \phi_c(s) = \phi_c(s');$
- $\forall s \in S . s \xrightarrow{\tau_c} \phi_c(s).$

The first condition ensures that equivalent states are mapped to the same representative, and the second makes sure that every representative is in a TSCC. If c is a probabilistically confluent set of τ-transitions, the second condition and Theorem 19 immediately imply that $s \leftrightarrow_{\mathrm{bp}} \phi_c(s)$ for every state s.

The next proposition states that for finite-state PAs and probabilistically confluent sets c, there always exists a representation map. As τ_c-transitions are required to always have a deterministic distribution, probabilities are not involved and the proof is identical to the proof for the non-probabilistic case [3].

Proposition 22. *Let $\mathcal{A} = \langle S, s^0, L, \Delta \rangle$ be a PA and c a probabilistically confluent subset of its τ-transitions. Moreover, let S be finite. Then, there exists a function $\phi_c \colon S \to S$ such that ϕ_c is a representation map for $\mathcal{A}$ under c.*

We can now define a PA modulo a representation map ϕ_c. The set of states of such a PA consists of all representatives. When originally $s \xrightarrow{a} \mu$ for some state s, in the reduced system $\phi_c(s) \xrightarrow{a} \mu'$ where μ' assigns a probability to each representative equal to the probability of reaching any state that maps to this representative in the original system. The system will not have any τ_c-transitions.

Definition 23 ($\mathcal{A}/\phi_c$). *Let $\mathcal{A} = \langle S, s^0, L, \Delta \rangle$ be a PA and c a set of τ-transitions. Moreover, let ϕ_c be a representation map for $\mathcal{A}$ under c. Then, we write $\mathcal{A}/\phi_c$ to denote the PA $\mathcal{A}$ modulo ϕ_c. That is,*

$$\mathcal{A}/\phi_c = \langle \phi_c(S), \phi_c(s^0), L, \Delta_{\phi_c} \rangle,$$

where $\phi_c(S) = \{\phi_c(s) \mid s \in S\}$, and $\Delta_{\phi_c} \subseteq \phi_c(S) \times L \times \mathrm{Distr}(\phi_c(S))$ such that $s \xrightarrow{a}_{\phi_c} \mu$ if and only if $a \neq \tau_c$ and there exists a transition $t \xrightarrow{a} \mu'$ in $\mathcal{A}$ such that $\phi_c(t) = s$ and $\forall s' \in \phi_c(S) . \mu(s') = \mu'(\{s'' \in S \mid \phi_c(s'') = s'\})$.

From the construction of the representation map it follows that $\mathcal{A}/\phi_c \leftrightarrow_{\mathrm{bp}} \mathcal{A}$ if c is (strongly) probabilistically confluent.

Theorem 24. *Let $\mathcal{A}$ be a PA and c a probabilistically confluent set of τ-transitions. Also, let ϕ_c be a representation map for $\mathcal{A}$ under c. Then, $(\mathcal{A}/\phi_c) \leftrightarrow_{\mathrm{bp}} \mathcal{A}$.*

Using this result, state space generation of PAs can be optimised in exactly the same way as has been done for the non-probabilistic setting [4]. Basically, every state visited during the generation is replaced on-the-fly by its representative. In the absence of τ-loops this is easy; just repeatedly follow confluent τ-transitions until none are enabled anymore. When τ-loops are present, a variant of Tarjan's algorithm for finding SCCs can be applied (see [3] for the details).

6   Symbolic Detection of Probabilistic Confluence

Before any reductions can be obtained in practice, probabilistically confluent τ-transitions need to be detected. As our goal is to prevent the generation of large state spaces, this has to be done symbolically.

We propose to do so in the framework of prCRL and LPPEs [10], where systems are modelled by a process algebra and every specification is *linearised* to an intermediate format: the LPPE (linear probabilistic process equation). Basically, an LPPE is a process X with a vector of global variables $\boldsymbol{g}$ of type $\boldsymbol{G}$ and a set of *summands*. A summand is a symbolic transition that is chosen nondeterministically, provided that its guard is enabled (similar to a guarded command). Each summand i is of the form

$$\sum_{\boldsymbol{d_i}:\boldsymbol{D_i}} c_i(\boldsymbol{g},\boldsymbol{d_i}) \Rightarrow a_i(\boldsymbol{g},\boldsymbol{d_i}) \sum_{\boldsymbol{e_i}:\boldsymbol{E_i}} f_i(\boldsymbol{g},\boldsymbol{d_i},\boldsymbol{e_i}) : X(\boldsymbol{n_i}(\boldsymbol{g},\boldsymbol{d_i},\boldsymbol{e_i})).$$

Here, $\boldsymbol{d_i}$ is a (possibly empty) vector of local variables of type $\boldsymbol{D_i}$, which is chosen nondeterministically such that the condition c_i holds. Then, the action $a_i(\boldsymbol{g},\boldsymbol{d_i})$ is taken and a vector $\boldsymbol{e_i}$ of type $\boldsymbol{E_i}$ is chosen probabilistically (each $\boldsymbol{e_i}$ with probability $f_i(\boldsymbol{g},\boldsymbol{d_i},\boldsymbol{e_i})$). Then, the next state is set to $\boldsymbol{n_i}(\boldsymbol{g},\boldsymbol{d_i},\boldsymbol{e_i})$.

The semantics of an LPPE is given as a PA, whose states are precisely all vectors $\boldsymbol{g} \in \boldsymbol{G}$. For all $\boldsymbol{g} \in \boldsymbol{G}$, there is a transition $\boldsymbol{g} \xrightarrow{a} \mu$ if and only if for at least one summand i there is a choice of local variables $\boldsymbol{d_i} \in \boldsymbol{D_i}$ such that

$$c_i(\boldsymbol{g},\boldsymbol{d_i}) \wedge a_i(\boldsymbol{g},\boldsymbol{d_i}) = a \wedge \forall \boldsymbol{e_i} \in \boldsymbol{E_i} . \ \mu(\boldsymbol{n_i}(\boldsymbol{g},\boldsymbol{d_i},\boldsymbol{e_i})) = \sum_{\substack{\boldsymbol{e_i'} \in \boldsymbol{E_i} \\ \boldsymbol{n_i}(\boldsymbol{g},\boldsymbol{d_i},\boldsymbol{e_i})=\boldsymbol{n_i}(\boldsymbol{g},\boldsymbol{d_i},\boldsymbol{e_i'})}} f_i(\boldsymbol{g},\boldsymbol{d_i},\boldsymbol{e_i'}).$$

Example 25. As an example of an LPPE, observe the following specification:

$$X(pc : \{1,2\}) = \sum_{n:\{1,2,3\}} pc = 1 \Rightarrow \text{output}(n) \sum_{i:\{1,2\}} \tfrac{i}{3} : X(i) \qquad (1)$$

$$+ \qquad pc = 2 \Rightarrow \text{beep} \sum_{j:\{1\}} 1 : X(j) \qquad (2)$$

The system has one global variable pc (which can be either 1 or 2), and consists of two summands. When $pc = 1$, the first summand is enabled and the system non-deterministically chooses n to be 1, 2 or 3, and outputs the chosen number. Then, the next state is chosen probabilistically; with probability $\tfrac{1}{3}$ it will be $X(1)$, and with probability $\tfrac{2}{3}$ it will be $X(2)$. When $pc = 2$, the second summand is enabled, making the system beep and deterministically returning to $X(1)$.

In general, the conditions and actions may depend on both the global variables (in this case pc) and the local variables (in this case n for the first summand), and the probabilities and expressions for determining the next state may additionally depend on the probabilistic variables (in this case i and j). □

Instead of designating *individual* τ-transitions to be probabilistically confluent, we designate *summands* to be so in case we are sure that *all* transitions they might generate are probabilistically confluent. For a summand i to be confluent, clearly $a_i(\boldsymbol{g},\boldsymbol{d_i}) = \tau$ should hold for all possible values of $\boldsymbol{g}$ and $\boldsymbol{d_i}$. Also, the next state of each of the transitions it generates should be unique: for every possible valuation of $\boldsymbol{g}$ and $\boldsymbol{d_i}$, there should be a single $\boldsymbol{e_i}$ such that $f_i(\boldsymbol{g},\boldsymbol{d_i},\boldsymbol{e_i}) = 1$.

Moreover, a confluence property should hold. For efficiency, we detect a strong variant of strong probabilistic confluence. Basically, a confluent τ-summand i has to commute properly with every summand j (including itself). More precisely, when both are enabled, executing one should not disable the other and the order of their execution should not influence the observable behaviour or the final state. Additionally, i commutes with itself if it generates only one transition. Formally:

$$\big(c_i(g, d_i) \wedge c_j(g, d_j)\big) \rightarrow \big(i = j \wedge n_i(g, d_i) = n_j(g, d_j)\big) \vee$$
$$\begin{pmatrix} c_j(n_i(g, d_i), d_j) \wedge c_i(n_j(g, d_j, e_j), d_i) \\ \wedge\, a_j(g, d_j) = a_j(n_i(g, d_i), d_j) \\ \wedge\, f_j(g, d_j, e_j) = f_j(n_i(g, d_i), d_j, e_j) \\ \wedge\, n_j(n_i(g, d_i), d_j, e_j) = n_i(n_j(g, d_j, e_j), d_i) \end{pmatrix} \quad (3)$$

where g, d_i, d_j and e_j universally quantify over G, D_i, D_j, and E_j, respectively. We used $n_i(g, d_i)$ to denote the unique target state of summand i given global state g and local state d_i (so e_i does not need to appear).

As these formulas are quantifier-free and in practice often trivially false or true, they can easily be solved using an SMT solver for the data types involved. For n summands, n^2 formulas need to be solved; the complexity of this depends on the data types. In our experiments, all formulas could be checked with fairly simple heuristics (e.g., validating them vacuously by finding contradictory conditions or by detecting that two summands never use or change the same variable).

Theorem 26. *Let X be an LPPE and $\mathcal{A}$ its PA. Then, if for a summand i we have $\forall g \in G, d_i \in D_i \,.\, a_i(g, d_i) = \tau \wedge \exists e_i \in E_i \,.\, f_i(g, d_i, e_i) = 1$ and formula (3) holds, the set of transitions generated by i is strongly probabilistically confluent.*

7 Case Study

To illustrate the power of probabilistic confluence reduction, we applied it on two leader election protocols. We implemented a prototype tool in Haskell for confluence detection using heuristics and state space generation based on confluence information, relying on Theorem 26 and Theorem 24. The results were obtained on a 2.4 GHz, 2 GB Intel Core 2 Duo MacBook[1].

First, we analysed the leader election protocol introduced in [10]. This protocol, between two nodes, decides on a leader by having both parties throw a die and compare the results. In case of a tie the nodes throw again, otherwise the one that threw highest will be the leader. We hid all actions needed for rolling the dice and communication, keeping only the declarations of leader and follower. The complete model in LPPE format can be found in [17].

In [10] we showed the effect of dead-variable reduction [18] on this system. Now, we apply probabilistic confluence reduction both to the LPPE that was already reduced in this way (basicReduced) and the original one (basicOriginal).

The results are shown in Table 1; we list the size of the original and reduced state space, as well as the number of states and transitions that were visited

---

[1] The implementation, case studies and a test script can be downloaded from
http://fmt.cs.utwente.nl/~timmer/prcrl/papers/TACAS2011

Table 1. Applying confluence reduction to two leader election protocols

	Original		Reduced		Visited		Runtime (sec)	
Specification	States	Trans.	States	Trans.	States	Trans.	Before	After
basicOriginal	3,763	6,158	631	758	3,181	3,290	0.45	0.22
basicReduced	1,693	2,438	541	638	1,249	1,370	0.22	0.13
leader-3-12	161,803	268,515	35,485	41,829	130,905	137,679	67.37	31.53
leader-3-15	311,536	515,328	68,926	80,838	251,226	264,123	145.17	65.82
leader-3-18	533,170	880,023	118,675	138,720	428,940	450,867	277.08	122.59
leader-3-21	840,799	1,385,604	187,972	219,201	675,225	709,656	817.67	211.87
leader-3-24	1,248,517	2,055,075	280,057	326,007	1,001,259	1,052,235	1069.71	333.32
leader-3-27	out of memory		398,170	462,864	1,418,220	1,490,349	–	503.85
leader-4-5	443,840	939,264	61,920	92,304	300,569	324,547	206.56	75.66
leader-4-6	894,299	1,880,800	127,579	188,044	608,799	655,986	429.87	155.96
leader-4-7	1,622,682	3,397,104	235,310	344,040	1,108,391	1,192,695	1658.38	294.09
leader-4-8	out of memory		400,125	581,468	1,865,627	2,005,676	–	653.60
leader-5-2	208,632	561,630	14,978	29,420	97,006	110,118	125.78	30.14
leader-5-3	1,390,970	3,645,135	112,559	208,170	694,182	774,459	1504.33	213.85
leader-5-4	out of memory		472,535	847,620	2,826,406	3,129,604	–	7171.73

during its generation using confluence. Probabilistic confluence reduction clearly has quite an effect on the size of the state space, as well as the running time. Notice also that it nicely works hand-in-hand with dead-variable reduction.

Second, we analysed several versions of a leader election protocol that uses asynchronous channels and allows for more parties (Algorithm $\mathcal{B}$ from [6]). We denote by `leader-i-j` the variant with i parties each throwing a j-sided die, that was already optimised using dead-variable reduction. Confluence additionally reduces the number of states and transitions by 77% – 92% and 84% – 94%, respectively. Consequently, the running times more than halve. With probabilistic POR, relatively smaller reductions were obtained for similar protocols [9].

For each experiment, linearisation and confluence detection only took a fraction of time. For the larger state spaces swapping occured, explaining the growth in running time. Confluence clearly allows us to do more before reaching this limit.

8   Conclusions

This paper introduced three new notions of confluence for probabilistic automata. We first established several facts about these notions, most importantly that they identify branching probabilistically bisimilar states. Then, we showed how probabilistic confluence can be used for state space reduction. As we used representatives in terminal strongly connected components, these reductions can even be applied to systems containing τ-loops. We discussed how confluence can be detected in the context of a probabilistic process algebra with data by proving formulas in first-order logic. This way, we enabled on-the-fly reductions when generating the state space corresponding to a process-algebraic specification. A case study illustrated the power of our methods.

References

[1] Baier, C., D'Argenio, P.R., Größer, M.: Partial order reduction for probabilistic branching time. In: Proc. of the 3rd Workshop on Quantitative Aspects of Programming Languages (QAPL). ENTCS, vol. 153(2), pp. 97–116 (2006)

[2] Baier, C., Größer, M., Ciesinski, F.: Partial order reduction for probabilistic systems. In: Proc. of the 1st International Conference on Quantitative Evaluation of Systems (QEST), pp. 230–239. IEEE Computer Society, Los Alamitos (2004)

[3] Blom, S.C.C.: Partial τ-confluence for efficient state space generation. Technical Report SEN-R0123, CWI, Amsterdam (2001)

[4] Blom, S.C.C., van de Pol, J.C.: State space reduction by proving confluence. In: Brinksma, E., Larsen, K.G. (eds.) CAV 2002. LNCS, vol. 2404, pp. 596–609. Springer, Heidelberg (2002)

[5] D'Argenio, P.R., Niebert, P.: Partial order reduction on concurrent probabilistic programs. In: Proc. of the 1st International Conference on Quantitative Evaluation of Systems (QEST), pp. 240–249. IEEE Computer Society, Los Alamitos (2004)

[6] Fokkink, W., Pang, J.: Simplifying Itai-Rodeh leader election for anonymous rings. In: Proc. of the 4th International Workshop on Automated Verification of Critical Systems (AVoCS). ENTCS, vol. 128(6), pp. 53–68 (2005)

[7] Giro, S., D'Argenio, P.R., Ferrer Fioriti, L.M.: Partial order reduction for probabilistic systems: A revision for distributed schedulers. In: Bravetti, M., Zavattaro, G. (eds.) CONCUR 2009. LNCS, vol. 5710, pp. 338–353. Springer, Heidelberg (2009)

[8] Groote, J.F., Sellink, M.P.A.: Confluence for process verification. Theoretical Computer Science 170(1-2), 47–81 (1996)

[9] Größer, M.: Reduction Methods for Probabilistic Model Checking. PhD thesis, Technische Universität Dresden (2008)

[10] Katoen, J.-P., van de Pol, J.C., Stoelinga, M.I.A., Timmer, M.: A linear process-algebraic format for probabilistic systems with data. In: Proc. of the 10th International Conference on Application of Concurrency to System Design (ACSD), pp. 213–222. IEEE Computer Society, Los Alamitos (2010)

[11] De Nicola, R., Vaandrager, F.W.: Action versus state based logics for transition systems. In: Guessarian, I. (ed.) LITP 1990. LNCS, vol. 469, pp. 407–419. Springer, Heidelberg (1990)

[12] Pace, G.J., Lang, F., Mateescu, R.: Calculating τ-confluence compositionally. In: Hunt Jr., W.A., Somenzi, F. (eds.) CAV 2003. LNCS, vol. 2725, pp. 446–459. Springer, Heidelberg (2003)

[13] Peled, D.: All from one, one for all: on model checking using representatives. In: Courcoubetis, C. (ed.) CAV 1993. LNCS, vol. 697, pp. 409–423. Springer, Heidelberg (1993)

[14] Segala, R.: Modeling and Verification of Randomized Distributed Real-Time Systems. PhD thesis, Massachusetts Institute of Technology (1995)

[15] Segala, R., Lynch, N.A.: Probabilistic simulations for probabilistic processes. Nordic Journal of Computation 2(2), 250–273 (1995)

[16] Stoelinga, M.I.A.: Alea jacta est: verification of probabilistic, real-time and parametric systems. PhD thesis, University of Nijmegen (2002)

[17] Timmer, M., Stoelinga, M.I.A., van de Pol, J.C.: Confluence reduction for probabilistic systems (extended version). Technical Report 1011.2314, ArXiv e-prints (2010)

[18] van de Pol, J.C., Timmer, M.: State space reduction of linear processes using control flow reconstruction. In: Liu, Z., Ravn, A.P. (eds.) ATVA 2009. LNCS, vol. 5799, pp. 54–68. Springer, Heidelberg (2009)

[19] van Glabbeek, R.J., Weijland, W.P.: Branching time and abstraction in bisimulation semantics. Journal of the ACM 43(3), 555–600 (1996)

Model Repair for Probabilistic Systems

Ezio Bartocci[1], Radu Grosu[2], Panagiotis Katsaros[3],
C.R. Ramakrishnan[2], and Scott A. Smolka[2]

[1] Department of Applied Math and Statistics, Stony Brook University
Stony Brook, NY 11794-4400, USA
[2] Department of Computer Science, Stony Brook University
Stony Brook, NY 11794-4400, USA
[3] Department of Informatics, Aristotle University of Thessaloniki
54124 Thessaloniki, Greece

Abstract. We introduce the problem of Model Repair for Probabilistic Systems as follows. Given a probabilistic system M and a probabilistic temporal logic formula ϕ such that M fails to satisfy ϕ, the Model Repair problem is to find an M' that satisfies ϕ and differs from M only in the transition flows of those states in M that are deemed controllable. Moreover, the cost associated with modifying M's transition flows to obtain M' should be minimized. Using a new version of parametric probabilistic model checking, we show how the Model Repair problem can be reduced to a nonlinear optimization problem with a minimal-cost objective function, thereby yielding a solution technique. We demonstrate the practical utility of our approach by applying it to a number of significant case studies, including a DTMC reward model of the Zeroconf protocol for assigning IP addresses, and a CTMC model of the highly publicized Kaminsky DNS cache-poisoning attack.

Keywords: Model Repair, Probabilistic Model Checking, Nonlinear Programming.

1 Introduction

Given a model M and a temporal logic formula ϕ, the *Model Checking problem* is to determine if $M \models \phi$, i.e. does M satisfy ϕ? In the case of a positive result, a model checker returns true and may also provide further diagnostic information if *vacuity checking* is enabled [9]. In the case of a negative result, a model checker returns false along with a counterexample in the form of an execution path in M leading to the violation of ϕ. One can then use the counterexample to debug the system model (assuming the problem lies within M as opposed to ϕ) and ultimately *repair* the model so that revised version satisfies ϕ.

Even in light of model checking's widespread success in the hardware, software, and embedded systems arenas (see, e.g., [6]), one can argue that existing model checkers *do not go far enough* in assisting the user in repairing a model that fails to satisfy a formula. Automating the repair process is the aim of the *Model Repair problem*, which we consider in the context of *probabilistic systems*

P.A. Abdulla and K.R.M. Leino (Eds.): TACAS 2011, LNCS 6605, pp. 326–340, 2011.

such as discrete-time Markov chains (DTMCs), continuous-time Markov chains (CTMCs), and Markov decision processes (MDPs).

The Model Repair problem we consider can be stated as follows. Given a probabilistic system M and a probabilistic temporal logic formula ϕ such that M fails to satisfy ϕ, the *probabilistic Model Repair problem* is to find an M' that satisfies ϕ and differs from M only in the transition flows[1] of those states in M that are deemed *controllable*. Moreover, the *cost* associated with modifying M's transition flows to obtain M' should be minimized. A related but weaker version of the Model Repair problem was first considered in [5], in the (non-probabilistic) context of Kripke structures and the CTL temporal logic. See Section 8 for a discussion of related work.

Our main contributions to the probabilistic Model Repair problem can be summarized as follows:

- Using a new version of *parametric probabilistic model checking* [7,13], we show how the Model Repair problem can be reduced to a *nonlinear optimization* problem with a minimal-cost objective function, thereby yielding a solution technique.
- We consider related solution feasibility and optimality conditions, and provide an implementation of our solution technique using the PARAM tool for parametric model checking [15] and the Ipopt open-source software package for large-scale nonlinear optimization [2].
- We also consider a *Max-Profit* version of the Model Repair problem for reward-based systems, where profit is defined as the difference between the expected reward and the cost of model repair.
- We also provide a control-theoretic characterization of the probabilistic Model Repair problem, and in the process establish a formal link between model repair and the controller-synthesis problem for linear systems.
- We demonstrate the practical utility of our approach by applying it to a number of significant case studies, including a DTMC reward model of the Zeroconf protocol for assigning IP addresses, and a CTMC model of the highly publicized Kaminsky DNS cache-poisoning attack [1].

The rest of the paper develops along the following lines. Section 2 provides background on parametric model checking. Section 3 contains our formulation of the probabilistic Model Repair problem, while Section 4 characterizes Model Repair as a nonlinear optimization problem. Section 5 considers related feasibility and optimality conditions. Section 6 examines the link between model repair and optimal controller synthesis. Section 7 presents our case studies, while Section 8 discusses related work. Section 9 offers our concluding remarks and directions for future work.

2   Parametric Probabilistic Model Checking

In this section, we show how the model-checking problem for parametric DTMCs can be reduced to the evaluation of a multivariate rational function. The

---

[1] For a DTMC, each row of the probability transition matrix represents the *transition flow* out of the corresponding state.

definition of a parametric DTMC is from [13] and the definition of the finite state automaton derived from a parametric DTMC is from [7].

Definition 1. *A (labeled) Discrete-Time Markov Chain (DTMC) is a tuple $D = (S, s_0, \mathbf{P}, L)$ where:*

- *S is a finite set of states*
- *$s_0 \in S$ is the initial state*
- *$\mathbf{P} : S \times S \to [0, 1]$ is a function such that $\forall s \in S$, $\sum_{s' \in S} \mathbf{P}(s, s') = 1$*
- *$L : S \to 2^{AP}$ is a labeling function assigning to each state a set of atomic propositions from the denumerable set of atomic propositions AP.*

Probabilistic model checking is based on the definition of a probability measure over the set of paths that satisfy a given property specification [20]. In the PCTL [17] temporal logic, property specifications are of the form $\mathcal{P}_{\sim b}(\psi)$, with $\sim \in \{<, \leq, >, \geq\}, 0 \leq b \leq 1$, and ψ a path formula defined using the X (next) and $\mathcal{U}^{\leq h}$ (bounded/unbounded until) operators for $h \in \mathbb{N} \cup \{\infty\}$. A state s satisfies $\mathcal{P}_{\sim b}(\psi)$, denoted as $s \models \mathcal{P}_{\sim b}(\psi)$, if $\mathbb{P}(Path_s(s, \psi)) \sim b$; i.e. the probability of taking a path from s that satisfies ψ is $\sim b$.

Definition 2. *Let $V = \{v_1, \cdots, v_r\}$ be a set of real variables and let $\mathbf{v} = (v_1, \ldots, v_r)$. A multivariate rational function f over V is a function of the form*

$$f(\mathbf{v}) = \frac{f_1(\mathbf{v})}{f_2(\mathbf{v})}$$

where f_1, f_2 are two polynomials in $\mathbf{v}$.

Let $\mathcal{F}_V(\mathbf{v})$ be the field of real-valued rational functions. Given $f \in \mathcal{F}_V$ and an evaluation function $u : V \to \mathbb{R}$, we denote by $f[V/u]$ the value obtained by substituting each occurrence of $v \in V$ with $u(v)$.

Definition 3. *A parametric DTMC (PDTMC) is a tuple $\mathcal{D} = (S, s_0, \mathbf{P}, V)$ where S, s_0 are as in Def. 1 and $\mathbf{P} : S \times S \to \mathcal{F}_V$, where $V = \{v_1, \cdots, v_r\}$ is a finite set of parameters.*

Given a PDTMC $\mathcal{D}$ over parameters V, an evaluation u of V is said to be *valid* for $\mathcal{D}$ if the induced probability transition matrix $\mathbf{P}_u : S \times S \to [0, 1]$ is such that $\sum_{s' \in S} \mathbf{P}_u(s, s')[V/u] = 1, \forall s \in S$.

Definition 4. *For a PDTMC $\mathcal{D}$ and PCTL formula $\phi = \mathcal{P}_{\sim b}(\psi)$, the derived finite state automaton (dFSA) $\mathcal{A}_{\mathcal{D},\psi}$ is given by $\mathcal{A}_{\mathcal{D},\psi} = \{S, \Sigma, s_0, \delta, S_f\}$, where:*

- *S is the same set of states of $\mathcal{D}$*
- *$\Sigma = \{f \in \mathcal{F}_V \mid \exists i, j, \mathbf{P}(i, j) = f(\mathbf{v}) \neq 0\}$ is the finite alphabet*
- *s_0 is $\mathcal{D}$'s initial state*
- *$\delta : S \times \Sigma \mapsto 2^S$ is the transition function derived from $\mathbf{P}$ such that $\delta(s, f) = Q$ implies $\forall q \in Q$, $\mathbf{P}(s, q) = f(\mathbf{v})$*
- *$S_f \subseteq S$ is the set of final states and depends on ψ.*

The set $\mathcal{R}(\Sigma)$ of regular expressions over alphabet Σ can be translated into a multivariate rational function. The composition function $comp : \mathcal{R} \mapsto \mathcal{F}_V$ is defined inductively by the following rules:

$$comp(f) = f(\mathbf{v}) \qquad\qquad comp(x|y) = comp(x) + comp(y)$$
$$comp(x.y) = comp(x) \cdot comp(y) \quad comp(x^*) = \frac{1}{1-comp(x)}$$

It can be proved that $comp(\alpha)$ yields the probability measure of the set $\bigcup_{s_f \in S_f} Paths(s_0, s_f)$ of paths in $\mathcal{A}_{\mathcal{D},\psi}$ from s_0 to some state s_f in S_f. In [7], Daws characterizes the set of paths satisfying an unbounded formula $\phi = \mathcal{P}_{\sim b}(\psi)$, but without nested probabilistic quantifiers, as a dFSA $\mathcal{A}_{\mathcal{D},\psi}$, and proves:

Proposition 1. *For a PDTMC $\mathcal{D}$ and PCTL formula $\phi = \mathcal{P}_{\sim b}(\psi)$, with ψ a path formula, let α be the regular expression for $\mathcal{L}(\mathcal{A}_{\mathcal{D},\psi})$. Then,*

$$s_0 \models \phi \text{ iff there is an evaluation } u \text{ s.t. } u \text{ is valid for } \mathcal{D} \text{ and } comp(\alpha) \sim b$$

Given a PDTMC $\mathcal{D}$ and bounded reachability property ϕ, Hahn [14] presents a simple recursive algorithm for deriving the multivariate function f that computes the probability by which $\mathcal{D}$ satisfies ϕ. Since the only arithmetic operators appearing in f are addition and multiplication, $f \in \mathcal{F}_V$, as $\mathcal{F}_V$ is a field.

3   The Model Repair Problem

Given a set of parameters V, we write $span(V)$ for the set of linear combinations of the elements in V. A n-state DTMC D can be turned into a *controllable DTMC $\tilde{D}$* by pairing it with an $n \times n$ matrix $\mathbf{Z}$ that specifies which states of D are controllable, and how the transitions out of these states are controlled using elements of $span(V)$.

Definition 5. *A controllable DTMC over a set of parameters V is a tuple $\tilde{\mathcal{D}} = (S, s_0, \mathbf{P}, \mathbf{Z}, L)$ where $(S, s_0, \mathbf{P}, L)$ is a DTMC and $\mathbf{Z} : S \times S \rightarrow span(V)$ is an $|S| \times |S|$ matrix such that $\forall s \in S$, $\sum_{t \in S} \mathbf{Z}(s,t) = 0$. A state $s \in S$ is a controllable state of $\tilde{\mathcal{D}}$ if $\exists t \in S$ such that $\mathbf{Z}(s,t) \neq 0$.*

Matrix $\mathbf{Z}$ can be understood as a strategy for altering or controlling the behavior of a DTMC, typically for the purpose of repair; i.e. forcing a particular property to hold for the DTMC. The constraint on $\mathbf{Z}$ implies that the control strategy embodied in $\mathbf{Z}$ should neither change the structure of the DTMC nor its stochasticity. Which states of the DTMC are controllable depends on the model parameters that can be tuned. In general, a model may be repaired by a number of different strategies.

Example 1. Fig. 1 shows a DTMC in (a), the controllable DTMC in (b), with s_0, s_2 controllable, and the associated matrix $\mathbf{Z}$ in (c).

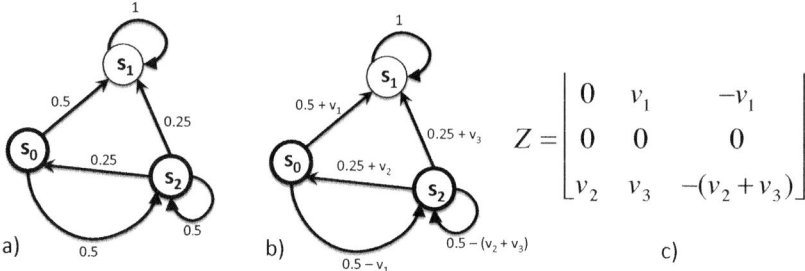

Fig. 1. A DTMC (a) and the controllable DTMC (b), with s_0, s_2 controllable by $\mathbf{Z}$ (c)

Definition 6. *Let $\tilde{\mathcal{D}} = (S, s_0, \mathbf{P}, \mathbf{Z}, L)$ be a controllable DTMC over parameters V, $D = (S, s_0, \mathbf{P}, L)$ the DTMC underlying $\tilde{\mathcal{D}}$, ϕ a PCTL formula for which $D, s_0 \nvDash \phi$, and $g(\mathbf{v})$ a possibly nonlinear cost function, which is always positive, continuous, and differentiable. The* Model Repair problem *is to find a DTMC $D' = (S, s_0, \mathbf{P}' = \mathbf{P} + \mathbf{Z}[V/u], L)$, where $u : V \to \mathbb{R}$ is an evaluation function satisfying the following conditions:*

$$u = \arg\min g \tag{1}$$

$$D', s_0 \models \phi \tag{2}$$

$$\mathbf{P}(i,j) = 0 \text{ iff } \mathbf{P}'(i,j) = 0, 1 \leq i, j \leq |S| \tag{3}$$

The repair process seeks to manipulate the parameters of $\tilde{\mathcal{D}}$ to obtain a DTMC D' such that $D', s_0 \models \phi$ and the cost of deriving probability transition matrix $\mathbf{P}'$ from $\mathbf{P}$ is minimized. A typical cost function is $g(\mathbf{v}) = w_1 v_1^2 + \cdots + w_r v_r^2$, $\mathbf{w} \in \mathbb{R}_+^r$: a weighted sum of squares of the parameters with weights $w_k, 1 \leq k \leq r$, specifying that some parameters affect the model to a greater extent than others. For $\mathbf{w} = 1_r$, g is the square of the L2-norm $\|\mathbf{v}\|_2^2$. The insertion of new transitions and the elimination of existing ones are not allowed by Condition (3).

The repair process as specified by Def. 6 is robust in the following sense.

Proposition 2. *A controllable DTMC $\tilde{\mathcal{D}}$ and its repaired version D' are ϵ-bisimilar, where ϵ is the largest value in the matrix $\mathbf{Z}[V/u]$.*

Proof. The result immediately follows from Def. 5 and the definition of ϵ-bisimulation [11].

Example 2. As an example of the repair process, consider the Die problem proposed by Knuth et al. in [19], where a fair coin is tossed three or more times to obtain the face of a die; see Fig. 2(a). We wish to repair the model so that the formula $\mathcal{P}_{\leq 1/8} F[die = 1]$ is satisfied, thereby imposing a probability bound of $\frac{1}{8}$ to eventually obtain a 1. In the original model, the probability to eventually reach any face of the die is $\frac{1}{6}$, and the property is not satisfied.

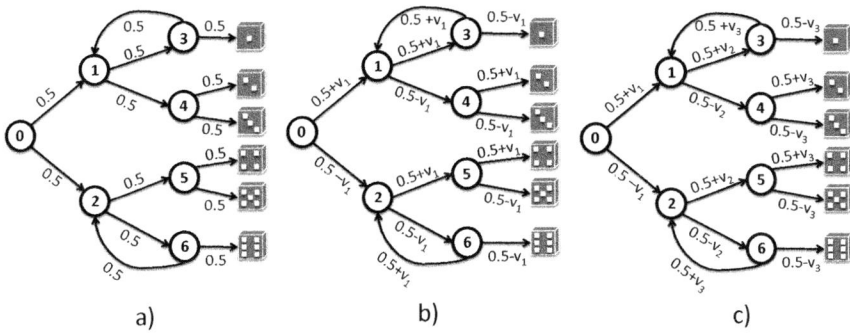

Fig. 2. Two different model-repair strategies for the Knuth Die problem

To repair the model, we need to decide by how much the coin should be biased each time it is tossed. Figs. 2(b) and (c) provide two different model-repair strategies. In Fig. 2(b), we use a single biased coin (controllable DTMC with one parameter), while in Fig. 2(c), we use three differently biased coins (three parameters). In the latter case, the third coin can be tossed again if state 3 or 6 is reached. The case with three parameters gives us the opportunity to prioritize the parameters according to the impact they should have on the repaired model's state space using the weighted sum-of-squares cost function described above.

3.1   The Max-Profit Model Repair Problem

Definition 7. *A controllable Markov Reward Model (MRM) is a tuple $\mathcal{R} = (\hat{\mathcal{D}}, \rho, \iota)$ where $\hat{\mathcal{D}} = (S, s_0, \mathbf{P}, \mathbf{Z}, L)$ is a controllable DTMC, $\rho : S \to \mathbb{R}_{\geq 0}$ a state reward function, and $\iota : S \times S \to \mathbb{R}_{\geq 0}$ a transition reward function.*

During the repair process, the Max-Profit Model Repair problem seeks to maximize the expected profit, as measured in terms of the difference between the expected reward and the expected cost.

Definition 8. *Let $\mathcal{R} = (\tilde{\mathcal{D}}, \rho, \iota)$ be a controllable MRM with $\tilde{\mathcal{D}} = (S, s_0, \mathbf{P}, \mathbf{Z}, L)$ the associated controllable DTMC and $D = (S, S_0, \mathbf{P}, L)$ $\tilde{\mathcal{D}}$'s underlying DTMC. Also, let ϕ be a PCTL formula such that $D, s_0 \nvDash \phi$, $g(\mathbf{v})$ a possibly nonlinear cost function, and $e(\mathbf{v})$ an expected reward function, which, using ρ and ι, measures the expected reward accumulated along any path in $\mathcal{R}$ that eventually reaches a state in a given set of states B, $B \subseteq S$ [13]. We assume that $e(\mathbf{v}) - g(\mathbf{v})$ is always positive, continuous, and differentiable. The* Max-Profit Model Repair *problem is to find a DTMC $D' = (S, s_0, \mathbf{P}' = \mathbf{P} + \mathbf{Z}[V/u], L)$ with $u : V \to \mathbb{R}$ an evaluation function that satisfies the Conditions (2) and (3) of Def. 6, and the following condition:*

$$u = \arg \max\ e - g \qquad (4)$$

4   Model Repair as a Nonlinear Programming Problem

The controllable DTMC $\tilde{\mathcal{D}} = (S, s_0, \mathbf{P}, \mathbf{Z}, L)$ over set of parameters V corresponds to a PDTMC with probability matrix $\mathbf{P} + \mathbf{Z}$. If $\tilde{\mathcal{D}} \nvDash \mathcal{P}_{\sim b}(\psi)$, by parametric model checking, we derive the nonlinear constraint $f(\mathbf{v}) \sim b$, where f is a multivariate rational function. Let Y be the set of linear combinations in V defined as $Y = \{y(\mathbf{v}) = \mathbf{P}(i,j) + \mathbf{Z}(i,j) : \mathbf{Z}(i,j) \neq 0, 1 \leq i, j \leq |S|\}$.

Proposition 3. *A solution to the Model Repair problem of Def. 6 satisfies the constraints of the following nonlinear program (NLP):*

$$\mathbf{min}\ g(\boldsymbol{v}) \tag{5}$$

$$f(\boldsymbol{v}) \sim b \tag{6}$$

$$\forall y \in Y : 0 < y(\mathbf{v}) < 1 \tag{7}$$

Proof. Cost function (5) ensures that if there is a solution to the NLP, this yields an evaluation u such that the resulting DTMC is the closest to $\tilde{\mathcal{D}}$ with the desired property, as implied by Condition (1) of Def. 6. Constraint (6) enforces requirement (2) of Def. 6. Constraint (7), along with the selection of $\mathbf{Z}$ in Def. 5, assure that evaluation u is valid and preserves the stochasticity of the new transition probability matrix, while the strict inequalities also enforce constraint (3) of Def. 6. If the NLP has no solution, then model repair for $\tilde{\mathcal{D}}$ is infeasible. For the solution of the Max Profit repair problem, we only have to replace *cost function* (5) with *profit function* (4).

Example 3. To find a solution for the Model Repair problem of Example 2 with the strategy shown in Fig. 2(b), the associated NLP is formulated as follows:

$$\mathbf{min}\ w_1 v_1^2 + w_2 v_2^2 + w_3 v_3^2$$

$$\frac{8v_1 v_2 v_3 - 4(v_2 v_1 - v_2 v_3 - v_1 v_3) - 2(v_1 + v_2 - v_3) - 1}{8v_2 v_3 + 4v_2 + 4v_3 - 6} - \frac{1}{8} \leq 0$$

$$\forall i \in \{1, \cdots, 3\}, -0.5 < v_i < 0.5$$

Fig. 3 shows two different solutions based on the choice of the vector $\mathbf{w}$ of weights associated with the cost function. In Fig. 3(a), we consider the same cost for changing each transition probability, i.e. $\mathbf{w} = [1\ 1\ 1]$. In Fig. 3(b), we provide a solution where the cost of v_1 is ten times the cost of v_3 and two times the cost of v_2, i.e. $\mathbf{w} = [10\ 5\ 1]$.

We denote by $D_{NLP} = Dom(g) \cap Dom(f) \cap \bigcap_{y \in Y} Dom(y)$ the domain of the NLP problem. A solution $\mathbf{v}^* \in D_{NLP}$ is a *local minimizer* if there is $\epsilon > 0$ such that $g(\mathbf{v}^*) \leq g(\mathbf{v})$ for all $|\mathbf{v} - \mathbf{v}^*| < \epsilon$ with $\mathbf{v} \in D_{NLP}$. If $g(\mathbf{v}^*) < g(\mathbf{v})$, $\mathbf{v}^*$ is a *strict* local minimizer. A local minimizer $\mathbf{v}^* \in D_{NLP}$ is a *globally optimal solution* if $g(\mathbf{v}^*) \leq g(\mathbf{v})$ for all $\mathbf{v} \in D_{NLP}$.

All nonlinear optimization algorithms search for a locally feasible solution to the problem. Such a solution can be found by initiating the search from the point $\mathbf{0}^{1 \times r}$, representing the no-change scenario. If no solution is found, the problem is locally infeasible and the analyst has to initiate a new search from another point or else to prove that the problem is infeasible.

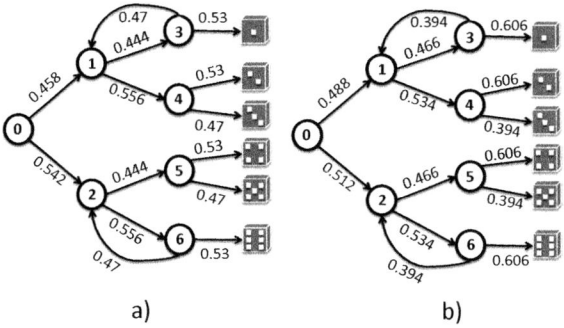

Fig. 3. Model Repair solutions for Knuth Die problem

5   Model Repair Feasibility and Optimality

For a Model Repair problem with property constraint $f(\mathbf{v}) \sim b$, we denote:

$$h(\mathbf{v}) = \begin{cases} f(\mathbf{v}) - b, & \text{if } \sim \in \{<, \leq\} \\ b - f(\mathbf{v}), & \text{if } \sim \in \{>, \geq\} \end{cases}$$

such that the constraint is written as $h(\mathbf{v}) \leq 0$, with the inequality becoming strict if $\sim \in \{<, >\}$.

Definition 9. *For the NLP problem, the* Lagrangian function $\mathcal{X} : D_{NLP} \times \mathbb{R}^{2|Y|+1} \to \mathbb{R}$ *is defined as:*

$$\mathcal{X}(\mathbf{v}, \lambda) = g(\mathbf{v}) + \lambda_0 \cdot h(\mathbf{v}) + \sum_{k=1}^{|Y|} \lambda_k \cdot (y_k(\mathbf{v}) - 1) - \sum_{k=|Y|+1}^{2|Y|} \lambda_k \cdot y_{k-|Y|}(\mathbf{v})$$

The vector λ is the Lagrange multiplier vector. *The* Lagrange dual function

$$w(\lambda) = \inf_{\mathbf{v} \in D_{NLP}} (\mathcal{X}(\mathbf{v}, \lambda))$$

yields the minimum of the Lagrangian function over $\mathbf{v}$.

It is easy to verify that for $\lambda \geq 0$, the Lagrangian dual function w yields lower bounds on the optimal cost $g(\mathbf{v}^*)$ of the NLP problem, since $h(\mathbf{v}^*) \leq 0$, $y(\mathbf{v}^*) - 1 < 0$, and $y(\mathbf{v}^*) > 0$, for all $y \in Y$.

Proposition 4. The best lower bound *on the the NLP cost function is the cost for the solution of the following Lagrangian dual problem:*

$$\mathbf{max} \ \ w(\lambda)$$

$$\lambda \geq 0$$

$$\lambda_k \neq 0, \ k = 1, \dots, 2 \cdot |Y|$$

The constraints for λ_k correspond to the strict inequalities in constraint (7), and if $\sim \in \{<, >\}$, we eventually have $\lambda > 0$.

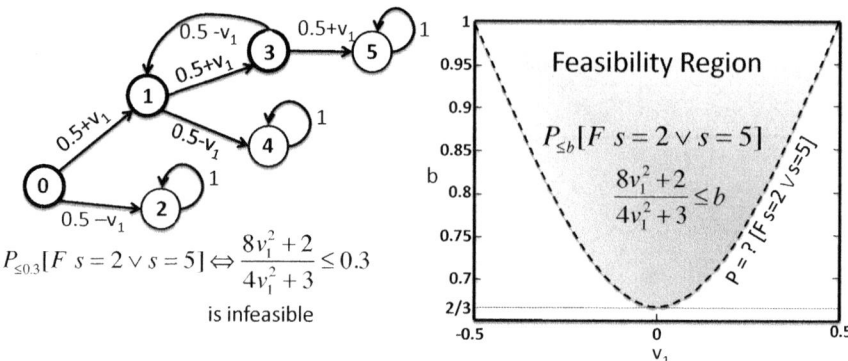

Fig. 4. Feasibility analysis for a probabilistic Model Repair problem

In [4], the authors developed a criterion for analyzing the feasibility of nonlinear optimization programs, based on the theory of Lagrange duality. More precisely, the system of inequalities in (6) (Prop. 3) is feasible when the NLP_f

$$\textbf{min } 0 \tag{8}$$

$$h(\mathbf{v}) \sim 0 \tag{9}$$

$$\forall y \in Y : y(\mathbf{v}) - 1 < 0 \tag{10}$$

$$\forall y \in Y : -y(\mathbf{v}) < 0 \tag{11}$$

is also feasible. The optimal cost for the NLP_f is:

$$p^* = \begin{cases} 0 & \text{if } NLP_f \text{ is feasible} \\ \infty & \text{if } NLP_f \text{ is infeasible} \end{cases} \tag{12}$$

The Lagrangian dual function w_f for this program with zero cost is:

$$w_f(\lambda) = \inf_{\mathbf{v} \in D_{NLP}} (\mathcal{X}(\mathbf{v}, \lambda) - g(\mathbf{v})) \tag{13}$$

We note that w_f is positive homogeneous in λ, i.e. $\forall \alpha > 0$, $w_f(\alpha \cdot \lambda) = \alpha \cdot w_f(\lambda)$. The associated Lagrange dual problem is to maximize $w_f(\lambda)$ subject to the constraints $\lambda \geq 0$ and $\lambda_k \neq 0$, $k = 1, \dots, 2 \cdot |Y|$. The optimal cost for the homogeneous w_f in the dual problem is:

$$d^* = \begin{cases} \infty & \text{if } \lambda \geq 0, \ \lambda_k \neq 0, \ w_f(\lambda) > 0 \text{ is feasible} \\ 0 & \text{if } \lambda \geq 0, \ \lambda_k \neq 0, \ w_f(\lambda) > 0 \text{ is infeasible} \end{cases} \tag{14}$$

for $k = 1, \dots, 2 \cdot |Y|$. By taking into account the property

$$d^* \leq p^*$$

which is known as *weak duality* [22], we conclude:

Proposition 5. *If the Lagrange dual problem of NLP_f, with the NLP constraints and cost is feasible, then the NLP for model repair is infeasible. Conversely, if NLP is feasible, then the Lagrange dual problem of NLP_f is infeasible.*

Example 4. For the controllable DTMC $\tilde{\mathcal{D}}$ of Fig. 4, we show that Model Repair is not always feasible. For path formula $\psi = F[s = 2 \lor s = 5]$, from Prop. 1 we have:

$$\mathcal{D}, s_0 \models \mathcal{P}_{\leq b}\psi \quad \text{iff} \quad \frac{8v_1^2 + 2}{4v_1^2 + 3} \leq b$$

The Lagrangian dual function for the NLP_f program is:

$$w_f(\lambda) = \inf_{v_1 \in]-0.5, 0.5[} \left(\lambda_0 \left(\frac{8v_1^2 + 2}{4v_1^2 + 3} - b \right) + \lambda_1(v_1 - 1) + \lambda_2(-v_1) \right)$$

where $\lambda_0 \geq 0$ and $\lambda_1, \lambda_2 > 0$. The rational function for ψ is minimized in $v_1 = 0$ (Fig. 4) and therefore

$$w_f(\lambda) = \lambda_0 \left(\frac{2}{3} - b \right) - \lambda_1$$

The nonlinear program in (14) becomes feasible when $b < 2/3$ and from Prop. 5 the NLP for repairing $\tilde{\mathcal{D}}$ is infeasible for these values of b.

From [4], if $\mathbf{v}^*$ is a local minimizer that also fulfills certain constraint qualifications, then the Karush-Kuhn-Tucker (KKT) conditions are satisfied. On the other hand, if for some feasible point $\mathbf{v}'$ there is a Lagrange multiplier vector λ^* such that the KKT conditions are satisfied, sufficient conditions are provided which guarantee that $\mathbf{v}'$ is a strict local minimizer. Since all the parameters are bounded it is possible to check global optimality by using an appropriate constraint solver, such as RealPaver [12].

6   Model Repair and Optimal Control

We show how the Model Repair problem is related to the optimal-controller synthesis problem for linear systems. Given a (right-)linear system

$$x(n+1) = x(n)A + u(n)B, \quad x(0) = x_0$$

where x is the state vector, u is a vector of inputs, and A and B are matrices representing the system dynamics, the synthesis of an optimal controller is typically performed in a compositional way as follows: (1) Synthesize a linear quadratic regulator; (2) Add a Kalman filter state estimator; (3) Add the reference input.

For (1), the input can be written as $u(n) = x(n)K$, where K is the controller matrix to be synthesized. Then:

$$x(n+1) = x(n)(A + KB)$$

An optimal K is then obtained by minimizing the cost function

$$\mathcal{J} = 1/2 \sum_{k=0}^{\infty} (x^T(k)Qx(k) + u^T(k)Ru(k))$$

where Q and R are nonnegative definite symmetric weighting matrices to be selected by the designer, based on the relative importance of the various states x_i and controls u_j.

The purpose of (2) is to add an optimal state estimator, which estimates the current state based on the previous input and the previously measured output. This is typically done in a recursive fashion with the help of a Kalman filter.

Finally, in (3), the reference input is added, which drives the overall behavior of the controlled system. Typically, the addition of the reference input is complemented with an integral control.

Although not immediately apparent, all the above steps are related to the Model Repair problem (Definition 6). First, matrix $\mathbf{Z}$ contains in each entry a linear combination of the parameters in V. This linear combination can be obtained by decomposing $\mathbf{Z}$ as $\mathbf{Z} = KB$, where K is a matrix of parameters (to be synthesized) and B is a matrix defining the contribution of each parameter to the next state of the system.

Note however, that the above optimization problem for $\mathcal{J}$ does not encode the stochasticity requirement of $\mathbf{Z}$. Hence, using the available tools for optimal-controller synthesis, one cannot impose this requirement.

Second, the reference input can be related to the PCTL formula ϕ. Adding the reference input to the system is, in many ways, the same as imposing the satisfaction of ϕ. Note, however, that the reference input is added *after* K is synthesized. In contrast, we synthesize the parameters in $\mathbf{Z}$ with ϕ included as a constraint of the nonlinear optimization problem.

Finally, the variables occurring in the PCTL formula can be seen as the observables of the system, i.e., as the outputs of the system. In the model repair case, we assume that we have full access to the state variables and the model precisely captures the output.

Overall, our approach seems to be a generalization of the optimal-controller synthesis problem for the class of linear systems represented by DTMCs. We also perform a nonlinear multivariate optimization which is by necessity more powerful than the one used for linear systems.

7   Applications

We present two Model Repair applications: (i) the Kaminsky DNS cache-poisoning attack, along with the proposed fix (repair), and (ii) the Zeroconf protocol as a Max-Profit problem. Model Repair was performed using the PARAM tool for parametric model checking [15] and the Ipopt software package for large-scale nonlinear optimization [2].

Example 5. The Kaminsky DNS attack makes clever use of cache poisoning, so that when a victim DNS server is asked to resolve URLs within a non-malicious domain, it replies with the IP address of a malicious web server. The proposed fix is to randomize the UDP port used in name-resolution requests. As such, an intruder can corrupt the cache of a DNS server with a falsified IP address for a URL, only if it manages to guess a 16-bit source-port id, in addition to the 16-bit query id assigned to the name-resolution request.

Our CTMC model for the Kaminsky attack [1] implements a victim DNS server that generates `times_to_request_url` queries to resolve one or more resource names within some domain. While the victim waits for a legitimate response to its query, the intruder tries with rate `guess` to provide a fake response that, if correctly matching the query id, will be accepted by the victim, thus corrupting its cache.

The only parameter the victim can control is the range of port-id values used by the proposed fix, which affects the rate at which correct guesses arrive at the victim. Other parameters that affect the rate of correct guesses, but are not controlled by the victim are the `popularity` of the requested names, and the rate at which `other_legitimate_requests` arrive at the victim. If the fix is disabled, the number of port ids is one, and experiments show that for `guess` ≥ 200, the attack probability is greater than 0.9 if `times_to_request_url` ≥ 6.

By applying model repair on the controllable embedded DTMC, we determined the minimum required range of port ids such that $\mathcal{P}_{\leq 0.05}F$ cache_poisoned. While the value of `times_to_request_url` determines the size of the state space, we observed that nonlinear optimization with Ipopt is not affected by state-space growth. This is not the case, however, for the parametric model-checking times given in Table 1 (`popularity`=3,`guess`=150,`other_legitimate_requests`=150). The model was successfully repaired for all values of `times_to_request_url` from 1 to 10.

Example 6. According to the Zeroconf protocol for assigning IP addresses in a network, when a new host joins the network it randomly selects an IP address among K possible ones. With m hosts in the network, the collision probability is $q = m/K$. A new host asks the other hosts whether the randomly selected IP address is already used and waits for an answer. The probability that the new host does not get any answer is p, in which case it repeats the query. If after n tries there is still no answer, the host will erroneously consider the chosen address as valid.

We used Max Profit model repair on the DTMC model of [7] to determine the collision probability q that optimizes the trade-off between (a) the expected number of tries until the algorithm terminates, and (b) the cost of changing q from its default value. The change applied to q is the only parameter used in our Max Profit model; all other transition probabilities were maintained as constants as in the original model. For $n = 3$, $p = 0.1$, and initial $q = 0.6$, we determined the optimal q to be 0.5002, which reduced the expected number of steps to termination from 6.15 to 5.1.

Table 1. Model Repair of the Kaminsky CTMC

times_to_request	States	Transitions	CPU_PARAM	PORT_ID P=?	[F cache_poisoned]
1	10	13	0m0.390s	5	0.04498
2	60	118	0m0.430s	10	0.04593
3	215	561	0m0.490s	14	0.04943
4	567	1759	0m1.750s	19	0.04878
5	1237	4272	0m15.820s	24	0.04840
6	2350	8796	1m56.650s	28	0.0498
7	4085	16163	10m55.150s	33	0.0494
8	6625	27341	47m21.220s	38	0.0491
9	10182	43434	167m58.470s	42	0.0499
10	14992	65682	528m32.720s	47	0.0496

8   Related Work

Prior work has addressed a related version of the Model Repair problem in the nonprobabilistic setting. In [5], abductive reasoning is used to determine a suitable modification of a Kripke model that fails to satisfy a CTL formula. Addition and deletion of state transitions are considered, without taking into account the cost of the repair process. The problem of automatically revising untimed and real-time programs with respect to UNITY properties is investigated in [3], such that the revised program satisfies a previously failed property, while preserving the other properties. A game-based approach to the problem of automatically fixing faults in a finite-state program is considered in [18]. The game consists of the product of a modified version of the program and an automaton representing an LTL specification, such that every winning finite-state strategy corresponds to a repair. In [23], the authors introduce an algorithm for solving the parametric real-time model-checking problem: given a real-time system and temporal formula, both of which may contain parameters, and a constraint over the parameters, does every allowed parameter assignment ensure that the real-time system satisfies the formula?

In related work for probabilistic models, a Bayesian estimator based on runtime data is used in [10] to address the problem of model evolution, where model parameters may change over time. The authors of [21] consider parametric models for which they show that finding parameter values for a property to be satisfied is in general undecidable. In [8], a model checker together with a genetic algorithm drive the parameter-estimation process by reducing the distance between the desired behavior and the actual behavior. The work of [16] addresses the parameter-synthesis problem for parametric CTMCs and time-bounded properties. The problem is undecidable and the authors provide an approximation method that yields a solution in most cases.

9   Conclusions

We have defined, investigated, implemented, and benchmarked the Model Repair problem for probabilistic systems. Ultimately, we show how Model Repair can be seen as both a nontrivial extension of the parametric model-checking problem for probabilistic systems and a nontrivial generalization of the controller-synthesis problem for linear systems. In both cases, its solution requires one to solve a nonlinear optimization problem with a minimal-cost (or maximal-profit) objective function.

The problem we considered is one of *offline* model repair. As future work, we would like to investigate the *online* version of the problem, where an online controller runs concurrently with the system in question, appropriately adjusting its parameters whenever a property violation is detected. Meeting this objective will likely require a better understanding of the similarities between the model repair and controller synthesis problems.

Acknowledgements. We thank the anonymous referees for their valuable comments. Research supported in part by NSF Grants CCF-0926190, CCF-1018459, CNS 0831298, CNS 0721665, ONR grant N00014-07-1-0928, and AFOSR Grant FA0550-09-1-0481. The research of Professor Katsaros was conducted while on Sabbatical leave at Stony Brook University.

References

1. Alexiou, N., Deshpande, T., Basagiannis, S., Smolka, S.A., Katsaros, P.: Formal analysis of the kaminsky DNS cache-poisoning attack using probabilistic model checking. In: Proceedings of the 12th IEEE International High Assurance Systems Engineering Symposium, pp. 94–103. IEEE Computer Society, Los Alamitos (2010)
2. Biegler, L.T., Zavala, V.M.: Large-scale nonlinear programming using IPOPT: An integrating framework for enterprise-wide dynamic optimization. Computers & Chemical Engineering 33(3), 575–582 (2009)
3. Bonakdarpour, B., Ebnenasir, A., Kulkarni, S.S.: Complexity results in revising UNITY programs. ACM Trans. Auton. Adapt. Syst. 4(1), 1–28 (2009)
4. Boyd, S., Vandenberghe, L.: Convex Optimization. Camb. Univ. Press, Cambridge (2004)
5. Buccafurri, F., Eiter, T., Gottlob, G., Leone, N.: Enhancing model checking in verification by AI techniques. Artif. Intell. 112(1-2), 57–104 (1999)
6. Clarke, E.M., Emerson, E.A., Sifakis, J.: Model checking: Algorithmic verification and debugging. Communications of the ACM 52(11), 74–84 (2009)
7. Daws, C.: Symbolic and parametric model checking of discrete-time Markov chains. In: Liu, Z., Araki, K. (eds.) ICTAC 2004. LNCS, vol. 3407, pp. 280–294. Springer, Heidelberg (2005)
8. Donaldson, R., Gilbert, D.: A model checking approach to the parameter estimation of biochemical pathways. In: Heiner, M., Uhrmacher, A.M. (eds.) CMSB 2008. LNCS (LNBI), vol. 5307, pp. 269–287. Springer, Heidelberg (2008)
9. Dong, Y., Sarna-Starosta, B., Ramakrishnan, C.R., Smolka, S.A.: Vacuity checking in the modal mu-calculus. In: Kirchner, H., Ringeissen, C. (eds.) AMAST 2002. LNCS, vol. 2422, pp. 147–162. Springer, Heidelberg (2002)

10. Epifani, I., Ghezzi, C., Mirandola, R., Tamburrelli, G.: Model evolution by run-time parameter adaptation. In: ICSE 2009: Proceedings of the 31st International Conference on Software Engineering, pp. 111–121. IEEE Computer Society Press, Washington, DC, USA (2009)

11. Giacalone, A., Chang Jou, C., Smolka, S.A.: Algebraic reasoning for probabilistic concurrent systems. In: Proc. of the IFIP TC2 Working Conference on Programming Concepts and Methods, pp. 443–458. North-Holland, Amsterdam (1990)

12. Granvilliers, L., Benhamou, F.: RealPaver: an interval solver using constraint satisfaction techniques. ACM Trans. Math. Softw. 32, 138–156 (2006)

13. Hahn, E., Hermanns, H., Zhang, L.: Probabilistic reachability for parametric Markov models. International Journal on Software Tools for Technology Transfer, 1–17 (April 2010)

14. Hahn, E.M.: Parametric Markov model analysis. Master's thesis, Saarland University (2008)

15. Hahn, E.M., Hermanns, H., Wachter, B., Zhang, L.: PARAM: A Model Checker for Parametric Markov Models. In: Touili, T., Cook, B., Jackson, P. (eds.) CAV 2010. LNCS, vol. 6174, pp. 660–664. Springer, Heidelberg (2010)

16. Han, T., Katoen, J.-P., Mereacre, A.: Approximate parameter synthesis for probabilistic time-bounded reachability. In: IEEE International Real-Time Systems Symposium, pp. 173–182 (2008)

17. Hansson, H., Jonsson, B.: A logic for reasoning about time and reliability. Formal Aspects of Computing 6, 102–111 (1994)

18. Jobstmann, B., Griesmayer, A., Bloem, R.: Program repair as a game. In: Etessami, K., Rajamani, S.K. (eds.) CAV 2005. LNCS, vol. 3576, pp. 226–238. Springer, Heidelberg (2005)

19. Knuth, D., Yao, A.: The complexity of nonuniform random number generation. In: Algorithms and Complexity: New Directions and Recent Results. Academic Press, London (1976)

20. Kwiatkowska, M.Z., Norman, G., Parker, D.: Stochastic model checking. In: Bernardo, M., Hillston, J. (eds.) SFM 2007. LNCS, vol. 4486, pp. 220–270. Springer, Heidelberg (2007)

21. Lanotte, R., Maggiolo-Schettini, A., Troina, A.: Parametric probabilistic transition systems for system sesign and analysis. Formal Aspects of Computing 19(1), 93–109 (2007)

22. Sinha, S.M.: Duality in nonlinear programming. In: Mathematical Programming, pp. 423–430. Elsevier Science, Burlington (2006)

23. Zhang, D., Cleaveland, R.: Fast on-the-fly parametric real-time model checking. In: Proceedings of the 26th IEEE International Real-Time Systems Symposium, pp. 157–166. IEEE Computer Society, Los Alamitos (2005)

Boosting Lazy Abstraction for SystemC with Partial Order Reduction

Alessandro Cimatti, Iman Narasamdya, and Marco Roveri

Fondazione Bruno Kessler — Irst
{cimatti,narasamdya,roveri}@fbk.eu

Abstract. The SystemC language is a de-facto standard for the description of systems on chip. A promising technique, called ESST, has recently been proposed for the formal verification of SystemC designs. ESST combines *Explicit* state techniques to deal with the SystemC *Scheduler*, with *Symbolic* techniques, based on lazy abstraction, to deal with the *Threads*. Despite its relative effectiveness, this approach suffers from the potential explosion of thread interleavings.

In this paper we propose the adoption of partial order reduction (POR) techniques to alleviate the problem. We extend ESST with two complementary POR techniques (persistent set, and sleep set), and we prove the soundness of the approach in the case of safety properties. The extension is only seemingly trivial: the POR, applied to the scheduler, must be proved not to interfere with the lazy abstraction of the threads.

We implemented the techniques within the software model checker KRATOS, and we carried out an experimental evaluation on benchmarks taken from the SystemC distribution and from the literature. The results showed a significant improvement in terms of the number of visited abstract states and run times.

1   Introduction

SystemC is widely used for the design of systems on chip. Executable models written in SystemC are amenable for high-speed simulation before synthesizing the RTL hardware description. Formal verification of SystemC designs can help to pinpoint errors, preventing their propagation down to the hardware, but can also help to reveal errors in the specifications.

Despite its importance, however, formal verification of SystemC is a very hard challenge. Indeed, a SystemC design is a very complex entity. In addition to rich data, SystemC features a form of multi-threading, where scheduling is cooperative and carried out according to a specific set of rules [20], and the execution of threads is mutually exclusive.

A promising technique, called ESST [7], has recently been proposed for the verification of SystemC designs. ESST combines *Explicit* state techniques to deal with the SystemC *Scheduler*, with *Symbolic* techniques, based on lazy abstraction [2], to deal with the *Threads*. Despite its relative effectiveness, this technique requires the exploration of a large number of thread interleavings, many of which are redundant, with subsequent degradations in the run time performance and high memory consumption.

Partial-order reduction (POR) [11,18,22] is a well known model checking technique that tackles the state explosion problem by exploring only representative subset of all

P.A. Abdulla and K.R.M. Leino (Eds.): TACAS 2011, LNCS 6605, pp. 341–356, 2011.
© Springer-Verlag Berlin Heidelberg 2011

possible schedules. In general, POR exploits the commutativity of concurrent transitions that result in the same state when they are executed in different orders. POR techniques have successfully been integrated in explicit-state software model checkers like SPIN [13] and VERISOFT [10], and also applied in symbolic model checking [15,23,1].

In this paper we boost ESST with two complementary POR techniques [11], *persistent set* and *sleep set*. The POR techniques are used in the ESST algorithm to limit the expansion of the transitions in the explicit scheduler, while the nature of the symbolic search of the threads, based on lazy abstraction, remains unchanged. Notice that the application of POR in ESST algorithm is only seemingly trivial, because POR could in principle interact negatively with the lazy abstraction used for the search within the threads. In fact, we prove that the pruning carried out by POR in the abstract space preserves the reachability in the concrete space, which yields the soundness of the approach in the case of safety properties.

We implemented these POR techniques within the KRATOS software model checker. KRATOS implements the ESST algorithm, and is at the core of the tool chain described in [7], which also includes a SystemC front-end derived from PINAPA [17]. We perform an experimental evaluation on the benchmark set used in [7], that includes problems from the SystemC distribution and from the literature. The results show that POR techniques can yield substantial improvements on the performance of the ESST algorithm in terms of the number of visited abstract states and run times.

This paper is structured as follows. In Sec. 2 we briefly introduce SystemC and we briefly describe the ESST algorithm. In Sec. 3 we show by means of an example the possible state explosion problem that may arise. In Sec. 4 we show how to lift POR techniques to the ESST algorithm. In Sec. 5 we revise the related work. Finally, in Sec. 7 we draw some conclusions and we outline future work.

2   Background

The SystemC language. SystemC is a C++ library that consists of (1) a core language that allows one to model a System-on-Chip (SoC) by specifying its components and architecture, and (2) a simulation kernel (or scheduler) that schedules and runs processes (or threads) of components. SoC components are modeled as SystemC modules that communicate through channels (that are bound to the ports specified in the modules).

A module consists of one or more threads that describe the parallel behavior of the SoC design. SystemC provides general-purpose events as a synchronization mechanism between threads. For example, a thread can suspend itself by waiting for an event or by waiting for some specified time. A thread can perform immediate notification of an event or delayed notification.

The SystemC scheduler is a cooperative non-preempting scheduler that runs at most one thread at a time. During a simulation, the status of a thread changes from sleeping, to runnable, and to running. A running thread will only give control back to the scheduler by suspending itself. The scheduler runs all runnable threads, one at a time, in a single delta cycle, while postponing the channel updates made by the threads. When there are no more runnable threads, the scheduler materializes the channel updates, and wakes up all sleeping threads that are sensitive to the updated channels. If, after this

```
1  SC_MODULE( numgen ) {                        31  SC_MODULE( stage2 ) {
2    sc_out<int> o;  // output port.            32    sc_in<int>   i;  // input port.
3    sc_in<bool> ck;  // input port for clock.  33    sc_in<bool> ck;  // input port for clock.
4                                                34
5    // Reads input from environment            35    // Method for checking value.
6    void gen() { int x = read_input(); o.write(x); }  36    void check() { int x = i.read(); assert(x == 0); }
7                                                37
8    SC_CTOR( numgen ) {                         38    SC_CTOR( stage2 ) {
9      // declare "gen" as a method thread.      39      // declare "check" as a method thread
10     SC_METHOD( gen );                         40      SC_METHOD( check );
11     dont_initialize();                        41      dont_initialize();
12     sensitive << ck.pos();                    42      sensitive << ck.pos();
13   }                                           43    }
14 }                                             44  }
15                                               45
16 SC_MODULE( stage1 ) {                         46  int sc_main() {
17   sc_in<int>  i;  // input port.              47    sc_signal<int> gen_to_s1 , s1_to_s2;
18   sc_out<int> o;  // output port.             48    sc_signal<bool> ck;
19   sc_in<bool> ck;  // input port for clock.   49
20                                               50    numgen N; N.o(gen_to_s1); N.ck(ck);
21   // Pass value from input port to output port.  51
22   void pass() { int x = i.read(); o.write(x); }  52    stage1 S1; S1.i(gen_to_s1); S1.o(s1_to_s2); S1.ck(ck);
23                                               53
24   SC_CTOR( stage1 ) {                         54    stage2 S2; S2.i(s1_to_s2); S2.ck(ck);
25     // declare "pass" as a method thread      55
26     SC_METHOD( pass );                        56    sc_start(0);
27     dont_initialize();                        57    for (int i=0; i < 3; ++i) {
28     sensitive << ck.pos();                    58      ck.write(1); sc_start(0); ck.write(0); sc_start(0);
29   }                                           59    }
30 }                                             60  }
```

Fig. 1. Example of SystemC design

step, there are some runnable threads, then the scheduler moves to the next delta cycle. Otherwise, it accelerates the simulation time to the nearest time point where a sleeping thread or an event can be woken up. The scheduler quits when there are no more runnable threads after time acceleration.

An example of SystemC design is shown in Fig. 1. It consists of three modules: numgen, stage1, and stage2. In the sc_main function we create an instance for each module, and we connect them such that the instances of numgen and stage1 are connected by the signal gen_to_s1 and the instances of stage1 and stage2 are connected by the signal s1_to_s2. The thread gen of numgen reads an integer value from the environment and sends it to stage1 through the signal gen_to_s1. The thread pass of stage1 simply reads the value from the signal gen_to_s1 and sends it to stage2 through the signal s1_to_s2. The thread check of stage2 reads the value from s1_to_s2 and asserts that the read value equals 0. All threads are made sensitive to the positive edge of the clock ck (modeled as a boolean signal): they become runnable when the value of the ck changes from 0 to 1. The function dont_initialize makes the most-recently declared thread sleep initially. The clock cycle is controlled by the loop in the sc_main function. The function sc_start runs a simulation until there are no more runnable threads. The property that we want to check is that the value read by numgen from the environment reaches stage2 in three clock cycle.

The Explicit Scheduler + Symbolic Threads (ESST) approach. The ESST technique [7] is a counter-example guided abstraction refinement [8] based technique that combines explicit-state technique with lazy predicate abstraction [2]. In the same way as the classical lazy abstraction, the data path of the threads is analyzed by means of predicate abstraction, while the flow of control of each thread and the state of the scheduler are analyzed with explicit-state techniques.

We assume that the SystemC design has been translated into a threaded C program [7] in which each SystemC thread is represented by a C function. Each function corresponding to a thread is represented by a *control-flow automaton* (CFA), which is a pair (L, G), where L is the set of control locations and $G \subseteq L \times Ops \times$ is the set of edges such that each edge is labelled by an operation from the set Ops of operations.

Threads in a threaded C program communicate with each other by means of shared global variables, and use primitive functions and events as synchronization mechanism. For SystemC we have the following primitive functions: `wait_event(e)`,`wait_time(t)`, `notify_event(e)`, `notify_event_at_time(e,t)`, and `cancel_event(e)` for an event e and a time unit t.

The ESST algorithm is based on the construction and analysis of an *abstract reachability forest* (ARF) that describes the reachable abstract states of the threaded program. An ARF consists of connected *abstract reachability trees* (ART's), each of which is obtained by unwinding the CFA corresponding to the running thread.

For a threaded program with n threads, a *node* in an ARF is a tuple $(\langle l_1, \varphi_1 \rangle, \ldots, \langle l_n, \varphi_n \rangle, \varphi, S)$, where l_i and φ_i are, respectively, the program location and the region of thread i, φ is the region of global variables, and S is the state of the scheduler. Regions are formulas describing the values of variables. Information maintained in the scheduler state S includes the status of threads and events, the events that sleeping threads are waiting for their notifications, and the delays of event notifications.

An ARF is constructed by unwinding the CFA's of threads, and by executing the scheduler. Unwinding a CFA involves computing the *abstract strongest post-condition* $SP^\pi(\varphi, op)$ of a region φ with respect to the operation op labelling the unwound CFA edge and the precision π. In the ESST approach, the precision π can contain a set of predicates that are tracked for the global region and for the thread regions.

Unwinding the CFA by executing primitive functions is performed by the function SEXEC that takes as inputs a scheduler state and a call to a primitive function p, and returns the updated scheduler state obtained from executing p. For example, the state $S' = \text{SEXEC}(S, \text{wait\_event(e)})$ is obtained from the state S by changing the status of the running thread to sleep, and noting that the now sleeping thread is waiting for the event e.

We implement the scheduler by the function SCHED that, given a scheduler state where all threads are sleeping, outputs a set of scheduler states representing all possible schedules such that each of the output scheduler states has exactly one running thread. We expand a node $(\langle l_1, \varphi_1 \rangle, \ldots, \langle l_n, \varphi_n \rangle, \varphi, S)$ by means of the following rules [7]:

E1. If there is a running thread i in S such that the thread performs an operation op and (l_i, op, l'_i) is an edge of the CFA of thread i, then

  – If op is *not* a call to primitive function, then the successor node is $(\langle l_1, \varphi_1 \rangle, \ldots, \langle l'_i, \varphi'_i \rangle, \ldots, \langle l_n, \varphi'_n \rangle, \varphi', S)$, where $\varphi'_i = SP^\pi(\varphi_i \wedge \varphi, op)$, $\varphi'_j = SP^\pi(\varphi_j \wedge \varphi, \text{HAVOC}(op))$) for $j \neq i$, and $\varphi' = SP^\pi(\varphi, op)$. The function HAVOC collects all global variables possibly updated by op, and builds a new operation where these variables are assigned with fresh variables.

  – If op is a primitive function, then the successor node is $(\langle l_1, \varphi_1 \rangle, \ldots, \langle l'_i, \varphi_i \rangle, \ldots, \langle l_n, \varphi_n \rangle, \varphi, \text{SEXEC}(S, op))$.

E2. If there is no running thread in S, then, for each $S' \in \text{SCHED}(S)$, we create a successor node $(\langle l_1, \varphi_1 \rangle, \ldots, \langle l_n, \varphi_n \rangle, \varphi, S')$, such that the node becomes the root node of a new ART that is then added to the ARF. Such a connection between two nodes is called *ARF connector*.

In the construction of an ARF, one stops expanding a node if the node is covered by other nodes or if the conjunction of all its thread regions and the global region is unsatisfiable. We say that a node $(\langle l_1, \varphi_1 \rangle, \ldots, \langle l_n, \varphi_n \rangle, \varphi, S)$ is *covered* by a node $(\langle l_1', \varphi_1' \rangle, \ldots, \langle l_n', \varphi_n' \rangle, \varphi', S')$ if (1) $l_i = l_i'$ for $i = 1, \ldots, n$, (2) $S = S'$, and (3) $\varphi \Rightarrow \varphi'$ and $\bigwedge_{i=1,\ldots,n}(\varphi_i \Rightarrow \varphi_i')$ are valid. An ARF is *complete* if it is closed under the expansion of the above rules. An ARF is *safe* if it is complete and, for every node $(\langle l_1, \varphi_1 \rangle, \ldots, \langle l_n, \varphi_n \rangle, \varphi, S)$ in the ARF such that $\varphi \wedge \bigwedge_{i=1,\ldots,n} \varphi_i$ is satisfiable, none of the locations $l_1, \ldots, l_n$ are error locations.

The construction of an ARF starts with a single ART representing reachable states of the main function. In the root node of that ART all regions are initialized with $True$, all thread locations are set to the entries of the corresponding threads and the only running thread in the scheduler state is the main function. The main function then suspends itself by calling a primitive function that starts the simulation. We expand the ARF using the rules E1 and E2 until either the ARF is complete or we reach a node where one of the thread's location is an error location. In the latter case we build a counterexample consisting of paths in the trees of the ARF and check if the counterexample is feasible. If it is feasible, then we have found a real counterexample witnessing that the program is unsafe. Otherwise, we use the spurious counterexample to discover predicates to refine the ARF. We refer to [7] for further details.

3   The Problem of Multiple Interleavings

The ESST algorithm often has to explore a large number of possible schedules. However, some of them might be redundant because the order of interleavings of some threads is irrelevant.

Fig. 2 depicts, for the example in Fig. 1, the ARF constructed by the ESST algorithm in every delta cycle. This figure clearly shows that there are 6=3! possible schedules. These threads communicate with each other using signals. A signal s in SystemC can be viewed as a pair (s_{old}, s_{new}) of variables such that writing to s is modeled by writing to s_{new} while reading from s is modeled by reading from s_{old}. Thus, these threads access disjoint sets of variables: gen writes to gen_to_s1_{new}, pass reads from gen_to_s1_{old} and writes to s1_to_s2_{new}, and check reads from s1_to_s2_{old}. Therefore, when all the threads become runnable, the order of running them is irrelevant. Consequently, instead of exploring all 3! possible schedules, it is sufficient to explore only one of them.

Partial order reduction techniques (POR) [11,18,22] can be used to avoid exploring redundant schedules. However, we need to ensure that in the construction of the ARF the partial order reduction does not remove all counter-example paths witnessing that the SystemC design is unsafe. In the following we will see how to extend the the ESST algorithm to exploit POR in the construction of the ARF guaranteeing that the above condition is satisfied.

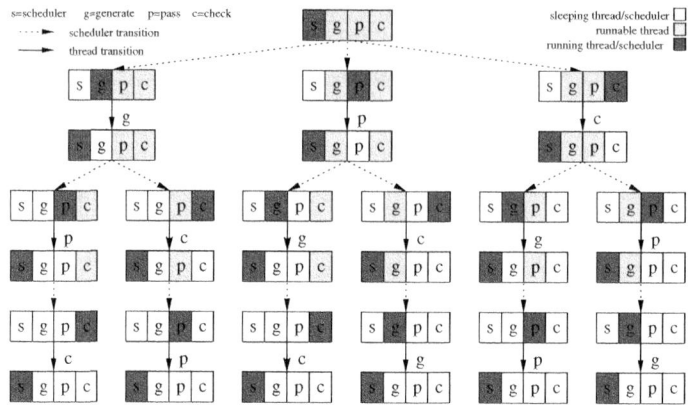

Fig. 2. The 6 possible interleavings for threads of the example in Fig. 1

4 Reduction Algorithms in ESST

Partial-order reduction (POR) [11,18,22] is a model checking technique that is aimed at combating the state explosion by exploring only representative subset of all possible interleavings. In this paper we apply POR to the ESST technique.

4.1 Partial-Order Reduction Techniques

For presentation of POR in this section we follow the standard notions and notations used in other literature [11,9]. We represent a concurrent program as a transition system $M = (S, S_0, T)$, where S is the finite set of states, $S_0 \subset S$ is the set of initial states, and T is a set of transitions such that for each $\alpha \in T$, we have $\alpha \subset S \times S$. We say that $\alpha(s, s')$ holds and often writes it as $s \xrightarrow{\alpha} s'$ if $(s, s') \in \alpha$. A state s' is a successor of a state s if $s \xrightarrow{\alpha} s'$ for some transition $\alpha \in T$. In the following we will only consider deterministic transitions A transition α is *enabled* in a state s if there is a state s' such that $\alpha(s, s')$ holds. The set of transitions enabled in a state s is denoted by $enabled(s)$. A *path* from a state s in a transition system is a finite or infinite sequence $s_0 \xrightarrow{\alpha_0} s_1 \xrightarrow{\alpha_1} \cdots$ such that $s = s_0$ and $s_i \xrightarrow{\alpha_i} s_{i+1}$ for all i. A path is empty if the sequence consists only of a single state. The length of a finite path is the number of transitions in the path.

Let $M = (S, S_0, T)$ be a transition system, we denote by $Reach(S_0, T) \subseteq S$ the set of states reachable from the states in S_0 by the transitions in T. In this work we are interested in verifying safety properties in the form of program assertion. To this end, we assume that there is a set $T_{err} \subseteq T$ of *error transitions* such that the set

$$E_{M, T_{err}} = \{ s \in S \mid \exists s' \in S. \exists \alpha \in T_{err}.\ \alpha(s', s) \text{ holds} \}$$

is the set of *error states* of M with respect to T_{err}. A transition system M is *safe with respect to the set* $T_{err} \subseteq T$ *of error transitions* iff $Reach(S_0, T) \cap E_{M, T_{err}} = \emptyset$.

Selective search in POR exploits the commutativity of concurrent transitions. The concept of commutativity of concurrent transitions can be formulated by defining an independence relation on pairs of transitions.

DEFINITION 4.1 (INDEPENDENCE RELATION, INDEPENDENT TRANSITIONS). An *independence relation* $I \subseteq T \times T$ is a symmetric, anti-reflexive relation such that for each state $s \in S$ and for each $(\alpha, \beta) \in I$ the following conditions are satisfied: (***Enabledness***) If α is in $enabled(s)$, then β is in $enabled(s)$ iff β is in $enabled(\alpha(s))$. (***Commutativity***) If α and β are in $enabled(s)$, then $\alpha(\beta(s)) = \beta(\alpha(s))$.

We say that two transitions α and β are *independent* of each other if for every state s they satisfy the enabledness and commutativity conditions. We also say that two transitions α and β are *independent in a state s* of each other if they satisfy the enabledness and commutativity conditions in s. □

In the sequel we will use the notion of valid dependence relation to select a representative subset of transitions that need to be explored.

DEFINITION 4.2 (VALID DEPENDENCE RELATION). A *valid dependence relation* $D \subseteq T \times T$ is a symmetric, reflexive relation such that for every $(\alpha, \beta) \notin D$, the transitions α and β are independent of each other. □

The Persistent Set approach. To reduce the number of possible interleavings, in every state visited during the state space exploration one only explores a representative subset of transitions that are enabled in that state. However, to select such a subset we have to avoid possible dependencies that can happen in the future. To this end, we appeal to the notion of persistent set [11].

DEFINITION 4.3 (PERSISTENT SET). A set $P \subseteq T$ of enabled transitions in a state s is *persistent* in s if for every finite nonempty path $s = s_0 \xrightarrow{\alpha_0} s_1 \cdots s_n \xrightarrow{\alpha_n} s_{n+1}$ such that $\alpha_i \notin P$ for all $i = 0, \ldots, n$, we have α_n independent of any transition in P in s_n. □

Note that the persistent set in a state is not unique. To guarantee the existence of successor state, we impose the *successor-state* condition on the persistent set: the persistent set in s is empty iff so is $enabled(s)$. For simplicity, in the sequel whenever we speak about persistent sets, then the sets satisfy the successor-state condition. We say that a state s is *fully expanded* if the persistent set in s equals $enabled(s)$. It is easy to see that, for any transition α not in the persistent set P in a state s, the transition α is disabled in s or independent of any transition in P.

We denote by $Reach_{red}(S_0, T) \subseteq S$ the set of states reachable from the states in S_0 by the transitions in T such that, during the state space exploration, in every visited state we only explore the transitions in the persistent set in that state. It is easy to see that $Reach_{red}(S_0, T) \subseteq Reach(S_0, T)$.

To preserve safety properties of a transition system we need to guarantee that the reduction by means of persistent set does not remove all interleavings that lead to an error state. To this end, we impose the so-called *cycle condition* on $Reach_{red}(S_0, T)$ [9,18]: a cycle is not allowed if it contains a state in which a transition α is enabled, but α is never included in the persistent set of any state s on the cycle.

THEOREM 4.4. *A transition system $M = (S, S_0, T)$ is safe w.r.t. a set $T_{err} \subseteq T$ of error transitions iff $Reach_{red}(S_0, T)$ that satisfies the cycle condition does not contains any error state from $E_{M, T_{err}}$.* □

The Sleep Set approach. We consider also the *sleep set* POR technique. This technique exploits independencies of enabled transitions in the current state. For example, suppose that in some state s there are two enabled transitions α and β, and they are independent of each other. Suppose further that the search explores α first from s. Then, when the search explores β from s such that $s \xrightarrow{\beta} s'$ for some state s', we associate with s' a sleep set containing only α. From s' the search only explores transitions that are not in the sleep set of s'. That is, although the transition α is still enabled in s', it will not be explored. Both persistent set and sleep set techniques are orthogonal and complementary, and thus can be applied simultaneously.

Note that the sleep set technique only removes transitions, and not states. Thus Theorem 4.4 still holds when the sleep set technique is applied.

4.2   Applying POR to ESST

Applying POR to the ESST algorithm is not trivial. The ESST algorithm is based on the construction of an ARF that describes the reachable abstract states of the threaded program, while the description of POR in Sec. 4.1 is based on the analysis of reachable concrete states. One then needs to guarantee that the original ARF is safe iff. the reduced ARF, obtained by applying POR in the construction of ARF, is safe. That is, we have to ensure that the selective search performed during the construction of ARF does not remove all non-spurious paths that lead to error locations. In particular, the construction of reduced ARF has to check if the cycle condition is satisfied in its concretization.

To integrate POR techniques into the ESST algorithm we first need to identify transitions in the threaded program. In the above description of POR the execution of a transition is atomic. We introduce the notion of atomic block as the notion of transition in the threaded program. Intuitively an atomic block is a block of operations between calls to primitive functions that can suspend the thread. For simplicity, let us call such primitive functions *wait functions*.

An *atomic block* of a thread is a rooted sub-graph of the CFA satisfying the following conditions: (1) its unique entry is the entry of the CFA or the location that immediately follows a call to a wait function; (2) its exit is the exit of the CFA or the location that immediately follows a call to a wait function; and (3) there is no call to a wait function in any CFA path from the entry to an exit except the one that precedes the exit. Note that an atomic block has a unique entry, but can have multiple exits. We often identify an atomic block by its entry.

For example, consider the thread code of Fig. 3. One atomic block starts from the entry of the thread and ends at the label `lab` and at the exit of the thread. The other atomic block starts from the label `lab` and ends at the label `lab` too and at the exit of the thread. In the sequel we will use the terms transition and atomic block interchangeably.

```
void thread_t() {
  while (...) {
    ...
    wait_event(e);
    lab: ...
  }
}
```

Fig. 3. Fragment of code

We use static analysis techniques to compute a valid dependence relation. In particular, a pair (α, β) of atomic blocks are in a valid dependence relation if one of the following criteria is satisfied: (1) The atomic block α contains a write to a shared (or global) variable g, and the atomic block β contains a write or a read to g. (2) The atomic block α contains an immediate

Algorithm 1. Persistent set.

Input: a set T_E of enabled atomic blocks.
Output: a persistent set P.
1. Let $T_P := \{\alpha\}$, where $\alpha \in T_E$.
2. For each atomic block $\alpha \in T_P$:
   (a) If $\alpha \in T_E$ (α is enabled): Add into T_P every atomic block β such that $(\alpha, \beta) \in D$.
   (b) If $\alpha \notin T_E$ (α is disabled): Add into T_P a set of atomic blocks whose executions guarantee α to become enabled.
3. Repeat step 2 until no more transition can be added into T_P.
4. $P := T_P \cap T_E$.

notification of an event e, and the atomic block β contains a wait for e. (3) The atomic block α contains a delayed notification of an event e, and the atomic block β contains a cancellation of a notification of e.

Persistent sets are computed using a valid dependence relation. Let D be a valid dependence relation. Algorithm 1 computes persistent sets. It is easy to see that the persistent set computed by this algorithm satisfies the successor-state condition.

Algorithm 1 is a variant of the stubborn set algorithm presented in [11], that is, we use a valid dependence relation as the interference relation described in [11].

We apply POR to the ESST algorithm by modifying the ARF node expansion rule E2 (see Section 2) by first computing a persistent set from a set of scheduler states output by the function SCHED; and then by ensuring that the cycle condition is satisfied by the concretization of the constructed ARF.

First, we assume that a valid dependence relation D has been produced by static analysis on the threaded program. Second, we introduce the function PERSISTENT that computes a persistent set of a set of scheduler states. PERSISTENT takes as inputs an ARF node and a set S of scheduler states, and outputs a subset S' of S. The input ARF node keeps track of the thread locations, which are used to identify atomic blocks, while the input scheduler states keep track of the status of the threads. From the ARF node and the set S, the function PERSISTENT extracts the set T_E of enabled atomic blocks. PERSISTENT then computes a persistent set P from T_E using Algorithm 1. Finally PERSISTENT constructs back a subset S' of the input set S of scheduler states from the persistent set P.

Let $A = (\langle l_1, \varphi_1 \rangle, \dots, \langle l_n, \varphi_n \rangle, \varphi, S)$ be an ARF node that is going to be expanded. We replace the rule E2 in the following way: instead of creating a new ART for each state $S' \in$ SCHED(S), we create a new ART whose root is the node $(\langle l_1, \varphi_1 \rangle, \dots, \langle l_n, \varphi_n \rangle, \varphi, S')$ for each state $S' \in$ PERSISTENT$(A,$ SCHED$(S))$ (rule E2').

To guarantee the preservation of safety properties, following [9], instead of checking that the cycle condition is satisfied we check a stronger condition: at least one state along the cycle is fully expanded.

In the ESST algorithm a *potential* cycle occurs if an ARF node is covered by one of its predecessors in the ARF. Let $A = (\langle l_1, \varphi_1 \rangle, \dots, \langle l_n, \varphi_n \rangle, \varphi, S)$ be an ARF node. We say that the scheduler state S is *running* if there is a running thread in S. Recall our assumption that there is at most one running thread in the scheduler state. We also say

---

Algorithm 2. ARF expansion algorithm for non-running node.

Input: a non-running ARF node A that contains no error locations.
1. Let $NonRunning(ARFPath(A, F))$ be $A_0, \ldots, A_m$ such that $A = A_m$
2. If there exists $i < m$ such that A_i covers A:
   (a) Let $A_{m-1} = (\langle l'_1, \varphi'_1 \rangle, \ldots, \langle l'_n, \varphi'_n \rangle, \varphi', S')$.
   (b) If $\text{PERSISTENT}(A_{m-1}, \text{SCHED}(S')) \subset \text{SCHED}(S')$:
      – For all $S'' \in \text{SCHED}(S') \setminus \text{PERSISTENT}(A_{m-1}, \text{SCHED}(S'))$:
         • Create a new ART with root node $(\langle l'_1, \varphi'_1 \rangle, \ldots, \langle l'_n, \varphi'_n \rangle, \varphi', S'')$.
3. If A is covered: Mark A as covered.
4. If A is not covered: Expand A by rule E2'.

---

that the node A is *running* if its scheduler state S is. Note that during ARF expansion the input of SCHED is always a non-running scheduler state.

A path in an ARF can be represented as a sequence $A_0, \ldots, A_m$ of ARF nodes such that for all i, we have A_{i+1} is a successor of A_i in the same ART or there is an ARF connector from A_i to A_{i+1}. Given an ARF node A of ARF F, we denote by $ARFPath(A, F)$ the ARF path $A_0, \ldots, A_m$ such that (1) A_0 has neither a predecessor ARF node nor an incoming ARF connector, and (2) $A_m = A$. Let Π be an ARF path, we denote by $NonRunning(\Pi)$ the *maximal* sub-sequence of non-running node in Π.

Algorithm 2 shows how a non running ARF node A (containing no error locations) is expanded in the presence of POR. We fully expand the immediate non-running predecessor node of A when a potential cycle is detected. Otherwise the node is expanded as usual.

Our POR technique differs from that one described in [9], and implemented in SPIN [14]. The technique in [9] tries to select a persistent set that does not create a cycle, and if it does not succeed, then it fully expands the node. In the context of ESST such a technique is expensive. To detect a cycle one has to expand a node by a transition. As explained before, a transition or an atomic block in our case can span over multiple operations in CFA. Thus, a cycle detection often requires expensive computations of abstract strongest post condition.

In addition to coverage check, in the above algorithm one can also check if the detected cycle is spurious or abstract. We only fully expand a node iff the detected cycle is not spurious. As cycles are rare, the benefit of POR can be defeated by the price of generating and solving the constraints that encode the cycle.

In SystemC cycles can occur because there is a chain of wait and immediate notifications of events. For SystemC cycle detection can be optimized by only considering the sequence of non-running ARF node that belongs to the same delta-cycle.

POR based on sleep set can also be applied to ESST by extending the ARF node to include a sleep set. The application of sleep set technique into the ESST algorithm is similar to that of in the case of explicit-state model checking. Due to lack of space we refer to the appendix for a detailed discussion on the application of sleep set to ESST.

4.3   Correctness of Reduction in ESST

The correctness of POR with respect to verifying program assertions of transition systems has been shown by Theorem 4.4. The correctness proof relies on the enabledness and commutativity of independent transitions. However, the proof is applied in concrete state space of the transition system, while the ESST algorithm works in abstract state space represented by an ARF. The following observation shows that two transition that are independent in the concrete state space may not commute in the abstract state space.

For simplicity of presentation, we represent an abstract state by a formula representing a region. Let g_1, g_2 be global variables, and p, q be predicates such that $p \Leftrightarrow (g_1 < g_2)$ and $q \Leftrightarrow (g1 = g2)$. Let α be the transition $g_1 \; := \; g_1 \; - \; 1$ and β be the transition $g_2 \; := \; g_2 \; - \; 1$. It is obvious that α and β are independent of each other. However, Fig. 4 shows that the two transition do no commute when we start from an abstract state A_1 such that $A_1 \Leftrightarrow p$. The edges in the figure represent the computation of abstract strongest post condition of the corresponding abstract states and transitions.

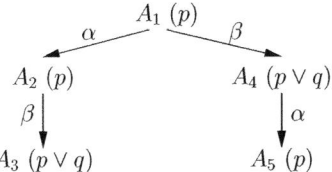

Fig. 4. Independent transitions do not commute in abstract state space

Even though two independent transitions do not commute in abstract state space, they still commute in the concrete state space over-approximated by the abstract state space, as shown by the lemma below. For a concrete state s and an abstract state A, we write $s \models A$ iff A holds in s.

LEMMA 4.5. Let α and β be transitions that are independent of each other such that for concrete states s_1, s_2, s_3 and abstract state A we have $s_1 \models A$, and both $\alpha(s_1, s_2)$ and $\beta(s_2, s_3)$ holds. Let A' be the abstract successor state of A by applying the abstract strongest post operator to A and β, and A'' be the abstract successor state of A' by applying the abstract strongest post operator to A' and α. Then, there are concrete states s_4 and s_5 such that: (1) $\beta(s_1, s_4)$ holds, (2) $s_4 \models A'$, (3) $\beta(s_4, s_5)$ holds, (4) $s_5 \models A''$, and (5) $s_3 = s_5$.   □

The above lemma shows that POR can be applied in abstract state space. The following theorem states the correctness of the reduction in the ESST technique.

THEOREM 4.6. *Let ARF F_1 be obtained by node expansion without POR and ARF F_2 be obtained with POR. F_1 is safe iff F_2 is safe.*   □

5   Related Work

It is discussed in [3] an approach aiming at the generation of a SystemC scheduler from a SystemC design that at run time will use race analysis information to speed up the simulation by reducing the number of possible interleavings. The race condition itself is formulated as a guarded independence relation. The independence relation used in [3] is more precise than the one defined in this paper. In our work two transitions are independent if they are independent in any state. While in [3] two transitions are independent in some state if the guard associated with the pair of transitions is satisfied by the state. The guards of two transitions are computed using model checking techniques. The soundness of the synthesized scheduler relies on the assumption that the SystemC threads cannot enable each other during the evaluation phase. Therefore, immediate event notifications are not allowed. Our work does not have such an assumption. More-over, unlike our work, the search in [3] is stateless and it is bounded on the number of simulation steps: cycle condition cannot be detected, and thus this approach cannot be used to verify program assertions.

Dynamic POR techniques for scalable testing of SystemC designs are described in [16,12]. Unlike other works on dynamic POR that collect and analyze run time in-formation to determine dependency of transitions, the work in [16] uses information obtained by static analysis to construct persistent set. For SystemC the criteria that they use to determine dependency of transitions are similar to the criteria we used in this paper. However, similar to [3], the simulation is bounded and stateless, and thus cannot be used to verify program assertions.

POR has been successfully implemented in explicit-state model checkers, like SPIN [13,14,19] and VERISOFT [10]. Despite the inability of explicit-state model checkers to handle non-deterministic inputs, there have been several attempts to en-code SystemC designs in PROMELA, the input language of SPIN. Examples of such attempts include the works described in [21,5]. The aim of those work is to take ad-vantages of search optimizations provided by SPIN, including POR. However, those works cannot fully benefit from the POR implemented in SPIN due to the limitations of the encodings themselves and the limitations of SPIN with respect to the intrinsic structure of SystemC designs. The encoding in [21] is unaware of atomic blocks in the SystemC design. The PROMELA encoding in [5] groups each atomic block in the Sys-temC design as an indivisible transition using SPIN atomic facility. Such an encoding can reduce the number of visited states, and thus relieves the state explosion problem. It is also shown in that paper that the application of POR reduces the number of visited states. However, the reductions that SPIN can achieve is not at the level of the atomic blocks of threads: SPIN still explores redundant interleavings. Moreover, to be able to handle cycle conditions similarly to Algorithm 2, one has to modify SPIN itself.

There have been some works on applying POR to symbolic model checking tech-niques as shown in [15,23,1]. In these works POR during state-space exploration is obtained by statically adding constraints describing the reduction technique into the en-coding of the program. The work in [1] applies POR to symbolic BDD-based invariant checking. The work in [23] can be considered as symbolic sleep-set based technique. The work also introduces the notion of guarded independence relation, where a pair of transitions are independent of each other if certain conditions specified in the pair's guards are satisfied. Such an independence relation is used in [3] for race analysis. Our work in this paper can be extended to use guarded independence relation by exploiting the thread and global regions, but we leave it as future work. The work in [15] consid-ers patterns of lock acquisition to refine the notion of independence transition, which subsequently yields better reductions.

6   Experiments

We have implemented the persistent and sleep sets based POR within the tool chain for SystemC verification described in [7]. It consists of a SystemC front-end, and a model checker called KRATOS which implements the ESST approach. The front-end

Table 1. Results of the experimental evaluation

Name	V	Visited ARF nodes				Run time (.sec)			
		No-POR	P-POR	S-POR	PS-POR	No-POR	P-POR	S-POR	PS-POR
toy	S	896	856	624	592	1.900	1.900	1.890	1.900
toy-bug-1	U	834	794	562	530	1.800	1.800	1.800	1.700
toy-bug-2	U	619	589	415	391	0.600	0.600	0.600	0.600
token-ring-1	S	60	60	60	60	0.000	0.010	0.010	0.000
token-ring-2	S	161	153	127	119	0.090	0.090	0.100	0.090
token-ring-3	S	417	285	259	195	0.190	0.090	0.100	0.090
token-ring-4	S	1039	480	538	296	0.400	0.200	0.200	0.200
token-ring-5	S	2505	1870	961	742	0.800	0.690	0.390	0.400
token-ring-6	S	5883	1922	2114	556	2.000	0.600	0.800	0.300
token-ring-7	S	13533	4068	4324	948	4.600	1.200	1.500	0.400
token-ring-8	S	30623	7781	8264	1293	12.400	2.390	2.800	0.600
token-ring-9	S	68385	17779	15938	3262	40.190	5.600	5.690	1.300
token-ring-10	S	151075	41517	30192	4531	155.390	16.500	12.600	1.900
token-ring-11	S	330789	78229	59310	9564	595.640	43.590	34.590	3.800
token-ring-12	S	M.O.	119990	121616	11106	M.O.	85.990	107.590	4.490
token-ring-13	S	M.O.	M.O.	230479	108783	M.O.	M.O.	344.360	96.280
transmitter-1	U	32	32	32	32	0.010	0.010	0.010	0.010
transmitter-2	U	83	45	67	45	0.010	0.010	0.010	0.010
transmitter-3	U	209	66	114	66	0.010	0.010	0.010	0.010
transmitter-4	U	509	176	254	98	0.090	0.010	0.010	0.010
transmitter-5	U	1205	88	478	88	0.190	0.010	0.090	0.010
transmitter-6	U	2789	483	918	172	0.400	0.100	0.100	0.010
transmitter-7	U	6341	307	2124	121	1.000	0.000	0.300	0.010
transmitter-8	U	14213	3847	4540	337	2.690	0.700	0.690	0.100
transmitter-9	U	31493	1209	7964	214	8.390	0.200	1.500	0.010
transmitter-10	U	69125	2053	13943	290	32.890	0.400	2.800	0.100
transmitter-11	U	150533	3298	36348	289	130.690	0.690	9.690	0.100
transmitter-12	U	M.O.	9784	50026	640	M.O.	2.200	21.390	0.200
transmitter-13	U	M.O.	4234	108796	334	M.O.	1.000	75.590	0.100
pipeline	S	25347	7135	8568	7135	205.270	54.690	61.000	54.690
kundu	S	1004	1004	1004	1004	1.190	1.200	1.180	1.200
kundu-bug-1	U	221	221	221	221	0.390	0.400	0.390	0.390
kundu-bug-2	U	866	866	866	866	1.090	1.100	1.200	1.200
bistcell	S	305	305	305	305	0.500	0.500	0.490	0.500
pc-sfifo-1	S	152	152	152	152	0.290	0.300	0.300	0.290
pc-sfifo-2	S	197	197	197	197	0.300	0.300	0.290	0.300
mem-slave-1	S	556	556	556	556	3.390	3.400	3.400	3.400
mem-slave-2	S	992	992	992	992	15.000	15.090	15.390	15.190
mem-slave-3	S	1414	1414	1414	1414	181.980	173.080	183.770	193.180
mem-slave-4	-	T.O.	T.O.	T.O.	T.O.	T.O.	T.O.	T.O.	T.O.
mem-slave-5	-	T.O.	T.O.	T.O.	T.O.	T.O.	T.O.	T.O.	T.O.

translates SystemC designs into sequential C programs and into threaded C programs. KRATOS can verify the sequential C programs using the classical lazy abstraction algorithm, or it can verify the threaded C programs using the ESST algorithm (extended with the two POR techniques). KRATOS is built on top of an extended version of NUSMV [6] that integrates the MATHSAT SMT solver [4] and provides advanced predicate abstraction techniques by combining BDDs and SMT formulas.

We have performed an experimental evaluation on the same benchmarks used in [7]. We run the experiments on an Intel-Xeon DC 3GHz running Linux, equipped with 4GB RAM. We fixed the time limit to 1000 seconds and the memory limit to 2GB. We experimented without any POR technique (No-POR), enabling only the persistent set reduction (P-POR), enabling only the sleep set reduction (S-POR), and finally enabling both the persistent and the sleep set reductions (PS-POR).

The results of the experimental evaluation are reported in Table 1. The first column lists the name of the benchmarks. The second column shows the status of the verification: S for safe, U for unsafe, - for unknown. Then we report for each experimented technique the number of visited ARF nodes, and the run time. On the table we indicate out of time with T.O., and out of memory with M.O. All the material to reproduce the experiments can be found at http://es.fbk.eu/people/roveri/tests/tacas2011.

The table clearly shows that with POR we can verify more benchmarks. Without POR the verifications of token-ring-x, and, transmitter-x, with $x \in \{12, 13\}$ resulted in out of memory. For some benchmarks, like bist-cell, kundu, mem-slave-x, the POR defined in this work is not applicable. In such benchmarks atomic blocks of threads access the same global variables, and thus are dependent on each other. When POR is not applicable, we can see from the table that there is no reduction in the number of visited ARF nodes. However, the times spent for the verification are almost identical. It means that the time spent for the computation of persistent and sleep sets when POR is enabled is negligible.

The table also shows that, on the pipeline, token-ring-x and transmitter-x POR results in a significant reduction in the number of visited ARF nodes, and in a reduction of run time. Moreover, the combination of persistent and sleep sets gives the best results in terms of visited nodes and run time.

With POR enabled, we are still not able to verify mem-slave-4 and mem-slave-5 given the resource limits. This is because KRATOS employs a precise predicate abstraction for expanding ARF nodes. Such an abstraction is expensive when there are a large number of predicates involved. For verifying mem-slave-3, we already discovered 65 predicates associated with the global region and 37 predicates associated with the thread regions, with an average of 5 predicates per location associated with the thread CFA.

POR, in principle, could interact negatively with the ESST algorithm. The construction of ARF in ESST is sensitive to the explored scheduler states and to the tracked predicates. POR prunes some scheduler states that ESST has to explore. However, exploring such scheduler states can yield a smaller ARF than if they are omitted. In particular, for an unsafe benchmark, exploring omitted scheduler states can lead to the shortest counter-example path. Furthermore, exploring the omitted scheduler states could lead to spurious counter-example ARF paths that yield predicates that allow ESST to perform less refinements and construct a smaller ARF.

Regardless of the fact that there is no guarantee that POR always boosts the ESST algorithm, Table 1 shows that POR can be useful in improving the performance of the ESST algorithm. In the future we will investigate the effectiveness of POR in the ESST algorithm by using randomization on deciding the set of omitted scheduler states to improve the quality of the analysis.

7   Conclusion and Future Work

In this paper we have shown how to extend the ESST approach with POR techniques for the verification of SystemC designs. We proved the correctness of the approach for the verification of program assertions, implemented the approach in the KRATOS model checker, and experimentally evaluated the approach. The proposed techniques significantly reduces the number of visited nodes and the verification time, and allows for the verification of designs that could not be verified without POR.

As future work, we will investigate how to extend the ESST approach to deal with symbolic primitive functions. This requires a generalization of the scheduler exploration with a hybrid (explicit-symbolic) approach, and the use of AllSMT techniques to enumerate all possible next states. We will also apply the ESST techniques to the verification of concurrent C programs from other application domains (e.g. robotics, railways), where different scheduling policies have to be taken into account.

References

1. Alur, R., Brayton, R.K., Henzinger, T.A., Qadeer, S., Rajamani, S.K.: Partial-order reduction in symbolic state-space exploration. Formal Methods in System Design 18(2), 97–116 (2001)
2. Beyer, D., Henzinger, T.A., Jhala, R., Majumdar, R.: The software model checker Blast. STTT 9(5-6), 505–525 (2007)
3. Blanc, N., Kroening, D.: Race analysis for SystemC using model checking. In: ICCAD, pp. 356–363. IEEE, Los Alamitos (2008)
4. Bruttomesso, R., Cimatti, A., Franzén, A., Griggio, A., Sebastiani, R.: The MATHSAT 4 SMT solver. In: Gupta, A., Malik, S. (eds.) CAV 2008. LNCS, vol. 5123, pp. 299–303. Springer, Heidelberg (2008)
5. Campana, D., Cimatti, A., Narasamdya, I., Roveri, M.: SystemC verification via an encoding in Spin (submitted for publication)
6. Cimatti, A., Clarke, E.M., Giunchiglia, F., Roveri, M.: NuSMV: A New Symbolic Model Checker. STTT 2(4), 410–425 (2000)
7. Cimatti, A., Micheli, A., Narasamdya, I., Roveri, M.: Verifying SystemC: a Software Model Checking Approach. In: FMCAD 2010 (to appear)
8. Clarke, E.M., Grumberg, O., Jha, S., Lu, Y., Veith, H.: Counterexample-guided abstraction refinement for symbolic model checking. J. ACM 50(5), 752–794 (2003)
9. Clarke, E.M., Grumberg, O., Peled, D.A.: Model Checking. The MIT Press, Cambridge (1999)
10. Godefroid, P.: Software Model Checking: The VeriSoft Approach. F. M. in Sys. Des. 26(2), 77–101 (2005)
11. Godefroid, P.: Partial-Order Methods for the Verification of Concurrent Systems - An Approach to the State-Explosion Problem. LNCS, vol. 1032. Springer, Heidelberg (1996)
12. Helmstetter, C., Maraninchi, F., Maillet-Contoz, L., Moy, M.: Automatic generation of schedulings for improving the test coverage of systems-on-a-chip. In: FMCAD, pp. 171–178. IEEE Computer Society, Los Alamitos (2006)
13. Holzmann, G.J.: Software model checking with SPIN. Advances in Computers 65, 78–109 (2005)
14. Holzmann, G.J., Peled, D.: An improvement in formal verification. In: 7th IFIP WG6.1 Int. Conf. on Formal Description Techniques VII, London, UK, UK, pp. 197–211 (1995)
15. Kahlon, V., Gupta, A., Sinha, N.: Symbolic model checking of concurrent programs using partial orders and on-the-fly transactions. In: Ball, T., Jones, R.B. (eds.) CAV 2006. LNCS, vol. 4144, pp. 286–299. Springer, Heidelberg (2006)
16. Kundu, S., Ganai, M.K., Gupta, R.: Partial order reduction for scalable testing of systemC TLM designs. In: DAC, pp. 936–941. ACM, New York (2008)
17. Moy, M., Maraninchi, F., Maillet-Contoz, L.: Pinapa: an extraction tool for SystemC descriptions of systems-on-a-chip. In: EMSOFT, pp. 317–324. ACM, New York (2005)

18. Peled, D.: All from one, one for all: on model checking using representatives. In: Courcoubetis, C. (ed.) CAV 1993. LNCS, vol. 697, pp. 409–423. Springer, Heidelberg (1993)
19. Peled, D.: Combining partial order reductions with on-the-fly model-checking. Formal Methods in System Design 8(1), 39–64 (1996)
20. Tabakov, D., Kamhi, G., Vardi, M.Y., Singerman, E.: A Temporal Language for SystemC. In: FMCAD, pp. 1–9. IEEE, Los Alamitos (2008)
21. Traulsen, C., Cornet, J., Moy, M., Maraninchi, F.: A SystemC/TLM Semantics in PROMELA and Its Possible Applications. In: Bošnački, D., Edelkamp, S. (eds.) SPIN 2007. LNCS, vol. 4595, pp. 204–222. Springer, Heidelberg (2007)
22. Valmari, A.: Stubborn sets for reduced state generation. In: Rozenberg, G. (ed.) APN 1990. LNCS, vol. 483, pp. 491–515. Springer, Heidelberg (1991)
23. Wang, C., Yang, Z., Kahlon, V., Gupta, A.: Peephole partial order reduction. In: Ramakrishnan, C.R., Rehof, J. (eds.) TACAS 2008. LNCS, vol. 4963, pp. 382–396. Springer, Heidelberg (2008)

Modelling and Verification of Web Services Business Activity Protocol

Anders P. Ravn, Jiří Srba*, and Saleem Vighio**

Department of Computer Science, Aalborg University,
Selma Lagerlöfs Vej 300, DK-9220 Aalborg East, Denmark
{apr,srba,vighio}@cs.aau.dk

Abstract. WS-Business Activity specification defines two coordination protocols in order to ensure a consistent agreement on the outcome of long-running distributed applications. We use the model checker UPPAAL to analyse the Business Agreement with Coordination Completion protocol type. Our analyses show that the protocol, as described in the standard specification, violates correct operation by reaching invalid states for all underlying communication media except for the perfect FIFO. Based on this result, we propose changes to the protocol. A further investigation of the modified protocol suggests that messages should be received in the same order as they are sent so that a correct protocol behaviour is preserved. Another important property of communication protocols is that all parties always reach their final states. Based on the verification with different communication models, we prove that our enhanced protocol satisfies this property for asynchronous, unreliable, order-preserving communication whereas the original protocol does not.

1 Introduction

Numerous protocols from the web services protocol stack [9] are currently in active development in order to support communication schemes that guarantee consistent and reliable executions of distributed transactions. As applications depend on the correctness of these protocols, guarantees about their functionality should be given prior to the protocols being put into industrial use. However, design and implementation of these protocols is an error-prone process, partly because of the lack of details provided in their standards [7,17]. Therefore, formal approaches provide a valuable supplement during the discussion and clarification phases of protocol standards. The advantage of formal methods is that automatic tools like UPPAAL [3] and TLC [7] can be applied to analyse protocol behaviours and verify general correctness criteria.

In this paper we consider the WS-Coordination framework [12] which, among others, includes the WS-Atomic Transaction (WS-AT) [10] and WS-Business

---

* The author is partially supported by the Ministry of Education of Czech Republic, grant no. MSM 0021622419.
** The author is supported by Quaid-e-Awam University of Engineering, Science, and Technology, Nawabshah, Pakistan, and partially by the Nordunet3 project COSoDIS.

P.A. Abdulla and K.R.M. Leino (Eds.): TACAS 2011, LNCS 6605, pp. 357–371, 2011.

Activity (WS-BA) [11] standards. The WS-AT specification provides protocols used for simple short-lived activities, whereas WS-BA provides protocols used for long-lived business activities. The WS-AT protocol has recently been in focus in the formal methods community and its correctness has been verified using both the TLC model checker [7] where the protocol was formalized in the TLA^+ [8] language as well as using the UPPAAL tool and networks of communicating timed automata [15]. In [15], we discussed the key aspects of the two approaches, including the characteristics of the specification languages, the performances of the tools, and the robustness of the specifications with respect to extensions.

In the present work we analyse the WS-BA standard which (to the best of our knowledge) has not yet been automatically verified in the literature. It consists of two coordination protocols: Business Agreement with Participant Completion (BAwPC) and Business Agreement with Coordinator Completion (BAwCC). We focus on BAwCC in our analysis. It is more complex in its behaviour and has a larger number of states, transitions and messages than BAwPC. We develop several UPPAAL [3] models related to the WS-BA protocols based on the state-tables provided in the standard specification (see [11] or the appendix for the complete tables). We use with advantage the C-like constructs available in UPPAAL and the model of the BAwCC protocol contains more than 600 lines of C code. Our tool supported analysis unexpectedly reveals several problems. The *safety* property, that the protocol never enters an invalid state, is checked for a range of communication mechanisms. The main result is that the property is violated by all considered communication mechanisms but perfect FIFO (queue).

Based on a detailed analysis of the error traces produced by UPPAAL, we suggest fixes to the protocol. Moreover, in contrast to [7,15], we do not limit our analyses to only one type of asynchronous communication policy where messages can be reordered, lost and duplicated, but study different communication mechanisms (see e.g. [1]). This fact appears crucial as even the fixed protocol behaves correctly only for some types of communication media, whereas for others it still violates the correctness criteria.

Another important property of web services applications is that they should terminate in consistent end states, irrelevant of the actual behaviour of the other participating parties [6]. This kind of property is usually called *liveness* and for most nontrivial protocols it cannot be established without some fairness assumptions, such that if a particular transition is infinitely often enabled then it is also executed. In our setting we use a more engineering-like approach by introducing tire-outs (delays before an alternative action is chosen, essentially the "execution delay" of ATP [13]) on the resubmission of messages, as this is a likely way this situation is handled in practice. UPPAAL enables us to specify the timing information in a simple and elegant way and our verification results show that under suitable timing constraints used for tire-outs, we can guarantee the termination property for the fixed protocol, at least for the communication policies where the protocol is correct.

The rest of the paper is organized as follows. In Section 2, we give an overview of the web services business activity protocol and discuss different types of communication policies. Section 3 introduces the UPPAAL modeling approach used in the case study. Properties of the original and the fixed protocols are discussed in Sections 4 and 5. Section 6 describes the termination property and its verification. Finally, Section 7 gives a summary and suggestions for the future research. The appendix contains a full overview of the state-transition tables of the original and modified BAwCC protocol.

2 WS-Business Activity Protocol

WS-Business Activity (WS-BA) [11] and WS-Atomic Transaction (WS-AT) [10] both built on top of WS-Coordination specification [12] form the Web Services Transaction Framework (WSTF). WS-Coordination describes an extensible framework for coordinating transactional web services. It enables an application service to create a context needed to propagate an activity to other services and to register for coordination protocols. These coordination protocols are described in WS-AT and WS-BA specifications. WS-AT provides protocols based on the ACID (atomicity, consistency, isolation, durability) principle [5] for simple short-lived activities, whereas WS-BA provides protocols used for long-lived business activities with relaxation of ACID properties.

WS-BA [11] describes two coordination types: AtomicOutcome and Mixed-Outcome. In AtomicOutcome the coordinator directs all participants to the same outcome, i.e. either to close or to cancel/compensate. In MixedOutcome some participants may be directed to close and others to cancel/compensate. Each of these coordination types can be used in two coordination protocols: WS-Business Agreement with Participant Completion (BAwPC) and WS-Business Agreement with Coordination Completion (BAwCC) that we shall focus on. A participant registers for one these two protocols, which are managed by the coordinator of the activity.

2.1 Business Agreement with Coordination Completion

A state-transition diagram for BAwCC is shown in Figure 1. Note that the figure depicts a combined view and the concrete coordinator and participant states are abstracted away. The complete transition tables are listed in the appendix.

A participant registered for this protocol is informed by its coordinator that it has received all requests to perform its work and no more work will be required. In this version of the protocol the coordinator decides when an activity is terminated, so completion notification comes from the coordinator: It sends a Complete message to the participant to inform it that it will not receive any new requests within the current business activity and it is time to complete the processing. The Complete message is followed by the Completed message by the participant, provided it can successfully finish its work. This protocol also introduces a new Completing state between Active and Completed states. Once the coordinator reaches the Completed state, it can reply with either a Close or a Compensate message. A Close message informs the participant that the activity

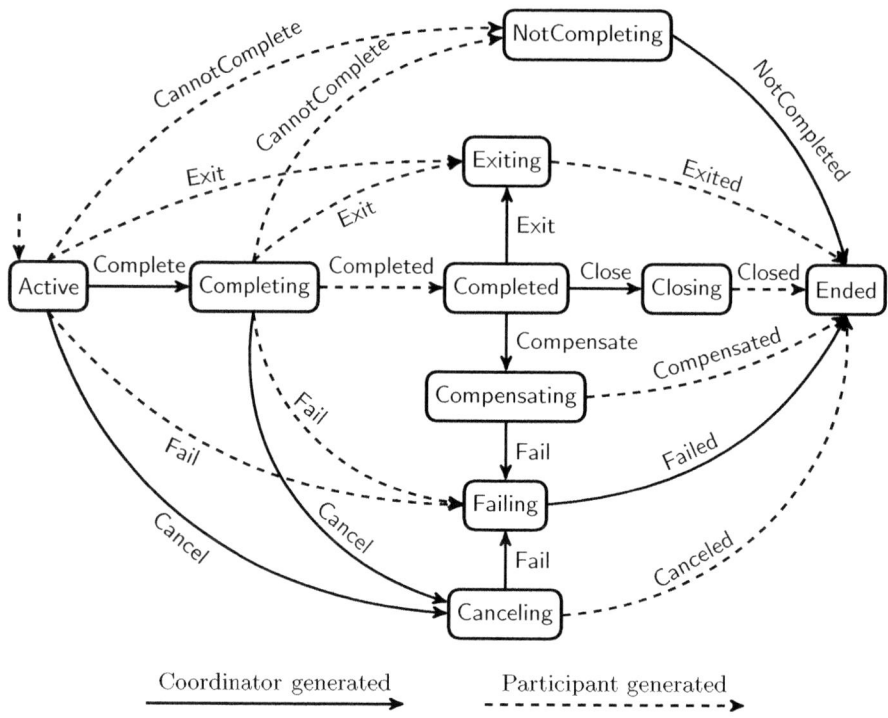

Fig. 1. Business Agreement with Coordinator Completion

has completed successfully. A participant then sends a Closed notification and forgets about the activity. Upon receipt of a Closed notification the coordinator knows that the participant has successfully completed its work and forgets about the participant's state.

A Compensate message, on the other hand, instructs the participant to undo the completed work and to restore the recorded data to its initial state. A participant in response can either send a Compensated or a Fail notification. The Compensated message informs the coordinator that the participant has successfully compensated its work for the business activity, the participant then forgets about the activity and the coordinator forgets about the participant. Upon receipt of a Fail message, the coordinator knows that the participant has encountered a problem and has failed during processing of the activity. The coordinator then replies with a Failed message and forgets about the state of the participant. The participant in turn also forgets about the activity. A participant can also send CannotComplete or Exit messages while being in Active, or Completing states. A CannotComplete notification informs the coordinator that the participant can not successfully complete its work and any pending work will be discarded and completed work will be canceled. The coordinator replies with a NotCompleted message and forgets about the state of the participant. The participant also forgets about the activity in turn. In case of an Exit message the coordinator knows that the participant will no longer engage in the business activity and the pending work will be discarded and any work performed will be canceled. The coordinator will reply with the Exited message and will forget about the participant. The participant will also forget about the activity. In Active and Completing states the coordinator can end a transaction by sending a Cancel message. A participant can either reply with a Canceled or a Fail notification. A Canceled message informs the coordinator that the work has been successfully canceled and then the participant forgets about the activity.

2.2   Communication Policies

The WS-BA specification is not explicit about the concrete type of communication medium for exchanging messages apart from implicitly expecting that the communication is asynchronous. In [7] the authors (two of them were designers of the specification) studied WS-AT and agreed that one should consider asynchronous communication where messages can be lost, duplicated and reordered. Indeed, the WS-AT protocol was proved correct in this setting. It seems natural to adopt the same communication assumptions also for WS-BA,

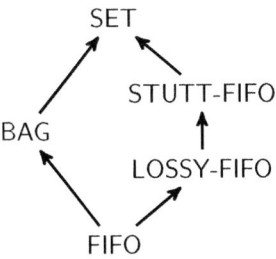

Fig. 2. Communication media

however, as we show later on, the BAwCC protocol is not correct under such a liberal communication policy. We therefore consider a hierarchy of five different communication policies for asynchronous message passing in our study.

– *Unreliable Unordered Asynchronous Communication.* In this type of asynchronous communication the messages may arrive in different order than they were sent and the communication medium is assumed to be unreliable as messages can be lost and duplicated. It corresponds well with the elementary UDP protocol of TCP/IP. As argued in [7], this kind of policy is conveniently implemented as a pool of messages mathematically represented by a set. Adding more messages of the same sort to a set has no additional effect and as our correctness property is a safety property, lossiness is implicitly included by the fact that protocol participants are not in any way forced to read messages contained in the pool (see [7,15] for further discussion on this issue). In the rest of the paper we call this kind of communication implementation SET.
– *Reliable Unordered Asynchronous Communication.* This kind of communication still does not preserve the order of messages but it is a completely reliable medium where a message can only be received as many times as it was sent. Therefore we have to keep track of the number of messages of the same type currently in transit. We can model this communication medium as a multiset (also called a bag) of messages. We refer to this particular implementation of the communication medium as BAG.

- *Reliable Ordered Asynchronous Communication.* This type of communication channel represents the perfect communication medium where messages are delivered according to the FIFO (first in, first out) policy and they can be neither duplicated nor lost. The problem with this medium is that for most nontrivial protocols there is no bound on the size of the communication buffer storing the queue of messages in transit (thanks to the asynchronous nature of the communication) and automatic verification of protocols using this communication policy is often impossible due to the infinite state-space of possible protocol configurations. We refer to this communication as FIFO. It is essentially implemented by the FTP protocol of TCP/IP.
- *Lossy Ordered Asynchronous Communication.* Here we assume an order preserving communication policy like in FIFO but messages can now be also lost before their delivery. The problem with unbounded size of this communication channel remains for most of interesting protocols. We call this policy LOSSY-FIFO.
- *Stuttering Ordered Asynchronous Communication.* In order to overcome the infinite state-space problem mentioned in the FIFO and LOSSY-FIFO communication policies, we introduce an abstraction that ignores stuttering, i.e. repetition of the same message inside of an ordered sequence of messages. We can also consider it as a lossy and duplicating medium which, however, preserves the order among different types of messages. In practice this means that if a message is sent and the communication buffer contains the same message as the most recently sent one, then the message will be ignored. Symmetrically, if a message is read from the buffer, it can be read as many times as required providing it is of the same type. This means that the communication buffer can remain finite even if the protocol includes retransmission of messages, as e.g. both protocols from WS-BA specification do. We call this communication type STUTT-FIFO.

Figure 2 shows the relationship among the different communication media. The arrows indicate the inclusions (in the sense of possible behaviours) of the presented media. Hence any protocol execution with the FIFO communication policy is possible also in any other communication type above it. This means that if we can introduce the validity of any safety property for e.g. the SET medium, this result will hold also for any other medium below it and finding an error trace in the protocol with e.g. the FIFO medium implies the presence of such a trace also in any other medium above it.

While the communication policies SET, BAG, FIFO and LOSSY-FIFO are well studied, the STUTT-FIFO communication we introduce in this paper is nonstandard and not implemented in any of industrial applications that we are aware of. Although, as remarked above, FTP will work this way if the application level avoids retransmission of data. The main reason why we consider this kind of communication is that it allows us to validate the protocols in question while preserving the finiteness of the state-space. Hence we can establish safety guarantees also for the FIFO and LOSSY-FIFO communication policies, which would be otherwise impossible as the size of such channels is not bounded in our setting.

a) WS-BA (BAwCC):

Coordinator View for Inbound Events (received messages).

Inbound Events		States			
	...	Closing	...	Exiting	...
⋮	⋮	⋮	...	⋮	⋮
Complete	...	resend Close goto Closing	...	...	...
⋮	⋮	⋮	...	⋮	⋮
CannotComplete	...	goto Invalid-State	...	...	...

c) Uppaal encoding with SET Communication (simplified):

Send_Msg() and Receive_Msg() functions for the coordinator.

```
typedef int[0,6] MsgsTC;          typedef int[0,6] MsgsP;
const MsgsTC CANCEL_TC = 0;       const MsgsP EXIT_P = 0;
const MsgsTC COMPLETE_TC = 1;     const MsgsP COMPLETED_P = 1;
const MsgsTC CLOSE_TC = 2;        const MsgsP FAIL_P = 2;
const MsgsTC COMPENSATE_TC =3;    const MsgsP CANNOT_COMPLETE_P = 3;
const MsgsTC FAILED_TC = 4;       const MsgsP CANCELED_P = 4;
const MsgsTC EXITED_TC = 5;       const MsgsP CLOSED_P = 5;
const MsgsTC NOT_COMPLETED_TC = 6; const MsgsP COMPENSATED_P = 6;
bool msgTC[MsgsTC];               bool msgP[MsgsP];

void Send_Msg(MsgsTC s) {         bool Receive_Msg(MsgP r) {
 msgTC[s] = true; }                 return msgP[r]; }
```

Behaviour of transaction coordinator upon the receipt of the message Complete is modelled by a loop transition with the following guard and update (action).

```
bool guard() {
 return Receive_Msg(COMPLETE_P) && stTC == TC_CLOSING;  }

void action() {  Send_Msg(CLOSE_TC); stTC = TC_CLOSING; }
```

Behaviour of transaction coordinator upon the receipt of the message CannotComplete is modelled by a transition to a new error-state called INVALID with the following guard and with no update.

```
bool guard() {
 return Receive_Msg(CANNOT_COMPLETE_P) && stTC == TC_CLOSING; }
```

Fig. 3. Implementation of selected WS-BA rules in UPPAAL

3 Formal Modelling of BAwCC in UPPAAL

The WS-BA standard [11] provides a high-level description of the WSwCC protocol. It is essentially a collection of protocol behaviours described in English accompanied by diagrams like the graph shown in Figure 1 and state-transition tables for the parties involved in the protocol. See Figure 3 a) for a fragment of such a table and the appendix for a complete collection of the tables.

Figure 3 a) describes how the transaction coordinator, being in its internal state Closing, handles the message Complete arriving from the participant. It will simply resend to the participant the message Close and remain in the state Closing. The table also describes that while being in the state Closing, the coordinator does not expect to receive the message CannotCompensate from the participant, and should this happen, it will enter an invalid state.

The UPPAAL implementation of this behaviour is given in Figure 3 b). The syntax should be readable even without any prior knowledge of the tool, but we refer the interested reader to [3] for a thorough introduction to UPPAAL. The code in the figure first lists the names of constants that represent messages sent from the transaction coordinator to the participant and vice versa. Then it defines two functions Send_Msg and Receive_Msg that take care of sending and receiving of messages via the bit-vectors msgTC and msgP. The code is shown only for the simplest SET implementation. For BAG, FIFO, LOSSY-FIFO and STUTT-FIFO the code is more complex but implemented in a standard way. The only complication is that the data structures representing these four types of communication are in general unbounded, so to ensure automatic verification we introduce a constant upper bound on the buffer size and we register a buffer overflow in a boolean variable called overflow.

The transitions described in the state tables are then implemented in the expected way as shown by the two examples in Figure 3 b). The final timed automata model then consists of a process for the coordinator with two locations (normal execution and invalid state) and a similar process for the participant running in parallel with the coordinator process. All data management (states, buffer content, etc.) is performed via C-like data structures, as this is an efficient and manageable way to handle this relatively large model. In total the C part of the implementation contains more than 600 nonempty lines of code. The complete UPPAAL model can be downloaded at [14].

4   Analysis of BAwCC

As already noted, WS-BA relaxes the ACID principles and allows for a mixed outcome of a transaction. Therefore, we cannot expect that all parties of the protocol agree on the outcome, as it was the case for WS-AT protocols [7,15]. Instead, we focus on the analysis of the actual state-transition tables w.r.t. reachability of invalid states. Invalid states appear in the tables both for inbound and outbound messages. The meaning of these states is not clearly stated in WS-BA specification but we contacted the designers via their discussion forum and received (citing [16]):

> "For outbound events, an Invalid State cell means that this is not
> a valid state for the event to be produced. ... For inbound events, an
> Invalid State cell means that the current state is not a valid state for the

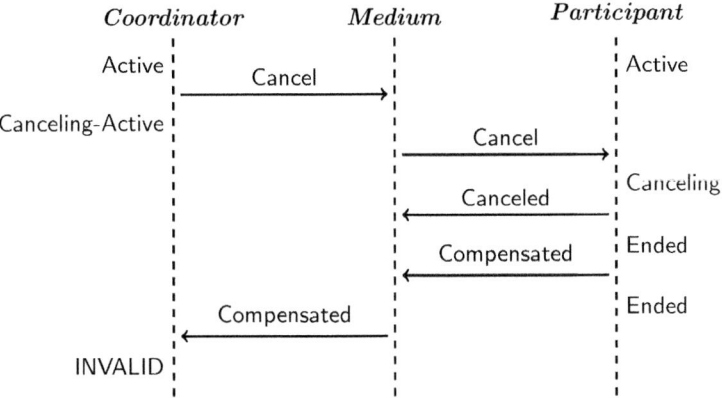

Fig. 4. Error trace in BAwCC leading to an invalid state

inbound message. For example, for Participants in BusinessAgreement-
WithCoordinationCompletion (table B.3) the Canceling state is not a
valid state for receiving a Close message. There are no circumstances
where a Participant in this state should ever receive a Close message,
indicating an implementation error in the Coordinator which sent the
message. This is a protocol violation ..."

This means that in the tables for outbound events, messages that lead to invalid states are never sent (and hence omitted in the UPPAAL model) and for
inbound events the possibility to enter an invalid state is a protocol violation.
This requirement is easily formulated in the UPPAAL query language (a subset
of TCTL) as follows.

`E<> (tc.INVALID || par.INVALID) && !overflow`

This is a safety property asking whether there is a protocol execution in which
either the transaction coordinator (process called `tc`) or the participant (process
called `par`) enters the state INVALID while at the same time there was no buffer
overflow. We have checked this property for all five communication policies we
consider and the property surprisingly turned out to be true for all of them except
for FIFO. The tool automatically generated an error trace, seen in Figure 4.
It is easy to see that this trace is executable both for LOSSY-FIFO and BAG
communication (and hence also for any other above them in the hierarchy in
Figure 2). The main point in this trace is that the message Canceled that is sent
by the participant is either lost (possible in LOSSY-FIFO) or reordered with the
message Compensated (possible in BAG).

It is also clear that this error trace cannot be executed in the perfect FIFO
communication policy. For FIFO we were able to verify that the protocol is correct
for up to six messages in transit (three from coordinator to participant and three
in the opposite direction). As perfect FIFO communication is known to have the

full Turing power [4], there is no hope to establish the correctness of the protocol with unbounded FIFO communication in a fully automatic way.

Another interesting question we can ask about the protocol is whether the communication medium is bounded for BAwCC or not. This can be done by asking the following UPPAAL query.

```
E<> overflow
```

Verification results show that all communication media except for SET can always reach a buffer overflow for any given buffer size that we were able to verify (up to 20 messages in transit). This is a good indication that the communication buffer is indeed unbounded and a simple (manual) inspection of the protocol confirms this fact.

5   Enhanced BAwCC

Given the verification results in the previous section, we found the BAwCC protocol not completely satisfactory as even a simple relaxation of the perfect communication policy results in incorrect behaviour. Taking into account that the protocols in WS-AT avoided invalid states even under the most general SET communication, we shall further analyze the protocol and suggest an improvement.

The error trace in Figure 4 hints at the source of problems. Once a participant reaches the Ended state, it it instructed to forget all state information and just send the last message by which the transition to the Ended state was activated. The problem is that there are three different reasons for reaching the Ended state, but BAwCC allows for the retransmission of all three messages at the same time, whenever the participant is in the state Ended. As seen in Figure 4, the participant after receiving the message Cancel correctly answers with the message Canceled, but then sends the message Compensated. This causes confusion on the coordinator side. A similar problem can occur in a symmetric way.

In our proposed fix to the BAwCC protocol, we introduce three additional end states, both for the participant as well as for the coordinator, in order to avoid the confusion. The complete state tables of the enhanced protocol are given in the appendix. We modelled and verified the enhanced protocol in UPPAAL and the results are as follows.

Under the STUTT-FIFO communication, the medium is bounded with no overflow, so all verification results are conclusive. We also established that there is no execution of the modified protocol that leads to an invalid state. As this is a safety property, the positive result holds automatically also for LOSSY-FIFO and FIFO.

However, when considering the media BAG and SET representing a communication where messages can be reordered, the tool still returns error traces like the one depicted in Figure 5. This problem is more inherent to the protocol design and the reason for the confusion is the fact that the messages Canceled and Fail sent be the participant are delivered in the opposite order.

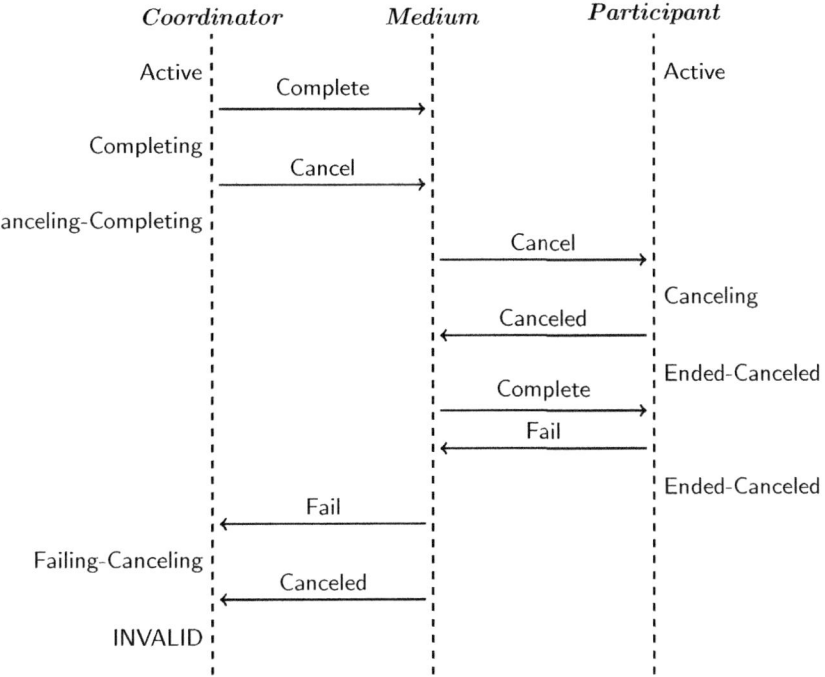

Fig. 5. Error trace in enhanced BAwCC leading to an invalid state

To conclude, our enhanced protocol, unlike the original one, is immune to lossiness and duplication of messages (stuttering) as long as their order is preserved. Making the protocol robust w.r.t. reordering of messages would, in our opinion, require a substantial and nontrivial redesign of the BAwCC protocol.

6   Termination under Fairness

In this section we shall turn our attention to another important property of distributed protocols, namely the *termination* property. Termination means that as long as the communication parties follow the protocol, any concrete execution will always bring them to their end states. In UPPAAL this property for our protocol can be formulated as follows.

$$A<> stTC == TC\_ENDED \&\& stP == P\_ENDED$$

The semantics is that in any maximal computation of the protocol, we will eventually reach a situation where the states of the transaction coordinator as well as the participant are TC_ENDED and P_ENDED, respectively. Termination is hence a liveness property.

It is clear that the original BAwCC fails to satisfy termination as we can reach invalid states from which there is no further continuation. This is true for all types of communication, except for FIFO, where on the other hand we cannot prove termination due to the unboundedness of the medium. We shall therefor focus on our enhanced BAwCC protocol and the communication medium STUTT-FIFO where the protocol is correct and the medium bounded. A positive result will imply termination also for LOSSY-FIFO and FIFO.

A quick query about termination in UPPAAL shows that it fails the property and the tool returns error traces that reveal the reason: there is no bound on the number of retransmissions of messages and this can create infinite process executions where the same message is retransmitted over and over. This is to be expected for any nontrivial protocol and in theory the issue is handled by imposing an additional assumption on *fairness* of the protocol execution. This can for example mean that we require that whenever during an infinite execution some action is infinitely often enabled then it has to be also executed. Such assumptions will guarantee that there is a progress in the protocol execution and are well studied in the theory (see e.g. [2]).

The complication is that fairness concerns infinite executions and is therefore difficult to implement in concrete applications. Software engineers would typically use only a limited number of retransmissions within a fixed time interval and give up resending messages after a certain time has passed.

So far, we have used UPPAAL only for verification of discrete systems, but the tool allows us to specify also *timed* automata models and supports their automatic verification. By introducing the timing aspects into the protocol behaviour, we will be able to argue about fairness properties like termination.

We model the retransmission feature using tire-outs. A tire-out imposes a progress in the model and as already outlined in the introduction it is essentially the "execution delay" of ATP [13]. In our model we introduce two clocks x and y local both for the coordinator and the participant. We also assume two global constants MIN-DELAY and TIRE-OUT, representing the minimal possible delay between two retransmissions and a tire-out time after which the protocol will not attempt to retransmit the message any more. Figure 6 shows the implementation of this feature in the protocol model. We already explained that the rules of the protocol are modelled using loops in UPPAAL automata and the discrete data are handled using guards and updates (not shown in the illustration). In the figure we can separate all transitions into two categories: progress transitions and retransmission transitions. Retransmission transitions retransmit a message and remain in the same state, while progress transitions change the state of the participant or the coordinator. The clock x represents the time delay since the last progress transition occurred (it is reset to 0 by any progress transition) and clock y represents the time elapsed since the last retransmission. These two clocks restrict the behaviour of the retransmission transitions so that they are enabled only if at least the minimal delay has passed since last retransmission and the clock x has not exceeded the tire-out limit. The presence of the invariant $x \leq$ TIRE-OUT then ensures a progress.

Using the tire-out modeling as described above we were able to automatically verify that the enhanced BAwCC protocol with the STUTT-FIFO communication

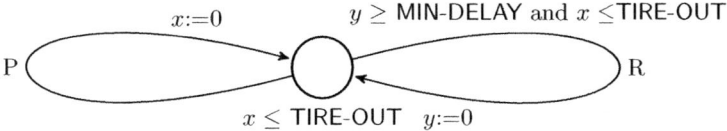

$x:=0$ $y \geq$ MIN-DELAY and $x \leq$ TIRE-OUT

P   R

$x \leq$ TIRE-OUT $y:=0$

Fig. 6. Tire-outs modelling; P is a progress transition, R is a retransmission transition

Buffer Type	Properties	BAwCA Protocol	
		Original	Enhanced
SET	Correctness	No	No
	Boundedness	Yes	Yes
	Termination	No	No
BAG	Correctness	No	No
	Boundedness	No	No
	Termination	No	No
STUTT-FIFO	Correctness	No	Yes
	Boundedness	No	Yes
	Termination	No	Yes
LOSSY-FIFO	Correctness	No	Yes
	Boundedness	No	No
	Termination	No	Yes
FIFO	Correctness	Yes?	Yes
	Boundedness	No	No
	Termination	Yes?	Yes

Fig. 7. Overview of verification results for BAwCC and enhanced BAwCC

policy satisfies the termination property for suitable constants MIN-DELAY and TIRE-OUT where, for example, the minimal delay is set to one time unit and the tire-out deadline to 30 time units. By changing the two constants we can experiment with different timing options while making (automatically) sure that the termination property is preserved.

7   Conclusion and Future Work

We provided a formal UPPAAL model of the Business Agreement with Coordinator Completion (BAwCC) protocol from the WS-BA specification. The model is based on the state-transition tables provided in the specification. We also introduced several ways to model the communication medium, starting with perfect

FIFO channels and ending up with lossy, duplicating and orderless medium. We have verified that the protocol may enter invalid states for all communication policies apart from the FIFO. For FIFO we verified that no invalid states are reachable for up to six messages in transit (three in each direction), however, this is not a guarantee that the protocol is correct for any size of the FIFO buffer.

Based on the analysis of the protocol in UPPAAL, we suggested an enhanced protocol which distinguishes among three different ways of entering the ended states. This protocol is correct also for all imperfect media based on FIFO but may still reach invalid states if more liberal communication is assumed. By introducing timing constraints (tire-outs) to the protocol behaviour, we were also able to verify the termination property for imperfect FIFO communication. Figure 7 gives the summary of the results for all five communication policies and the original and enhanced protocols. Correctness stands for the absence of invalid states in protocol executions, boundedness describes whether the communication channels have bounded size and termination guarantees that during any protocol behaviour, all parties eventually reach their final (ended) states.

To conclude, the BAwCC protocol seems correct for the perfect FIFO communication as provided e.g. by the FTP of TCP/IP. We assume that the protocol was also mainly tested in this setting and hence the tests did not discover any problematic behaviour. On the other hand, the protocol contains a number of message retransmissions, which would not be necessary for the perfect medium. This signals that the designers planned to extend the applicability of the protocol also to frameworks with unreliable communication but as we demonstrated, some fixes have to be applied to the protocol in order to guarantee the correct operation also in this case. In any case, WS-BA specification is not explicit about the assumptions on the communication medium, but this should be perhaps considered for the future design of protocols.

Finally, the manual creation of UPPAAL models for WS-BA protocols was a long and time demanding process and in our future work we will try to automate the process of creating timed automata templates directly from the state-transition tables. For widely used, standardized protocols, this is probably not going to find defects. Yet, in concrete implementations some optimizations and specializations may be included, and here a tool support may assist in validating the effect of presumably small innocent changes.

Acknowledgement. The authors are grateful to the anonymous reviewers for their comments on the perspective of this work.

References

1. Afek, Y., Attiya, H., Fekete, A., Fischer, M., Lynch, N., Mansour, Y., Wang, D.-W., Zuck, L.: Reliable communication over unreliable channels. J. ACM 41(6), 1267–1297 (1994)
2. Apt, K.R., Francez, N., Katz, S.: Appraising fairness in languages for distributed programming. Distributed Computing 2, 226–241 (1988)

3. Behrmann, G., David, A., Larsen, K.G.: A tutorial on UPPAAL. In: Bernardo, M., Corradini, F. (eds.) SFM-RT 2004. LNCS, vol. 3185, pp. 200–236. Springer, Heidelberg (2004)
4. Brand, D., Zafiropulo, P.: On communicating finite-state machines. Journal of the ACM 30(2), 323–342 (1983)
5. Gray, J., Reuter, A.: Transaction Processing: Concepts and Techniques. Morgan Kaufmann, San Francisco (1993)
6. Greenfield, P., Kuo, D., Nepal, S., Fekete, A.: Consistency for web services applications. In: VLDB 2005: Proceedings of the 31st International Conference on Very Large Data Bases, pp. 1199–1203. VLDB Endowment (2005)
7. Johnson, J.E., Langworthy, D.E., Lamport, L., Vogt, F.H.: Formal specification of a web services protocol. Journal of Logic and Algebraic Programming 70(1), 34–52 (2007)
8. Lamport, L.: Specifying Systems. Addison-Wesley, Reading (2003)
9. Mathew, B., Juric, M., Sarang, P.: Business Process Execution Language for Web Services, 2nd edn. Packt Publishing (2006)
10. Newcomer, E., Robinson, I. (chairs): Web services atomic transaction (WS-atomic transaction) version 1.2 (2009),
http://docs.oasis-open.org/ws-tx/wstx-wsat-1.2-spec.html
11. Newcomer, E., Robinson, I. (chairs): Web services business activity (WS-businessactivity) version 1.2 (2009),
http://docs.oasis-open.org/ws-tx/wstx-wsba-1.2-spec-os/
wstx-wsba-1.2-spec-os.html
12. Newcomer, E., Robinson, I. (chairs): Web services coordination (WS-coordination) version 1.2 (2009),
http://docs.oasis-open.org/ws-tx/wstx-wscoor-1.2-spec-os/
wstx-wscoor-1.2-spec-os.html
13. Nicollin, X., Sifakis, J.: The algebra of timed processes, ATP: Theory and application. Information and Computation 114(1), 131–178 (1994)
14. Ravn, A.P., Srba, J., Vighio, S.: UPPAAL model of the WS-BA protocol, Available in the UPPAAL example section at http://www.uppaal.org
15. Ravn, A.P., Srba, J., Vighio, S.: A formal analysis of the web services atomic transaction protocol with uppaal. In: Margaria, T., Steffen, B. (eds.) ISoLA 2010. LNCS, vol. 6416, pp. 579–593. Springer, Heidelberg (2010)
16. Robinson, I.: Answer in WS-BA discussion forum, July 14 (2010),
http://markmail.org/message/wriewgkboaaxw66z
17. Vogt, F.H., Zambrovski, S., Gruschko, B., Furniss, P., Green, A.: Implementing web service protocols in SOA: WS-coordination and WS-businessactivity. In: Proceedings of the Seventh IEEE International Conference on E-Commerce Technology Workshops (CECW 2005), pp. 21–28. IEEE Computer Society, Los Alamitos (2005)

CADP 2010: A Toolbox for the Construction and Analysis of Distributed Processes*

Hubert Garavel, Frédéric Lang, Radu Mateescu, and Wendelin Serwe

INRIA Laboratoire d'Informatique de Grenoble, VASY team
655, avenue de l'Europe, 38330 Montbonnot St Martin, France
{Hubert.Garavel,Frederic.Lang,Radu.Mateescu,Wendelin.Serwe}@inria.fr

Abstract. CADP (*Construction and Analysis of Distributed Processes*) is a comprehensive software toolbox that implements the results of concurrency theory. Started in the mid 80s, CADP has been continuously developed by adding new tools and enhancing existing ones. Today, CADP benefits from a worldwide user community, both in academia and industry. This paper presents the latest release CADP 2010, which is the result of a considerable development effort spanning the last four years. The paper first describes the theoretical principles and the modular architecture of CADP, which has inspired several other recent model checkers. The paper then reviews the main features of CADP 2010, including compilers for various formal specification languages, equivalence checkers, model checkers, performance evaluation tools, and parallel verification tools running on clusters and grids.

1 Introduction

Among all the scientific issues related to the reliability of computer systems, concurrency has a major place, because the design of parallel systems is a complex, error-prone, and largely unmastered activity. Thirty years after the first attempts at building automated verification tools for concurrent systems, the problem is still there; it has even gained in relevance because system complexity has increased, and because concurrency is now ubiquitous, from multicore microprocessors to massively parallel supercomputers.

To ensure the reliability of a concurrent system under design, it is understood that the first step consists in establishing a precise model of the system behavior, this model usually consisting of several concurrent processes, together with a description of the data types, constants, variables, and functions manipulated by these processes. This opens the debate on the most appropriate languages to express system models, with a large choice of candidates ranging from semiformal to formal languages.

* This work has been partly funded by Bull, by the French National Agency for Research (project OpenEmbedd), by the French Ministry of Economics and Industry (Aerospace Valley project Topcased), and by the *Conseil Général de l'Isère* (Minalogic project Multival).

P.A. Abdulla and K.R.M. Leino (Eds.): TACAS 2011, LNCS 6605, pp. 372–387, 2011.

Once a precise, if not formal, model is available, one needs automated methods to prove the correctness of the system with respect to its specification or, at least, to search for the presence of certain mistakes. Without neglecting recent progresses in theorem proving and static analysis, state space exploration techniques (among which reachability analysis and model checking) remain the most successful approaches for dealing with complex concurrent systems, especially during the design phase, when system specifications are evolving frequently.

State space exploration techniques are usually grouped in two classes: *enumerative* (or *explicit state*) techniques consider each state of the system separately, whereas *symbolic* (or *implicit state*) techniques manipulate sets of states represented using decision diagrams (BDDs and their variants) or logical formulas whose satisfiability is determined using SAT and SMT solvers. In this paper, we will use the term enumerative instead of explicit-state in order to avoid possible confusions with the terminology about explicit and implicit models (see Sect. 2). Enumerative techniques are based on a forward exploration of the transition relation between states (*post* function), making them suitable for the on-the-fly verification of specifications written in languages with arbitrary data types. Although they enable exploration of *a priori* fewer states than their symbolic counterparts, enumerative techniques prove to be adequate for the analysis of asynchronous parallel systems containing complex data structures. Among the enumerative model checkers developed in the 80s, SPIN [33] and CADP are the two oldest that are still available on the latest 64-bit architectures and being used in an industrial setting. The principles underlying these two model checkers are a source of inspiration for other recent verification tools based on similar concepts.

CADP (*Construction and Analysis of Distributed Processes*)[1] is a toolbox for verifying asynchronous concurrent systems. The toolbox, whose development started in 1986, is at the crossroads between several branches of computer science: concurrency theory, formal methods, and computer-aided verification. Initially, CADP consisted of only two tools: CÆSAR [14], a compiler and explicit state space generator for the LOTOS language, and ALDÉBARAN [11,13], an equivalence checker based on bisimulation minimization. Over the past 25 years, CADP has been continuously improved and extended [12,19,20]. This paper presents the latest release, CADP 2010 *"Zurich"*, which currently contains 45 tools.

CADP offers now a comprehensive set of functionalities covering the entire design cycle of asynchronous systems: specification, interactive simulation, rapid prototyping, verification, testing, and performance evaluation. For verification, it supports the three essential approaches existing in the field: model checking, equivalence checking, and visual checking. To deal with complex systems, CADP implements a wide range of verification techniques (reachability analysis, on-the-fly verification, compositional verification, distributed verification, static analysis) and provides scripting languages for describing elaborated verification scenarios. In addition, CADP 2010 brings deep changes with respect to previous releases, especially the support for many different specification languages.

---

[1] `http://vasy.inria.fr/cadp`

This paper gives an overview of CADP 2010, highlighting new tools and recent enhancements. It is organized as follows. Sect. 2 presents the core semantic models of CADP. Sect. 3 describes the three languages now supported by CADP and lists translations developed for other languages. Sect. 4, 5, 6, and 7 respectively present the CADP tools for model checking, equivalence checking, performance evaluation, and distributed verification. Finally, Sect. 8 summarizes the achievements and indicates directions for future work.

2   Architecture and Verification Technology

Compared to other explicit-state model checkers, CADP has the following principles and distinctive features (some of which were already present in inspiring tools rooted in concurrency theory, such as CWB [8] and CWB-NC [7]):

– CADP supports both high-level languages with a formal semantics (process calculi) and lower level formalisms (networks of communicating automata); it also accepts connections from informal or semi-formal languages that have a means to compute the *post* transition function.
– Contrary to most model checkers supporting only scalar types, CADP has from the outset supported concurrent programs with complex and/or dynamic data structures (records, unions, lists, trees, etc.) provided that these data structures are not shared by concurrent processes.
– CADP relies on *action-based* (rather than *state-based*) semantic models inherited from concurrency theory, in which one can only refer to the observable actions performed by a system instead of the internal contents of states, which are supposed to be hidden and implementation dependent, and thus are not abstract enough. This encompasses the classical concepts of LTSs (for verification), discrete- and continuous-time Markov chains (for performance evaluation), and extended Markovian models, such as *Interactive Markov Chains* (IMCs) [31], which combine LTSs and Markov chains.
– Relying on action-based models enables equivalence checking, i.e., the comparison of specifications for equality or inclusion; this corresponds to the notions of bisimulations for LTSs and aggregation/lumpability for Markov chains. Also, the possibility of replacing a state space by an equivalent but smaller one is fundamental in compositional verification.
– As a consequence, the model checkers of CADP are based on branching-time (rather than linear-time) logics, which are adequate with bisimulation reductions.
– CADP is equipped with an original software architecture, which has widely inspired recent competing model checkers developed in the 2000s. Early model checkers were "monolithic" in the sense that they tightly combined (1) the source language used to describe the concurrent system under verification and the compiling algorithms used to generate/explore the state space of the concurrent system, and (2) the temporal logic language used to specify correctness formulas and the verification algorithms that evaluate these formulas over the state space. CADP took a different approach and adopted

a modular architecture with a clear separation between language-dependent and language-independent aspects. Different verification functionalities are implemented in different tools, which can be reused for several languages and which are built upon well-specified interfaces that enable code factoring.

- CADP 2010 can manage state spaces as large as 10^{10} explicit states; by employing compositional verification techniques on individual processes, much larger state spaces can be handled, up to sizes comparable to those reached using symbolic techniques, such as BDDs.

CADP can be seen as a rich set of powerful, interoperating software components for manipulating automata and Markov chains. All these tools are integrated in two ways: for interactive use, a graphical user-interface (named EUCALYPTUS) with contextual menus is provided; for batch-mode use, a scripting language named SVL [18] was designed, with user-friendly syntax and powerful verification strategies that make of SVL a unique feature of CADP.

Explicit state spaces. In the terminology of CADP, an *explicit* state space is a state-transition graph defined *extensively*, meaning that the sets of states and transitions are entirely known, because they have been already computed.

In the early 90s, most verification tools represented explicit state spaces using textual file formats, which were only adequate for small graphs but would not scale satisfactorily, e.g., to millions of states. To solve this issue, CADP was equipped in 1994 with BCG (*Binary-Coded Graphs*), a portable file format for storing LTSs. BCG is a binary format, which was designed to handle large state spaces (up to 10^8 states and transitions initially — this limit was raised to 10^{13} in CADP 2010 to take into account 64-bit machines). Because the BCG format is not human readable, it comes with a collection of code libraries and utility programs for handling BCG files.

Two key design goals for BCG are file compactness and the possibility to encode/decode files quickly and dynamically (i.e., without requiring knowledge of the entire state space in advance); these goals are achieved using dedicated compression techniques that give significant results: usually, two bytes per transition on average, as observed on VLTS (*Very Large Transition Systems*)[2], a benchmark suite used in many scientific publications. A third design goal is the need to preserve in BCG files the source-level information (identifiers, line numbers, types, etc.) present in the source programs from which BCG files are generated, keeping in mind that these programs could be written in different languages.

Implicit state spaces. In the terminology of CADP, an *implicit* state space is a state-transition graph defined *comprehensively*, meaning that only the initial state and the *post* transition function are given, such that (a fragment of) the graph is progressively explored and discovered on demand, depending on the verification goals. Handling implicit state spaces properly is a prerequisite for on-the-fly verification.

In addition to BCG, which only applies to explicit state spaces, CADP provides OPEN/CÆSAR [16], a software framework for implicit state spaces,

---

[2] http://vasy.inria.fr/cadp/resources/benchmark_bcg.html

which enforces modularity by clearly separating language-dependent aspects (i.e., compiler algorithms) from language-independent aspects (i.e., verification algorithms). OPEN/CÆSAR is organized around three components: the *graph module* (which encapsulates all language-dependent aspects, typically code generated from a high-level source program to compute states and transitions), the *library module* (which provides useful generic data structures, e.g., stacks, tables, hash functions, etc.), and the *exploration module* (which gathers language-independent aspects, typically verification and state exploration algorithms). All the internal details of the graph module are hidden behind a programming interface, which provides an abstraction for states and transition labels (making them available as opaque types) and implements the transition relation by means of a higher-order iterator.

Since the introduction of the OPEN/CÆSAR architecture in 1992, each of its three modules has been progressively extended. Regarding the graph module, only LOTOS was supported at first, but support for more languages has been added, either by VASY or other research teams. Regarding the library module, its data structures and algorithms have been continuously optimized and enriched. Regarding the exploration module, many OPEN/CÆSAR tools have been developed for simulation, random execution, model checking, equivalence checking, and test case generation.

Boolean equation systems (BESs [39]). These are a useful low-level formalism for expressing analysis problems on LTSs, i.e., model checking, equivalence checking, partial order reductions, test case generation, and behavioral adaptation. A BES is a collection of equation blocks, each defining a set of Boolean variables (left-hand sides) by propositional formulas (right-hand sides). All equations in a block have the same fixed point sign: either minimal (μ) or maximal (ν). BESs can be represented as *Boolean graphs* [1] and are closely related to game graphs [51] and parity games [50].

The CÆSAR_SOLVE library [43] of OPEN/CÆSAR contains a collection of linear-time algorithms for solving alternation-free BESs using various exploration strategies of its underlying Boolean graph (depth-first search, breadth-first search, etc.). The resolution works on the fly, the BES being constructed (e.g., from the evaluation of a temporal logic formula on an LTS, or from the comparison of two LTSs) at the same time it is solved, new equations being added to the BES and solved as soon as they are discovered. All algorithms of CÆSAR_SOLVE can generate diagnostics, i.e., compute a minimal (in the sense of graph inclusion) Boolean subgraph explaining why a given Boolean variable is true or false [42].

New strategies have been added to CADP 2010 for solving conjunctive BESs (arising from equivalence checking) and disjunctive BES (arising from model checking), keeping in memory only the vertices (and not the edges) of the Boolean graphs. Currently, CADP 2010 offers nine resolution strategies, which can solve BESs containing 10^7 variables in ten minutes. Recently, a new linear-time algorithm generalizing the detection of accepting cycles in Büchi automata was added [47], which serves for model checking fairness properties. For testing and benchmarking purposes, CADP 2010 provides the new BES_SOLVE tool, which can evaluate BESs entirely constructed and stored in (gzipped) files, or built on the fly randomly according to fourteen parameters (number of variables, equation length, percentage of disjunctive and conjunctive operators, etc.).

Parameterized Boolean equation systems. CADP also uses internally the PBES (Parameterized BES) model [41], which extends the BES model by adding typed data parameters and arbitrary Boolean expressions over these parameters. The PBES model was originally invented as a means to represent the model checking of MCL formulas (μ-calculus extended with typed data), implemented in the EVALUATOR 4.0 model checker now available in CADP 2010 (see Sect. 4). Recently, this model received much attention from the model checking community, which investigates two approaches: symbolic resolution or instantiations towards BESs followed by on-the-fly resolution, the latter being somehow close to SAT-solving. Beyond verification, PBESs can express other problems such as evaluation of parameterized Horn clauses or DATALOG queries over data bases.

3   Specification Languages

A major difference of CADP 2010 compared with earlier versions is the support for several specification languages, while previously only LOTOS was supported.

3.1   Support for the LOTOS Language

LOTOS [34] is a formal specification language standardized by ISO to describe communication protocols. It is composed of two different languages in one: a data part, based on algebraic abstract data types, and a control part, which is a process calculus combining the best features of CCS, CSP, and CIRCAL. For this reason, CADP provides two LOTOS compilers, both sharing a common front-end.

Compiling the data part. The CÆSAR.ADT compiler [15,28] translates the data part of a LOTOS program (i.e., a collection of sorts, constructors, and functions defined by algebraic equations) into executable C code. The translation aims at verification efficiency, by first optimizing memory (which is essential for state space exploration, where every bit counts), then time. The compiler automatically recognizes certain classes of usual types (natural numbers, enumerations, tuples, etc.), which are implemented optimally. The algebraic equations of LOTOS are translated using a pattern-matching compilation algorithm for rewrite systems with priority. Amusingly, most of the compiler is written using LOTOS abstract data types, so that CÆSAR.ADT is used to bootstrap itself.

The version of CÆSAR.ADT included in CADP 2010 allows values of complex types (such as tuples, unions, lists, trees, strings, sets, etc.) to be represented "canonically", meaning that these values are stored in tables, represented in normal form as table indexes and thus are stored only once in memory. A technical challenge was to make this feature optional: the user can selectively store certain types in tables, while other types remain implemented as before.

Compiling the control part. The CÆSAR compiler [25,24] translates an entire LOTOS program (reusing the C code generated by CÆSAR.ADT) into C code that can be used either for generating an explicit LTS (encoded in the BCG format) or an implicit LTS (represented using the OPEN/CÆSAR programming interface), or for rapid prototyping (using the EXEC/CÆSAR programming interface, which enables the connection with a real-world environment). The subset of LOTOS accepted by CÆSAR must obey certain constraints, which forbid unbounded dynamic creation of processes and non-terminal recursion in process calls; practically, these constraints are acceptable in most cases.

The translation is done using several intermediate steps, so as to perform, for efficiency reasons, as many computations as possible at compile-time. The LOTOS program is first translated into a simplified language named SUBLOTOS, then into a (hierarchical) Petri net extended with atomic transitions, typed local/global variables, and arbitrary combinations of conditions and actions attached to Petri net transitions. This Petri net is then simplified by applying a collection of optimizations on its control and data flows, and finally transformed into a C program, which is then compiled and executed.

In addition to various bug fixes, the version of CÆSAR included in CADP 2010 delivers increased performance, especially by introducing dynamically resizable state tables and by optimizing the generated C code for the amount of physical memory available. Also, the reduction techniques based on data flow analysis [24], which typically reduce state spaces by several orders of magnitude, have been enhanced by applying data-flow optimizations iteratively, following the hierarchical structure of the Petri net: for 22% of the benchmarks, the number of states is divided by 2.4 on average (on certain benchmarks, it is divided by 25).

3.2 Support for the FSP Language

FSP (*Finite State Process*) is a concise algebraic notation for concurrent processes [40], supported by the LTSA (*Labelled Transition System Analyser*) verification tool designed at Imperial College (London, United Kingdom). FSP and LTSA are particularly suited for students to practice with academic examples.

Although FSP and LOTOS share many fundamental concepts, they differ slightly in their expressiveness. On the one hand, FSP provides a priority operator that has no equivalent in LOTOS. On the other hand, LOTOS allows abstract data types to be defined by the user, while FSP provides Booleans, integers, labels, and predefined numeric functions only. Also, LOTOS allows sequential and parallel composition operators to be combined with only few restrictions, while FSP imposes a strict separation between sequential and parallel processes, so that parallel processes cannot be composed in sequence.

CADP 2010 supports the FSP language, following the translation approach of [38], implemented in two new tools. The FSP2LOTOS tool translates each sequential FSP process into a LOTOS process, and each parallel FSP process into an EXP.OPEN [37] network of communicating processes with priorities. The FSP.OPEN tool provides a transparent interface between FSP and the OPEN/CÆSAR environment.

For the FSP user community, CADP 2010 brings the following advantages: it can handle FSP programs with non-guarded process recursion; it can handle larger FSP programs than LTSA, due to the particular attention to performance issues in CADP and to the support of 64-bit architectures, whereas LTSA suffers from Java's 32-bit limitations; finally, CADP offers many tools that complement the functionalities provided by LTSA.

3.3 Support for the LOTOS NT Language

A major new feature of CADP 2010 is the support of LOTOS NT [5], a specification language derived from the ISO standard E-LOTOS [35]. LOTOS NT is an attempt [17] at merging the most salient features of process calculi (concurrency, abstraction, congruence results) into mainstream programming languages (imperative and functional languages for sequential programming). Contrary to LOTOS, which gathers two different languages into one, LOTOS NT exhibits a single unified language, in which the data part can be seen as a subset of the control part (i.e., functions are a particular case of processes): absence of such a nice symmetry in LOTOS is a drawback and a cause for its steep learning curve.

LOTOS NT has convenient features that LOTOS is lacking: it has a set of predefined data types (Booleans, natural numbers, integers, reals, characters, and strings); it provides short-hand notations for lists, sets, and arrays; it eases the definition of inductive types by automatically generating common operations (equality and order relations, field accessors, etc); it enables typing of communication channels; it introduces the notion of modules. Similar to the LOTOS compilers of CADP, LOTOS NT can import hand-written, external C code that implements LOTOS NT types and functions; under some conditions, it is also possible to combine LOTOS and LOTOS NT code into the same specification.

The feedback received about LOTOS NT from both academia and industry is highly positive: it is observed that people quickly start writing meaningful LOTOS NT specifications without the need for a long prior training. As of January 2010, the VASY team has switched from LOTOS to LOTOS NT for all its modeling activities, and LOTOS NT is used internally in companies such as Bull, CEA/Leti, and STMicroelectronics.

CADP 2010 includes a set of tools (LPP preprocessor, LNT2LOTOS translator, and LNT.OPEN connector to OPEN/CÆSAR) that implement LOTOS NT by translation to LOTOS, which enables one to reuse the CÆSAR and CÆSAR.ADT compilers to analyze and execute LOTOS NT specifications. To reduce the translation complexity, many semantic checks are deferred to the CÆSAR.ADT and CÆSAR compilers that will run on the generated, possibly incorrect LOTOS code.

The translation of LOTOS NT data part into LOTOS (which is, to some extent, the reverse of the translation performed by CÆSAR.ADT) requires compilation of functions defined in imperative-style into rewrite systems with priorities. It reuses an existing algorithm [48] for translating a subset of the C language into Horn clauses, but largely extends this algorithm to handle reference-passing parameters, pattern matching ("**case**" statements), loop interruptions ("**break**" statements), multiple "**return**" statements within function bodies, uncatchable exceptions ("**raise**" statements), and overloading of function names.

The translation of the LOTOS NT control part into LOTOS process algebraic terms borrows from a prior translation of CHP into LOTOS [23], which was adapted and optimized for LOTOS NT. The translation is tricky because LOTOS is much less "regular" than LOTOS NT for certain aspects (sequential composition, functionality typing for process termination) and because LOTOS lacks certain concepts (graphical parallel composition [26], type checking for communication channels). Surprisingly, the state spaces generated from LOTOS NT programs are in general not larger than those generated from "equivalent" LOTOS programs, due to the precise analysis and sharing of continuations during the translation.

3.4   Support for Other Languages

Numerous other languages have been connected to CADP 2010. Fig. 1 gives a global picture; dark grey boxes indicate the languages and software components included in CADP 2010; light grey boxes indicate the languages for which VASY has developed translators and connections to CADP 2010, these translators being distributed separately from CADP 2010; arcs are labeled with bibliographic references; arcs without labels correspond to work in progress.

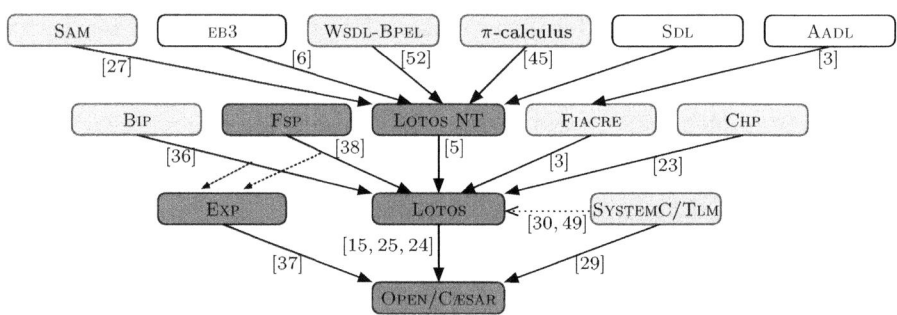

Fig. 1. Connection of the input languages of CADP 2010

4   Model Checking

CADP contains three model checkers operating on explicit and implicit LTSs. XTL (*eXecutable Temporal Language*) [44] is a functional language dedicated to the exploration and querying of an explicit LTS encoded in the BCG format. XTL handles (sets of) states, labels, and transitions as basic data types, enabling temporal logic operators to be implemented using their fixed point characterizations. Temporal logic operators can be mixed with non-standard properties (e.g., counting states, transitions, etc.) and, more generally, with arbitrary computations described as recursive functions exploring the LTS. XTL specifications can include reusable libraries of operators (15 such libraries are available in CADP) and can also be interfaced with external C code for specific computations.

EVALUATOR 3.x [46] evaluates formulas of RAFMC (the regular alternation-free μ-calculus) on an implicit LTS on the fly. RAFMC incorporates the PDL modalities containing regular formulas over transition sequences, which are much more concise and intuitive than their fixed point counterparts: for instance, safety properties are simply expressed using the modality $[R]$ **false**, which forbids the transition sequences characterized by the regular formula R. The tool works by reformulating the model checking problem as a BES resolution, which is performed using the linear-time local algorithms of the CÆSAR_SOLVE library [43]. According to the shape of the formula, the most memory-efficient algorithm of the library is selected automatically. The tool produces examples and counterexamples, which are general LTS subgraphs (i.e., may contain branches and/or cycles), and also enables the definition of reusable libraries of property patterns.

EVALUATOR 4.0 [47] is a new model checker handling formulas written in MCL (*Model Checking Language*), which conservatively extends RAFMC with two kinds of features. First, MCL adds data-handling mechanisms to parse and exploit structured transition labels (containing a channel/gate name and a list of values exchanged), generated from value-passing specification languages. MCL contains action predicates with value extraction, fixed point operators parameterized with data values, quantifiers over finite data domains, regular formulas extended with counters, and constructs inspired from functional programming languages ("**let**", "**if-then-else**", "**case**", "**while**", "**repeat**", etc.).

Second, MCL adds fairness operators, inspired from those of PDL-Δ, which characterize complex, unfair cycles consisting of infinite repetitions of regular subsequences. These operators belong to $L\mu_2$, the μ-calculus fragment of alternation depth two and were shown to subsume CTL*. Although $L\mu_2$ has, in the worst case, a quadratic model checking complexity, the fairness operators of MCL are evaluated in linear-time using an enhanced resolution algorithm of CÆSAR_SOLVE [47].

5   Equivalence Checking

Equivalence checking is useful to guarantee that some properties verified on one graph are also satisfied by another. Alternatively, equivalence checking can be used to minimize a graph by collapsing its equivalent states. Concurrency theory produced many graph equivalence relations, including strong bisimulation, branching bisimulation, as well as stochastic/probabilistic extensions of strong and branching bisimulations (which take into account the notion of *lumpability*) for models combining features from LTSs and Markov chains. From the beginning, equivalence checking has been a key feature of CADP, first with the ALDÉBARAN tool [11,13] and, since 1999, with the BCG_MIN 1.0 tool for minimization of explicit graphs using various partition-refinement algorithms. The functionalities of these two tools have been progressively subsumed by improved tools, namely BCG_MIN 2.0 and BISIMULATOR, available in CADP 2010.

BCG_MIN 2.0 enables an explicit LTS to be minimized according to various equivalence relations. It implements partition-refinement algorithms based on the notion of state signature [4]. Intuitively, the signature of a state is the set of all couples "(transition label, block of the target state)" of the outgoing transitions (possibly following some compressed sequence of internal transitions in the case of branching bisimulation). Refinement of the state partition consists in dispatching states with different signatures to different blocks until the fixpoint has been reached, each block thus corresponding to a class of equivalent states. BCG_MIN 2.0 extends this algorithm to the stochastic/probabilistic extensions of strong and branching bisimulations, by incorporating lumpability in the computation of signatures.

For strong and branching bisimulations, tests on more than 8000 BCG graphs show that BCG_MIN 2.0 is 20 times faster and uses 1.3 times less memory than BCG_MIN 1.0. For stochastic/probabilistic bisimulations, BCG_MIN 2.0 is more than 500 (occasionally, 8500) times faster and uses 4 times less memory. Large graphs of more than 10^8 states and 10^9 transitions have been minimized in a few hours, using less than 100 Gbytes RAM.

BISIMULATOR [2,43] compares an implicit LTS (usually, describing a *protocol*) with an explicit LTS (usually, describing the expected *service*) on the fly, by encoding the problem as a BES, which is solved using the linear-time local algorithms of the CÆSAR_SOLVE [43] library of CADP. This encoding generalizes and, due to optimizations applied on the fly depending on the LTS structure, outperforms the pioneering on-the-fly equivalence checking algorithms [13]. For typical cases (e.g., when the service LTS is deterministic and/or τ-free, τ denoting the hidden/invisible action), the tool automatically chooses an appropriate memory-efficient BES resolution algorithm, which stores only the states, and not the transitions.

BISIMULATOR implements seven equivalence relations (strong, weak, branching, $\tau^*.a$ [13], safety, trace, and weak trace) and their associated preorders, and is one of the richest on-the-fly equivalence checkers available. For non-equivalent LTSs, the tool can generate a counterexample, i.e., a directed acyclic graph containing the minimal set of transition sequences that, if traversed simultaneously in the two LTSs, lead to couples of non-equivalent states. Minimal-depth counterexamples can be obtained using breadth-first strategies for BES resolution. The tool is also equipped with reductions modulo τ-compression (collapse of τ-cycles) and τ-confluence (elimination of redundant interleavings), which preserve branching equivalence and can improve performance by several orders of magnitude.

6   Performance Evaluation

During the last decade, CADP has been enhanced for performance evaluation operating on extended Markovian models encoded in the BCG format (see details in [9]). Besides BCG_MIN, the EXP.OPEN tool [37] now supports also the parallel composition of extended Markovian models, implementing maximal progress of internal transitions in choice with stochastic transitions. New tools have been added, namely DETERMINATOR [32], which eliminates stochastic nondeterminism in extended Markovian models on the fly using a variant of the algorithm presented in [10], and the BCG_STEADY and BCG_TRANSIENT tools, which compute, for each state s of an extended Markovian model, the probability of being in s either on the long run (i.e., in the "steady state") or at each time instant t in a discrete set provided by the user.

More recently, the new CUNCTATOR on-the-fly steady-state simulator for extended Markovian models has been added to CADP. The tool explores a random execution sequence in the model until a non-Markovian transition or a deadlock state is found, or the sequence length or virtual time (obtained by summing up the Markovian information present on transitions) reaches a maximum value specified by the user, or the user interactively halts the simulation. Upon termination, the throughputs of labeled transitions chosen by the user are displayed, together with information such as the number of τ-transitions encountered and the presence of nondeterminism (i.e., states with more than one outgoing τ-transition). The context of a simulation can be saved and restored for starting subsequent simulations, enabling one to implement convergence criteria (e.g., based on confidence intervals) by executing series of increasingly long simulations in linear time. For nondeterministic models, CUNCTATOR selects between conflicting τ-transitions according to one of three scheduling policies (the first, the last, or a randomly chosen transition). Thus, launching simulations using different scheduling policies provides more insight about the stochastic behavior of the model. Compared to BCG_STEADY, which computes exact throughputs, CUNCTATOR consumes less memory but achieving the same accuracy may require more time.

7   Parallel and Distributed Methods

Verification algorithms based on state space exploration have high computing and memory requirements and, thus, are often limited by the capabilities of one single sequential machine. However, the limits can be pushed forward by new algorithms capable of exploiting processing resources offered by networks of workstations, clusters, grids, etc.

CADP was among the first toolboxes to release tools for distributed model checking. The first step was to parallelize the state space construction, which is a bottleneck for verification because storing all reachable states requires a considerable amount of memory. For this purpose, the DISTRIBUTOR and BCG_MERGE tools [22,21] split the generation of an LTS across several machines, each machine building only a fragment of the entire LTS. Interestingly, essential DISTRIBUTOR features, such as the PBG (*Partitioned* BCG *Graph*) format and the graphical monitor that displays in real-time the progress of generation across all the machines, have been replicated in competing verification toolsets.

The second step was the integration into CADP 2010 of a collection of new software tools (PBG_CP, PBG_MV, PBG_RM, and PBG_OPEN) to manipulate an LTS in the PBG format, and their connection to OPEN/CÆSAR.

The third step was the parallelization of on-the-fly verification itself. Therefore we designed a distributed version of the CÆSAR_SOLVE library to solve Boolean equation systems on the fly using several machines, thus enabling the development of parallel model and equivalence checkers.

8   Conclusion

Concurrency theory is now 40-year old; formal methods are 35-year old; model checking verification is nearly 30-year old. To push theoretical ideas into reality and to obtain new scientific results, significant effort must be put into software development and confrontation with industrial applications.

This was indeed the case with CADP 2010 which, besides all aforementioned new tools and major enhancements, also required large amounts of programming work: porting to various processors (Itanium, PowerPC, Sparc, x86, x64), operating systems (Linux, MacOS X, Solaris, Windows) and C compilers (gcc 3, gcc 4, Intel, and Sun); careful code cleanup to remove all compiler and lint warnings, not only in the C code of the CADP tools themselves, but also in the C code that they may generate (this ensures that all compiler warnings received by end-users are related to some mistakes in their LOTOS or LOTOS NT code); significant documentation effort; intensive nonregression testing using thousands of LOTOS and LOTOS NT programs, BCG files, temporal logic formulas, Boolean equation systems, etc. together with a new tool named CONTRIBUTOR that will allow CADP users to send such test cases to the VASY team.

The relevance of these efforts and the maturity of CADP can be estimated from its dissemination and impact figures. As of December 2010, academic and commercial licences have been signed with more than 435 universities, public research institutes, and global corporations; 137 case-studies have been tackled using CADP; 58 research software applications have been developed using CADP; numerous academic courses are using CADP to teach concurrency; the CADP user forum gathers more than 150 registered members with 1000 messages exchanged.

Regarding future work, we plan to develop a native LOTOS NT compiler, to connect even more concurrent languages to CADP, and add new verification tools that exploit massively parallel computing platforms. The latter research area is especially difficult, because it superposes the algorithmic complexities of verification and distributed programming; yet this is the only way to exploit parallel computing resources, which are becoming pervasive.

References

1. Andersen, H.R.: Model Checking and Boolean Graphs. TCS 126(1), 3–30 (1994)
2. Bergamini, D., Descoubes, N., Joubert, C., Mateescu, R.: BISIMULATOR: A Modular Tool for On-the-Fly Equivalence Checking. In: Halbwachs, N., Zuck, L.D. (eds.) TACAS 2005. LNCS, vol. 3440, pp. 581–585. Springer, Heidelberg (2005)
3. Berthomieu, B., Bodeveix, J.-P., Farail, P., Filali, M., Garavel, H., Gaufillet, P., Lang, F., Vernadat, F.: FIACRE: An Intermediate Language for Model Verification in the TOPCASED Environment. In: ERTS (2008)

4. Blom, S., Orzan, S.: Distributed state space minimization. STTT 7, 280–291 (2005)
5. Champelovier, D., Clerc, X., Garavel, H., Guerte, Y., McKinty, C., Lang, F., Serwe, W., Smeding, G.: Reference Manual of the LOTOS NT to LOTOS Translator (Version 5.1). Tech. Report INRIA/VASY, 117 pages (2010)
6. Chossart, R.: Évaluation d'outils de vérification pour les spécifications de systèmes d'information. Mémoire maître ès sciences, Univ. de Sherbrooke, Canada (2010)
7. Cleaveland, R., Li, T., Sims, S.: The Concurrency Workbench of the New Century (Version 1.2). User's Manual (2000)
8. Cleaveland, R., Parrow, J., Steffen, B.: The Concurrency Workbench. In: Sifakis, J. (ed.) CAV 1989. LNCS, vol. 407. Springer, Heidelberg (1990)
9. Coste, N., Garavel, H., Hermanns, H., Lang, F., Mateescu, R., Serwe, W.: Ten Years of Performance Evaluation for Concurrent Systems Using CADP. In: Margaria, T., Steffen, B. (eds.) ISoLA 2010. LNCS, vol. 6416, pp. 128–142. Springer, Heidelberg (2010)
10. Deavours, D.D., Sanders, W.H.: An Efficient Well-Specified Check. In: 8th International Workshop on Petri Nets and Performance Models, PNPM 1999 (1999)
11. Fernandez, J.-C.: ALDEBARAN : un système de vérification par réduction de processus communicants. Thèse de Doctorat, Univ. J. Fourier, Grenoble (1988)
12. Fernandez, J.-C., Garavel, H., Kerbrat, A., Mateescu, R., Mounier, L., Sighireanu, M.: CADP (CÆSAR/ALDÉBARAN Development Package): A Protocol Validation and Verification Toolbox. In: Alur, R., Henzinger, T.A. (eds.) CAV 1996. LNCS. vol. 1102, Springer, Heidelberg (1996)
13. Fernandez, J.-C., Mounier, L.: "On the Fly" Verification of Behavioural Equivalences and Preorders. In: Larsen, K.G., Skou, A. (eds.) CAV 1991. LNCS, vol. 575, Springer, Heidelberg (1992)
14. Garavel, H.: Compilation et vérification de programmes LOTOS. Thèse de Doctorat, Univ. J. Fourier, Grenoble (1989)
15. Garavel, H.: Compilation of LOTOS Abstract Data Types. In: FORTE (1989)
16. Garavel, H.: OPEN/CAESAR: An Open Software Architecture for Verification, Simulation, and Testing. In: Steffen, B. (ed.) TACAS 1998. LNCS, vol. 1384, p. 68. Springer, Heidelberg (1998)
17. Garavel, H.: Reflections on the Future of Concurrency Theory in General and Process Calculi in Particular. In: LIX Colloquium on Emerging Trends in Concurrency Theory. ENTCS, vol. 209 (2008)
18. Garavel, H., Lang, F.: SVL: a Scripting Language for Compositional Verification. In: FORTE. IFIP (2001)
19. Garavel, H., Lang, F., Mateescu, R.: An Overview of CADP 2001. EASST Newsletter 4, 13–24 (2002)
20. Garavel, H., Mateescu, R., Lang, F., Serwe, W.: CADP 2006: A toolbox for the construction and analysis of distributed processes. In: Damm, W., Hermanns, H. (eds.) CAV 2007. LNCS, vol. 4590, pp. 158–163. Springer, Heidelberg (2007)
21. Garavel, H., Mateescu, R., Bergamini, D., Curic, A., Descoubes, N., Joubert, C., Smarandache, I., Stragier, G.: DISTRIBUTOR and BCGMERGE: Tools for distributed explicit state space generation. In: Hermanns, H. (ed.) TACAS 2006. LNCS, vol. 3920, pp. 445–449. Springer, Heidelberg (2006)
22. Garavel, H., Mateescu, R., Smarandache, I.: Parallel state space construction for model-checking. In: Dwyer, M.B. (ed.) SPIN 2001. LNCS, vol. 2057, p. 217. Springer, Heidelberg (2001)
23. Garavel, H., Salaün, G., Serwe, W.: On the Semantics of Communicating Hardware Processes and their Translation into LOTOS for the Verification of Asynchronous Circuits with CADP. SCP 74(3), 100–127 (2009)

24. Garavel, H., Serwe, W.: State Space Reduction for Process Algebra Specifications. TCS 351(2), 131–145 (2006)
25. Garavel, H., Sifakis, J.: Compilation and Verification of LOTOS Specifications. In: PSTV. IFIP (1990)
26. Garavel, H., Sighireanu, M.: A Graphical Parallel Composition Operator for Process Algebras. In: FORTE/PSTV (1999)
27. Garavel, H., Thivolle, D.: Verification of GALS systems by combining synchronous languages and process calculi. In: Păsăreanu, C.S. (ed.) Model Checking Software. LNCS, vol. 5578, pp. 241–260. Springer, Heidelberg (2009)
28. Garavel, H., Turlier, P.: CÆSAR.ADT : un compilateur pour les types abstraits algébriques du langage LOTOS. In: Actes du CFIP (1993)
29. Helmstetter, C.: TLM.Open: a SYSTEMC/TLM Front-End for the CADP Verification Toolbox, http://hal.archives-ouvertes.fr/hal-00429070/
30. Helmstetter, C., Ponsini, O.: A Comparison of Two SystemC/TLM Semantics for Formal Verification. In: MEMOCODE (2008)
31. Hermanns, H.: Interactive Markov Chains and the Quest for Quantified Quality. In: Hermanns, H. (ed.) Interactive Markov Chains. LNCS, vol. 2428, p. 57. Springer, Heidelberg (2002)
32. Hermanns, H., Joubert, C.: A Set of Performance and Dependability Analysis Components for CADP. In: Garavel, H., Hatcliff, J. (eds.) TACAS 2003. LNCS, vol. 2619, pp. 425–430. Springer, Heidelberg (2003)
33. Holzmann, G.J.: The SPIN Model Checker: Primer and Reference Manual. Addison-Wesley, Reading (2003)
34. ISO/IEC. LOTOS— A Formal Description Technique Based on the Temporal Ordering of Observational Behaviour. International Standard 8807, International Organization for Standardization, Geneva (1989)
35. ISO/IEC. Enhancements to LOTOS(E-LOTOS). International Standard 15437:2001, International Organization for Standardization, Geneva (2001)
36. Khan, A.M.: Connection of Compositional Verification Tools for Embedded Systems. Mémoire master 2 recherche, Univ. J. Fourier, Grenoble (2006)
37. Lang, F.: Exp.Open 2.0: A flexible tool integrating partial order, compositional, and on-the-fly verification methods. In: Romijn, J.M.T., Smith, G.P., van de Pol, J. (eds.) IFM 2005. LNCS, vol. 3771, pp. 70–88. Springer, Heidelberg (2005)
38. Lang, F., Salaün, G., Hérilier, R., Kramer, J., Magee, J.: Translating FSP into LOTOS and Networks of Automata. FACJ 22(6), 681–711 (2010)
39. Mader, A.: Verification of Modal Properties Using Boolean Equation Systems. Bertz, Berlin (1997)
40. Magee, J., Kramer, J.: Concurrency: State Models and Java Programs. Wiley, Chichester (2006)
41. Mateescu, R.: Vérification des propriétés temporelles des programmes parallèles. Thèse de Doctorat, Institut National Polytechnique de Grenoble (April 1998)
42. Mateescu, R.: Efficient Diagnostic Generation for Boolean Equation Systems. In: Graf, S. (ed.) TACAS 2000. LNCS, vol. 1785, p. 251. Springer, Heidelberg (2000)
43. Mateescu, R.: CÆSAR_SOLVE: A Generic Library for On-the-Fly Resolution of Alternation-Free Boolean Equation Systems. STTT 8(1), 37–56 (2006)
44. Mateescu, R., Garavel, H.: XTL: A Meta-Language and Tool for Temporal Logic Model-Checking. In: STTT. BRICS (1998)
45. Mateescu, R., Salaün, G.: Translating Pi-Calculus into LOTOS NT. In: Méry, D., Merz, S. (eds.) IFM 2010. LNCS, vol. 6396, pp. 229–244. Springer, Heidelberg (2010)

46. Mateescu, R., Sighireanu, M.: Efficient On-the-Fly Model-Checking for Regular Alternation-Free Mu-Calculus. SCP 46(3), 255–281 (2003)
47. Mateescu, R., Thivolle, D.: A Model Checking Language for Concurrent Value-Passing Systems. In: Cuellar, J., Sere, K. (eds.) FM 2008. LNCS, vol. 5014, pp. 148–164. Springer, Heidelberg (2008)
48. Ponsini, O., Fédèle, C., Kounalis, E.: Rewriting of imperative programs into logical equations. SCP 56(3), 363–401 (2005)
49. Ponsini, O., Serwe, W.: A Schedulerless Semantics of TLM Models Written in SystemC Via Translation into LOTOS. In: Cuellar, J., Sere, K. (eds.) FM 2008. LNCS, vol. 5014, pp. 278–293. Springer, Heidelberg (2008)
50. Schewe, S.: Solving Parity Games in Big Steps. In: Arvind, V., Prasad, S. (eds.) FSTTCS 2007. LNCS, vol. 4855, pp. 449–460. Springer, Heidelberg (2007)
51. Stevens, P., Stirling, C.: Practical Model-Checking Using Games. In: Steffen, B. (ed.) TACAS 1998. LNCS, vol. 1384, p. 85. Springer, Heidelberg (1998)
52. Thivolle, D.: Langages modernes pour la vérification des systèmes asynchrones. PhD thesis, Univ. J. Fourier Grenoble and Polytechnic. Univ. of Bucharest (2011)

GameTime: A Toolkit for Timing Analysis of Software

Sanjit A. Seshia and Jonathan Kotker

EECS Department, UC Berkeley
{sseshia,jamhoot}@eecs.berkeley.edu

Abstract. Timing analysis is a key step in the design of dependable real-time embedded systems. In this paper, we present GameTime, a toolkit for execution time analysis of software. GameTime is based on a combination of game-theoretic online learning and systematic testing using satisfiability modulo theories (SMT) solvers. In contrast with many existing tools for timing analysis, GameTime can be used for a range of tasks, including estimating worst-case execution time, predicting the distribution of execution times of a task, and finding timing-related bugs in programs. We describe key implementation details of GameTime and illustrate its usage through examples.

1 Introduction

Timing properties of embedded systems are determined by the behavior of both the control software and the platform the software executes on. The verification of such properties is made difficult by their heavy dependence on characteristics of the platform, including details of the processor and memory hierarchy.

Several kinds of timing analysis problems arise in practice. First, for hard real-time systems, a classic problem is to estimate the worst-case execution time (WCET) of a terminating software task. Such an estimate is relevant for verifying if deadlines or timing constraints are met as well as for use in scheduling strategies. Second, for soft real-time systems, it can be useful to estimate the distribution of execution times exhibitable by a task. Third, it can be very useful to find a test case on which the program exhibits anomalous timing behavior; e.g., a test case causing a task to miss its deadline. Finally, in "software-in-the-loop" simulation, the software implementation of a controller is simulated along with a model of the continuous plant it controls, with the simulations connected using execution time estimates. For scalability, such simulation must be performed on a workstation, not on the target embedded platform. Consequently, during the workstation-based simulation, it is necessary to predict the timing of the program along a particular execution path on the target platform.

All of the problems mentioned in the preceding paragraph are instances of *predicting* a particular execution time property of a terminating software task. In this paper, we present GAMETIME, a toolkit for timing analysis of software. In contrast with existing tools for timing analysis (see, e.g., [4]), GAMETIME can predict not only extreme-case behavior, but also certain execution time statistics (e.g., the distribution) as well as a program's timing along particular execution paths. Additionally, it is measurement-based, making it easy to port to new platforms. The GAMETIME approach, along with an exposition of theoretical and experimental results, including comparisons with other

P.A. Abdulla and K.R.M. Leino (Eds.): TACAS 2011, LNCS 6605, pp. 388–392, 2011.

methods, is described in existing papers [5,6]. The goal of this paper is to describe the overall tool flow along with aspects of the implementation not described in detail in those papers. We also illustrate, with a running example, how GAMETIME can be used to make various execution time predictions.

2 Running Example

We consider programs P where loops have statically-known finite loop bounds and function calls have known finite recursion depths. Thus P can be unrolled to an equivalent program Q where every execution path in the (possibly cyclic) control-flow graph of P is mapped 1-1 to a path in the acyclic control-flow graph of Q. Our running example is the modular exponentiation code given in Figure 1(a). Modular exponentiation is a necessary primitive for implementing public-key encryption and decryption. The unrolled version of this code for a 2-bit exponent is given in Figure 1(b).

```
1   modexp(base, exponent) {
2     result = 1;
3     for(i=EXP_BITS; i>0; i--) {
4       // EXP_BITS = 2
5       if ((exponent & 1) == 1) {
6         result = (result * base) % p;
7       }
8       exponent >>= 1;
9       base = (base * base) % p;
10    }
11    return result;
12  }
```
(a) Original code P

```
1   modexp_unrolled(base, exponent) {
2     result = 1;
3     if ((exponent & 1) == 1) {
4       result = (result * base) % p;
5     }
6     exponent >>= 1;
7     base = (base * base) % p;
8     // unrolling below
9     if ((exponent & 1) == 1) {
10      result = (result * base) % p;
11    }
12    exponent >>= 1;
13    base = (base * base) % p;
14    return result;
15  }
```
(b) Unrolled code Q

Fig. 1. Modular exponentation. Both programs compute the value of $base^{exponent}$ modulo p

3 The GAMETIME Approach

We begin with a brief overview of the approach taken by GAMETIME and a description of one of the core components, the generation of basis paths of a program (Sec. 3.1). Sec. 3.2 gives a sample experimental result on the running example described in Sec. 2.

Figure 2 depicts the operation of GAMETIME. As shown in the top-left corner, the process begins with the generation of the control-flow graph (CFG) corresponding to the program, where all loops have been unrolled to the maximum loop bound, and all function calls have been inlined into the top-level function. The CFG is assumed to have a single source node (entry point) and a single sink node (exit point); if not, dummy source and sink nodes are added. The next step is a critical one, where a subset of program paths, called *basis paths* are extracted. These basis paths are those that form a basis for the set of all paths, in the standard linear algebra sense of a basis. A satisfiability modulo theories (SMT) solver is invoked to ensure that the generated basis paths are feasible. We discuss this step in more detail in Sec. 3.1.

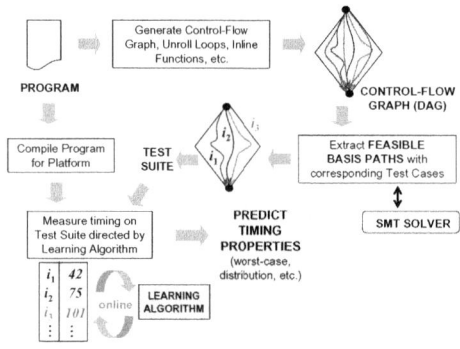

Fig. 2. GAMETIME **overview**

The basis paths are generated along with the corresponding test cases that drive execution down those paths. The program is then compiled for the target platform, and executed on these test cases. In the basic GAMETIME algorithm (described in [5,6]), the sequence of tests is randomized, with basis paths being chosen uniformly at random to be executed. The overall execution time of the program is recorded for each test case. From these end-to-end execution time measurements, GAMETIME's learning algorithm generates a weighted graph model that is used to make predictions about timing properties of interest. The predictions hold with high probability; see the previous papers on GAMETIME [5,6] for details.

In principle, GAMETIME can be set up to use any compiler front-end to generate the CFG and perform test generation along basis paths using an SMT solver; we have experimented with using both CIL [2] and the Microsoft Phoenix compiler front-end [1]. Similarly, any SMT solver for the combination of bit-vector arithmetic and arrays can be used. The core GAMETIME algorithms (involving linear algebra to generate basis paths and the learning algorithm) are implemented separately in Python.

3.1   Generating Basis Paths

In the CFG extracted from a program, nodes correspond to program counter locations, and edges correspond to basic blocks or branches.

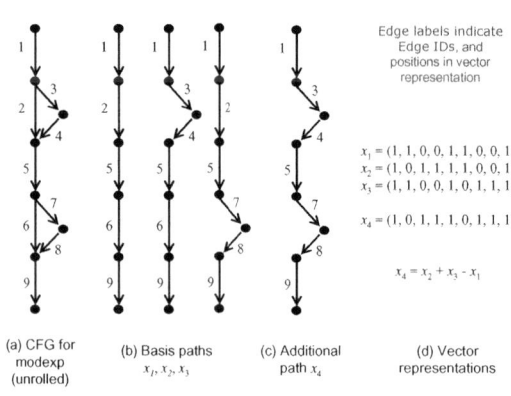

Fig. 3. CFG and Basis Paths for Code in Fig. 1(b)

Figure 3(a) denotes the control-flow graph for the code in Figure 1(b). Each source-sink path in the CFG can be represented as a 0-1 vector with m elements, where m is the number of edges. The interpretation is that the ith entry of a path vector is 1 iff the ith edge is on the path (and 0 otherwise). For example, in the graph of Fig. 3(a), each edge is labeled with its index in the vector representation of the path. For example, edge 2 and 3 correspond to the else (0th bit of exponent = 0) and then branches of the condition statements at lines 3 and 9 respectively in the code, while edge 5 corresponds to the basic block comprising lines 6 and 7. We denote by $\mathcal{P}$ the subset of $\{0,1\}^m$ corresponding to valid program paths. Note that this set can be exponentially large in m.

A key feature of GameTime is the ability to exploit correlations between paths so as to be able to estimate program timing along any path by testing a relatively small subset of paths. This subset is a basis of the path-space $\mathcal{P}$, with two valuable properties: any path in the graph can be written as a linear combination of the paths in the basis, and the coefficients in this linear combination are bounded in absolute value. The first requirement says that the basis is a good representation for the exponentially-large set of possible paths; the second says that timings of some of the basis paths will be of the same order of magnitude as that of the longest path. These properties enable us to repeatedly sample timings of the basis paths to reconstruct the timings of all paths. As GameTime constructs each *basis path*, it ensures that it is feasible by formulating and checking an SMT formula that encodes the semantics of that path; a satisfying assignment yields a test case that drives execution down that path.

Fig. 3(b) shows the basis paths for the graph of Fig. 3(a). Here x_1, x_2, and x_3 are the paths corresponding to exponent taking values 00, 10, and 01 respectively. Fig. 3(c) shows the fourth path x_4, expressible as the linear combination $x_2 + x_3 - x_1$ (see Fig. 3(d)).

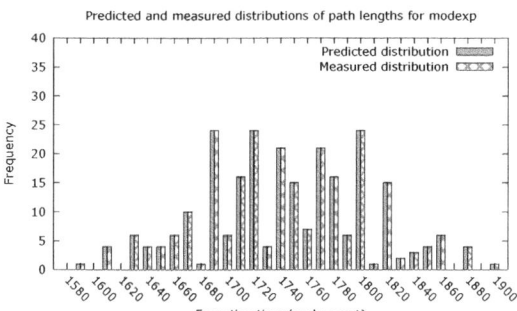

Fig. 4. Actual (striped) and Predicted (shaded) Execution Times for Code in Fig. 1.

The number of feasible basis paths b is bounded by $m - n + 2$ (where n is the number of CFG nodes). Note that our example graph has a "2-diamond" structure, with 4 feasible paths, any 3 of which make up a basis. In general, an "N-diamond" graph with 2^N feasible paths has at most $N + 1$ basis paths.

Computing tests for all basis paths can be viewed as a structural test coverage criterion. Covering all basis paths (for any basis) gives full statement and branch coverage, but not full path coverage. Also, generating tests for basis paths can be viewed as a way of exploiting the structure of the program's CFG for systematic test generation. We have found basis path coverage to be valuable for the prediction of timing properties of a program.

3.2 Sample Experimental Result

We used GameTime to estimate the distribution of execution times of modexp function for an 8-bit exponent (256 program paths) by testing only the (9) basis paths. The experiments were performed for the StrongARM-1100 processor – which implements the ARM instruction set with a 5-stage pipeline and both data and instruction caches – using the SimIt-ARM cycle-accurate simulator [3]. Fig. 4 shows the predicted and actual distribution of execution times – we see that GAMETIME predicts the distribution perfectly. Also, GAMETIME correctly predicts the WCET (and produces the corresponding test case: `exponent=255`).

Acknowledgments. This work was supported in part by NSF grants CNS-0644436 CNS-0627734, and CNS-1035672, an Alfred P. Sloan Research Fellowship, and the Multiscale Systems Center (MuSyC), one of six research centers funded under the Focus Center Research Program (FCRP), a Semiconductor Research Corporation entity. We also acknowledge the contributions of Andrew Chan, Sagar Jain, and Min Xu to the development of GAMETIME.

References

1. Phoenix software optimization and analysis framework,
   `https://connect.microsoft.com/Phoenix`
2. Necula, G., et al.: CIL - infrastructure for C program analysis and transformation,
   `http://manju.cs.berkeley.edu/cil/`
3. Qin, W., Malik, S.: Simit-ARM: A series of free instruction-set simulators and microarchitecture simulators,
   `http://embedded.eecs.berkeley.edu/mescal/forum/2.html`
4. Wilhelm, R., et al.: The Determination of Worst-Case Execution Times—Overview of the Methods and Survey of Tools. ACM Transactions on Embedded Computing Systems, TECS (2007)
5. Seshia, S.A., Rakhlin, A.: Game-theoretic timing analysis. In: Proc. IEEE/ACM International Conference on Computer-Aided Design (ICCAD), pp. 575–582 (2008)
6. Seshia, S.A., Rakhlin, A.: Quantitative analysis of systems using game-theoretic learning. ACM Transactions on Embedded Computing Systems (TECS) (to appear)

Author Index

Alglave, Jade 41
Annapureddy, Yashwanth 254

Barbot, Benoît 128
Bartocci, Ezio 326
Bertrand, Nathalie 96
Buckl, Christian 258
Burnim, Jabob 11

Chamarthi, Harsh Raju 291
Chang, Jinn-Shu 262
Chang, Yi-Wen 262
Chatterjee, Krishnendu 267
Chen, Taolue 128
Cheng, Chih-Hong 258
Cimatti, Alessandro 341
Codish, Michael 189
Conchon, Sylvain 45
Contejean, Evelyne 45

Dillinger, Peter 291

Ehlers, Rüdiger 272

Fainekos, Georgios 254
Fekete, Yoav 189
Forejt, Vojtěch 112
Fränzle, Martin 158
Fuhs, Carsten 189

Garavel, Hubert 372
Garbervetsky, Diego 65
Garg, Pranav 26
Gopinath, Divya 173
Gorín, Daniel 65
Griggio, Alberto 143
Grosu, Radu 326

Han, Hyojung 239
Han, Tingting 128
Henzinger, Thomas A. 267
Hoder, Kryštof 60
Holzmann, Gerard J. 1
Howar, Falk 220

Iguernelala, Mohamed 45

Jéron, Thierry 96
Jobstmann, Barbara 267
Jung, Yungbum 205

Katoen, Joost-Pieter 128
Katsaros, Panagiotis 326
Khurshid, Sarfraz 173
Kiefer, Stefan 296
Knoll, Alois 258
Kotker, Jonathan 388
Kovács, Laura 60
Krichen, Moez 96
Kroening, Daniel 81
Kupferschmid, Sebastian 276
Kwiatkowska, Marta 112

Lang, Frédéric 372
Le, Thi Thieu Hoa 143
Lee, Wonchan 205
Liu, Che 254
Luttenberger, Michael 258

Madhusudan, P. 26
Malik, Muhammad Zubair 173
Manolios, Panagiotis 291
Maranget, Luc 41
Margaria, Tiziana 220
Mateescu, Radu 372
Mereacre, Alexandru 128
Merten, Maik 220

Narasamdya, Iman 341
Neisen, Ariel 65
Norman, Gethin 112

Parker, David 112
Podelski, Andreas 3

Qu, Hongyang 112

Ramakrishnan, C.R. 326
Ravn, Anders P. 357
Roveri, Marco 341
Rybalchenko, Andrey 3

Sankaranarayanan, Sriram   254
Sarkar, Susmit   41
Schneider-Kamp, Peter   189
Sebastiani, Roberto   143
Sen, Koushik   11
Serwe, Wendelin   372
Seshia, Sanjit A.   388
Sewell, Peter   41
Sharygina, Natasha   81
Singh, Rohit   267
Smolka, Scott A.   326
Srba, Jiří   357
Stainer, Amélie   96
Steffen, Bernhard   220
Stergiou, Christos   11
Stoelinga, Mariëlle   311

Talupur, Muralidhar   239
Teige, Tino   158

Timmer, Mark   311
Tsai, Ming-Hsien   262
Tsay, Yih-Kuen   262
Tsitovich, Aliaksei   81

van de Pol, Jaco   311
Vighio, Saleem   357
Voronkov, Andrei   60
Vroon, Daron   291

Wang, Bow-Yaw   205
Wehrle, Martin   276
Wimmel, Harro   224
Wintersteiger, Christoph M.   81
Wojtczak, Dominik   296
Wolf, Karsten   224

Yi, Kwangkuen   205